Methods in Molecular Biology

For further volumes:
http://www.springer.com/series/7651

Antimicrobial Peptides

Methods and Protocols

Edited by

Paul R. Hansen

Department of Drug Design and Pharmacology, Faculty of Health and Medical Sciences,
University of Copenhagen, Copenhagen, Denmark

Humana Press

Editor
Paul R. Hansen
Department of Drug Design and Pharmacology
Faculty of Health and Medical Sciences
University of Copenhagen
Copenhagen, Denmark

ISSN 1064-3745 ISSN 1940-6029 (electronic)
Methods in Molecular Biology
ISBN 978-1-4939-6735-3 ISBN 978-1-4939-6737-7 (eBook)
DOI 10.1007/978-1-4939-6737-7

Library of Congress Control Number: 2016955825

Printed on acid-free paper

This Humana Press imprint is published by Springer Nature
The registered company is Springer Science+Business Media LLC
The registered company address is: 233 Spring Street, New York, NY 10013, U.S.A.

Preface

Antibiotics have saved millions of lives and helped shape modern medicine. However, the emergence of multidrug-resistant pathogens has become a global health problem, and novel approaches to antimicrobial therapy are urgently needed.

Since their discovery in the early 1980s, antimicrobial peptides (AMPs) have attracted considerable interest as novel antibiotics. AMPs are produced by most living organisms and constitute the first line of defense against invading pathogens. AMPs display broad-spectrum antimicrobial activity, paired with rapid and selective bactericidal activity. Furthermore, they can display anti-biofilm, anticancer, anti-virulence, lipopolysaccharide-neutralizing, and immunomodulatory activities.

The aim of this volume of *Methods in Molecular Biology: Antimicrobial Peptides* is to provide basic and advanced laboratory protocols which would interest both the nonspecialist researcher and the experienced researcher.

To this end, we have asked experts in the field to share some protocols used in their own laboratories. As a result, this book is divided in three parts: (1) inserted properties, design, synthesis and characterization of AMPs, (2) studying the interaction of AMPs with model systems or bacteria, and (3) assaying selected biological activities of AMPs.

Part one contains an excellent overview of antimicrobial peptides followed by protocols on design, synthesis, and characterization of AMPs by HPLC, NMR, as well as MS.

Part two describes a number of different techniques for studying the interaction of AMPs with model systems or bacteria, including fluorescence spectroscopy methods, circular dichroism, NMR, isothermal titration calorimetry, atomic force microscopy, patch-clamp techniques, molecular dynamics, PCR, and many more.

Part three describes protocols for determining selected biological activities of AMPs, including minimum inhibitory concentration, anti-biofilm activity, anticancer activity, lipopolysaccharide and lipoteichoic acid neutralization, anti-virulence, anti-keratitis, and hemolytic activity. AMP prodrugs are also covered.

I thank all authors for their excellent contributions, the series editor John M. Walker, and the editorial staff at Springer.

Finally, I'd like to thank past and present collaborators and students.

Copenhagen, Denmark *Paul R. Hansen*

Contents

Contributors

BIRGITTA AGERBERTH • *Department of Laboratory Medicine, Karolinska Institutet, Huddinge, Sweden*
FERNANDO ALBERICIO • *Catalysis and Peptide Research Unit, School of Health Sciences, University of KwaZulu-Natal, Durban, South Africa; School of Chemistry and Physics, University of KwaZulu-Natal, Durban, South Africa; CIBER-BBN, Networking Centre on Bioengineering, Biomaterials and Nanomedicine, Barcelona Science Park, Barcelona, Spain; Department of Organic Chemistry, University of Barcelona, Barcelona, Spain*
ZAHRAA AL-SHARSHAHI • *Department of Pharmaceutical and Medicinal Chemistry, Centre for Synthesis and Chemical Biology, Royal College of Surgeons in Ireland, Dublin, Ireland; Department of Clinical Microbiology, Royal College of Surgeons in Ireland, Education and Research Centre, Dublin, Ireland*
THOMAS L. ANDRESEN • *Department of Micro- and Nanotechnology, DTU Nanotech, Technical University of Denmark, Kgs. Lyngby, Denmark; Center for Nanomedicine and Theranostics, Technical University of Denmark, Kgs. Lyngby, Denmark*
SHIMA ARASTEH • *Department of Chemistry, Center for Biophysics and Computational Biology, Institute for Computational Molecular Science, Temple University, Philadelphia, PA, USA*
MAURICIO ARIAS • *Biochemistry Research Group, Department of Biological Sciences, University of Calgary, Calgary, AB, Canada*
TEERAKUL ARPORNSUWAN • *Department of Medical Technology, Faculty of Allied Health Sciences, Thammasat University, Pathumtani, Thailand*
CONCETTA AVITABILE • *Institute of Biostructure and Bioimaging CNR, Naples, Italy*
MOJTABA BAGHERI • *Peptide Chemistry Laboratory, Institute of Biochemistry and Biophysics, University of Tehran, Tehran, Iran*
HASNA BAIDOURI • *The Ocular Surface Institute, College of Optometry, University of Houston, Houston, TX, USA*
KRISTIN RENEE BAKER • *Department of Biomedical Sciences, Ross University School of Veterinary Medicine, Basseterre, St Kitts and Nevis, West Indies*
MARA BALDRY • *Department of Veterinary Disease Biology, Faculty of Health and Medical Sciences, University of Copenhagen, Copenhagen, Denmark*
MONICA BENINCASA • *Department of Life Sciences, University of Trieste, Trieste, Italy*
PETER BERGMAN • *Department of Laboratory Medicine, Karolinska Institutet, Huddinge, Sweden*
MARTIN SAXTORPH BOJER • *Department of Veterinary Disease Biology, Faculty of Health and Medical Sciences, University of Copenhagen, Copenhagen, Denmark*
VALERIE BOOTH • *Department of Physics and Physical Oceanography, Memorial University of Newfoundland and Labrador, St. John's, NL, Canada; Department of Biochemistry, Memorial University of Newfoundland and Labrador, St. John's, NL, Canada*
FLORIANA CAPPIELLO • *Department of Biochemical Sciences, Istituto Pasteur Italia-Fondazione Cenci Bolognetti, Sapienza University of Rome, Rome, Italy*

Bruno Casciaro • *Department of Biochemical Sciences, Istituto Pasteur Italia-Fondazione Cenci Bolognetti, Sapienza University of Rome, Rome, Italy*
Luca Domenico D'Andrea • *Institute of Biostructure and Bioimaging CNR, Naples, Italy*
Marc Devocelle • *Department of Pharmaceutical and Medicinal Chemistry, Centre for Synthesis and Chemical Biology, Royal College of Surgeons in Ireland, Dublin, Ireland*
Deirdre Fitzgerald-Hughes • *Department of Clinical Microbiology, Royal College of Surgeons in Ireland, Dublin, Ireland*
Peter Fojan • *Department of Physics and Nanotechnology, Aalborg University, Aalborg, Denmark*
Éanna B. Forde • *Department of Pharmaceutical and Medicinal Chemistry, Centre for Synthesis and Chemical Biology, Royal College of Surgeons in Ireland, Dublin, Ireland; Department of Clinical Microbiology, Royal College of Surgeons in Ireland, Dublin, Ireland*
Henrik Franzyk • *Department of Drug Design and Pharmacology, Faculty of Health and Medical Sciences, University of Copenhagen, Copenhagen, Denmark*
Renato Gennaro • *Department of Life Sciences, University of Trieste, Trieste, Italy*
Sanne Gottschalk • *Department of Veterinary Disease Biology, Faculty of Health and Medical Sciences, University of Copenhagen, Copenhagen, Denmark*
Thavendran Govender • *Catalysis and Peptide Research Unit, School of Health Sciences, University of KwaZulu-Natal, Durban, South Africa*
Luca Guardabassi • *Department of Biomedical Sciences, Ross University School of Veterinary Medicine, Basseterre, St Kitts and Nevis, West Indies; Department of Veterinary Disease Biology, Faculty of Health and Medical Sciences, University of Copenhagen, Copenhagen, Denmark*
Gudmundur H. Gudmundsson • *Institute of Biology and Biomedical Center, University of Iceland, Reykjavík, Iceland*
Leonid Gurevich • *Department of Physics and Nanotechnology, Aalborg University, Aalborg, Denmark*
Robert E.W. Hancock • *Department of Microbiology and Immunology, Center for Microbial Diseases and Immunity Research, University of British Columbia, Vancouver, BC, Canada*
Evan F. Haney • *Department of Microbiology and Immunology, Center for Microbial Diseases and Immunity Research, University of British Columbia, Vancouver, BC, Canada*
Paul R. Hansen • *Department of Drug Design and Pharmacology, Faculty of Health and Medical Sciences, University of Copenhagen, Copenhagen, Denmark*
Jonas R. Henriksen • *Department of Chemistry, DTU Chemistry, Technical University of Denmark, Kgs. Lyngby, Denmark; Center for Nanomedicine and Theranostics, Technical University of Denmark, Kgs. Lyngby, Denmark*
Andreas Herrmann • *Theoretische und Molekulare Biophysik, Humboldt-Universität zu Berlin, Berlin, Germany*
Rickey P. Hicks • *Department of Chemistry and Physics, College of Science and Mathematics, Augusta University, Augusta, GA, USA*
David W. Hoskin • *Department of Microbiology and Immunology, Dalhousie University, Halifax, NS, Canada; Department of Pathology, Dalhousie University, Halifax, NS, Canada; Department of Surgery, Dalhousie University, Halifax, NS, Canada*
Hanne Ingmer • *Department of Veterinary Disease Biology, Faculty of Health and Medical Sciences, University of Copenhagen, Copenhagen, Denmark*

Bimal Jana • *Department of Biomedical Sciences, Ross University School of Veterinary Medicine, Basseterre, St Kitts and Nevis, West Indies*
Jianping Jiang • *Chengdu Institute of Biology, Chinese Academy of Sciences, Chengdu, China*
Davor Juretić • *Mediterranean Institute for Life Sciences, Split, Croatia*
Graeme Kelly • *Department of Pharmaceutical and Medicinal Chemistry, Centre for Synthesis and Chemical Biology, Royal College of Surgeons in Ireland, Dublin, Ireland*
Edda Klipp • *Theoretische und Molekulare Biophysik, Humboldt-Universität zu Berlin, Berlin, Germany*
Satya Sree Kolar • *The Ocular Surface Institute, College of Optometry, University of Houston, Houston, TX, USA*
Kasper Kristensen • *Department of Micro- and Nanotechnology, DTU Nanotech, Technical University of Denmark, Kgs. Lyngby, Denmark; Center for Nanomedicine and Theranostics, Technical University of Denmark, Kgs. Lyngby, Denmark*
Hendrik G. Kruger • *Catalysis and Peptide Research Unit, School of Health Sciences, University of KwaZulu-Natal, Durban, South Africa*
Jiongyu Liu • *Chengdu Institute of Biology, Chinese Academy of Sciences, Chengdu, China*
Laurence Madera • *Department of Microbiology and Immunology, Dalhousie University, Halifax, Canada*
Hisham Makki • *Department of Pharmaceutical and Medicinal Chemistry, Centre for Synthesis and Chemical Biology, Royal College of Surgeons in Ireland, Dublin, Ireland; Department of Clinical Microbiology, Royal College of Surgeons in Ireland, Education and Research Centre, Dublin, Ireland*
Maria Luisa Mangoni • *Department of Biochemical Sciences, Istituto Pasteur Italia-Fondazione Cenci Bolognetti, Sapienza University of Rome, Rome, Italy*
Sarah C. Mansour • *Department of Microbiology and Immunology, Center for Microbial Diseases and Immunity Research, University of British Columbia, Vancouver, BC, Canada*
Mario Mardirossian • *Department of Life Sciences, University of Trieste, Trieste, Italy*
Alison M. McDermott • *The Ocular Surface Institute, College of Optometry, University of Houston, Houston, TX, USA*
Michael R. Morrow • *Department of Physics and Physical Oceanography, Memorial University of Newfoundland and Labrador, St. John's, NL, Canada*
Lena Münzker • *Department of Drug Design and Pharmacology, Faculty of Health and Medical Sciences, University of Copenhagen, Copenhagen, Denmark*
Frank Nylen • *Department of Laboratory Medicine, Karolinska Institutet, Huddinge, Sweden*
Alberto Oddo • *Department of Drug Design and Pharmacology, Faculty of Health and Medical Sciences, University of Copenhagen, Copenhagen, Denmark*
Elmar J. Prenner • *Biochemistry Research Group, Department of Biological Sciences, University of Calgary, Calgary, AB, Canada*
Suhas Ramesh • *Catalysis and Peptide Research Unit, School of Health Sciences, University of KwaZulu-Natal, Durban, South Africa*
Gabriel Del Rio • *Biochemistry and structural Biology Department, Instituto de Fisiología Celular, Universidad Nacional Autónoma de México, México D.F, México*
Giorgio Rispoli • *Department of Life Science and Biotechnology, University of Ferrara, Ferrara, Italy*

ALESSANDRA ROMANELLI • *Institute of Biostructure and Bioimaging CNR, Naples, Italy; Department of Pharmacy, University of Naples "Federico II", Naples, Italy*
SITTIRUK ROYTRAKUL • *Genome Institute, National Center for Genetic Engineering and Biotechnology, Pathumthani, Thailand Science Park, Thailand*
GIULIA RUNTI • *Department of Life Sciences, University of Trieste, Trieste, Italy*
NURY P. SANTISTEBAN • *Department of Physics and Physical Oceanography, Memorial University of Newfoundland and Labrador, St. John's, NL, Canada*
VIJAYALEKSHMI SAROJINI • *School of Chemical Sciences, The University of Auckland, Auckland, New Zealand*
MARCO SCOCCHI • *Department of Life Sciences, University of Trieste, Trieste, Italy*
LI-AV SEGEV-ZARKO • *Department of Biological Chemistry, Weizmann Institute of Science, Rehovot, Israel*
YECHIEL SHAI • *Department of Biological Chemistry, Weizmann Institute of Science, Rehovot, Israel*
SARAH LINE SKOVBAKKE • *Department of Drug Design and Pharmacology, Faculty of Health and Medical Sciences, University of Copenhagen, Copenhagen, Denmark*
HATHAITIP SRITANAUDOMCHAI • *Department of Oral Biology, Faculty of Dentistry, Mahidol University, Bangkok, Thailand*
WIMOLPAK SRIWAI • *Department of Medical Technology, Faculty of Allied Health Sciences, Thammasat University, Pathumthani, Thailand*
LINE ELNIF THOMSEN • *Department of Veterinary Disease Biology, Faculty of Health and Medical Sciences, University of Copenhagen, Copenhagen, Denmark*
BEATRIZ G. DE LA TORRE • *Catalysis and Peptide Research Unit, School of Health Sciences, University of KwaZulu-Natal, Durban, South Africa*
ALESSANDRO TOSSI • *Department of Life Sciences, University of Trieste, Trieste, Italy*
LARS ERIK UGGERHØJ • *Department of Chemistry and Bioscience, University of Aalborg, Aalborg, Denmark*
HANS J. VOGEL • *Biochemistry Research Group, Department of Biological Sciences, University of Calgary, Calgary, AB, Canada*
DAMIR VUKIČEVIĆ • *Department of Mathematics, Faculty of Science, University of Split, Split, Croatia*
REINHARD WIMMER • *Department of Chemistry and Bioscience, University of Aalborg, Aalborg, Denmark*

Part I

Properties, Design, Synthesis and Characterization of AMPs

Chapter 1

Antimicrobial Peptides: An Introduction

Evan F. Haney*, Sarah C. Mansour*, and Robert E.W. Hancock

Abstract

The "golden era" of antibiotic discovery has long passed, but the need for new antibiotics has never been greater due to the emerging threat of antibiotic resistance. This urgency to develop new antibiotics has motivated researchers to find new methods to combat pathogenic microorganisms resulting in a surge of research focused around antimicrobial peptides (AMPs; also termed host defense peptides) and their potential as therapeutics. During the past few decades, more than 2000 AMPs have been identified from a diverse range of organisms (animals, fungi, plants, and bacteria). While these AMPs share a number of common features and a limited number of structural motifs; their sequences, activities, and targets differ considerably. In addition to their antimicrobial effects, AMPs can also exhibit immunomodulatory, anti-biofilm, and anticancer activities. These diverse functions have spurred tremendous interest in research aimed at understanding the activity of AMPs, and various protocols have been described to assess different aspects of AMP function including screening and evaluating the activities of natural and synthetic AMPs, measuring interactions with membranes, optimizing peptide function, and scaling up peptide production. Here, we provide a general overview of AMPs and introduce some of the methodologies that have been used to advance AMP research.

Key words Antimicrobial peptides, Host defense peptides, Immunomodulatory function, Antibiofilm activity

1 Introduction

When two cationic antimicrobial peptides (AMPs), cecropins A and B, were identified in the hemolymph of silk moths in the early 1980s [1], they were considered to be the primary mode of insect defense against invading pathogens. The subsequent discovery of magainins in *Xenopus* frogs [2] changed our view of these natural antibiotics, revealing that they are produced by vertebrates and other higher organisms. It has now been demonstrated that AMPs are produced by virtually every living organism [3] and they are considered an important first line of defense against invading pathogens [4, 5]. Generally speaking, these peptides have modest

*Both authors are contributed equally to this work.

Paul R. Hansen (ed.), *Antimicrobial Peptides: Methods and Protocols*, Methods in Molecular Biology, vol. 1548,
DOI 10.1007/978-1-4939-6737-7_1, © Springer Science+Business Media LLC 2017

direct antimicrobial activity under physiological conditions although there are exceptions (e.g., pig protegrin [6]), and synthetic derivatives can be made with much greater potency. In mammals, natural AMPs are often produced by epithelial cells and they can be found in tissues, fluids, and at body surfaces such as the mucosa, which coincidentally are the same sites within the body that encounter a multitude of bacteria, both commensals and pathogens, on a daily basis [7]. In addition, AMPs are produced by various cells of the immune system, especially phagocytes, where they are present in granules. At sites of infection or inflammation, these phagocytic cells release their AMPs at concentrations that are likely high enough to directly target bacterial cells in the immediate vicinity [8].

While a diverse range of organisms produce AMPs, there are a number of common features that characterize these peptides. In general, AMPs are ribosomally synthesized polypeptide sequences, often produced as inactive pro-peptides that are processed into active antibacterial forms that typically range in size from 12 to 50 residues [9]. A number of proteins (100–300 residues) also have the ability to inhibit bacterial growth or kill bacterial cells [7], and peptides with anionic sequences can be active under specific conditions [10], but for the purposes of this overview, we will focus on the shorter cationic and amphipathic AMP sequences. The primary amino acid sequences of these AMPs are diverse and varied; however, they are characterized by an abundance of cationic residues (i.e., Arg and Lys) that confer a positive charge on these molecules at neutral pH [11]. In addition, AMPs typically have a high proportion (up to 50% or more) of hydrophobic amino acids [11]. These two features allow AMPs to fold (e.g., after interaction with membranes) into amphipathic secondary structures with the hydrophobic residues localizing on one side of the molecule and the cationic and polar residues appearing on the opposite face [12].

Online databases have been established as repositories of peptides sequences with antibacterial properties [13–15]. The most extensive of these is the antimicrobial peptide database (APD) that was established in 2004 as a manually curated repository of naturally occurring AMPs [16]. Since that time, more than 2600 AMP sequences from across all six kingdoms of life have been added [17]. The identification of new AMP sequences from natural sources remains an avenue of active research and reports of novel AMPs continue to appear in the literature. In addition, it has become widely appreciated that AMPs have other biological activities including antiviral [18], anticancer [19], antibiofilm [20], and immunomodulatory activities [21, 22]. These additional roles for AMPs have stimulated continued interest into this class of molecule, and many of the techniques and methods used to study these processes are explored and discussed in chapters included in this book.

1.1 Synthesis of AMPs

Early studies of AMPs relied on purifying peptides from natural sources and assessing the antibacterial activity of these extracted samples. The disadvantage of this method is that it necessitates a large amount of raw biological sample to obtain modest quantities of peptide. For instance, to purify dermaseptin from the skin of *Phyllomedusa sauvagii* frogs, 1 g of dried skin was required to obtain only 40 μg of pure peptide [23]. Additionally, natural AMPs are often translated as larger precursor proteins, which are then proteolytically cleaved to generate the active AMP. For example, the human cathelicidin hCAP-18 is synthesized and stored intracellularly as a larger preprotein [24] which is processed by protease-3 during secretion [25] to release the active form of the LL-37 peptide. Interestingly, LL-37 can also be degraded by proteases at other sites in the body to generate alternative active forms of the peptide [26, 27]. Therefore, this posttranslational processing of AMPs makes it exceedingly difficult to purify these sequences from natural sources as the desired peptide may not be present in a unique and/or active form.

Fortunately, chemical synthesis of peptides has quickly become the preferred method for obtaining AMPs at high yield and purity. Solid-phase peptide synthesis is widely used to generate large quantities of peptide using primarily Fmoc protecting groups on the elongating polypeptide chain [28]. Following cleavage from the solid-phase support, the AMP of interest is purified using reversed-phase liquid chromatography, and the identity of the AMP can be confirmed by mass spectrometry. Since the activity of a peptide is directly related to its sequence, the potency of synthetic peptides is identical to natural peptides, and large quantities of highly pure AMPs can be quickly obtained at a reasonable price.

Synthesis of peptides offers several additional advantages to purification from natural sources. Since each amino acid is added sequentially to the AMP of interest, this allows researchers to precisely modify AMP sequences to modulate their antibacterial potency and investigate structure-activity relationships. In addition, synthetic peptides are not limited to the 20 naturally occurring amino acids, and virtually any nonnatural amino acid can be inserted or mutated in an AMP to improve biological activity and stability. For example, substitution of two of the Arg residues in oncocin (an AMP derived from *Oncopeltus* antibacterial peptide-4) to ornithines increased the serum half-life of the peptide and improved the antibacterial potency against many Gram-negative pathogens [29]. Furthermore, synthetic chemistry provides the opportunity to use peptidomimetics such as all D-forms of AMPs [30] or sequences consisting of β-amino acids [31]. Peptidomimetics often retain the antimicrobial activity of the parent AMP molecule since general characteristics like positive charge and amphipathicity are preserved, but they are impervious to proteolytic degradation by enzymes [32].

In addition to synthetic peptides, many research groups have used molecular cloning technology to recombinantly express and purify AMPs. Conceptually, genetically engineering microorganisms to synthesize peptides that kill them would seem to be a contradictory proposition. However, this can be overcome by synthesizing AMPs with a large fusion partner that masks the antimicrobial activity (and cleavability by intracellular proteases) of the recombinant peptide within the expressing cell. This fusion partner can then be cleaved from the AMP of interest and removed during subsequent purification. Examples of fusion partners that have been used to generate recombinant AMPs include SUMO [33], calmodulin [34], thioredoxin [35], and maltose-binding protein [36]. Another strategy involves genetically engineering eukaryotic yeast cells, such as *Pichia pastoris*, to produce the AMP of interest [37] since many AMPs are selectively toxic toward bacteria.

1.2 Identification and Design of Novel AMPs

Various strategies have been used to identify novel AMP sequences or design peptides with enhanced activities. Some natural AMPs are actually the result of proteolytic cleavage of larger proteins which release the active peptide sequence. For instance, bovine lactoferricin is released from the iron-binding protein lactoferrin upon pepsin digestion in the gut [38], and pepsin hydrolysis of whey protein also releases AMPs with antimicrobial activity [39]. These studies suggest that AMPs could be generated in large quantities as by-products from the food industry. This observation has led groups to search for AMP sequences in various other large protein molecules resulting in the identification of AMPs such as lactoferrampin from lactoferrin [40] and puroindoline from wheat endosperm [41]. Interestingly, AMPs have even been identified within larger antimicrobial proteins such as a helix-loop-helix peptide that was proteolytically cleaved from hen egg white lysozyme [42]. Additionally, segments of larger proteins can serve as structural scaffolds to design novel AMPs from otherwise inactive sequences. For example, the membrane proximal region of the HIV glycoprotein, gp41, is a Trp-rich segment of the protein that interacts with the viral membrane at the interfacial region of the bilayer. Manipulation of this membrane-associated peptide scaffold to increase the overall positive charge and amphipathicity resulted in a new AMP sequence with antibacterial activity, whereas the parent sequence was completely inactive [43].

Numerous reports aimed at optimizing AMPs for their antimicrobial potency have been reported in the literature (recently reviewed in [44, 45]). Often a small library of AMP derivatives are synthesized focusing on altering certain peptide characteristics such as increasing positive charge, hydrophobicity, or overall amphipathicity. This strategy has been applied numerous times with moderate success to incrementally improve the antibacterial activity of AMPs. Recent examples include AMPs based on tryptophan zipper-like β-hairpins [46], and mastoparan X derivatives [47]. Various computational tools have also been developed to help design AMPs, some of which are described in chapters in this book including the design of amphipathic α-helical peptides or orally active AMPs.

SPOT-synthesis of peptide arrays on cellulose membranes has allowed researchers to generate large libraries of synthetic peptides at a reasonable cost [48, 49]. To illustrate their usefulness, SPOT-synthesized peptide arrays were used to generate a complete amino acid substitution library of a bovine bactenecin derivative [50]. All of these peptide derivatives were then screened against *P. aeruginosa* to determine which substitutions improved the antibacterial activity and optimized peptides were generated which combined the most active substitutions resulting in AMPs with MICs as low as 0.5 μg/ml [50]. An extension of these methods involves the generation of quantitative structure activity relationships (QSAR) models that take into account hundreds of peptide sequences (based on so-called "descriptors" derived from the primary amino acid sequence that provide surrogates for secondary structure) and their antibacterial activities to generate computational models that accurately predict the antibacterial activity of peptides in silico. These models are then used to compute the antibacterial activity of thousands of virtual peptides to quite accurately identify novel AMPs with predicted enhanced antibacterial activity. This strategy was successfully employed to identify two novel AMPs, HHC-10 and HHC-36, with potent and broad-spectrum activity against a wide range of multidrug-resistant bacterial isolates, as well as systemic activity in mouse model infections [51].

1.3 Structure

AMPs are typically unstructured in aqueous solution but adopt an amphipathic conformation in the presence of a biological membrane or membrane mimetics [12]. This ability to fold into an amphipathic structure is attributed to the abundance of cationic and hydrophobic amino acids that make up AMPs, and the specific membrane bound conformation is largely dictated by the primary sequence of the peptide. Numerous AMP conformations have been described, including α-helical, β-sheet, and extended and looped structures [52]. Some larger AMPs are stabilized by intramolecular disulfide bonds or by cyclization of the peptide backbone which limits the structural flexibility and maintains these peptides in a relatively restricted conformation [53]. Examples of AMPs from different structural classes are shown in Fig. 1.

Various methods have been used to study the structural characteristics of peptides. Circular dichroism spectroscopy is a valuable method that provides information on the general conformation and structure of an AMP in solution as well as upon binding to micelles or vesicles. Recently it was demonstrated that CD spectroscopy could also be used to study the conformation of AMPs bound to whole

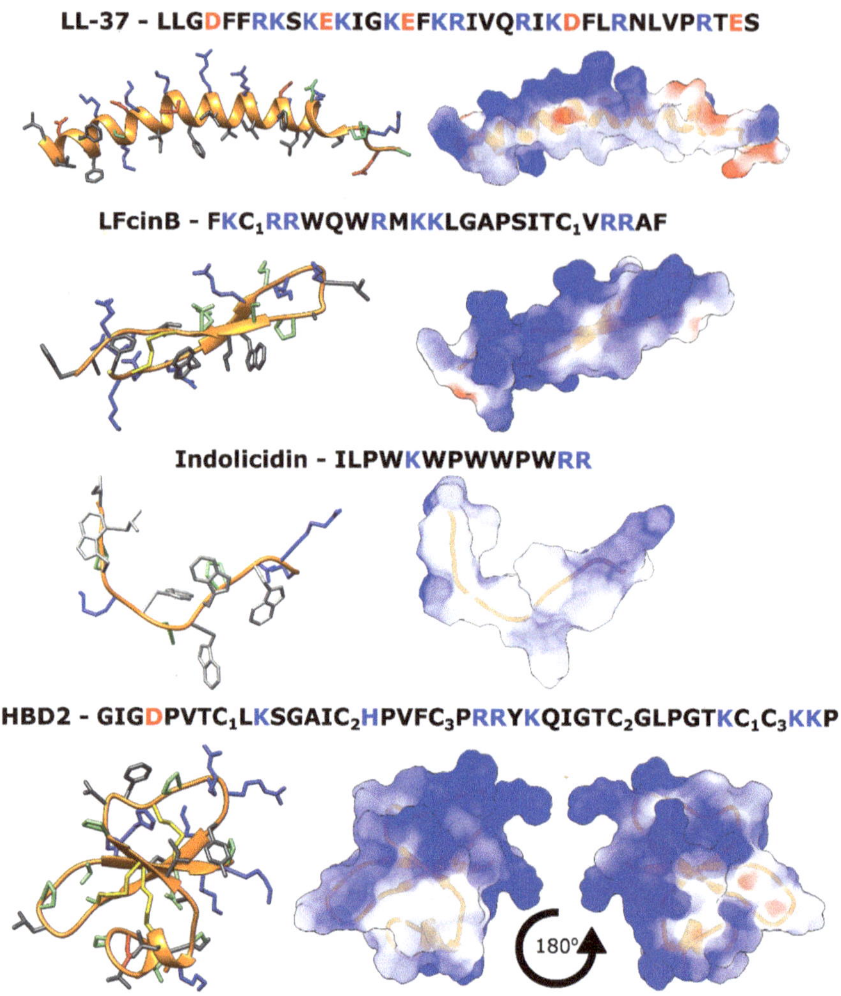

Fig. 1 Sequences and three-dimensional structures of selected AMPs from different structural classes. The human cathelicidin, LL-37, adopts an amphipathic α-helical conformation when bound to sodium dodecyl sulfate micelles [114]. Bovine lactoferricin (LFcinB), an AMP that is enzymatically released from the larger iron-binding protein, lactoferrin, adopts a β-sheet structure in solution [115]. Bovine indolicidin is a short 13-residue AMP that adopts an elongated turn structure in the presence of dodecylphosphocholine (DPC) micelles [116]. Human beta-defensin-2 (HBD2) forms a triple stranded β-sheet with an α-helical N-terminal region [117]. In the primary amino acid sequences, cationic residues are highlighted in *blue* and anionic residues in *red* and disulfide-bonded Cys residues are denoted by a *numerical subscript. Ribbon diagrams* highlight cationic residue in *blue*, anionic residues in *red*, hydrophobic side chains in *dark gray* and *cysteine* and disulfide bonds in *gold.* All other residues are colored in *light green.* The surface charge distribution represents the coulombic surface potential with the positively charged regions highlighted in *blue*, anionic regions in *red* and uncharged and/or hydrophobic regions appearing *white.* All the images were generated with Chimera molecular modeling software [118] version 1.10.2. PDB IDs for each peptide structure shown are 2K60 (LL-37), 1LFC (LFcinB), 1G89 (indolicidin) and 1FQQ (HBD2)

bacterial cells [54]. Oriented CD methods have been described which use macroscopically oriented membranes to determine the alignment of a peptide within a bilayer and provide important insights into the interactions between AMPs and lipid molecules [55].

To determine the high-resolution structure of polypeptides, two experimental techniques are available to researchers: nuclear magnetic resonance (NMR) spectroscopy and X-ray crystallography. Unfortunately, most AMP sequences do not readily crystallize due to their small size and inherent flexibility in aqueous solution. On the other hand, NMR spectroscopy has proven to be an invaluable tool for studying the three-dimensional structures of AMPs and has provided important insights into how these peptides fold and interact with biological membranes [56]. NMR is an ideal method to determine the high-resolution structure of a given peptide since they are relatively small molecules and their conformation can be solved using standard homonuclear proton NMR techniques of unlabeled peptide samples. The classical approach exploits nuclear overhauser effect spectroscopy (NOESY) to determine the inter-proton distance of the hydrogen atoms within a given peptide sample. If necessary, incorporation of NMR active nuclei such as N^{15} and C^{13} extends the structural information that can be obtained from NMR by providing probes for the nitrogen and carbon atoms within the peptide backbone. These can then be used in heteronuclear NMR experiments to probe the conformation of a peptide bound to larger lipid vesicles or they can give information regarding the orientation of a folded AMP within an oriented bilayer. In addition, NMR has provided structural insights into the molecular interactions between AMPs and bacterial cell components, such as lipopolysaccharides (LPS) [57].

1.4 Mechanisms of Action

AMPs exhibit a great number of fundamentally different activities, implying that there is no single target site or mechanism of action that is common to all AMPs. For bacterial species, the initial contact with AMPs relies on an electrostatic attraction between the anionic molecules on the bacterial cell surface and the cationic residues within the AMPs. Specifically, AMPs are attracted to LPS found on the outer membrane of Gram-negative species and lipoteichoic acids from the cell wall of Gram-positive species. Once AMPs pass the bacterial cell wall, they interact preferentially with the negatively charged lipids found in bacterial cytoplasmic membranes such as phosphatidylglycerol and cardiolipin [45]. AMPs displace the divalent cations that stabilize these phospholipids, resulting in membrane perturbation [58]. Due to their amphipathic composition, AMPs also insert themselves within the bilayer, resulting in alterations to membrane structure, such as membrane thinning, changes in membrane curvature, and disruption of the bilayer permeability barrier [59].

1.5 Membrane-Disrupting Models

Several membrane-disrupting models have been described in the literature, such as the barrel-stave, toroidal-pore, aggregate, and carpet models. Briefly, in the "barrel-stave model," AMPs insert within the bilayer by orienting their hydrophobic regions in the lipid core of the bilayer, forming a barrel-like transmembrane pore [60]. While this type of pore is widely described in the literature, there is very little evidence favoring this model for the cationic amphipathic peptides, and only the fungal peptide alamethicin is thought to form such a structure [60, 61]. In the "toroidal-pore model," AMPs are proposed to insert within the membrane and cause the lipids within the bilayer to bend, forming a peptide-lined pore in which the AMPs associate with the polar head groups of the phospholipids. Magainins, protegrins, and melittin have been proposed to form pores using the toroidal-pore model [60], but again there is only modest support for the formation of such a toroidal structure, and molecular dynamics simulations of the interaction of these peptides with the membrane do not favor a formal peptide-lined toroidal channel but rather a series of discrete lipid-associated peptide molecules [62, 63]. The "aggregate model," which our lab proposed [64], is a less formal model that indicates the formation of transient peptide-lipid aggregates that enable observed rapid current fluctuations in transmembrane conductance as well as translocation of peptides across membranes. In the "carpet model" (and a variation termed the "detergent model"), AMPs fold into an amphipathic conformation and accumulate on the membrane surface forming a carpet-like cover [65]. At a critically high concentration of AMP, the bilayer structure becomes disrupted (by either electrostatic or detergent effects) leading to the disintegration of membranes into smaller fragments or micelles [66]. Cecropins are insect-derived peptides that have been proposed to cause lysis via carpet-like mechanisms [67], but again there is little support for a sudden disintegration of biological (cf. model) membranes, and in fact cecropin A is known to affect intracellular processes and alter transcription in *E. coli* at sub-inhibitory concentrations [68]. Overall we still lack a cohesive model that explains how peptides alter the barrier properties of biological membranes.

While the initial peptide-AMP interactions rely on electrostatic forces, the effect of the peptides on the membrane differs depending on the biophysical properties of the peptide (secondary structure, amphipathicity, size, net charge, and oligomerization) along with the composition of the membrane [66]. In this context, a variety of techniques (many of which are highlighted in this book) can be used to better define AMP-membrane interactions, specifically peptide attachment, insertion, orientation, and translocation. Many of these techniques rely on model membrane systems ranging from simple detergent micelles to vesicles containing phospholipids of varying head group and acyl chain composition. Model membrane studies can reveal important membrane interactions of AMPs; however, they represent bacterial membranes as simple bilayers and do not account for other bacterial surface domains, such as lipid rafts, membrane proteins and glycolipids [66]. To study more complex AMP and membrane/cell wall interactions, confocal microscopy coupled with biotinylated or fluorescently labeled peptides can further uncover where AMPs localize in membranes and whether they possess bacterial penetrating properties. For instance, studies using biotinylated peptides have shown that magainin 2 binds to the cell surface, while buforin II accumulates within the bacterial cytoplasm [69]. Other microscopic techniques, such as electron microscopy [70] or atomic force microscopy [71], have been used to evaluate the effect of an AMP on the overall structure or the bacterial cell and/or evaluate morphological changes that might occur in the presence of peptide. Overall, these techniques provide insights into the involvement of specific membrane components in peptide-membrane interactions.

Often forgotten is the fact that the membrane is the site at which important events occur that are required for viability including cell wall biosynthesis and cell division. Thus, membrane perturbation by an AMP can have significant effects on these events. For example, lantibiotics, a class of post-transcriptionally modified AMP, are known to bind to a peptidoglycan precursor, lipid II, resulting in the inhibition of cell wall biosynthesis [72]. This is a known mechanism of action for several AMPs including the defensins. Nisin is also known to cause membrane perturbation in a lipid II-dependent manner and interferes with cell wall synthesis by stimulating *Staphylococcus* autolysin, an enzyme known to hydrolyze the cell wall [73].

1.6 Intracellular Targets

Many AMPs inhibit cellular functions at concentrations where they do not cause major damage to the cytoplasmic membrane but rather freely translocate across the bilayer [59]. It is therefore no coincidence that the signature compositions of AMPs are shared by so-called cell-penetrating peptides [74]. Once inside the bacterial cell, AMPs can interact with intracellular targets, causing interference with important cellular and metabolic processes such as the inhibition of DNA, RNA, and protein synthesis as well as the inhibition of cytosolic enzymatic activity [45, 75]. Certain peptides are known to target DNA and RNA. For example, fish peptide pleurocidin inhibits macromolecular synthesis, affecting RNA and protein synthesis at lower concentrations [76]. Buforin II [77] and puroindoline [78] also bind directly to nucleic acids, presumably due to their electrostatic attraction to the negatively charged phosphodiester bonds in the nucleic acid backbone.

As noted above, the targets of AMPs differ considerably, resulting in a diverse range of activities. Certain methods used for studying the biological activities of AMPs are discussed below and summarized in Fig. 2.

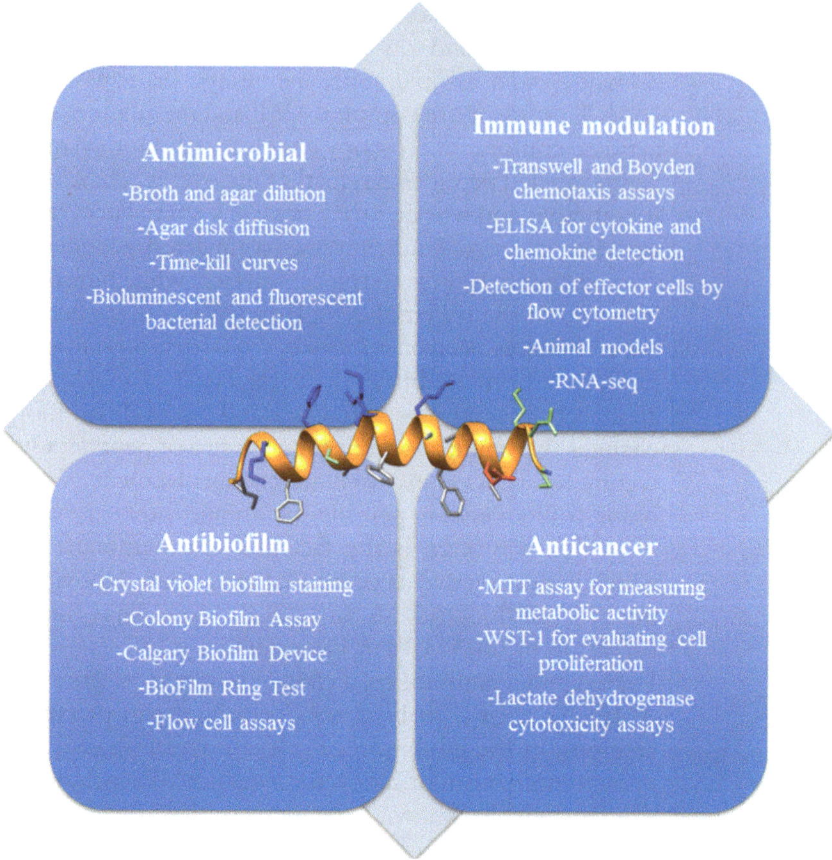

Fig. 2 Summary of methods mentioned for evaluating different biological activities of AMPs. Generally, AMPs have been studied for their antimicrobial activity, but their antibiofilm, immunomodulatory, and anticancer roles are becoming more widely appreciated. The NMR structure of magainin 2 bound to DPC micelles (PDB ID—2MAG [119]) is shown for illustrative purposes

1.7 Antimicrobial Activity

AMPs were first recognized as antimicrobial agents. However, unlike conventional antibiotics that target specific bacterial proteins or structures, AMPs have been proposed to have multiple targets, making them less likely to impose selective pressure on bacteria [59, 79]. Because of this, AMPs are considered to be excellent candidates for further development as a new class of antimicrobials to combat the emerging threat of multidrug-resistant bacteria. Many in vitro screening methods exist for evaluating the susceptibility of a given organism to an AMP. Determining the minimal inhibitory concentration (MIC), which is defined as the lowest concentration of the antimicrobial agent required to inhibit growth, is the gold standard method for determining and comparing the efficacy of antimicrobials. The MIC is normally measured using agar or broth dilution methods, whereby a standardized inoculum of bacteria is incubated in serial dilutions of the agent of interest and special modifications are used for measuring peptide

MICs [80]. Agar disk diffusion testing is also often used for antimicrobial susceptibility testing. In this method, bacteria are streaked onto agar to generate a bacterial lawn, and peptides are added to holes punched into the agar or to filter paper disks placed onto the agar surface. The AMP diffuses into the agar, inhibiting the growth of the bacteria which can be quantified as a "zone of inhibition" around the disk. While this method is cost-effective, it does not allow for the distinction of bacteriostatic and bactericidal agents, and the amount of AMP at the edge of the inhibition zone cannot be quantified.

While MIC measurements determine the susceptibility to an agent after longer incubation periods (16–20 h) [80], bacterial time-kill assays can determine short-term killing kinetics of a drug [81]. Thus, they can be used to determine concentration dependence as well as time-dependent bactericidal activity of AMPs [81]. Bioluminescent bacteria can also be used to screen real-time killing by AMPs [82]. Moreover, fluorescent dyes paired with fluorescent microscopy or flow cytometry can be used to determine bacterial viability after exposure to AMPs. For example, propodium iodide, a DNA intercalating agent, is often used to detect perforated cells [83]. While the use of fluorimetric assays reveals lysis and cellular damage, they are often more challenging as they require specific expensive equipment (e.g., flow cytometry and fluorescent microscope) and throughput is low; therefore, they are conducted less often.

It should be noted that the medium conditions used in these assays can dramatically alter the bacterial susceptibility to AMPs. For example, the presence of cations (Ca^{2+} and Mg^{2+}), physiological salt concentrations and serum can impede the activity of AMPs [84]. As a result, media that do not properly represent the host environment (e.g., phosphate buffer) may misrepresent the in vivo therapeutic potential of AMPs.

1.8 Antibiofilm Activity

Biofilms are structured aggregates of bacteria that are highly resistant to conventional antibiotics [85]. Biofilms represent a major growth state of bacteria in the environment and they contribute to as much as two thirds of all human infections [86]. Recently it has been shown that some AMPs possess antibiofilm activity that is independent of their antimicrobial activity against free-swimming (planktonic) cells. The concept that cationic peptides could serve as antibiofilm therapies first arose when LL-37, at one 16th its MIC, was shown to inhibit *Pseudomonas aeruginosa* biofilm formation as well as disperse preformed biofilms [87]. This led to an explosion of research aimed at identifying agents (both naturally occurring and synthetic) with antibiofilm potential [30, 88–90]. Many of the best antibiofilm peptides inhibit and dissipate biofilms at concentrations well below their MIC against planktonic cells, implying a mechanism of action that is distinct from the antimicrobial activity.

Mechanistic studies have shown that the antibiofilm activity exerted by synthetic bactenecin derivatives is due to the peptide targeting for degradation the stringent stress response signaling molecule, guanosine tetraphosphate (ppGpp) [89]. Under nutrient limiting conditions, bacteria produce ppGpp and this conserved response has been proposed as being required for the initiation and maintenance of protective biofilms. By targeting ppGpp, AMPs not only prevent biofilm formation and cause dispersal of preformed biofilms, but they also make bacteria more susceptible to conventional antibiotics [91]. Intriguingly, since the stringent response is highly conserved, antibiofilm peptides have very broad-spectrum activity.

Interest in antibiofilm compounds is intensifying since it is widely appreciated that biofilms contribute to a significant number of clinical infections. As a result, a number of techniques have been described to assess the antibiofilm activity of an AMP against biofilms formed by various pathogenic bacteria. The simplest method to quantify biofilm growth uses crystal violet dye to stain biofilm mass attached to plastic surfaces and evaluate the reduction in amount of biofilm in the presence of an AMP [92]. This approach can be used to determine the lowest concentration of a peptide required to inhibit biofilm formation or minimal biofilm inhibitory concentration (MBIC). Other static biofilm assays include the Calgary Biofilm Device and the BioFilm Ring test [92]. The Calgary Biofilm Device comprises of a 96-well dish with a lid consisting of 96 pegs to which biofilms adhere and grow [93]. Crystal violet can then be used to quantify the adherent biofilm mass on the pegs. In the BioFilm Ring test, bacterial biofilms are grown in the presence of magnetic beads ultimately immobilizing them, and the biofilm production can be measured by determining the mobility of the remaining beads [94]. However, all of these static methods measure attachment to and growth on surfaces that are limited by access to fresh growth medium, have high backgrounds (attached non-biofilm cells and cell components), and involve fairly immature biofilms, so they are not ideal for assessing biofilm inhibition. The colony biofilm assay [95] can also measure antibiofilm activity of AMPs by growing biofilms on peptide-infused agar, but there is some question as to the validity of bacterial colonies as a biofilm model. Antibiofilm activity can also be assessed using flow cell instruments coupled with confocal imaging which better visualizes biofilm attachment and growth in the presence of AMPs, but this has relatively low throughput [89].

1.9 Immunomodulatory Function

How antimicrobial peptides work in vivo has remained a subject of debate since their antimicrobial activity determined under in vitro conditions often does not reflect physiologically relevant conditions (divalent cations and polyanions found in blood and organs) which severely reduce antimicrobial activity [21]. For example, the activity of LL-37 is strongly dampened in media with ionic composition similar to that found within the human body [84]. Furthermore, while defensins have demonstrated membrane permeabilizing activity under in vitro model membrane assays, their antimicrobial activity is generally very weak under physiologically relevant salt conditions [59]. Despite these findings, the therapeutic efficacy of AMPs has been demonstrated in several animal models, inferring that AMPs may have a different mode of action in vivo. This concept is best illustrated by the activity of the synthetic peptide, IDR-1. This peptide offered protection in several murine Gram-negative and Gram-positive infection models, despite being an ineffective antimicrobial agent (MIC > 128 μg/mL) [96]. This protection was shown to be due to IDR-1 selectively enhancing monocyte chemoattraction while also reducing harmful proinflammatory responses [96].

Several other studies have shown that AMPs are capable of favorably modulating and enhancing the host immune response [21]. One major immunomodulatory activity of AMPs is their ability to stimulate cellular recruitment by promoting expression of chemokines by leukocytes [97]. Interestingly, AMPs and certain chemokines have strong similarities in that they are amphipathic cations and at sufficiently high concentrations, AMPs can exhibit direct chemokine activity [98]. Another important feature of AMPs is their ability to modulate proinflammatory responses by interfering with TLR-ligand-induced proinflammatory pathways [99]. Controlling inflammatory pathways may be beneficial during bacterial infections where excessive inflammation and cytokine production result in organ failure and septic shock syndrome [75]. Certain AMPs also possess wound healing capabilities by promoting the production of restructuring metalloproteinases and epithelial cell metabolism as well as the migration of epithelial and keratinocyte cells [100, 101]. AMPs can also promote cellular differentiation of macrophages and dendritic cells, resulting in the polarization of immune responses including activation of adaptive immunity [102, 103]. Interestingly, LL-37, despite its weak antimicrobial activity (but moderate antibiofilm activity), possesses all of these immunomodulatory activities and is often referred as a "host defense peptide" [84]. Similarly, the 12-amino acid innate defense regulator peptide IDR-1018 has optimized broad immunomodulatory activities, resulting in protection in numerous animal models, as well as excellent antibiofilm activity [21]. Because of their ability to modulate the host immune response, some postulate that this immunomodulatory activity is the most biologically relevant function of natural AMPs [104].

Peptide-induced chemotaxis can be investigated using Transwell or Boyden chamber assays that study leukocyte migration in vitro [105]. Furthermore, enzyme-linked immunosorbent assays (ELISAs) can be used to quantify specific chemokine and cytokine production profiles from various peptide-stimulated cell lines and primary cells [106]. Additionally, many of the signaling pathways and transcription factors dysregulated by peptides have been studied using transcriptomics (RNA-seq, microarrays, sophisticated bioinformatics) [96, 107]. Finally, chemical inhibitors [96] or small interfering RNAs (siRNAs) are often employed to silence suspected pathways and verify molecular targets.

1.10 Anticancer Activity

Along with the various host-targeted activities, certain AMPs are cytolytic towards tumor cells and, as a result, are also referred to as anticancer peptides (ACPs). Since tumor cells typically grow more rapidly than healthy cells, conventional chemotherapy treatments target rapidly dividing cells. Unfortunately, these drugs negatively impact all rapidly dividing cells throughout the body often resulting in hair loss, mucositis (inflammation of digestive tract), thrombocytopenia, and myelosuppression [66]. On the other hand, ACPs exhibit specific toxicity toward cancer cells, which makes them attractive as potential chemotherapeutics. The preferential killing of malignant cells is due to electrostatic interactions between ACPs and cancer cells. Malignant cells possess a strong overall negative charge on their cell surface due to the presence of anionic molecules (i.e., heparin sulfate, O-glycosylated mucins, sialylated gangliosides, and phosphatidylserine) [19, 66]. Conversely, normal healthy mammalian cells are very weakly anionic due to their high zwitterionic phospholipid content. Furthermore, while cancer cell membranes are exceptionally fluid, allowing ACPs to penetrate [108], healthy mammalian cell membranes contain a large amount of cholesterol, which makes their membranes more rigid and blocks the entry of ACPs [66]. Despite the specificity of ACPs for cancer cells however, some peptides can exert broad cytotoxic effects on mammalian cells as observed with melittin [109] and defensins [110].

Cytotoxicity can be assessed using a variety of colorimetric assays such as the tetrazolium dye methylthiazolyldipohenyl tetrazolium bromide (MTT) or water-soluble tetrazolium (WST)-1 and lactate dehydrogenase (LDH) assays. The WST-1 and MTT assays both measure metabolic activity by quantifying the activity of enzymes required for cellular proliferation and respiration, respectively [111]. The LDH assay measures cytolysis of eukaryotic cells by quantifying the release of cytosolic lactate dehydrogenase [112]. Hemolysis assays can also be employed to measure the hemolytic activity of an AMP on erythrocytes.

AMPs are thought to interact with cancer cells using analogous membranolytic mechanisms to those exerted on bacteria. Permeation via membranolytic events can also result in mitochondrial disruption resulting in swelling and release of cytochrome C, eventually leading to apoptosis [113]. Interestingly, not all AMPs target cancer cells and the properties that make certain AMPs active against malignant cells are unknown [66].

2 Concluding Remarks

The emergence of multidrug-resistant strains of pathogenic bacteria coupled with a dwindling supply of antibiotics presents an urgent need to develop new classes of antimicrobial agents. The potential of AMPs as a prospective solution to this impending health crisis has led many research groups to study these peptides and understand how they exert their antimicrobial and immunomodulatory activities. A number of experimental techniques have been used over the years to study AMPs which has dramatically improved our understanding of how these peptides exert their antibacterial effects. In this volume of Methods in Molecular Biology, a number of these methodologies are described covering a diverse range of topics including synthesis and design of AMPs, structural characterization of AMPs, assessment of antimicrobial activity, and elucidation of the mechanism of action for a given sequence. In addition, some AMPs and peptides related to AMPs exhibit additional biological activities including antibiofilm, anticancer and immune modulation, and experimental techniques employed to study these additional roles are also highlighted.

Acknowledgments

The authors acknowledge all the members of the Hancock Lab (both past and present) for their valuable input and discussions regarding antimicrobial, antibiofilm, and immunomodulatory peptides. This work was supported by the Canadian Institutes of Health Research [funding reference number MOP-74493]. REWH holds a Canada Research Chair.

References

1. Steiner H, Hultmark D, Engström Å et al (1981) Sequence and specificity of two antibacterial proteins involved in insect immunity. Nature 292:246–248
2. Zasloff M (1987) Magainins, A class of antimicrobial peptides from xenopus skin - isolation, characterization of 2 active forms, and partial cDNA sequence of a precursor. Proc Natl Acad Sci 84:5449–5453
3. Zasloff M (2002) Antimicrobial peptides of multicellular organisms. Nature 415:389–395
4. Mookherjee N, Hancock REW (2007) Cationic host defence peptides: innate immune regulatory peptides as a novel approach for treating infections. Cell Mol Life Sci 64:922–933
5. Hancock REW, Scott MG (2000) The role of antimicrobial peptides in animal defenses. Proc Natl Acad Sci 97:8856–8861
6. Lai JR, Huck BR, Weisblum B, Gellman SH (2002) Design of non-cysteine-containing antimicrobial β-hairpins: structure–activity relationship studies with linear protegrin-1 analogues. Biochemistry 41:12835–12842
7. Gallo RL, Hooper LV (2012) Epithelial antimicrobial defence of the skin and intestine. Nat Rev Immunol 12:503–516
8. Faurschou M, Borregaard N (2003) Neutrophil granules and secretory vesicles in inflammation. Microbes Infect 5:1317–1327
9. Jenssen H, Hamill P, Hancock REW (2006) Peptide antimicrobial agents. Clin Microbiol Rev 19:491–511
10. Phoenix DA, Dennison SR, Harris F (2013) Anionic antimicrobial peptides. In: Antimicrobial peptides. Wiley-VCH, GmbH & Co. KGaA, pp 83–113

11. Hancock REW, Sahl HG (2006) Antimicrobial and host-defense peptides as new anti-infective therapeutic strategies. Nat Biotechnol 24:1551–1557
12. Nguyen LT, Haney EF, Vogel HJ (2011) The expanding scope of antimicrobial peptide structures and their modes of action. Trends Biotechnol 29:464–472
13. Waghu FH, Barai RS, Gurung P, Idicula-Thomas S (2015) CAMPR3: a database on sequences, structures and signatures of antimicrobial peptides. Nucleic Acids Res 44:D1094–D1097
14. Piotto SP, Sessa L, Concilio S, Iannelli P (2012) YADAMP: yet another database of antimicrobial peptides. Int J Antimicrob Agents 39:346–351
15. Hammami R, Hamida JB, Vergoten G, Fliss I (2009) PhytAMP: a database dedicated to antimicrobial plant peptides. Nucleic Acids Res 37:D963–D968
16. Wang Z, Wang G (2004) APD: the antimicrobial peptide database. Nucleic Acids Res 32:D590–D592
17. Wang G, Li X, Wang Z (2015) APD3: the antimicrobial peptide database as a tool for research and education. Nucleic Acids Res 44:D1087–D1093
18. Skalickova S, Heger Z, Krejcova L et al (2015) Perspective of use of antiviral peptides against influenza virus. Viruses 7:5428–5442
19. Hoskin DW, Ramamoorthy A (2008) Studies on anticancer activities of antimicrobial peptides. Biochim Biophys Acta 1778:357–375
20. de la Fuente-Núñez C, Cardoso MH, de Souza Cândido E et al (2016) Synthetic antibiofilm peptides. Biochim Biophys Acta 1858:1061–1069
21. Mansour SC, Pena OM, Hancock REW (2014) Host defense peptides: front-line immunomodulators. Trends Immunol 35:443–450
22. Hilchie AL, Wuerth K, Hancock REW (2013) Immune modulation by multifaceted cationic host defense (antimicrobial) peptides. Nat Chem Biol 9:761–768
23. Mor A, Nguyen VH, Delfour A et al (1991) Isolation, amino acid sequence and synthesis of dermaseptin, a novel antimicrobial peptide of amphibian skin. Biochemistry (Mosc) 30:8824–8830
24. Durr UHN, Sudheendra US, Ramamoorthy A (2006) LL-37, the only human member of the cathelicidin family of antimicrobial peptides. Biochim Biophys Acta 1758:1408–1425
25. Sørensen OE, Follin P, Johnsen AH et al (2001) Human cathelicidin, hCAP-18, is processed to the antimicrobial peptide LL-37 by extracellular cleavage with proteinase 3. Blood 97:3951–3959
26. Sørensen OE, Gram L, Johnsen AH et al (2003) Processing of seminal plasma hCAP-18 to ALL-38 by gastricsin: a novel mechanism of generating antimicrobial peptides in vagina. J Biol Chem 278:28540–28546
27. Murakami M, Lopez-Garcia B, Braff M et al (2004) Postsecretory processing generates multiple cathelicidins for enhanced topical antimicrobial defense. J Immunol 172:3070–3077
28. Behrendt R, White P, Offer J (2016) Advances in Fmoc solid-phase peptide synthesis. J Pept Sci 22:4–27
29. Knappe D, Piantavigna S, Hansen A et al (2010) Oncocin (VDKPPYLPRPRPPRRIYNR-NH2): a novel antibacterial peptide optimized against gram-negative human pathogens. J Med Chem 53:5240–5247
30. de la Fuente-Núñez C, Reffuveille F, Mansour SC et al (2015) D-enantiomeric peptides that eradicate wild-type and multidrug-resistant biofilms and protect against lethal *Pseudomonas aeruginosa* infections. Chem Biol 22:196–205
31. Jahnsen RD, Frimodt-Moller N, Franzyk H (2012) Antimicrobial activity of peptidomimetics against multidrug-resistant Escherichia coli: a comparative study of different backbones. J Med Chem 55:7253–7261
32. Hilchie AL, Haney EF, Pinto DM et al (2015) Enhanced killing of breast cancer cells by a d-amino acid analog of the winter flounder-derived pleurocidin NRC-03. Exp Mol Pathol 99:426–434
33. Bommarius B, Jenssen H, Elliott M et al (2010) Cost-effective expression and purification of antimicrobial and host defense peptides in Escherichia coli. Peptides 31: 1957–1965
34. Arias M, Hoffarth ER, Ishida H et al (2016) Recombinant expression, antimicrobial activity and mechanism of action of tritrpticin analogs containing fluoro-tryptophan residues. Biochim Biophys Acta 1858:1012
35. Panteleev PV, Ovchinnikova TV (2015) Improved strategy for recombinant production and purification of antimicrobial peptide tachyplesin I and its analogs with high cell selectivity. Biotechnol Appl Biochem. doi: 10.1002/bab.1456
36. Li Y, Wang J, Yang J et al (2014) Recombinant expression, purification and characterization of antimicrobial peptide ORBK in Escherichia coli. Protein Expr Purif 95:182–187
37. Mulder KC, de Lima LA, Aguiar PS et al (2015) Production of a modified peptide clavanin in

Pichia pastoris: cloning, expression, purification and in vitro activities. AMB Express 5:46
38. Bellamy W, Takase M, Yamauchi K et al (1992) Identification of the bactericidal domain of lactoferrin. Biochim Biophys Acta 1121:130–136
39. Théolier J, Hammami R, Labelle P et al (2013) Isolation and identification of antimicrobial peptides derived by peptic cleavage of whey protein isolate. J Funct Foods 5:706–714
40. van der Kraan MIA, Nazmi K, Teeken A et al (2005) Lactoferrampin, an antimicrobial peptide of bovine lactoferrin, exerts its candidacidal activity by a cluster of positively charged residues at the C-terminus in combination with a helix-facilitating N-terminal part. Biol Chem 386:137–142
41. Jing W, Demcoe AR, Vogel HJ (2003) Conformation of a bactericidal domain of puroindoline a: structure and mechanism of action of a 13-residue antimicrobial peptide. J Bacteriol 185:4938–4947
42. Ibrahim HR, Thomas U, Pellegrini A (2001) A helix-loop-helix peptide at the upper lip of the active site cleft of lysozyme confers potent antimicrobial activity with membrane permeabilization action. J Biol Chem 276:43767–43774
43. Haney EF, Nguyen LT, Schibli DJ, Vogel HJ (2012) Design of a novel tryptophan-rich membrane-active antimicrobial peptide from the membrane-proximal region of the HIV glycoprotein, gp41. Beilstein J Org Chem 8:1172–1184
44. Haney EF, Hancock REW (2013) Peptide design for antimicrobial and immunomodulatory applications. Biopolymers 100:572–583
45. Fjell CD, Hiss JA, Hancock REW, Schneider G (2012) Designing antimicrobial peptides: form follows function. Nat Rev Drug Discov 11:37–51
46. Xu L, Chou S, Wang J et al (2015) Antimicrobial activity and membrane-active mechanism of tryptophan zipper-like β-hairpin antimicrobial peptides. Amino Acids 47:2385–2397
47. Henriksen JR, Etzerodt T, Gjetting T, Andresen TL (2014) Side chain hydrophobicity modulates therapeutic activity and membrane selectivity of antimicrobial peptide mastoparan-X. PLoS One 9:e91007
48. Hilpert K, Winkler DF, Hancock REW (2007) Peptide arrays on cellulose support: SPOT synthesis, a time and cost efficient method for synthesis of large numbers of peptides in a parallel and addressable fashion. Nat Protoc 2:1333–1349
49. Winkler DF, Hilpert K, Brandt O, Hancock REW (2009) Synthesis of peptide arrays using SPOT-technology and the CelluSpots-method. Methods Mol Biol 570:157–174
50. Hilpert K, Volkmer-Engert R, Walter T, Hancock REW (2005) High-throughput generation of small antibacterial peptides with improved activity. Nat Biotechnol 23:1008–1012
51. Cherkasov A, Hilpert K, Jenssen H et al (2009) Use of artificial intelligence in the design of small peptide antibiotics effective against a broad spectrum of highly antibiotic-resistant superbugs. ACS Chem Biol 4:65–74
52. Powers JPS, Hancock REW (2003) The relationship between peptide structure and antibacterial activity. Peptides 24:1681–1691
53. Taylor K, Barran PE, Dorin JR (2008) Structure–activity relationships in β-defensin peptides. J Pept Sci 90:1–7
54. Avitabile C, D'Andrea LD, Romanelli A (2014) Circular Dichroism studies on the interactions of antimicrobial peptides with bacterial cells. Sci Rep 4:4293
55. Bürck J, Wadhwani P, Fanghänel S, Ulrich AS (2016) Oriented circular dichroism: a method to characterize membrane-active peptides in oriented lipid bilayers. Acc Chem Res 49:184–192
56. Haney EF, Vogel HJ (2009) NMR of antimicrobial peptides. Ann Rep NMR Spectrosc 65:1–51
57. Bhunia A, Domadia PN, Torres J et al (2010) NMR structure of pardaxin, a pore-forming antimicrobial peptide, in lipopolysaccharide micelles: mechanism of outer membrane permeabilization. J Biol Chem 285:3883–3895
58. Peschel A, Sahl H-G (2006) The co-evolution of host cationic antimicrobial peptides and microbial resistance. Nat Rev Microbiol 4:529–536
59. Brogden KA (2005) Antimicrobial peptides: pore formers or metabolic inhibitors in bacteria? Nat Rev Microbiol 3:238–250
60. Yang L, Harroun TA, Weiss TM et al (2001) Barrel-Stave model or Toroidal model? A case study on melittin pores. Biophys J 81:1475–1485
61. Laver DR (1994) The barrel-stave model as applied to alamethicin and its analogs reevaluated. Biophys J 66:355–359
62. Leontiadou H, Mark AE, Marrink SJ (2006) Antimicrobial peptides in action. J Am Chem Soc 128:12156–12161
63. Sengupta D, Leontiadou H, Mark AE, Marrink S-J (2008) Toroidal pores formed by

antimicrobial peptides show significant disorder. Biochim Biophys Acta 1778:2308–2317
64. Wu M, Maier E, Benz R, Hancock REW (1999) Mechanism of interaction of different classes of cationic antimicrobial peptides with planar bilayers and with the cytoplasmic membrane of *Escherichia coli*. Biochemistry 38:7235–7242
65. Oren Z, Shai Y (1997) Selective lysis of bacteria but not mammalian cells by diastereomers of melittin: structure-function study. Biochemistry (Mosc) 36:1826–1835
66. Gaspar D, Veiga AS, Castanho MARB (2013) From antimicrobial to anticancer peptides. A review. Antimicrob Resist Chemother 4:294
67. Gazit E, Miller IR, Biggin PC et al (1996) Structure and orientation of the mammalian antibacterial peptide cecropin P1 within phospholipid membranes. J Mol Biol 258:860–870
68. Hong RW, Shchepetov M, Weiser JN, Axelsen PH (2003) Transcriptional profile of the Escherichia coli response to the antimicrobial insect peptide cecropin A. Antimicrob Agents Chemother 47:1–6
69. Park CB, Yi KS, Matsuzaki K et al (2000) Structure-activity analysis of buforin II, a histone H2A-derived antimicrobial peptide: the proline hinge is responsible for the cell-penetrating ability of buforin II. Proc Natl Acad Sci 97:8245–8250
70. Zhang X, Wang Y, Liu L et al (2016) Two-peptide bacteriocin PlnEF causes cell membrane damage to Lactobacillus plantarum. Biochim Biophys Acta 1858:274–280
71. Mularski A, Wilksch JJ, Wang H et al (2015) Atomic force microscopy reveals the mechanobiology of lytic peptide action on bacteria. Langmuir ACS J Surf Colloids 31:6164–6171
72. Martin NI, Breukink E (2007) Expanding role of lipid II as a target for lantibiotics. Future Microbiol 2:513–525
73. Bierbaum G, Sahl HG (1985) Induction of autolysis of staphylococci by the basic peptide antibiotics Pep 5 and nisin and their influence on the activity of autolytic enzymes. Arch Microbiol 141:249–254
74. Milletti F (2012) Cell-penetrating peptides: classes, origin, and current landscape. Drug Discov Today 17:850–860
75. Guilhelmelli F, Vilela N, Albuquerque P et al (2013) Antibiotic development challenges: the various mechanisms of action of antimicrobial peptides and of bacterial resistance. Front Microbiol 4:353
76. Patrzykat A, Friedrich CL, Zhang L et al (2002) Sublethal concentrations of pleurocidin-derived antimicrobial peptides inhibit macromolecular synthesis in *Escherichia coli*. Antimicrob Agents Chemother 46:605–614
77. Park CB, Kim HS, Kim SC (1998) Mechanism of action of the antimicrobial peptide buforin II: buforin II kills microorganisms by penetrating the cell membrane and inhibiting cellular functions. Biochem Biophys Res Commun 244:253–257
78. Haney EF, Petersen AP, Lau CK et al (2013) Mechanism of action of puroindoline derived tryptophan-rich antimicrobial peptides. Biochim Biophys Acta 1828:1802–1813
79. Friedrich CL, Moyles D, Beveridge TJ, Hancock REW (2000) Antibacterial action of structurally diverse cationic peptides on gram-positive bacteria. Antimicrob Agents Chemother 44:2086–2092
80. Wiegand I, Hilpert K, Hancock RE (2008) Agar and broth dilution methods to determine the minimal inhibitory concentration (MIC) of antimicrobial substances. Nat Protoc 3:163–175
81. Balouiri M, Sadiki M, Ibnsouda SK (2016) Methods for in vitro evaluating antimicrobial activity: a review. J Pharm Anal 6:71–79
82. Vojtek L, Dobes P, Büyükgüzel E et al (2014) Bioluminescent assay for evaluating antimicrobial activity in insect haemolymph. Eur J Entomol 111:335–340
83. Riccardi C, Nicoletti I (2006) Analysis of apoptosis by propidium iodide staining and flow cytometry. Nat Protoc 1:1458–1461
84. Bowdish DM, Davidson DJ, Lau YE et al (2005) Impact of LL-37 on anti-infective immunity. J Leukoc Biol 77:451–459
85. Lopez D, Vlamakis H, Kolter R (2010) Biofilms. Cold Spring Harb Perspect Biol 2:a000398
86. de la Fuente-Núñez C, Reffuveille F, Fernández L, Hancock RE (2013) Bacterial biofilm development as a multicellular adaptation: antibiotic resistance and new therapeutic strategies. Curr Opin Microbiol 16:580–589
87. Overhage J, Campisano A, Bains M et al (2008) Human host defense peptide LL-37 prevents bacterial biofilm formation. Infect Immun 76:4176–4182
88. Haney EF, Mansour SC, Hilchie AL et al (2015) High throughput screening methods for assessing antibiofilm and immunomodulatory activities of synthetic peptides. Peptides 71:276–285
89. de la Fuente-Núñez C, Reffuveille F, Haney EF et al (2014) Broad-spectrum anti-biofilm peptide that targets a cellular stress response. PLoS Pathog 10:e1004152

90. Wang G, Hanke ML, Mishra B et al (2014) Transformation of human cathelicidin LL-37 into selective, stable, and potent antimicrobial compounds. ACS Chem Biol 9:1997–2002
91. Reffuveille F, de la Fuente-Núñez C, Mansour S, Hancock REW (2014) A broad-spectrum anti-biofilm peptide enhances antibiotic action against bacterial biofilms. Antimicrob Agents Chemother 58:5363–5371
92. Lebeaux D, Chauhan A, Rendueles O, Beloin C (2013) From in vitro to in vivo models of bacterial biofilm-related infections. Pathogens 2:288–356
93. Ceri H, Olson ME, Stremick C et al (1999) The calgary biofilm device: new technology for rapid determination of antibiotic susceptibilities of bacterial biofilms. J Clin Microbiol 37:1771–1776
94. Olivares E, Badel-Berchoux S, Provot C et al (2016) The Biofilm Ring Test®: a rapid method for the routine analysis of P. aeruginosa biofilm formation kinetics. J Clin Microbiol 54:657
95. Merritt JH, Kadouri DE, O'Toole GA (2005) Growing and analyzing static biofilms. Curr Protoc Microbiol 22:1B11–1B118
96. Scott MG, Dullaghan E, Mookherjee N et al (2007) An anti-infective peptide that selectively modulates the innate immune response. Nat Biotechnol 25:465–472
97. Yang D, Chertov O, Oppenheim JJ (2001) Participation of mammalian defensins and cathelicidins in anti-microbial immunity: receptors and activities of human defensins and cathelicidin (LL-37). J Leukoc Biol 69:691–697
98. Tjabringa GS, Ninaber DK, Drijfhout JW et al (2006) Human cathelicidin LL-37 is a chemoattractant for eosinophils and neutrophils that acts via formyl-peptide receptors. Int Arch Allergy Immunol 140:103–111
99. Scott MG, Vreugdenhil ACE, Buurman WA et al (2000) Cutting edge: cationic antimicrobial peptides block the binding of lipopolysaccharide (LPS) to LPS binding protein. J Immunol 164:549–553
100. Steinstraesser L, Hirsch T, Schulte M et al (2012) Innate defense regulator peptide 1018 in wound healing and wound infection. PLoS One 7:e39373
101. Heilborn JD, Nilsson MF, Kratz G et al (2003) The cathelicidin anti-microbial peptide LL-37 is involved in re-epithelialization of human skin wounds and is lacking in chronic ulcer epithelium. J Invest Dermatol 120:379–389
102. van der Does AM, Joosten SA, Vroomans E et al (2012) The antimicrobial peptide hLF1-11 drives monocyte-dendritic cell differentiation toward dendritic cells that promote antifungal responses and enhance Th17 polarization. J Innate Immun 4:284–292
103. Davidson DJ, Currie AJ, Reid GS et al (2004) The cationic antimicrobial peptide LL-37 modulates dendritic cell differentiation and dendritic cell-induced T cell polarization. J Immunol 172:1146–1156
104. Mansour SC, de la Fuente-Núñez C, Hancock REW (2015) Peptide IDR-1018: modulating the immune system and targeting bacterial biofilms to treat antibiotic-resistant bacterial infections. J Pept Sci 21:323–329
105. Chen H-C (2005) Boyden chamber assay. Methods Mol Biol 294:15–22
106. Bowdish DME, Davidson DJ, Scott MG, Hancock REW (2005) Immunomodulatory activities of small host defense peptides. Antimicrob Agents Chemother 49:1727–1732
107. Hancock REW (2016) Bioinformatics: novel insights from genomic information. Nestlé Nutr Inst Workshop Ser 84:35–46
108. Kozłowska K, Nowak J, Kwiatkowski B, Cichorek M (1999) ESR study of plasmatic membrane of the transplantable melanoma cells in relation to their biological properties. Exp Toxicol Pathol 51:89–92
109. Eisenberg D, Terwilliger TC, Tsui F (1980) Structural studies of bee melittin. Biophys J 32:252–254
110. Lichtenstein A, Ganz T, Selsted ME, Lehrer RI (1986) In vitro tumor cell cytolysis mediated by peptide defensins of human and rabbit granulocytes. Blood 68:1407–1410
111. Riss TL, Moravec RA, Niles AL et al (2004) Cell viability assays. Assay Guidance Manual. Available from http://www.ncbi.nlm.nih.gov/books/NBK144065/
112. Smith SM, Wunder MB, Norris DA, Shellman YG (2011) A simple protocol for using a LDH-based cytotoxicity assay to assess the effects of death and growth inhibition at the same time. PLoS One 6:e26908
113. Mai JC, Mi Z, Kim S-H et al (2001) A proapoptotic peptide for the treatment of solid tumors. Cancer Res 61:7709–7712
114. Wang G (2008) Structures of human host defense cathelicidin LL-37 and its smallest antimicrobial peptide KR-12 in lipid micelles. J Biol Chem 283:32637–32643
115. Hwang PM, Zhou N, Shan X et al (1998) Three-dimensional solution structure of lactoferricin B, an antimicrobial peptide derived from bovine lactoferrin. Biochemistry 37:4288–4298
116. Rozek A, Friedrich CL, Hancock REW (2000) Structure of the bovine antimicrobial

peptide indolicidin bound to dodecylphosphocholine and sodium dodecyl sulfate micelles. Biochemistry 39:15765–15774

117. Sawai MV, Jia HP, Liu LD et al (2001) The NMR structure of human beta-defensin-2 reveals a novel alpha-helical segment. Biochemistry 40:3810–3816

118. Pettersen EF, Goddard TD, Huang CC et al (2004) UCSF Chimera--a visualization system for exploratory research and analysis. J Comput Chem 25:1605–1612

119. Gesell J, Zasloff M, Opella SJ (1997) Two-dimensional H-1 NMR experiments show that the 23-residue magainin antibiotic peptide is an alpha-helix in dodecylphosphocholine micelles, sodium dodecylsulfate micelles, and trifluoroethanol/water solution. J Biomol NMR 9:127–135

Chapter 2

Tools for Designing Amphipathic Helical Antimicrobial Peptides

Davor Juretić, Damir Vukičević, and Alessandro Tossi

Abstract

Methods are described for the design of amphipathic helical AMPs, to improve potency and/or increase selectivity with respect to host cells. One method is based on the statistical analysis of known helical AMPs to derive a sequence template and ranges of charge, hydrophobicity, and amphipathicity (hydrophobic moment) values that lead to broad-spectrum activity, but leaves optimization for selectivity to subsequent rounds of SAR determinations. A second method uses a small database of anuran AMPs with known potency (MIC values vs. *E. coli*) and selectivity (HC_{50} values vs. human erythrocytes), as well as the concept of longitudinal moment, to suggest sequences or sequence variations that can improve selectivity. These methods can assist in the initial design of novel AMPs with useful properties in vitro, but further development requires knowledge-based decisions and a sound prior understanding of how structural and physical attributes of this class of peptides affect their mechanism of action against bacteria and host cells.

Key words α-Helical AMPs, Anuran AMPs, Amphipathicity, Hydrophobic moment, Longitudinal moment, D-descriptor

1 Introduction

Given their potential for helping to meet the current antibiotics resistance crisis, a lot of effort has been spent on the redesign of natural or design of artificial antimicrobial peptides (AMPs) to improve their therapeutic applicability. Methods vary from simply tinkering with natural sequences to improve potency, selectivity, and/or bioavailability to de novo design using complex computational and QSAR methods analogous to those applied for small molecule drugs [1]. In this chapter, we present two methods that our groups have applied to derive novel amphipathic helical AMPs, based on the analysis of multiple sequences of natural AMPs that have maintained or converged to this structure in their evolution [2].

The first method is based on the use of a sequence template derived from the statistical analysis of aligned sequences of helical AMPs from widely different species. This analysis, apart from

Paul R. Hansen (ed.), *Antimicrobial Peptides: Methods and Protocols*, Methods in Molecular Biology, vol. 1548,
DOI 10.1007/978-1-4939-6737-7_2, © Springer Science+Business Media LLC 2017

allowing to extract conserved structural motifs, suggested appropriate value ranges for some simple physicochemical attributes (charge, hydrophobicity, amphipathicity). The template has guided the design of novel peptides using either natural or non-proteinogenic amino acid residues [2–4]. It does not rely on complex computational algorithms and is aimed at obtaining lead sequences with useful, broad-spectrum in vitro antimicrobial activities, which then need to be refined for other required attributes (selectivity, stability, bioavailability, etc.) in subsequent rounds of syntheses and SAR analysis [5, 6]. The second method complements the first, with a more sophisticated analysis of sequence characteristics, taking into account positional frequencies of residues with respect to one another and residue frequencies related to selectivity [7]. This method required the creation of a database of anuran AMPs with quantified antimicrobial and cytotoxic activities, also to allow predicting a selectivity index (SI). An algorithm based on the longitudinal asymmetry of residue hydrophobicity (longitudinal moment), which appears to be significantly related with selectivity, is used in this prediction process. This method can be used either to design AMPs de novo or suggest mutations to existing AMP sequences that might increase their selectivity [7–9].

The design of AMPs in any case necessitates knowledge-based decisions that in turn require a sound prior understanding of how structural and physical attributes of this class of peptides affect their mechanism of action against bacteria (mainly, but not only, membrane effects) and cytotoxic activity against host cells. In this respect, peptides of the amphipathic α-helical structural group are by far the most studied and best understood among AMPs, and within this group, those of anuran origin are preponderant.

2 Tools

AMP design requires the collection of various types of data: peptide sequences, quantitative measures of biological activities, and amino acid and peptide physical attributes. Databases that contain these data and the computational tools needed to extract them are often freely available online. Unfortunately, these resources are frequently subject to site changes or discontinuation, if not curated with constancy.

2.1 AMP Sequence Databases

There are several databases dedicated to AMP sequences, some specific to particular types of organisms, and others, more general.

1. A comprehensive list of these has been assembled [10] and a tool is provided for obtaining a set of nonredundant sequences from them (http://mobiosd-hub.com/doveranalyzer/). It provides some such sets, which include sequences from some of the older databases that are no longer available.

2. Among the more general AMP databases that are currently available, APD3 [11] provides over 2600 natural peptide sequences (http://aps.unmc.edu/AP/main.php), while DBAASP [12] provides sequence, structural, and functional information for thousands of natural and synthetic AMPs (http://dbaasp.org). The DADP database provides about 2000 sequences for anuran peptides [13].
3. Alternatively, one can assemble one's own database by searching the large institutional databases such as UniProtKB, using appropriate keywords. In this respect, rather than carrying out text searches using terms such as "antimicrobial," "antibacterial," "antifungal," "bactericidal," etc., one could first use the corresponding available UniProt keyword, such as "Antimicrobial [KW-0929]," "Fungicide [KW-0295]," "Amphibian defense peptide [0878]," and Defensin [0211], as this significantly focuses the search.

2.2 AMP Activity Data

Selection of sequences for deriving templates does not require that these be accompanied by quantitative activity data, if the design method aims to obtain a generic antimicrobial activity; their presence in a dedicated AMP database may be a sufficient criterion. If selectivity is required of the design method, then quantitative and comparable activity data should be associated with peptide sequences used to develop it.

1. While some biological activity data is provided in AMP sequence databases, it is generally necessary to follow the associated literature entries and carry out comprehensive literature surveys in PubMed to obtain these data from the original literature.
2. Data should be collected for a statistically useful number of sequences (at least 30 for each training and testing data sets) so it is advisable to seek the most common antimicrobial activity estimate; MIC is determined by the serial dilution method. Nonetheless, sufficient data may be available only for the most frequently used bacterium: Gram-negative *E. coli* (for the next most frequent, Gram-positive *S. aureus*, MIC data is significantly less abundant). Always verify that a comparable bacterial load (CFU/ml) and similar medium conditions were used to determine the MIC (*see* **Note 1**).
3. Cytotoxic activity is assessed in terms of HC_{50}, as the hemolysis assay is by far the most frequent assay used for AMPs. Even so, it is the limiting data item, as it is tenfold less abundant than *E. coli* MIC values. Always verify that comparable cell loads and conditions were used (*see* **Note 2**).
4. The ratio of hemolytic to antibacterial activity provides a commonly used selectivity index, $SI = HC_{50}/MIC$ (*see* **Note 3**).

2.3 Tools for Determining Peptide Attributes and Improving Selectivity

Helical AMP features most often used in SAR studies are net charge, mean (per residue) side-chain hydrophobicity (<H>), and mean hydrophobic moment (amphipathicity, <μH>). The amplitudes of the polar and hydrophobic sectors (angles subtended on a helical wheel projection; *see* Fig. 1a, c), which are related to the amphipathicity, may also be considered. Another relevant parameter, the propensity for helix formation (α), can be predicted, but then requires experimental verification (normally CD spectroscopy).

1. There are many different side-chain hydrophobicity index (H_i) scales, in the AAindex database (http://www.genome.jp/aaindex) [14] that can be used to calculate <H> and <μH>. Note that H_i values vary quite markedly in different scales. For

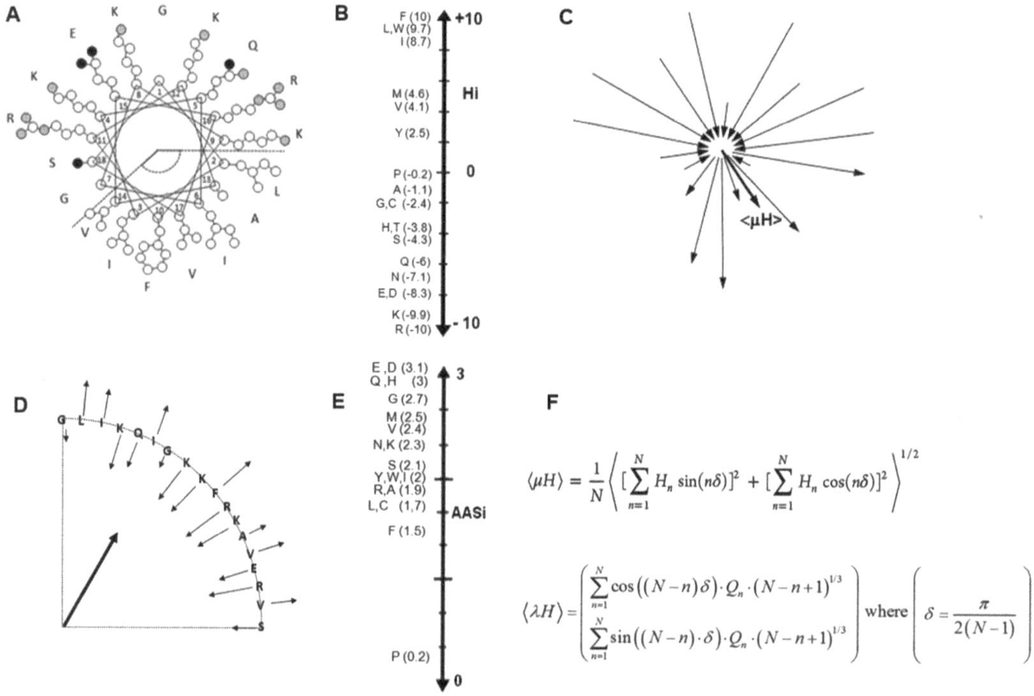

$$\langle \mu H\rangle = \frac{1}{N}\left\langle \left[\sum_{n=1}^{N} H_n \sin(n\delta)\right]^2 + \left[\sum_{n=1}^{N} H_n \cos(n\delta)\right]^2 \right\rangle^{1/2}$$

$$\langle \lambda H\rangle = \frac{\sum_{n=1}^{N} \cos((N-n)\delta)\cdot Q_n\cdot (N-n+1)^{1/3}}{\sum_{n=1}^{N} \sin((N-n)\cdot\delta)\cdot Q_n\cdot (N-n+1)^{1/3}} \quad \text{where} \quad \left(\delta = \frac{\pi}{2(N-1)}\right)$$

Fig. 1 Sequence and physicochemical attributes that can contribute to the potency and selectivity of helical AMPs. (**a**) Polar and hydrophobic sectors should be well separated on a helical wheel projection. The size of the sector is given by the angle it subtends (N^o if AA in the sector × 20°); (**b**) <H> and <μH> are determined by using hydrophobicity index (H_i) scales; the normalized consensus CCS scale is shown [15]. <H> = $\Sigma(H_i)i/N$ where N is the number of residues; (**c**) <μH> is the sum of vectors with scalar values equivalent to the H_i values for each side chain and orientation as on a helical wheel projection. Polar residues (negative H_i) point toward the helix axis, hydrophobic one away from it (see equation). (**d**) Method for calculating the longitudinal asymmetry moment <λH> [7] [*see* equation and Subheading 2.3, **item 5**]; (**e**) AA selectivity scale determined from a set of anuran helical AMPs with high SI values [7]; (**f**) equations for calculating <μH> and <λH>, Hn is the hydrophobicity index scalar value, N is total number of residues, n is the position, δ is the angle between vectors (100° for calculating <μH>); Qn is the averaged hydrophobicity of the sequence environment [7] (*see* Subheading 2.3, **item 5** and [**Notes 5** and **6**])

our initial sequence template, we used an ad hoc developed consensus scale based on 150 such scales (CCS scale; *see* Fig. 1b) [15].

2. Several online tools quantify <H> and <*μ*H> based on sequence and presumed helical or other conformations (e.g., HydroMCalc at http://splitbioinf.pmfst.hr/HydroMCalcV/HydroMCalc.html and HELIQUEST at http://heliquest.ipmc.cnrs.fr) [16] and provide helical wheel projections. Values depend markedly on the chosen H_i scale. The comparison of <H> or <*μ*H> determined with different scales is meaningful only for normalized scales (*see* **Note 4**). One can also use the equation in Fig. 1f to set up an Excel spreadsheet ad hoc to calculate <*μ*H> (*see* **Note 5**).
3. To obtain polar or hydrophobic sector angles, first define the sector size in terms of the number of continuous adjacent residues of the each type on a helical wheel projection and then multiply by 20° (*see* Fig. 1a).
4. Helix propensity can be predicted using applications such as Quark, which provides 3D models from peptide sequences [17]. Once peptides are synthesized, helix stability can be quantified in terms of the helix content as determined by CD spectroscopy under favorable conditions (e.g., in 50% trifluoroethanol): $\alpha = \left([\theta]^{TFE} - [\theta]^{H_2O}\right) / \left([\theta]\alpha - [\theta]^{H_2O}\right)$. $[\theta]^{H_2O}$ is the molar per/residue ellipticity in aqueous solution at 222 nm (predominantly random coil form), and $[\theta]\alpha$ is that of a perfectly helical peptide of the same size. Accounting for end effects, $[\theta]\alpha$ can be calculated by the method of Chen; $[\theta]\alpha = 39{,}000(1 - n/4)$ deg/cm/dmol where n is the residue number of residues and 39,000 molar per/residue ellipticity value with no end effects (perfect infinite helix) [18].
5. A prediction of AMP selectivity for anuran or anuran-like peptide sequences is made by the D-descriptor algorithm. It can be obtained as a first output of the Mutator tool (http://split4.pmfst.hr/mutator), which then also suggests single or double mutations to improve it (*see* Subheading 3.3). The underlying algorithm first uses a sliding window to calculate mean hydrophobicity *Qn* at each position (sequence environment) (*see* **Note 6**). It then bends the sequence into a 90° arc so that peptide N-terminal is at the *y*-axis and its C-terminal at the *x*-axis. A vector is associated with each amino acid so that its direction points away or toward the origin of the coordinate system depending on the arc and the *Qn* sign, while its length is equal to the *Qn* value (*see* Fig. 1d). Vector summation includes a weighting procedure giving greater weight to amino acids closer to N-terminus. This allows extracting a single vector, the longitudinal moment (<*λH*>), whose direction (the angle with respect to the *x*-axis) and size preserves

the lengthwise asymmetry information and encapsulates the sequence hydrophobicity profile. We found that SI correlates best with the cosine of the angle between longitudinal moments calculated using the scales of Janin and of Guy (*see* **Note** 7) [1, 7].

3 Methods

3.1 Basic Sequence Template Method

The template shown in Fig. 2a was originally developed from the alignment of over 150 helical AMP sequences from both vertebrate and invertebrate animals (mostly insects, frogs, and mammals) present in the AMSDb database (now discontinued). Many more are now available. Sequences were aligned simply from the N-terminus, based on the observation that >70% of helical AMP sequences had a conserved Gly residue at this position. With more focused sequence collections (such as related families of anuran AMPs [13] or mammalian helical cathelicidins [19]), internal conserved sequence motifs can also be used for pegging the alignment.

1. The template has been successfully used to design AMPs with as few as 13 residues. Normally, 18 residues were considered (a full helical wheel). Gly was placed at the C-terminus, and a further Tyr residue was added at the C-terminus to allow a more accurate spectrophotometric quantification without interference in CD spectra [3].

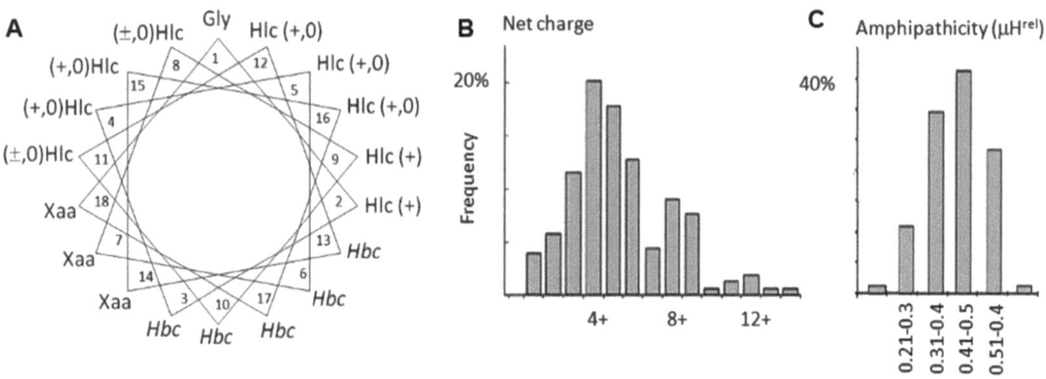

Fig. 2 Sequence template obtained from a set of over 150 helical AMPs of vertebrate and invertebrate animal origin. (**a**) The template is shown on a helical wheel projection. *Hlc*=*HydrophiLiC*, the type of polar residue most frequently found is indicated (cationic +, anionic− or neutral 0); *Hbc*=*HydrophoBiC*; *Xaa*=undefined/variable position, Gly and Pro are frequently found here; (**b**) net charge distribution among the analyzed helical AMPs; (**c**) relative amphipathicity (<μ**Hrel**>) for the analyzed AMPs. Amphipathicity was determined assuming that the considered peptide segment was perfectly helical, with no end effects or contribution for N- and C-terminals. Note that most AMPs fall between 0.3 and 0.6 range (30–60% the amphipathicity of a perfectly amphipathic peptide composed only of nine Arg and nine Phe, with the maximum and minimum H_i in the CCS scale, μH=6.4)

2. Charge is modulated by altering the ratio of cationic, anionic, and neutral polar residues in hydrophilic (*Hfc*) positions, as suggested by the template (*see* **Note 8**). This can be done without markedly altering μH, if residues with similar H_i values are used (e.g., *R*/*K*, *E*/*D*, and *Q*/*N* are all within the −10 to −6 range in the CCS scale; *see* Fig. 1b). Statistically, a charge of 4+/5+ was most frequent among considered AMPs (*see* Fig. 2b). The template allows for >9+, but in that case, electrostatic repulsion in the polar sector reduces helix-forming propensity (*see* **step 5**).
3. <H> is modulated either by altering nonpolar residues (H_i variation is much wider for these, ranging from −1 to +10 in the CCS scale; *see* Fig. 1b) or by altering the width of the hydrophobic sector using variable positions 7 and 14 (*Xaa*; *see* Fig. 2a). Natural helical AMPs tend to have about 50% hydro phobic residues, and <H> tends to be close to 0 or slightly negative [2, 4, 20].
4. Amphipathicity is also modulated mainly by altering the hydrophobic sector as described in **step 3**. Note that a relative amphipathicity (<μHrel>) of 0.5–0.6 is sufficient for good activity [4, 20] (*see* **Note 9**).
5. Helix-forming propensity can be modulated in several different ways. To increase it, one can place helix-favoring residues in the hydrophobic sector (aminoisobutyric acid is most often used but aminocylcopentanecarboxylic acid is even better [21]). This may be required to counteract the destabilizing effect of electrostatic repulsion in the polar sector, if the peptide's charge is required to be very high. To decrease helix stability/propensity, the helix-destabilizing residue Pro can be placed at one or more interface positions 7, 13, and 14, or d-amino acids can be placed in either or both the polar/hydrophobic sectors (*see* **Note 10**). Introducing Gly at one or more of the interface positions (e.g., 7 or 14) increases helix flexibility, but with differing effects on biological activity, depending on the position [6].

This template has been successfully used to design peptides with a potent and broad-spectrum antibacterial activity, which in some cases is extended to yeast and fungi. Peptides however tended to have a significant cytotoxicity. In general, by increasing stability, charge, and amphipathicity simultaneously, one could increase potency but at the expense of decreased selectivity. This seemed to depend strongly in the helix-forming propensity. It was also possible to use the template to identify novel natural AMPs in the sequence databases, and then suggest modifications to improve them [22].

3.2 Increasing Selectivity: The D-Descriptor Algorithm

As indicated in the previous section, the use of an adequate template generally assures a reasonable antimicrobial potency but not necessarily an acceptable selectivity toward host cells. One may need to tinker with the sequence in several subsequent rounds of SAR studies to achieve this, but it is generally quite difficult to reduce cytotoxic activity without affecting potency. In a second design method, analysis of the sequence of anuran helical AMPs with defined SI indicated other sequence features that could improve selectivity. The concept of lengthwise asymmetry (longitudinal moment $<\lambda H>$) in residue moments was also introduced as a relevant physical feature related to selectivity.

1. Analysis of anuran AMPs with high SI values resulted in a table of AA selectivity indices (*see* Fig. 1d). Peptide design took this into account so that at least a third of residues were from among five AA most frequently present in highly selective AMPs (*E*, *D*, *Q*, *H*, *G*).
2. This analysis also resulted in tables that help define motif regularity in AMP helices; for any given residue, the most frequent residue present in positions *i* (successor) and *i*+4 (adjacent on helix face), respectively, is defined (*see* [7]).
3. Indications in **steps 1** and **2** can then be implemented together with indications provided in Subheading 3.1, namely, Gly at position 1, separation of polar and nonpolar residues in the helical wheel projection, a charge of at least +4, and $<H>$ between −1.2 and 0, when designing novel AMPs.
4. A final prerequisite is that the SI index, calculated according to the D-descriptor algorithm (*see* **item 5** in Subheading 2.3), is as high as possible (*see* **Note 11**).

This method has been successfully used to design artificial AMPs called adepantins, with very low hemolytic activities [7, 8]. However, to train the D-descriptor algorithm, it was only possible to obtain a sufficient number of AMPs with high SI values relative to MIC against *E. coli*. As a consequence, although the resulting selectivity is good, activity is generally high only toward Gram-negative bacteria. Once a sufficient data set is found for *S. aureus*, activity may become more broad spectrum. Alternatively, one can use the D-descriptor algorithm to suggest modifications of AMPs already known to have good activity against Gram-positive bacteria, rather than to assist ab initio AMP design (*see* Subheading 3.3).

3.3 Increasing Selectivity of AMPs

The D-descriptor algorithm can be used independently of the design rules set out in Subheading 3.2. In fact, some peptides such as PGLa, which are predicted and confirmed to have high SI values, do not conform so well to the residue selectivity frequencies described above (*see* Subheading 3.2, **steps 1** and **2**). It is possible to use the algorithm to suggest possible point mutations to a known AMP sequence that might increase its selectivity. Variations should however remain compatible with typical helical AMP characteristics (*see* Subheadings 3.1 and 3.2), as this reduces the risk of losing potency while increasing selectivity.

1. The D-descriptor has been developed using anuran AMPs in the training and testing data sets. Its use is therefore not encouraged with AMPs of other origin. However, the underlying concept may be generally applicable to helical AMPs from other organisms, and it has been effective in improving the selectivity of teleost AMPs [23].
2. The "Mutator" algorithm (*see* Subheading 2.3, **step 5**) is based on a set of 26 AMPs with a potent antimicrobial activity and measured SI > 20 and uses the normalized H_i scales of Janin and Guy to calculate <λH> and then predict SI. It also uses Eisenberg's scale to determine <μH> as a filtering parameter [9]. It systematically carries out one or two mutations along the sequence but only reports those that predict an SI increase of at least 10 and which conform to the criteria set out in Subheading 3.2 (*see* **Note 12**).
3. The resulting sequences should in any case be inspected to determine compliance with typical AMP parameters (Subheading 3.1), such as sector separation on a helical wheel projection and q, <H>, and <μH> being within the expected ranges.
4. Adepantins, produced de novo by using the Designer algorithm [7], all had 23 amino acid residues. Shorter adepantins could be designed by inserting appropriate instruction in the source code; however they had weaker antimicrobial activity and decreased selectivity (lower SI), indicating that additional optimization is needed for computational design of such peptides.

4 Notes

1. Changing the bacterial load will result in an altered MIC, as for a given AMP concentration, the peptide/cell ratio changes. However, the relationship is not linear. It has been reported for a small helical AMP that an increase in CFU by a factor of 5 (e.g., $1 \times 10^5 \rightarrow 5 \times 10^5$) shifts the MIC by a factor of 2 [24]. Determination of MIC is described in Chapter 22.
2. Hemolysis assays are often reported using 0.5% RBC suspensions, but also 5–6%. The latter result in a significantly lower concentration-dependent lysis by AMPs. Furthermore, hemolysis data should only be considered if determined using erythrocytes from fresh human blood. The lytic effect of AMPs on erythrocytes from other animals (bovine, sheep, rat, or mouse

erythrocytes are sometimes used) can be quite different. A protocol for determination of hemolysis of AMPs against red blood cells is described in Chapter 31.

3. This is sometimes called the therapeutic index (TI), a rather inappropriate term, which is also used to indicate the active/lethal dose of a drug. Given that MICs are determined by serial dilution, the SI tends to vary in a discontinuous manner, and for AMPs with the same HC_{50} but different MICs, the SI will vary by multiples of a factor of 2.

4. Alternatively use the "relative hydrophobic moment" **μHrel** = μH/μ**Hmax**, where **μHmax** refers to a perfectly amphipathic 18-residue (one complete helix turn) peptide composed only of the most hydrophobic and hydrophilic amino acids in a given H_i scale. This is a sort of % maximum amphipathicity which may be more homogenous using different scales.

5. In this equation, N indicates a total residue number, n indicates the position, and $n\delta$ is the angle of successive side-chain H_i vectors, where $\delta = 100°$ for an α-helix and Hn is the H_i value at each position.

6. There are several different ways of assigning positional hydrophobicity. The simplest is to assign it the H_i value for the side chain of the AA in that position. Alternatively one can assign a mean hydrophobicity $\bar{H}_n$ calculated from a window of a given size associated with the central amino acid n, at any given position. Another option is to calculate the sequence environment Qn, omitting the central amino acid n. Mutator uses the last option.

7. An exhaustive examination was made of many possibilities (e.g., different arc bending angles and use of 1, 2, or 3 longitudinal moment vectors based on different H_i scales), but the final choice fell on the cosine of the angle between these two scales for an arc of 90° (D value). This resulted in the best correlation with the measured SI values of known anuran AMPs ($r^2 = 0.83$), where SI = 50–45D.

8. C-terminal amidation (often present in natural AMPs) increases charge by +1. Decide if histidine is to be considered cationic or neutral.

9. <μH> is normally calculated as a global property involving the whole sequence, as shown in Fig. 1c. However, natural AMPs are unlikely to conform well to this approximation. It could be more appropriate to use a sliding window of appropriate length (say, 11–12 residues) to obtain a μH profile. Furthermore, it is likely that the helical conformation will be somewhat distorted on membrane interaction, so that a certain flexibility could also be introduced in the projection angle mentioned in **Note 5** (say, $\delta = 80$–120°) [24, 25].

10. Single d-AA substitutions do not strongly destabilize membrane-bound helical peptides but can significantly decrease cytotoxicity [26]. Placing two successive d-AA residues in the sequence has a more marked effect [27].
11. On its own, the D-descriptor algorithm just indicates if a peptide is likely to be selective but not if it is likely to be a potent antimicrobial. That is why it is combined with the design rules listed in Subheading 3.2. Note that for predictions close to the maximum output value of the algorithm (~95), the measured SI value could be significantly higher.
12. If Mutator predicts SI > 85 for a starting peptide, it cannot suggest any substitution given the imposed improvement limit of at least 10 and the maximum possible SI of 95 (*see* Subheading 3.3 **step 2** and **Notes** 7 and **11**). One can sometimes get around this limit by first finding a substitution that reduces SI to <85, then keeping that position fixed and finding other substitutions that improve the SI back to the 85–95 range, and finally returning the first AA to the original.

Acknowledgments

Authors acknowledge funding from the Croatian Science Foundation project 8481 BioAmpMode.

References

1. Juretić D, Vukičević D, Petrov D et al (2011) Knowledge-based computational methods for identifying or designing novel, non-homologous antimicrobial peptides. Eur Biophys J 40:371–385
2. Tossi A, Sandri L, Giangaspero A (2000) Amphipathic, α-helical antimicrobial peptides. Biopolymers 55:4–30
3. Giangaspero A, Sandri L, Tossi A (2001) Amphipathic α helical antimicrobial peptides. Eur J Biochem 268:5589–5600
4. Zelezetsky I, Tossi A (2006) Alpha-helical antimicrobial peptides—using a sequence template to guide structure–activity relationship studies. Biochim Biophys Acta 1758:1436–1449
5. Zelezetsky I, Pacor S, Pag U et al (2005) Controlled alteration of the shape and conformational stability of α-helical cell-lytic peptides: effect on mode of action and cell specificity. Biochem J 390:177–188
6. Zelezetsky I, Pag U, Sahl H-G, Tossi A (2005) Tuning the biological properties of amphipathic α-helical antimicrobial peptides: rational use of minimal amino acid substitutions. Peptides 26:2368–2376
7. Juretić D, Vukičević D, Ilić N et al (2009) Computational design of highly selective antimicrobial peptides. J Chem Inf Model 49:2873–2882
8. Ilić N, Novković M, Guida F et al (2013) Selective antimicrobial activity and mode of action of adepantins, glycine-rich peptide antibiotics based on anuran antimicrobial peptide sequences. Biochim Biophys Acta 1828:1004–1012
9. Kamech N, Vukičević D, Ladram A et al (2012) Improving the selectivity of antimicrobial peptides from anuran skin. J Chem Inf Model 52:3341–3351
10. Aguilera-Mendoza L, Marrero-Ponce Y, Tellez-Ibarra R et al (2015) Overlap and diversity in antimicrobial peptide databases: compiling a non-redundant set of sequences. Bioinformatics 31:2553–2559

11. Wang G, Li X, Wang Z (2016) APD3: the antimicrobial peptide database as a tool for research and education. Nucleic Acids Res 44:D1087–D1093
12. Pirtskhalava M, Gabrielian A, Cruz P et al (2016) DBAASP v. 2: an enhanced database of structure and antimicrobial/cytotoxic activity of natural and synthetic peptides. Nucleic Acids Res 44:D1104–D1112
13. Novkovic M, Simunic J, Bojovic V et al (2012) DADP: the database of anuran defense peptides. Bioinformatics 28:1406–1407
14. Kawashima S, Pokarowski P, Pokarowska M et al (2008) AAindex: amino acid index database, progress report 2008. Nucleic Acids Res 36:D202–D205
15. Tossi A, Sandri L, Giangaspero A (2002) New consensus hydrophobicity scale extended to non-proteinogenic amino acids. Peptides 27:416–417
16. Gautier R, Douguet D, Antonny B, Drin G (2008) HELIQUEST: a web server to screen sequences with specific α-helical properties. Bioinformatics 24:2101–2102
17. Xu D, Zhang Y (2012) Ab initio protein structure assembly using continuous structure fragments and optimized knowledge-based force field. Proteins 80:1715–1735
18. Chen Y-H, Yang JT, Chau KH (1974) Determination of the helix and β form of proteins in aqueous solution by circular dichroism. Biochemistry (Mosc) 13:3350–3359
19. Xhindoli D, Pacor S, Benincasa M et al (2016) The human cathelicidin LL-37 - a pore-forming antibacterial peptide and host-cell modulator. Biochim Biophys Acta 1858:546–566
20. Pag U, Oedenkoven M, Sass V et al (2007) Analysis of in vitro activities and modes of action of synthetic antimicrobial peptides derived from an -helical “sequence template”. J Antimicrob Chemother 61:341–352
21. Tossi A, Scocchi M, Zahariev S, Gennaro R (2012) Use of unnatural amino acids to probe structure–activity relationships and mode-of-action of antimicrobial peptides. In: Pollegioni L, Servi S (eds) Unnatural amino acids. Humana Press, Totowa, NJ, pp 169–183
22. Zelezetsky I, Pag U, Antcheva N et al (2005) Identification and optimization of an antimicrobial peptide from the ant venom toxin pilosulin. Arch Biochem Biophys 434:358–364
23. Tessera V, Guida F, Juretić D, Tossi A (2012) Identification of antimicrobial peptides from teleosts and anurans in expressed sequence tag databases using conserved signal sequences. FEBS J 279:724–736
24. Uggerhøj LE, Poulsen TJ, Munk JK et al (2015) Rational design of alpha-helical antimicrobial peptides: do’s and don’ts. Chembiochem 16:242–253
25. Kozić M, Vukičević D, Simunić J et al (2015) Predicting the minimal inhibitory concentration for antimicrobial peptides with Rana-box domain. J Chem Inf Model 55:2275–2287
26. Oren Z, Shai Y (1997) Selective lysis of bacteria but not mammalian cells by diastereomers of melittin: structure-function study. Biochemistry (Mosc) 36:1826–1835
27. Dathe M, Wieprecht T (1999) Structural features of helical antimicrobial peptides: their potential to modulate activity on model membranes and biological cells. Biochim Biophys Acta 1462:71–87

Chapter 3

Chemical Synthesis of Antimicrobial Peptides

Lena Münzker, Alberto Oddo, and Paul R. Hansen

Abstract

Solid-phase peptide synthesis (SPPS) is the method of choice for chemical synthesis of peptides. In this nonspecialist review, we describe commonly used resins, linkers, protecting groups, and coupling reagents in 9-fluorenylmethyloxycarbonyl (Fmoc) SPPS. Finally, a detailed protocol for manual Fmoc SPPS is presented.

Key words 9-Fluorenylmethyloxycarbonyl (Fmoc) solid-phase peptide synthesis, DIC, Oxyma

1 Introduction

Solid-phase peptide synthesis (SPPS), introduced by Bruce Merrifield in 1963 [1], is a strategy for the synthesis of peptides which entails the repetitive coupling of the C-terminus of protected amino acids to an insoluble resin support. The attachment to a solid support via a cleavable linker enables the use of large excesses of reagents and facilitates an effective removal of by-products to obtain peptides of high purity. Moreover, the method permits the automation of the whole process and is, nowadays, the most commonly used strategy for peptide synthesis. For specialist reviews on SPPS, *see* [2–6] (Scheme 1).

1.1 General Concept: Transient/Semipermanent Protection Strategy

In the initial step of synthesis, the first amino acid is attached (typically by its carboxyl function) to the resin via a linker. To avoid polymerization, protecting groups shield the α-amino moiety, as well as reactive side-chain functionalities. The use of orthogonal protecting groups, such as acid and base labile, allows for the selective removal of the N^{α}-protecting group without affecting the side-chain protecting groups.

The peptide is synthesized stepwise until the desired length is achieved; however, researchers claim that only the synthesis of peptides of up to 50 amino acids in length is routinely feasible [7]. Depending on the linker and the protection strategy adopted, it is

Paul R. Hansen (ed.), *Antimicrobial Peptides: Methods and Protocols*, Methods in Molecular Biology, vol. 1548,
DOI 10.1007/978-1-4939-6737-7_3, © Springer Science+Business Media LLC 2017

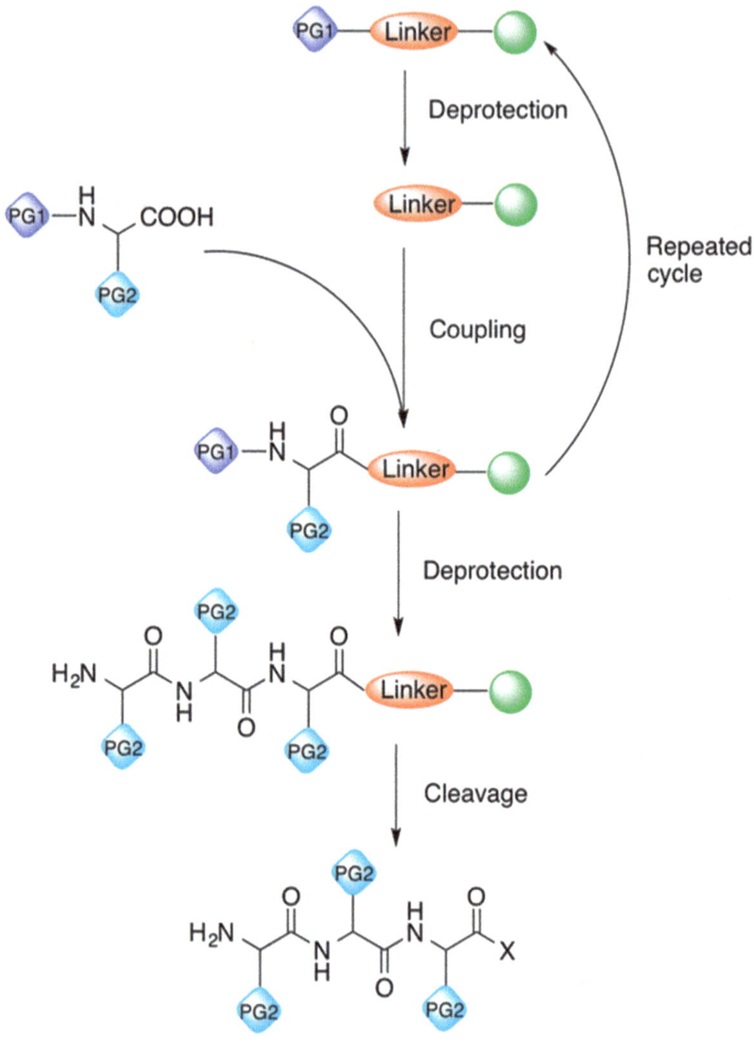

Scheme 1 Overview of solid-phase peptide synthesis. PG: protecting group, X = OH or NH_2

then possible to (a) selectively remove the side-chain protecting groups, or (b) release the (still protected) peptide from the solid support, or (c) do both at the same time to obtain the free peptide. For peptides longer than 50 residues, fragment condensation [8] or native chemical ligation may be considered [9].

1.2 Boc/Bzl-Fmoc/tBu Protection Schemes

Initially, Merrifield's synthesis featured Boc groups (**1**) on the N^{α}- and benzyl-based (**2**) side-chain protecting groups (Fig. 1). Although both groups are acid labile, this protection scheme can be applied to SPPS due to their differential acid lability: Boc groups can be removed with 50% trifluoroacetic acid (TFA), whereas the removal of benzyl-based protecting groups and the cleavage from the resin require very strong acids, such as hydrogen fluoride (HF)

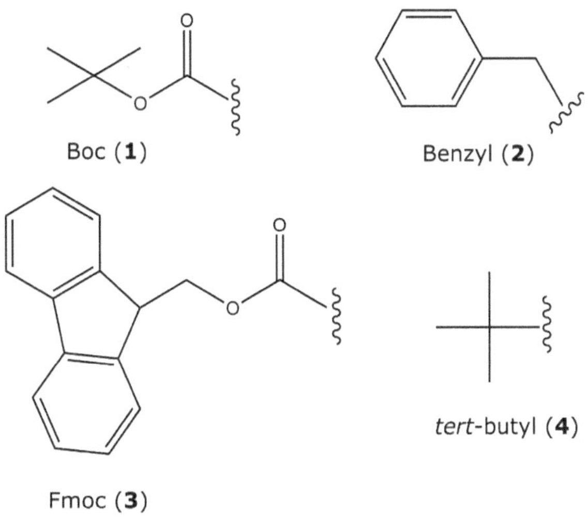

Fig. 1 Most important protecting groups used in Boc and Fmoc SPPS

and trifluoromethanesulfonic acid (TFMSA). Nevertheless, the use of these toxic and highly volatile reagents is hazardous and requires special equipment. For a recent review on Boc SPPS, *see* Pedersen et al. [10]. In comparison, the Fmoc strategy is truly orthogonal, using the base-labile Fmoc (**3**) for N^{α}-protection and acid-labile groups, e.g., *tert* butyl (**4**) for side-chain protection.

The removal of the Fmoc group via base-induced β-elimination results in the release of carbon dioxide and dibenzofulvene, a reactive electrophile. Secondary amines, such as piperidine, function both as bases and as scavengers for the released dibenzofulvene and prevent the irreversible reattachment of the latter to the amino group. Although Boc chemistry presents several advantages, such as increased solubility, less expensive building blocks, and compatibility with electrophilic side chains, Fmoc SPPS is widely preferred as it overcomes the application of HF.

For the remainder of this review, only Fmoc SPPS will be described.

1.3 Resins

A resin for SPPS needs to be chemically, mechanically, and physically stable. Apart from that, parameters such as the bead size, the degree of cross-linking, the particle size distribution, and the swelling capacity are considered, depending on the envisaged application. Most resins fall in one of three categories [11]: classic polystyrene (PS) resins, PS functionalized with polyethylene glycol (PEG) resins, and cross-linked PEG resins.

In polystyrene supports, the matrix is cross-linked with 1–2% divinylbenzene (DVB) which results in the best compromise between mechanical stability and swelling properties in low to middle polarity solvents, such as DMF (dimethylformamide) and DCM (dichloromethane). Polystyrene resins are mostly suitable for small peptides, since the highly hydrophobic environment of the resin amplifies the aggregative behavior of longer peptides, making the N-terminus less accessible during synthesis [12].

TentaGel (TG) resins are a hybrid composed of polyethylene glycol (PEG) coupled to a low cross-linked polystyrene matrix. They contain 50–70% (w/w) of PEG grafted to polystyrene that is cross-linked by an ether linkage [13]. PEG shows both hydrophobic and hydrophilic properties, and due to the high ratio of PEG in the resin, the graft copolymer's physicochemical properties are dependent on the PEG residues. TG resins facilitate the synthesis of longer peptide sequences, and their pressure stability allows for use in batch and continuous flow conditions.

PEGA (poly(ethylene glycol)-poly-(*N*,*N*-dimethylacrylamide) copolymer) resins, which were developed by Meldal and coworkers [14], have especially excellent swelling properties in water, as they do not contain any hydrophobic polystyrene. This makes them also applicable to protein interaction studies. However, their sticky nature only allows easy handling in the swollen state, and, once the resin is dry, the beads are easily damaged. In contrast to that, CM (ChemMatrix), which was first reported in 2006 [15], shows handling properties that are comparable to PS or TG resins. Moreover, as the resin only contains primary ether bonds, it offers high chemical stability.

1.4 Linkers

A linker is a bifunctional moiety that allows a cleavable connection between the peptide chain and the resin backbone—in most cases via the C-terminal α-carboxyl group. Alternatively, the peptide chain is anchored via a side chain or by a backbone amide linker (BAL). In the latter case, the first amino acid bears a carboxy-protecting group that can be removed to obtain the free α-carboxylic acid, enabling C-terminal modifications on solid phase.

Inevitably, the linker dictates the conditions for the peptide cleavage and therefore the overall synthetic strategy. Therefore, a large variety of linkers compatible with Fmoc SPPS have been developed. For the synthesis of peptide acids, the most popular are the 2-chlorotrityl (**5**) [16] and the Wang linker (**6**) [17]. The former may also be used for synthesis of protected peptides. The Rink amide linker (**7**) [18] is the most widely used for the generation of peptide amides, and the BAL linker (**8**) [19] may be used for C-modified [20] or cyclic peptides [21] (Fig. 2).

1.5 Coupling Reagents

Coupling reagents are necessary for the formation of an amide bond between the free N-terminus of one residue and the carboxyl function of another amino acid [22]. The reagent leads to the activation of the carboxy function by offering a favorable leaving group, which can be easily displaced by the resin-bound amino group. The activated species is most often an ester (**9**) (Fig. 3a).

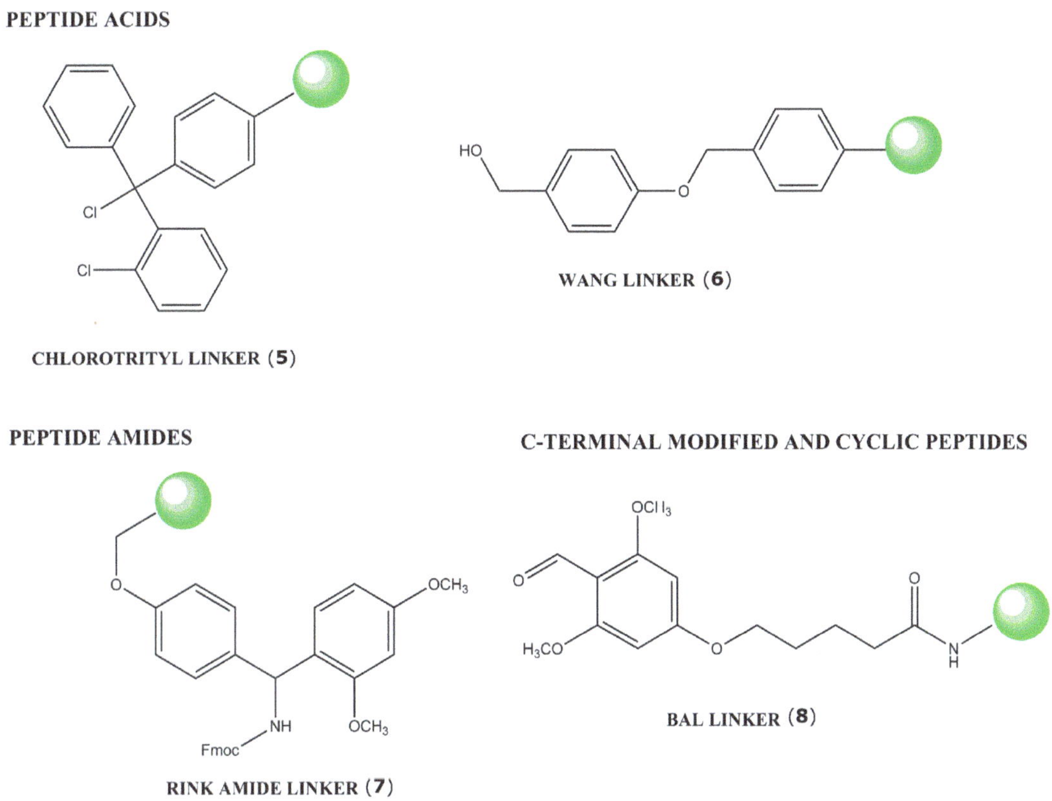

Fig. 2 The most commonly used linkers in Fmoc SPPS: 2-chlorotrityl, Wang, Rink, and backbone amide linker

However, activation of the carboxyl group can lead to racemization of a chiral α carbon via two major pathways, both of which are base catalyzed [23], direct enolization (**10–12**) (Fig. 3b), and 5(*4H*)-oxazolone formation (**13–18**) (Fig. 3c).

The most commonly used coupling reagents can be divided into three groups: carbodiimides, phosphonium salts, and aminium salts.

Carbodiimides, first described by Sheehan and Hess in 1955 [24], include DCC (**19**) and DIC (**20**). Generally, DIC is preferred over DCC since the urea by-product formed is soluble in DMF and less toxic. Carbodiimides lead to the formation of *O*-acylisourea, a reactive species which then undergoes aminolysis by the N-terminus of the resin-bound amino acid, thus resulting in an amide bond. Alternatively, (a) oxazolone formation can occur and result in racemization, or (b) the *O*-acylisourea intermediate can be rearranged to the unreactive *N*-acylurea, which terminates the chain elongation. Therefore, carbodiimides are commonly used with *N*-hydroxy derivatives, such as HOBt (**21**) [25], HOAt (**22**) [26], and Oxyma (**23**) [27] in order to suppress these undesired side reactions. Oxyma, recently introduced by El-Faham and Albericio, exhibits superior capacity to decrease racemization and avoids the risk of explosion of HOBt and HOAt.

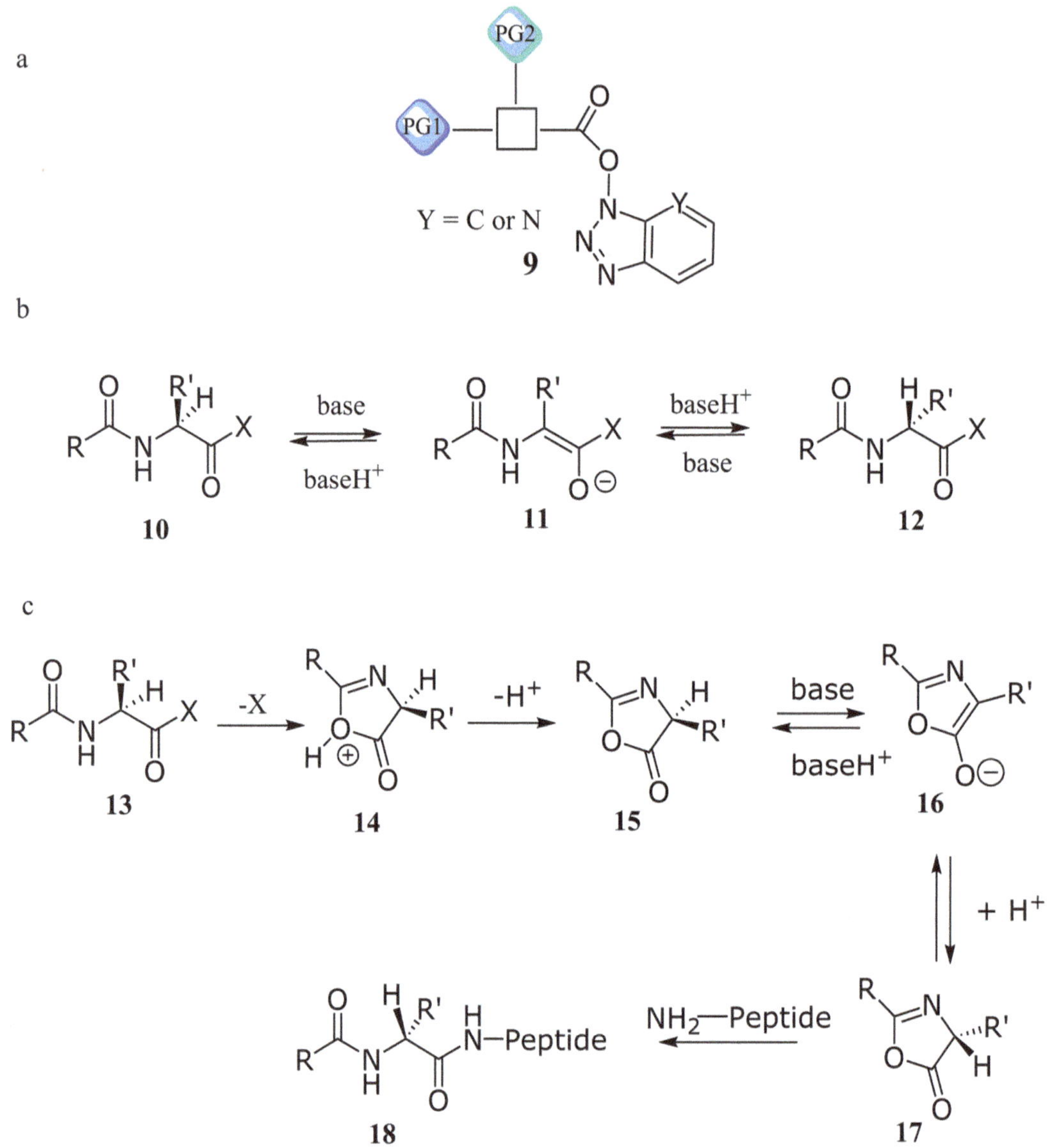

Fig. 3 (**a**) Activated ester. (**b**) Racemization via direct enolization. (**c**) Racemization via 5(*4H*)-oxazolone formation

Phosphonium-based coupling reagents constitute an efficient alternative to carbodiimides. BOP (**24**) was the first reagent of this type [28]. Nowadays, the application of BOP is widely displaced by other coupling reagents (such as PyBOP (**25**) [29], PyAOP (**26**) [30], and PyBroP (**27**) [31]) due to the release of the carcinogenic hexamethylphosphorotriamide (HMPA) as a by-product. Instead of a dimethylamine moiety, the compounds PyBOP, PyAOP, and PyBroP contain a pyrrolidine and react without releasing HMPA in the activation step. These reagents are typically used in combination with a non-nucleophilic tertiary amine, e.g., diisopropylethylamine (DIEA) or trimethylpyridine (TMP). The addition of *N*-hydroxy derivatives can further improve the coupling process and reduce the loss of configuration.

In aminium salts, such as HBTU (**28**) and TBTU (**29**) [32, 33], HATU (**30**) and TATU (**31**) [26], a positive nitrogen atom replaces the phosphonium residue. As for phosphonium salts, the coupling reaction requires the addition of base. El-Faham and Albericio recently described the morpholino-based aminium reagent COMU (**32**) [34, 35]. Increased efficiency compared to HATU/HBTU, compatibility with microwave-assisted SPPS, non-toxicity, and higher solubility in DMF make it an interesting alternative to the formers. However, COMU is not stable in DMF, thus being unsuitable for automated synthesis.

An overview of several different coupling reagents of various classes is given in Fig. 4.

1.6 Protecting Groups

Protecting groups are needed in order to shield the nucleophilic functions of the amino acid side chains [36]. The side chains of Ser, Thr, Tyr, Glu, and Asp are typically protected by the acid-labile *tert*-butyl (*t*Bu) group [37]. For Lys and Trp, the Boc group is preferred [38]. The 2,2,4,6,7-pentamethyl-dihydrobenzofuran-5-sulfonyl (Pbf) [39] is used for Arg. The trityl (Trt) group [40] is widely used for Cys, Asn, Gln, and His. Finally, a number of protecting groups are also available for special purposes and have been reviewed elsewhere [3, 36].

1.7 Cleavage Cocktails

In Fmoc SPPS, the peptide is typically released from the resin using a cocktail of acid and scavengers. The latters are necessary to capture reactive species produced during the acidolytic treatment. The acid is most often TFA in concentrations between 1 and 95 %.

Examples of established cleavage cocktails include:

Reagent K: TFA/thioanisole/H_2O/phenol/EDT (82.5:5:5:5:2.5) [41]

Reagent L: TFA/TIS/dithiothreitol/H_2O (88:2:5:5) [42]

Reagent R: TFA/thioanisole/ethandithiole/anisole (90:5:3:2) [43]

Reagent B: TFA/phenol/H_2O/triisopropylsilane (88:5:5:2) [44]

For protected peptide acids on a 2-chlorotrityl resin: acetic acid/TFE/dichloromethane [45] (20:20:60) or 1 % TFA in DCM

For most purposes, a combination of TFA/H_2O/triisopropylsilane (95:2.5:2.5) (v/v) works well in our hands [46]. However, thiols must be added to ensure peptides containing Cys remain in their reduced state. We recommend dithiothreitol (DTT, as in Reagent L) over ethanedithiol (EDT) and 2-mercaptoethanol; it is solid and far less obnoxious.

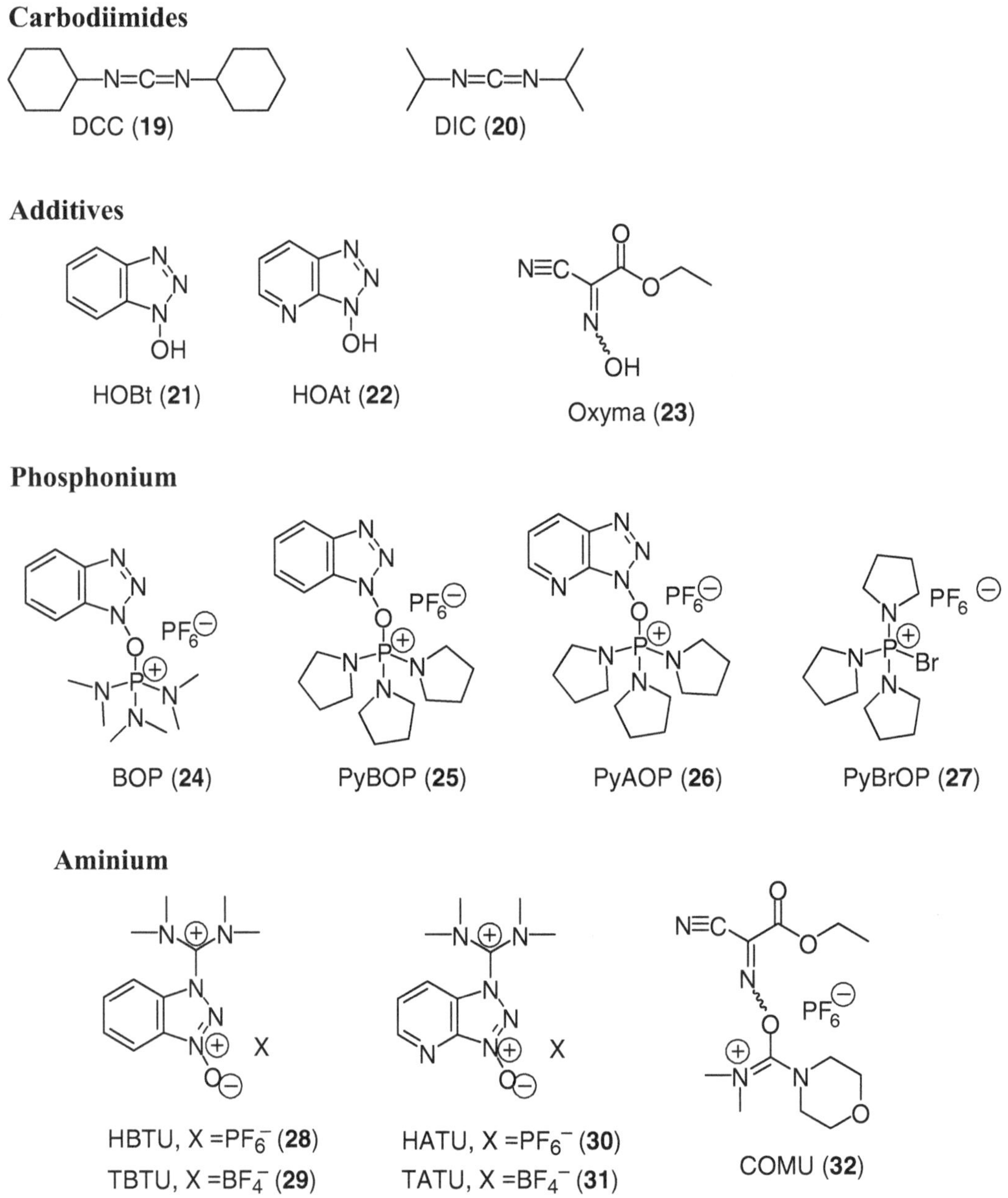

Fig. 4 The most commonly used coupling reagents

It may often be helpful to carry out a test cleavage run using less than 10 mg of resin to determine the best cleavage cocktail.

The cleavage solution is collected after 2 h, TFA reduced to a volume of approx. 200 μL by a stream of nitrogen, and the peptide precipitated in cold ether and centrifuged. The peptide is then washed twice with ether. After drying, the peptide is then dissolved in 10–20% CH_3CN in 0.1% aqueous TFA and lyophilized to give a fluffy white powder.

1.8 Analysis

LC-MS or analytical reverse-phase HPLC and MALDI-time-of-flight mass spectrometry are the methods of choice to analyze synthetic antimicrobial peptides. Analytical and preparative HPLC are most often done on a C-18 column using 0.1 % aqueous TFA and 0.1 % TFA in neat acetonitrile as buffers. An established matrix for MALDI-TOF-MS is based on α-cyano-4-hydroxycinnamic acid (10 mg/mL in $CH_3CN:H_2O$ 1:1, plus 0.1 % TFA). Alternatively, the HPLC UV and mass spectra may be generated by a LC-MS instrument. A detailed HPLC and LC-MS procedure is given in Chapter 5.

2 Materials and Preparations

A protocol for manual Fmoc SPPS protocol using DIC and Oxyma as coupling reagents is described.

2.1 Choosing the Resin

A preloaded 2-chlorotrityl or Wang resin is recommended if a free peptide acid is desired. A non-preloaded Tentagel®/Hypogel® resin functionalized with a RAM linker is widely used for the synthesis of peptide amides.

2.2 Calculation of the Amount of Resin Needed

Depending on how much peptide is desired, the following formula can be used to calculate the amount of resin needed:

$$\text{g of Resin} = \frac{\text{mmol of Peptide (wanted)}}{\text{Resin loading}\left(\frac{\text{mmol}}{\text{g}}\right)}$$

2.3 Syringes and Suction Device

The protocol described here is based on the use of 5 mL syringes equipped with a polytetrafluoroethylene (PTFE) filter as reactors for 50–200 mg of resin (synthesis scale of approximately 0.05 mmol) connected to a suction device (*see* **Notes 1** and **2**). The syringes are equipped with 200 μL pipette tips and placed in the suction plate (Fig. 5). However, the suction plate is not strictly necessary.

2.4 Oxyma and Amino Acid Solutions

Based on the synthesis scale calculated before, five equivalents of Oxyma are needed for each coupling.

It is helpful to calculate the total number of couplings and prepare a 0.4–0.6 M stock solution in DMF (the final volume should be sufficient to completely cover the resin; this, in turn, will depend on the resin loading):

$$\text{mg Oxyma} = 142.11\,\text{mg} / \text{mmol} \times 5 \times \left[\text{synthesis scale (mmol)}\right] \times \left[\text{number of couplings}\right]$$

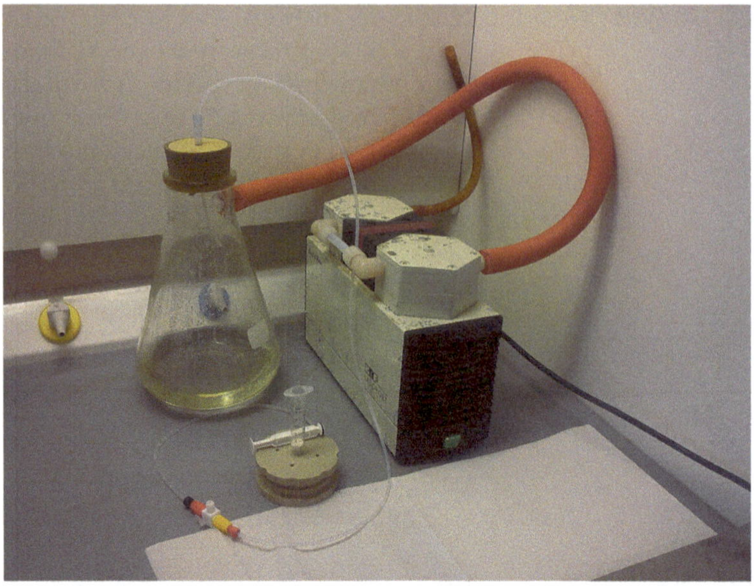

Fig. 5 Setup for manual SPPS

The total amount shall then be dissolved in DMF to achieve a 0.4–0.6 M concentration.

The total number of couplings dictates how many individual tubes the Oxyma solution should be divided in. The amino acid samples will be dissolved in the Oxyma solution as described in Subheading 2.5. We recommend using a spreadsheet to calculate the amounts automatically.

2.5 Amino Acid Solutions

The Oxyma stock solution prepared in Subheading 2.4 is used to prepare equimolar solutions of the amino acids by dissolving them in an appropriate volume. For each coupling one individual sample tube containing five equivalents of amino acid and Oxyma should be prepared.

The required amount of protected amino acid for each coupling should be weighed out in an individual tube according to the following formula:

$$\text{mg Amino acid} = \text{Molecular weight} \times 5 \times \left[\text{synthesis scale}\,(\text{mmol})\right]$$

Equally divide the Oxyma stock solution among the amino acid samples. Alternatively, prepare a stock solution of amino acid and Oxyma and then divide it into individual tubes.

The tubes can be stored in a refrigerator at +4 °C for up to 2 weeks. Let them warm up to room temperature before use.

DIC is added to the reagent solution just before coupling. The appropriate amount is calculated as follows:

$$\mu L\ DIC = \left[126.20\,mg/mmol \times 5 \times \left[synthesis\ scale\ (mmol)\right]\right] / 0.81\,mg/\mu L$$

2.6 Deprotection Solution to Remove the Fmoc Group

Prepare a 20% piperidine in DMF solution by measuring 20 mL piperidine and 80 mL of DMF and then mixing them in a glass bottle with a screw cap.

3 Methods

3.1 Synthesis Start

1. Remove the piston from the syringe, weigh out the resin directly inside the syringe (*see* **Note 3**), reinsert the piston, and push it just above the resin.
2. Draw 3 mL of DMF into the syringe from a beaker.
3. Use a pressure cap to close the bottom and let it swell in DMF for at least 30 min.

3.2 Quick Wash

1. Remove the piston and place the syringe on the suction plate.
2. If any resin residue is left on the pistonhead, wash it down with DMF.
3. Wash three times with DMF.

3.3 Amino Acid Coupling

1. Add DIC to the Oxyma/amino acid solution and mix.
2. Draw the coupling solution into the syringe.
3. Cover the syringe with tinfoil and leave on a shaker.
4. After 1.5 h drain the coupling solution.
5. If a recoupling is desired, do a *quick wash* (Subheading 3.2, steps 1–3) and then repeat **steps 1**–**4**.

3.4 Full Wash

1. Remove the piston and place the syringe on a suction plate.
2. If any resin residue is left on the pistonhead, wash it down with DMF.
3. Wash with DMF (3×), DCM (3×), and DMF again (5×).

3.5 Fmoc-Group Removal (See Note 4)

1. Transfer 2–3 mL of the 20% piperidine in DMF solution into the syringe.
2. After 4 min drain the solution and do a *quick wash* (see Subheading 3.2).
3. Repeat **steps 1** and **2** twice.
4. For peptides longer than 10 residues, we recommend extending the time of the last two deprotection cycles to 7 min each.
5. Put the piston back in place and change the syringe tip.

3.6 Full Wash (as in Subheading 3.4)

3.7 Reiteration of Coupling and Deprotection Cycle

1. For each subsequent amino acid, repeat Subheading 3.3, **steps 1–5**, until the sequence is completed.

3.8 Final Fmoc-Group Removal

1. Full wash as in Subheading 3.4, **steps 1–3**.
2. Fmoc deprotection as in Subheading 3.5, **steps 1–5**.
3. Full wash as in Subheading 3.4, **steps 1–3**.
4. Wash the resin five times with ethanol.
5. Insert the piston and push it halfway down, then discard the tip and leave the reactor in a lyophilizer overnight.

3.9 Cleavage

1. For each peptide, freshly prepare at least 6 mL of TFA/H_2O/triisopropylsilane (95:2.5:2.5) (v/v) cleavage cocktail (*see* **Note 5**).
2. Push the piston to the bottom of the syringe.
3. Transfer 3.5 mL of cleavage cocktail to a beaker and draw it into the syringe.
4. Cap the bottom of the syringe with a pressure cap.
5. Leave the syringe on a shaker for at least 2 h.
6. Minding possible pressure inside the syringe, carefully remove the bottom cap and use the piston to push the cleavage solution into a 5 mL cryotube.
7. Remove the piston.
8. Wash the resin twice with cleavage cocktail collecting the eluate in the cryotube.
9. Evaporate the solution using a gentle stream of N_2.
10. When less than 0.5 mL are left, add 4 mL of cold diethyl ether (*see* **Note 6**).
11. Put the cap on and shake gently.
12. Centrifuge for 4–6 min.
13. Remove the supernatant using a pipette.
14. Resuspend the solid into another 3–4 mL of cold ether.
15. Repeat **steps 12–14** for a total of three washes.
16. Let the residual ether evaporate by leaving the cryotube open in the fume hood.
17. Dissolve the crude product in 1–2 mL of 10–20% acetonitrile in H_2O.
18. Freeze-dry to obtain fluffy white crystals (*see* **Note 7**).

4 Notes

1. If a propylene 5 mL syringe equipped with PTFE filter is used, make sure that the piston does not have a rubber O-ring since it is labile to the chemicals used. Furthermore, we recommend pressure caps to close the narrow end of the syringe. As an alternative approach, 5 mL propylene reactors equipped with a PTFE filter are commercially available. If no suction device is available, syringes must be used by drawing and ejecting solutions using the piston.
2. The in-house-made Teflon® synthesis plate contains five holes which can accommodate syringes equipped with 200 μL pipette tips (Fig. 5). A 2.5 L Büchner flask with a sidearm is used to collect waste via a vacuum pump, Omnifit® fittings, and a three-way flow switch. The flask is equipped rubber ring adapter and with a cork stopper with a drilled hole.
3. When synthesizing multiple peptides in parallel, we recommend using tape of different colors to label the reactors.
4. Most resins (preloaded or not) are Fmoc protected (notable exceptions are 2-chlorotrityl resins).
5. TFA-containing cleavage cocktails and other caustic substances are highly corrosive and must be handled with care. All operations should be carried out in a fume hood, and safety goggles, coat, and gloves should be worn at all times.
6. Hexane may be used to precipitate short (≤6 AA) and lipophilic peptides which may be partly soluble in diethyl ether.
7. The peptide will be obtained as a TFA salt. If a hydrochloride is desired, *see* Chapter 10, **Note 1**.

References

1. Merrifield RB (1963) Solid phase peptide synthesis. I. The synthesis of a tetrapeptide. J Am Chem Soc 85:2149–2154
2. Tulla-Puche J, El-Faham A, Galanis AS, de Oliveira E, Zompra AA, Albericio F (2015) Methods for the peptide synthesis and analysis. In: Peptide chemistry and drug design. John Wiley & Sons Inc, New York, NY, pp 11–73
3. Hansen PR, Oddo A (2015) Fmoc solid-phase peptide synthesis. Methods Mol Biol 1348:33–50
4. Fields GB, Lauer-Fields JL, Liu R-Q, Barany G (2002) Principle and practice of solid-phase peptide synthesis. In: Synthetic peptides: a user's guide. Oxford University Press, New York, NY, pp 93–219
5. Jensen KJ (2013) Solid-phase peptide synthesis: an introduction. Methods Mol Biol 1047:1–21
6. Behrendt R, White P, Offer J (2016) Advances in Fmoc solid-phase peptide synthesis. J Pept Sci 22:4–27
7. Paradis-Bas M, Tulla-Puche J, Albericio F (2016) The road to the synthesis of "difficult peptides". Chem Soc Rev 45:631–654
8. Góngora-Benítez M, Tulla-Puche J, Albericio F (2013) Handles for Fmoc solid-phase synthesis of protected peptides. ACS Comb Sci 15:217–228
9. Dawson PE, Muir TW, Clark-Lewis I, Kent SBH (1994) Synthesis of proteins by native chemical ligation. Science 266:776–779

10. Pedersen SW, Armishaw CJ, Strømgaard K (2013) Synthesis of peptides using *tert* – butyloxycarbonyl (Boc) as the α-amino protection group. Methods Mol Biol 1047:65–80
11. Shelton P, Jensen KJ (2013) Linkers, resins, and general procedures for solid-phase peptide synthesis. Methods Mol Biol 1047:23–41
12. Meldal M (1997) Properties of solid supports. Methods Enzymol 289:83–104
13. Rapp W, Zhang L, Habich R, Bayer E (1989) Polystyrene-polyoxyethylene graftcopolymers for high speed peptide synthesis. Peptides 1988:199–201
14. Meldal M (1992) PEGA: a flow stable polyethylene glycol dimethyl acrylamide copolymer for solid phase synthesis. Tetrahedron Lett 33:3077–3080
15. García-Martín F, Quintanar-Audelo M, García-Ramos Y, Cruz LJ, Gravel C, Furic R, Côté S, Tulla-Puche J, Albericio F (2006) ChemMatrix, a poly(ethylene glycol)-based support for the solid-phase synthesis of complex peptides. J Comb Chem 8:213–220
16. Barlos K, Chatzi O, Gatos D, Stavropoulos G (1991) 2-Chlorotrityl chloride resin. Int J Pept Protein Res 37:513–520
17. Wang S-S, Gisin BF, Winter DP, Makofske R, Kulesha ID, Tzougraki C, Meienhofer J (1977) Facile synthesis of amino acid and peptide esters under mild conditions via cesium salts. J Org Chem 42:1286–1290
18. Rink H (1987) Solid-Phase synthesis of protected peptide fragments using a tri-Alkoxy-diphenyl-methylester resin. Tetrahedron Lett 28:3787–3790
19. Jensen KJ, Alsina J, Songster MF, Vágner J, Albericio F, Barany G (1998) Backbone Amide Linker (BAL) strategy for solid-phase synthesis of C-terminal-modified and cyclic peptides. J Am Chem Soc 123:5441–5452
20. Alsina J, Albericio F (2003) Solid-phase synthesis of C-terminal modified peptides. Biopolymers 71:454–477
21. Lambert JN, Mitchell JP, Roberts KD (2001) The synthesis of cyclic peptides. J Chem Soc Perk Trans 1:471–484
22. El-Faham A, Albericio F (2011) Peptide coupling reagents, more than a letter soup. Chem Rev 111:6557–6602
23. Romoff T (2004) Racemization assays. Houben-Weyl 22b:657–769
24. Sheehan JC, Hess GP (1955) A new method of forming peptide bonds. J Am Chem Soc 77:1067–1068
25. König W, Geiger R (1970) A new method for synthesis of peptides: activation of the carboxyl group with dicyclohexylcarbodiimide using 1-hydroxybenzotriazoles as additives. Chem Ber 103:788–798
26. Carpino LA (1993) 1-Hydroxy-7-azabenzotriazole. An efficient peptide additive. J Am Chem Soc 115:4397–4398
27. Subiros-Funosas R, Prohens R, Barbas R, El-Faham A, Albericio F (2009) Oxyma: an efficient additive for peptide synthesis to replace the benzotriazole-based HOBt and HOAt with a lower risk of explosion. Chemistry 15:9394–9403
28. Castro B, Dormoy JR, Evin G, Selve C (1975) Reactifs de couplage peptidique IV (1) - L'hexaflurophosphate de benzotriazolyl *N*-oxytrisdemethylamino phosphonium. Tetrahedron Lett 14:1219–1222
29. Coste J, LeNguyen D, Castro B (1990) PyBoP®: a new peptide coupling reagent devoid of toxic by-product. Tetrahedron Lett 31:205–208
30. Carpino LA, El-Faham A, Minor CA, Albericio A (1994) Advantageous applications of azabenzotriazole (triazolopyridine)-based coupling reagents to solid-phase peptide synthesis. J Chem Soc Chem Commun 25:201–203
31. Frérot E, Coste J, Pantaloni A, Dufour M-N, Jouin P (1991) PyBOP® and PyBroP: two reagents for the difficult coupling of the α, α-dialkyl amino acid, Aib. Tetrahedron 47:259–270
32. Knorr R, Trzcieak A, Bannwarth W, Gillesen D (1989) New coupling reagents in peptide chemistry. Tetrahedron Lett 30:1927–1930
33. Dourtoglou V, Gross B (1984) HBTU as coupling reagent for the synthesis of peptides of biological interest. Synthesis 573–574
34. El-Faham A, Subiros Funosas R, Prohens R, Albericio F (2009) COMU: a safer and more effective replacement for benzotriazole-based uronium coupling reagents. Chemistry 15:9404–9416
35. El-Faham A, Albericio F (2010) COMU: a third generation of uronium-type coupling reagents. J Pept Sci 16:6–9
36. Isidro-Llobet A, Alvarez M, Albericio F (2009) Amino acid-protecting groups. Chem Rev 109:2455–2504
37. Callahan FM, Anderson GW, Paul R, Zimmerman JE (1963) The tertiary butyl group as a blocking agent for hydroxyl, sulfhydryl and amido functions in peptide synthesis. J Am Chem Soc 85:201–207
38. McKay FC, Albertson NF (1957) New amine-masking groups for peptide synthesis. J Am Chem Soc 79:4686–4690

39. Carpino LA, Shroff H, Triolo SA, Mansour E-SME, Wenschuh H, Albericio F (1993) The 2,2,4,6,7-pentamethyldihydrobenzofuran-5-sulfonyl group (Pbf) as arginine side chain protectant. Tetrahedron Lett 34:7829–7832
40. Sieber P, Riniker B (1991) Protection of carboxamide functions by the trityl residue. Application to peptide synthesis. Tetrahedron Lett 32:739–742
41. King DS, Fields CG, Fields GB (1990) A cleavage method which minimizes side reactions following Fmoc solid phase peptide synthesis. Int J Pept Protein Res 36:254–266
42. Nielsen SL, Frimodt-Moller N, Kragelund BB, Hansen PR (2007) Structure activity study of the antibacterial peptide fallaxin. Protein Sci 16:1969–1976
43. Albericio F, Kneib-Cordonier N, Biancalana S, Gera L, Masada RI, Hudson D, Barany G (1990) Preparation and application of the PAL handle for the solid-phase peptide synthesis of C-terminal peptide amides under mild conditions. J Org Chem 55:3730–3743
44. Solé NA, Barany G (1992) Optimization of solid-phase peptide synthesis of [Ala^{8}]-dynorphin. J Org Chem 57:5399–5403
45. De Luca S, Bruno G, Fattorusso R, Isernia C, Pedrone C, Morelli G (1998) New synthetic tools for peptide-tetraphenylporphyrin derivatives. Lett Pept Sci 5:269–276
46. Oddo A, Thomsen TT, Kjelstrup S, Gorey C, Franzyk H, Frimodt-Møller N, Løbner-Olesen A, Hansen PR (2016) An all-D amphipathic undecapeptide shows promising activity against colistin-resistant strains of Acinetobacter baumannii and a dual mode of action. Antimicrob Agents Chemother 60:592–599

Chapter 4

Microwave-Assisted Synthesis of Antimicrobial Peptides

Suhas Ramesh, Beatriz G. de la Torre, Fernando Albericio, Hendrik G. Kruger, and Thavendran Govender

Abstract

Antimicrobial peptides (AMPs) are emerging as one of the unsurpassed therapeutic tools to treat various devastating diseases that are affecting millions of lives. Conventional synthesis of peptides requires longer times, and hence automated microwave technology could be regarded as an alternative implement which offers advantages like less reaction times and higher yields. In this sense, we herein describe a methodology to prepare AMPs through solid-phase peptide synthesis under microwave conditions. We have used LL37 as an example to discuss the synthetic protocol including the difficulties involved in the preparation of so-called long and difficult peptides and also remedial procedures to overcome these obstacles.

Key words Antimicrobial peptides, Microwave, Automated, LL37, SPPS

1 Introduction

One of the principal limitations to the use of therapeutic peptides is the cost factor [1–4]. There are three routes (i.e., liquid-phase, solid-phase, and recombinant technology) for the production of peptides and each with their pros and cons.

Liquid phase is useful for short peptides (<10 mer) and less expensive and has good purity profiles. The cycle times are relatively longer, with much more operational steps and often with lower yields [5].

Solid-phase synthesis is far more rapid, automatable, and scalable without the need to isolate intermediates resulting in faster production cycles. The downside to this approach is the cost of the raw materials and the probability of getting a complex impurity profile [5].

Recombinant technology has the advantage of creating peptides that are larger than 10 mer with high purities. It does suffer from higher development costs and unit operations [6]. Unlike the other two methods, it cannot tolerate any unnatural amino acids.

Paul R. Hansen (ed.), *Antimicrobial Peptides: Methods and Protocols*, Methods in Molecular Biology, vol. 1548, DOI 10.1007/978-1-4939-6737-7_4, © Springer Science+Business Media LLC 2017

Peptides up to 30 mer are normally more efficiently produced via chemical synthesis, while longer ones and proteins benefit from recombinant approaches. The costs may be lowered by a factor of a hundred, but the technical difficulties should not be underestimated. An estimated one million liters of fermentation mixture will be necessary to produce 100 kg of recombinant peptide. Pharmaceutical companies do not see this approach as feasible for the ton-scale production of AMPs required by developing countries.

The T-20 peptide (Fuzeon, Roche) is the most famous example of a multi-ton chemical synthesis of a peptide drug. It is synthesized through a combination of solid-phase and solution-phase methodologies. This single move by Roche has resulted in lower costs of synthesis reagents and making peptide synthesis more feasible [6]. The bottleneck in peptide synthesis has always been the purification step, and there is therefore a demand for higher-quality peptides.

The growing popularity of microwave-assisted peptide synthesis technology in combination with low-cost reagents is addressing this and modifying the current landscape even further [7]. Microwave heating is normally applied to the coupling and deprotection steps with significant reduction in reaction times and increased crude purities of the peptides. There are currently several manufacturers of dedicated microwave peptide synthesizers, and the instruments vary based on automation, reaction vessel sizes, and mode of mixing (Table 1).

AMPs are mainly categorized into two modes of action i.e. mainly cell wall disruption and secondarily intracellular targets. They are normally about 10–50 amino acids in length and comprised of

Table 1
A selection of microwave peptide synthesizers

Manufacturer	Instrument	Operation	Reaction vessels (mL)	Mixing mode
CEM	Discover Bio	Manual	4, 25	N_2 bubbling
CEM	Liberty Blue	Automated	30, 125	N_2 bubbling
Biotage	Initiator + SP Wave	Manual	2, 5, 10	Oscillation
Biotage	Initiator + Alstra	Automated	5, 10, 30	Oscillation
AAPPTEC	Infinity 2400	Automated	5, 25, 50, 100, 300	Oscillation and/or N_2 bubbling
Protein Technologies	Prelude X	Automated	10, 45	Oscillation and/or N_2 bubbling
Protein Technologies	Symphony X	Automated	10, 45	Oscillation and/or N_2 bubbling
Protein Technologies	Tribute	Automated	10, 45	Oscillation and/or N_2 bubbling

a significant percentage of lysine and arginine residues resulting in an overall positive charge [1–4]. This in combination with a substantial portion of hydrophobic residues results in peptides with amphipathic conformations to facilitate its cell disruption mode of action.

Several studies have also confirmed that certain classes of AMPs have intracellular targets. These peptides normally have prolines in combination with lysines and are able to interact with DNA and RNA.

Here we present a microwave-assisted synthesis of the antimicrobial peptide LL37.

2 Materials

Prepare all solutions (amino acids, coupling reagents, additives) in HPLC-grade *N*,*N*-dimethylformamide (DMF) freshly. Backflush the microwave peptide synthesizer using DMF before the launch of the synthesis. All connections are to be checked before starting the instrument. Also make sure that the cylinder has enough nitrogen before turning the instrument on. Perform the respective pressure and transfer volume calibration tests.

2.1 Chemicals

1. Fmoc amino acids.
2. HOBt (1-hydroxybenzotriazole).
3. NMP (*N*-methyl-2-pyrrolidone).
4. Dimethylformamide (DMF).
5. Acetonitrile.
6. Diisopropylamine (DIEA).
7. Piperidine.
8. α-Cyano-4-hydroxycinnamic acid (HCCA).
9. Trifluoroacetic acid (TFA).
10. Triisopropylsilane (TIS).
11. ChemMatrix® resin (*see* **Note 1**).

2.2 Solutions and Buffers

1. Coupling solution: 2 M diisopropylethyl amine (DIEA) in *N*-methyl-2-pyrrolidone (NMP) (*see* **Note 2**).
2. Fmoc deprotection solution: piperidine in DMF (1:4) and 0.1 M HOBt (*see* **Note 3**).
3. HPLC buffer A: 0.1 % trifluoroacetic acid (TFA) in water.
4. HPLC buffer B: 0.1 % TFA in acetonitrile. The same solvents hold good for analytical high performance liquid chromatography (HPLC).
5. MALDI-TOF-MS solution: HCCA in TA30 w/v (30 % ACN in 70 % of 0.1 % TFA in water) (1:9).

2.3 Instruments

1. CEM Liberty automated microwave peptide synthesizer.
2. Semi-preparative HPLC, C_{18} preparative column (250 × 10 × 21.2 mm).
3. Analytical HPLC, C_{18} column (3 μm × 4.6 × 50 mm).
4. MALDI-TOF-MS.
5. Sonicator.
6. Centrifuge.

3 Methods

Herein a peptide synthesis of 0.1 mmol scale is demonstrated.

3.1 Preparations and Setup of CEM Liberty Peptide Synthesizer

1. Weigh out the required quantity of Fmoc amino acids into each bottle and dissolve in the specified amount of DMF. Sonicate the solution to get uniform concentrations. The above holds good for the activating reagent (0.5 M).
2. Calibrate the instrument (pressure and volume). Adjust low and high pressure using the valve of the gauge present on the rear end of the instrument. Low and high pressure should be in between 2 and 4.5 psi and 15–16.5 psi, respectively. For volume calibration, verify "Add DMF-Top" and "Add DMF-Bottom" by specifying desired volume of dispensing.
3. Clean all the manifolds using DMF.
4. Initially load the amino acid sequence to the instrument by following the path shown below: (*see* **Note 4**)

Sequence → New folder → New sequence → create the sequence by a click on the desired amino acids → Save

5. Create a method as follows:

Methods → New root folder → New method → fill all the parameters like, loading, deprotection, C-terminal, cleavage, etc. → Apply to all → Save → Close (*see* **Note 5**)

6. Click on the method, verify for the correctness, and drag it to the loading position.

3.2 Peptide Synthesis and Final Cleavage

1. Weigh out 0.1 mmol of Rink amide-ChemMatrix resin (0.6 mmol/g loading) into the reaction vessel.
2. Swell the resin using DMF and drain (3×).
3. Treat the resin with 20% piperidine to remove Fmoc protecting group. Perform deprotection using 45 W microwave power at 90 °C for 65 s. This is followed by washings with DMF (Top-3× and Bottom-2×). Normally 7 mL DMF is used for washings at this scale (*see* **Note 6**).

4. Add amino acid (2.5 mL), activator (1.0 mL), and activator base (0.5 mL) (*see* **Notes 7** and **8**) and conduct the reaction at 90 °C for 120 s (*see* **Note 9**).
5. Wash the resin using DMF; Top-2× and Bottom-1×.
6. Repeat the **steps 4** and **5** till the completion of the sequence.
7. At the end of the synthesis, perform deprotection as shown in **step 3**.
8. Take the peptide-resin out of the reaction vessel and dry.
9. Treat the dried peptidyl-resin with 2 mL of cocktail solution (TFA:TIS:H_2O :: 95:2.5:2.5) and, for 2 h, add 5 mL cold diethyl ether to precipitate the peptide (*see* **Note 10**).
10. Centrifuge at 955 ×*g* for 10 min, discard the ether layer, rewash with ether, discard and dissolve the peptide in 10 mL water, filter and store at 4 °C before purification.

3.3 Identification of Problem Sequences

1. LCMS should be used to monitor peptides that are 6 amino acids or less while MALDI-TOF-MS is a better option for larger peptides.
2. Any mass less than your desired peptide represents deletion/s and any mass larger normally represents an incomplete removal of orthogonal protecting groups (bare in mind Na^+ and K^+ salts) (*see* **Notes 11–13**).

3.4 Purification and Characterization

1. Equip the semi-preparative HPLC for purification with C_{18} column at 220 nm detection (*see* **Note 14**).
2. Condition and wash the column prior to start purification with 95 % mobile phase B.
3. After the wash, stabilize the column with mobile phase B (*see* **Note 15**).
4. Set the method with a fixed flow and solvent gradient to run over 45 min.
5. As a trial run, inject a small amount of the sample (0.5 mL) and collect all peaks.
6. Identify the peak of interest by MALDI-TOF-MS (*see* **Note 16**) using HCCA as the matrix (*see* **Note 17**).
7. Once the peak of interest is identified, continue the purification and collect the peak of interest from each run.
8. Lyophilize the peptide.
9. Run analytical HPLC to confirm the purity of the sample using the appropriate conditions
10. Store at 4 °C until further use (*see* **Note 18**).

4 Notes

1. Polystyrene resins work just as good for up to about 20 mer sequences. Resins will have different loading capacities and swelling properties based on manufacturer.
2. 2 M DIEA is not miscible with DMF, NMP is used for this purpose. If NMP is not available, then use a 1 M solution in just DMF and ensure that the instrument now adds twice the volume of the standard settings.
3. HOBt is added in order to reduce the aspartimide formation [8] which is caused by aspartic acid present in the sequence.
4. Methods would vary between different manufacturers.
5. Some amino acids are hindered, less reactive or unstable. Hence, special conditions are required. The vendor would have special methods pre-programmed eg. 'double couplings' for Arginine and lower temperature for cysteine amino acids.
6. The standard methods normally involve two deprotection steps but this was found to be unnecessary as reported by Collins et al. [9] and confirmed by us [10]. The amount of power used is based on the what is required to reach the desired temperature in 20 s. This will vary amongst instrument manufacturers and even the age of the instrument. When evaluating this parameter, the concentrations of reagents must be as close as possible to mimic the dielectric constant of the reaction.
7. Adjust volume of DIEA depending on the concentration used as mentioned in **Note 1**.
8. Different activators can be employed (*see* Chapter 3 by Münzker et al.) depending on availability but the combinations of amino acid/HBTU/DIPEA in DMF (1:1:2) and using amino acid/DIC/OxymaPure in DMF (1:1:2) are currently most cost efficient.
9. Comparing the standard microwave heating protocols and the aggressive heating approach found that the latter works well for most amino acids. Cysteine and histidine are sensitive to epimerization during the coupling step. For cysteine, the DIC/Oxyma is preferred over the HBTU/DIEA activator. Aggressive heating should be avoided for histidine if the side chain protecting group is Trt. An alternative is to use the Mbom protecting group with a reaction time of 3 min and at 80 °C. Arginine normally needs double coupling but 25 min at room temperature followed by 2 min of microwave heating at 75 °C was found to be sufficient.
10. It is always desirable to verify the synthesis by means of 'mini-cleavage' in which a small amount of the peptidyl-resin (~2 mg) is treated with 0.1 mL cleavage cocktail solution, followed by addition of cold ether, centrifuged, discarded ether and dissolved

the peptide in a suitable solvent. If the peptide to be synthesized is too long, then mini-cleavage of the segments can be done to ensure successful synthesis.

11. A good example of the synthesis of a large AMP is the LL37. MALDI analysis revealed that there is a decrease in coupling efficiency from the 20th amino acid. Hence, segment-synthesis was undertaken as shown in Figs. 1 and 2. There is evidence for the formation of the product at 4493 m/z albeit in a small amount. There was no significant peak up to 2179 m/z leading us to conclude that the coupling of residues from 21 to 37 worked well. The rest of the peaks in this region indicate poor coupling efficiencies for the residues of positions 16–20 in the sequence. Thereafter, it seems to be a smooth run to the desired product masses.

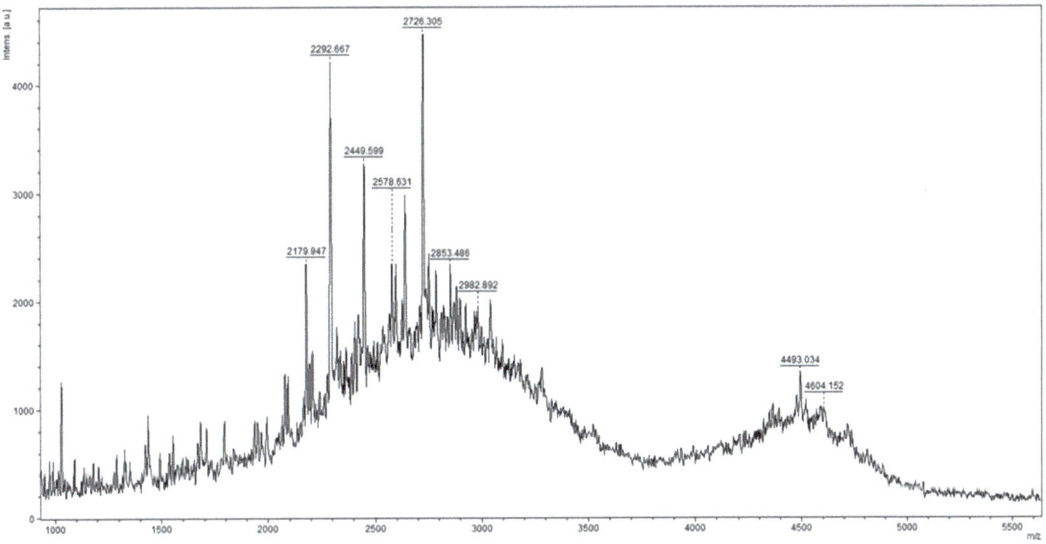

Fig. 1 MALDI-TOF MS showing non-specific products along with the desired mass *m/z* 4493.034

	1 10 20 30 37
I	P R T E S
II	V Q R I K D F L R N L V P R T E S
III	E F K R I V Q R I K D F L R N L V P R T E S
IV	E K I G K E F K R I V Q R I K D F L R N L V P R T E S
V.	L L G D F F R K S K E K I G K E F K R I V Q R I K D F L R N L V P R T E S

Fig. 2 Segment-based synthesis showing (*I*) fragment 33–37, (*II*) fragment 21–37, (*III*) fragment 16–37, (*IV*) fragment 11–37 and (*V*) fragment 1–37. Difficult coupling region/amino acids as found out experimentally are shown in *red*

12. Hence, it was decided to monitor the addition of the residues in this region with manual synthesis and without microwave heating. Since extended coupling times would be beneficial, the use of a more stable coupling reagent like DIC [11] is essential. Carbodiimides are used in combination with additives. Here a superior and less-risk additive like OxymaPure [12] is recommended over HOBt/HOAt [13]. A report has identified that THF with DIC/OxymaPure combination proved more efficient compared to DMF [14].
13. The isoleucine at amino acid residue 20 was concluded to be the main problem since it required double couplings for 1 h each to give a negative ninhydrin test. All other amino acids in this region only required single couplings.
14. Employing a guard column will increase the shelf life of the column.
15. The starting solvent composition will depend on the polarity or the peptide. AMPs can vary drastically in this regard and may need optimization on an analytical scale first to determine the correct buffer system and gradient. Try for conditions on the analytical HPLC until the maximum resolution is achieved since this is always worse when translated to a larger scale.
16. If its a high molecular weight peptide, it is better seen on MALDI-TOF MS. On the other hand, for peptides of low molecular weight LCMS can be used.
17. Depending on the molecular weight of the peptide, matrix has to be chosen for MALDI. Spotting of sample on the plate can be done using three ways: (1) sample first and matrix next, (2) matrix first and sample next and, (3) sandwich method, in which sample is spotted in between the layers of the matrix i.e., matrix-sample-matrix.
18. If the AMP synthesized is known, then its synthesis can also be verified by performing suitable biological activity and compare the results.

References

1. Seo M-D, Won H-S, Kim J-H, Mishig-Ochir T, Lee B-J (2012) Antimicrobial peptides for therapeutic applications: a review. Molecules 17:12276–12286
2. Phoenix DA, Dennison S, Harris F (2013) Antimicrobial peptides. Wiley-VCH, Weinheim
3. Giuliani A, Pirri G, Nicoletto SF (2007) Antimicrobial peptides: an overview of a promising class of therapeutics. Cent Eur J Biol 2:1–33
4. Paraje MG (2011) Antimicrobial resistance in biofilms. In: Mendez-Vilas A (ed) Science against microbial pathogens: communicating current research and technological advances. Formatex Research Center, Spain, pp 736–744
5. Bayer E, Mutter M (1972) Liquid phase synthesis of peptides. Nature 237:512–513
6. Thayer AM (2011) Improving peptides: making peptides at large scale. Chem Eng News 89:13–20
7. Pedersen SL, Tofteng AP, Malik L, Jensen KJ (2012) Microwave heating in solid-phase peptide synthesis. Chem Soc Rev 41: 1826–1844

8. Fields GB, Angeletti RH, Bonewald LF, Moore WT, Smith AJ, Stults JT, Williams LC (1995) Correlation of cleavage techniques with side - reactions following solid - phase peptide synthesis. In: Crabb J (ed) Techniques in protein chemistry IV. Academic, New York, NY, pp 539–548
9. Collins JM, Porter KA, Singh SK, Vanier GS (2014) High-efficiency solid phase peptide synthesis (HE-SPPS). Org Lett 16:940–943
10. Dutta J, Ramesh S, Radebe SM, Somboro AM, de la Torre BG, Kruger HG, Essack SY, Albericio F, Govender T (2015) Optimized microwave-assisted synthesis of LL37, a cathelicidin human antimicrobial peptide. Int J Pept Res Ther 21:13–20
11. Hachmann J, Lebl M (2006) Search for optimal coupling reagent in multiple peptide synthesizer. Biopolymers 84:340–347
12. Subiros-Funosas R, Prohens R, Barbas R, El-Faham A, Albericio F (2009) Oxyma: an efficient additive for peptide synthesis to replace the benzotriazole-based HOBt and HOAt with a lower risk of explosion. Chemistry 15:9394–9403
13. Wehrstedt KD, Wandrey PA, Heitkamp D (2005) Explosive properties of 1-hydroxybenzotriazoles. J Hazard Mater 126:1–7
14. Jad YE, Acosta GA, Khattab SN, de la Torre BG, Govender T, Kruger HG, El-Faham A, Albericio F (2015) Peptide synthesis beyond DMF: THF and ACN as excellent and friendlier alternatives. Org Bio Chem 13:2393–2398

Chapter 5

High-Performance Liquid Chromatography and Mass Spectrometry-Based Design of Proteolytically Stable Antimicrobial Peptides

Mojtaba Bagheri and Robert E.W. Hancock

Abstract

The emergence of multiresistant bacteria worldwide together with the shortage of effective antibiotics in the market emphasizes the need for the design and development of the promising agents for the treatment of superbug-associated infections. Antimicrobial peptides (AMPs) have been considered as excellent candidates to tackle this issue, and thousands of peptides of different lengths, amino acid compositions, and mode of action have been discovered and prepared to date. Nevertheless, it is of great importance to develop innovative formulation strategies for delivering these AMPs and to improve their low bioavailability and metabolic stability, particularly against proteases, if these peptides are to find applications in the clinic and administered orally or parenterally or used as dietary supplements. The purpose of this chapter is to describe basic experimental principles, based on analytical reversed-phase high-performance liquid chromatography (RP-HPLC) and mass spectrometry (MS), for the prospective design of orally bioavailable AMPs considering the structural characteristics of the peptides and the substrate specificity of proteases that abound in the body especially at sites of infection.

Key words Antimicrobial peptides, Bioavailability, Proteases, Reversed-phase high-performance liquid chromatography, Mass spectrometry

1 Introduction

AMPs are known as efficient endogenous components of the innate immune systems of many organisms that are involved in protection of the host against health-threatening multiresistant pathogens [1]. These multifaceted compounds can have direct killing effects (e.g., cell membrane-permeabilizing activities [2], dysfunction of membrane protein functions [3], disturbance of DNA synthesis/transcription and mRNA translation [4], inhibition of mitochondrial ATPase activity [5]), separate action against biofilm bacteria [6], and/or anti-infective and anti-inflammatory mediated by selective modulation of immune host defense responses [7]; they are often termed host defense peptides to encompass these other

Paul R. Hansen (ed.), *Antimicrobial Peptides: Methods and Protocols*, Methods in Molecular Biology, vol. 1548, DOI 10.1007/978-1-4939-6737-7_5,

activities. Many natural and synthetic AMPs with different mechanism of actions have been discovered or designed since the early 1980s [8]. However, to date none of these have achieved new drug approval; a few peptides have received statistically significant efficacy in Phase 3 clinical trials [8]. One main drawback to marketing these peptide antibiotics is their low bioavailability and metabolic stabilities for the systemic applications in clinics [9]. This contrasts with two highly modified natural cationic amphipathic peptides, the polymyxins and gramicidin S, which are highly used in treatment of infections [10]. Thus, the development of biocompatible formulation strategies of the AMPs, which are specifically suitable for noninvasive oral administration and increasing the plasma half-life of the peptides, is a high priority.

Although natural and synthetic nanocarrier systems (e.g., liposome or colloidal delivery systems), prodrug-based approaches (e.g., cell-penetrating peptide conjugates), and chemical modifications via site-specific polyethylene glycol (PEG)ylation, lipidation, and glycosylation have frequently been used in peptide drug development [11, 12], adverse consequences associated with these excipients, such as reduced bioactivity, aggregation and/or decreased solubility/stability of the formulated peptides, and toxicity-related issues are of concern [13, 14]. This becomes much more pronounced for the AMPs since there is a good correlation between active AMPs sequences and the content of amino acid residues in the aggregation-prone region of amyloid peptides leading to their consequent physiochemical properties [15].

Peptide structural stability has also been described to be important for the high activity of AMPs, especially, at physiological salt concentrations [16, 17]. Critically, the AMP sequences contain high numbers of residues sensitive to digestive serine and aspartate proteases [18], e.g., Lys and Arg as trypsin-sensitive cationic residues as well as hydrophobic residues such as Trp, Phe, Leu, Ile, Tyr, Ala, and Val that are sensitive to digestion by chymotrypsin, pepsin, and elastase proteases (Table 1); in particular these residues are functionally important for their activities and mediate peptide electrostatic attraction and hydrophobic insertion into the cell membrane [21]. Therefore, the strategies used to modify AMPs should result in minimal impact on peptides structure and tendency to aggregation and should improve the peptide stability against enzymatic action of endo- and exopeptidases.

Basically, substitution of the endoprotease-sensitive residues with noncanonical amino acids bearing the same cationicity as positively charged amino acids or the comparable hydrophobicity as the aromatic and aliphatic amino acids, but with different side-chain lengths or chemically modifications, provides one strategy for improving the proteolytic stabilities of AMPs and enables the prospective design of orally bioavailable peptides. Alternatives include the use of all-d-peptide isomers, including the retro-inverso

Table 1
Structural description and substrate specificity of human gastrointestinal enzymes affecting the bioavailability of antimicrobial peptides

Endopeptidase[a]	Catalytic dyad/triad	Optimal pH[b]	S1 primary structure[c]	Substrate specificity[d] S1	S1′
Pepsin	Asp-32, Asp-215	pH ~ 1.5	Val-71, Ser-72, Ile-73, **Thr-74**, Tyr-75, Gly-76, **Thr-77**, Gly-78, Ser-79, Met-80, Thr-81, Gly-82	Phe, Leu, (not Val, Ala, or Gly)	Trp, Phe, Tyr, Leu, Ile, Ala, other hydrophobic amino acids
Chymotrypsin	His-57, Asp-102, Ser-195	pH ~ 7.8–8.0	**Ser-189**, Ser-190, Cys-191, Met-192, Gly-193, Asp-194, Ser-195, Ser-214, Trp-215, **Gly-216**, Ser-217, Asp-218, Thr-219, Cys-220, Pro-225, **Gly-226**, Val-227, Tyr-228	Trp, Phe, Tyr	slight substrate specificity for hydrophilic amino acids (not Pro)
Trypsin	His-57, Asp-102, Ser-195	pH ~ 7.5–8.5	**Glu-189**, Ser-190, Cys-191, Gln-192, Gly-193, Asp-194, Ser-195, Ser-214, Trp-215, **Gly-216**, Asp-217, Gly-218, Cys-219, Pro-225, **Gly-226**, Val-227, Tyr-228	Arg, Lys	slight substrate specificity for hydrophobic amino acids (not Pro)
Elastase	His-57, Asp-102, Ser-195	pH ~ 7.0	**Ser-189**, Ser-190, Cys-191, Asn-192, Gly-193, Asp-194, Ser-195, Val-213, Ser-214, Phe-215, **Gly-216**, Ser-217, Arg-218, Leu-219, Gly-220, **Ser-226**, Val-227, Phe-228	Val, Ala (not Trp)	any amino acids (no substrate specificity)

[a]The endopeptidase is classified as aspartate proteases, e.g., pepsin, and serine proteases, e.g., chymotrypsin, trypsin, and elastase

[b]The pH at which the enzyme is most active

[c]The primary structure of the S1 binding site listed here belongs to human pepsin A and pancreatic chymotrypsin B, trypsin 1, and elastase 2A [19, 20]. The side chains of bold underlined residues point toward the S1 binding pocket and are in contact with the P1 side chain. Thus, they play important role in specificity determinants of the enzymes

[d]In the case of pepsin, only amino acids with high frequency in the AMPs are shown here as the substrate

sequences, and acetylation and amidation of the peptide N- and C- termini to, respectively, reduce susceptibility to amino- and carboxy-exopeptidases. For this, it is important to take into consideration the structural determinants of the enzyme binding pockets, particularly S1 site specificity and its preferences for the peptide side chains (P1) (Fig. 1), as well as the peptide structural characteristics (*see* **Note 1**). Cyclization to induce peptide structural constraints and creation of peptidomimetics such as α,β-/β-peptides and peptoids have also been used for the generation of metabolically stable AMPs [22]; however, these approaches are limited by our poor understanding of how such backbone modification impact on peptide structure in various environments (including the membrane) and consequently the amphiphilic arrangement of the side chains (or the subunits) and the antimicrobial and hemolytic activities of these compounds [23, 24].

Studies of the relationship between peptide structural elements and enzyme substrate specificity will be beneficial for the understanding of the modes of binding of AMPs to proteases and are very relevant to the establishment of a general approach for the design of AMPs with increased bioavailability [9]. Critically, HPLC and MS have tremendous potential for determining the peptide (plasma) half-lives and the site(s) of enzymatic cleavage, respectively, enabling an iterative peptide design process involving synthesis and evaluation of both the biological activities and proteolytic stability of peptides. Through the integration of the peak area in an HPLC spectrum and the mass profile analysis, the necessary information on enzymatic stability and the mode of AMP-protease binding is obtained within a few hours, providing an important advance in the design of the bioavailable peptides. In this chapter,

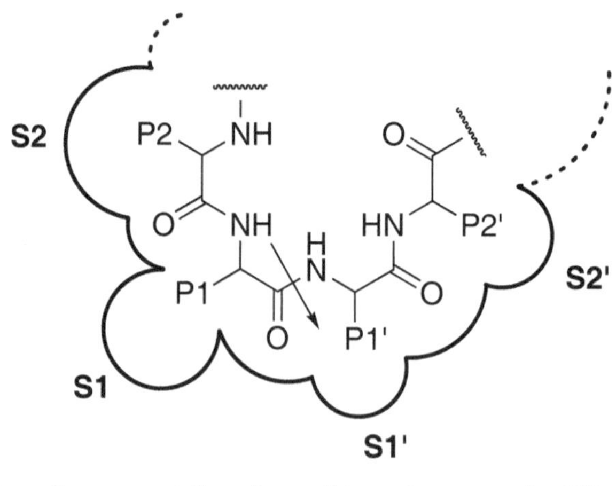

Fig. 1 Schematic representation of a peptide-protease complex. This is used to describe the enzyme substrate specificity according to Schechter's naming system. The *arrow* shows the scissile bond at which the peptide enzymatic hydrolysis occurs

the highly cationic Trp-rich ^{5Dab}HHC-36, an analog of active antibacterial HHC-36 peptide (Fig. 2) against pathogens, particularly, *Staphylococcus aureus* and *Pseudomonas aeruginosa* [25], is utilized as an example of a derivative that is less sensitive to serine endopeptidases (*see* **Note 2**) [9], and an efficient protocol for the analysis of its tryptic stability is described. The detailed protocol provided here is quite straightforward and can be applied to evaluate the stability of a variety of pharmaceutically interesting peptide sequences toward luminally secreted gastrointestinal proteases, thus addressing challenges in the design of orally and systemically administered peptides.

2 Materials

2.1 Enzyme, Peptide, Buffers, and Solvents

1. Trypsin; salt-free, lyophilized powder, ≥10,000 BAEE units/mg protein; molecular mass: ~ 24 kDa.
2. Lyophilized ^{5Dab}HHC-36: Dab-Dab-Trp-Trp-Dab-Trp-Trp-Dab-Dab-NH_2 (1261.8185 Da (*see* **Note 3**)).
3. Ammonium bicarbonate buffer: 0.1 M NH_4HCO_3.
4. Stock solution of trypsin from bovine pancreas: 0.02 mg/mL trypsin in 0.1 M NH_4HCO_3 buffer at pH = 8.2 (*see* **Note 4**).
5. Peptide stock solution: 600 μM peptide in 0.1 M NH_4HCO_3 buffer at pH = 8.2 (*see* **Note 5**).
6. Sodium hydroxide (NaOH).

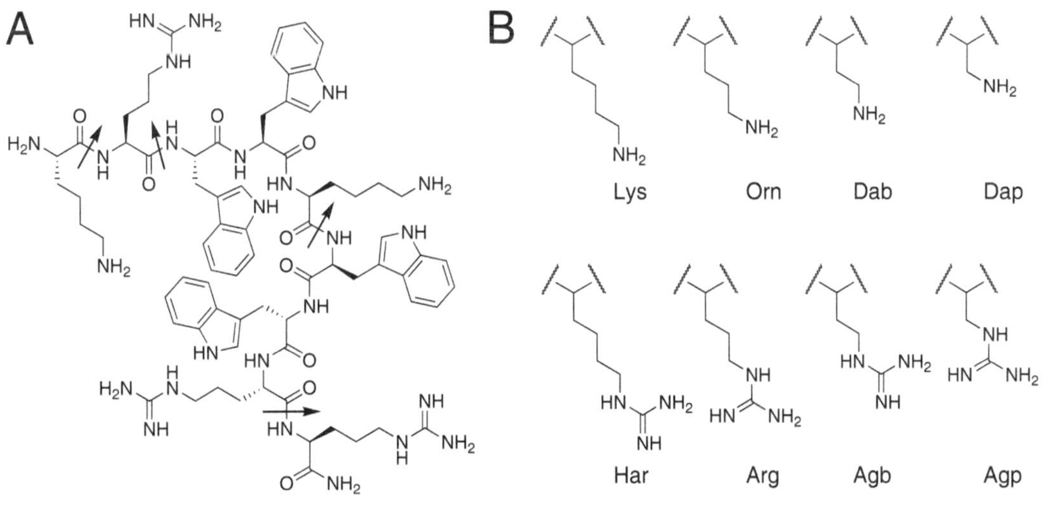

Fig. 2 Structure of HHC-36 peptide and amino acid substituents. (**a**) HHC-36 chemical structure and (**b**) example non-proteinogenic Lys and Arg analogs with varied methylene chains which can be substituted for the cationic residues to generate proteolytically stable AMPs. Lys analogs are: Orn, Dab, and Dap; Arg mimics are Har, Agb, and Agp. *Arrows* show the tryptic cleavage positions in the HHC-36

7. Hydrochloric acid (HCl).
8. Acetonitrile (ACN).
9. Acetic acid (HOAc).
10. Trifluoroacetic acid (TFA).
11. Quenching cocktail: 99.99 % HOAc and 0.1 % TFA.

2.2 HPLC System

Buffer A: H_2O containing 0.1 % TFA.

Buffer B: ACN/H_2O (4/1: v/v) containing 0.1 % TFA.

Waters analytical reversed-phase HPLC equipped with Breeze workstation software, autosampler with a diode-array ultraviolet/visible (UV/VIS) detector; reversed-phase Nova-Pak C18 Cartridge, 60 Å, 4 μm, 4.6 mm × 250 mm column.

The retention time (t_R) is determined using a linear gradient of 5–95 % of mobile phase B for 40 min at 23 °C and a flow rate of 1 mL/min. The injection volume is set to 200 μL, and the absorbance is detected at $\lambda = 220$ nm.

2.3 MS System

Agilent Technologies electrospray ionization-mass spectrometry (ESI-MS) in the positive ionization mode using an Agilent 6410 Atriple quadrupole LC/MS and equipped with MassHunter Workstation Software. Fragmentor and capillary voltages are set to 135 V and 4000 V, respectively (*see* **Note 6**). The other source parameters are set as follows: collision energy, 0 V; nebulizer pressure, 40 PSI; drying gas temperature, 325 °C; drying gas flow, 10 L/min; and MS heater temperature 100 °C. The mass spectra are recorded using the direct injection (*see* **Note 7**) in the full scan mode from the m/z 50 to m/z 2000 range (*see* **Note 8**). The sample injection volume is set to 20 μL.

2.4 Additional Materials

1. Thermomixer with 2 mL thermoblock.
2. pH meter.
3. NANOpure infinity UV/ultrafilter (UF) water purification system.
4. Measuring pipette and pipette tips and 1.5 mL and 2 mL vials.

3 Methods

3.1 Enzymatic Assessment and Data Analysis

1. Transfer 150 μL of the enzyme sample and 150 μL of the peptide stock solution into a 2 mL Eppendorf vial already filled with 1200 μL of the buffer.
2. Incubate the peptide-enzyme mixture in an Eppendorf thermomixer at 37 °C with gentle shaking.
3. Take a 200 μL aliquot from the reaction mixture at a specific time interval and transfer it to a 1.5 mL Eppendorf vial.

4. Add 50 μL of a quenching cocktail containing 99.99% HOAc and 0.1% TFA to the aliquot in **step 3** (*see* **Note 9**).
5. Analyze the quenched mixture using analytical RP-HPLC.
6. Use Breeze workstation software for HPLC peak detection, integration, and quantitation for the determination of the peptide half-life using Eq. 1 (*see* **Note 10**).
7. As a control repeat **steps 1–5** for samples in the absence of the enzyme. The final peptide concentration is the same as in **step 3** (60 μM).
8. Transfer 10 μL of the quenched enzymatic reaction mixture in **step 4** to a 1.5 mL Eppendorf vial prefilled with 900 μL of the buffer (100-fold dilution) and use it for the mass analysis.
9. Analyze the quenched mixture using the triple quadrupole MS instrument. Use the MassHunter Workstation Software for data analysis (*see* **Note 11**).

4 Notes

1. Schechter's model nomenclature system is used to describe the interaction between a peptidase and its substrate with respect to the enzyme binding subsites and the peptide side chains (Fig. 1) [26]. In this naming system, the scissile bond is taken as the reference, and the peptide side chains are continuously numbered as "P1, P2, …, Pn" and "P1′, P2′, …, Pn′" (where $n \geq 1$) toward the peptide's N-terminus and the C-terminus, respectively. Also, enzyme binding subsites fitted in by each individual residue's side chain are designated as "S1, S2, …, Sn" and "S1′, S2′, …, Sn′", where "Pn" (and/or "Pn′") is the substrate for "Sn" (and/or "Sn′") if "*n*" has the same value. This nomenclature system is widely used to describe the digestive enzyme substrate specificity.
2. Trypsin specifically cleaves peptides at the carboxyl side of arginine and lysine residues by an electrostatic interaction between Asp-189 in the S1 binding subsite (of trypsin) with the cationic residues (in the target peptide). Unlike S1, the other trypsin binding subsites have very broad substrate specificity [9]. Thus, the substitution of Arg and Lys residues with their noncanonical analogs enables the systematic design of trypsin-stable peptide sequences. HHC-36 (Lys-Arg-Trp-Trp-Lys-Trp-Trp-Arg-Arg-NH_2) is an Arg/Lys-rich sequence (Fig. 2). Using this template X_1-X_2-Trp-Trp-X_1-Trp-Trp-X_2-X_2-NH_2, where X_1 and X_2 are replaced by Arg and/or Lys analogs, a very large number of peptides can be made if every permutation and combination of Arg and Lys substitution is performed. Indeed, this requires a high-throughput method of generating

the varied peptide sequences to examine the effects of changing the side chain of the cationic residues on the peptide tryptic stability. To make the design process more cost-effective and faster, the noncanonical residues that are commercially available may be considered first, including l-ornithine (Orn), (2*S*)-2,4-di-amino butyric acid (Dab), (2*S*)-2,3-di-amino propionic acid (Dap), l-homoarginine (Har), (2*S*)-2-amino-4-guanidinobutyric acid (Agb), and (2*S*)-2-amino-3-(guanidino) propanoic acid (Agp) with the varied side-chain lengths (Fig. 2) compared to Arg and Lys. At this point, the easiest way is to use only one residue at each tryptic cleavage position, e.g., Dab-Dab-Trp-Trp-Dab-Trp-Trp-Dab-Dab-NH_2, etc. Once it is established whether it is better to elongate or shorten the cationic side chain and whether a guanidinium group or an amino group results in better antimicrobial activity, then combinations of Arg and Lys noncanonical analogs and the other conformationally restricted mimics of similar length can be included in the iterative peptide design process and checked for tryptic stability and bioactivity tests.

As a result of the stereoselective action of trypsin, it is also possible to make HHC-36 variants in which either the all-d-peptide isomers are constructed or the retro-inverso form (reversed sequence with d-amino acid residues in which all side chains will appear in the same position in 3D space as in the l-amino acid analog). Such a strategy will also result in trypsin-resistant peptide analogs.

The same systematic approach can be used for the design of peptide sequences that are resistant to other digestive enzymes, such as chymotrypsin, elastase, and pepsin, based on the enzyme binding subsites' substrate specificity (Table 1).

3. The purity of the peptide is recommended to be ≥95%. Small contaminations cause errors in the concentrations and thus create uncertainties in the peptide-enzyme binding studies, including for the proteolytic stability of the peptide.

4. The carbonate buffer is prepared by dissolving an appropriate amount of NH_4HCO_3 in pyrogen-free H_2O. The pH is adjusted to 8.2 by adding NaOH or HCl as assessed using a pH meter and is then filter sterilized. This buffer does not adsorb UV light at $\lambda = 220$ nm, which is used for the detection of peptides in the RP-HPLC. The carbonate buffer is, however, volatile and the pH changes due to the loss of carbon dioxide. As a result of this and to prevent microbial contamination, the buffer should be stored at 4 °C and renewed at least every 6 months after the first usage.

5. Peptides containing ≥50% hydrophobic residues may be insoluble in aqueous solutions and/or might self-aggregate at a desired concentration. In these circumstances, the stock solution of the

peptides should be dissolved in an appropriate amount of a dimethyl sulfoxide/H_2O (1/1: v/v) buffer until the desired concentration is reached.

6. High fragmentor and low capillary voltages are required to produce accurate masses with relatively good intensity for peptides with low molecular weights, particularly, during enzymatic peptide cleavage that produces fragments with short sequences.
7. As a result of harsh acidic condition of the quenched mixture of the trypsin-treated peptide sample, it is recommended to utilize direct injection in mass spectrometry where the reaction mixture is injected directly via a syringe injection autosampler, to prevent probable damage and consequent loss of performance of the column in the LC/MS instrument. In this mode, a single-run data collection takes ~3–5 min.
8. To detect the unknown peaks representing the peptide fragments in the sample, the MS system is used in the full scan mode. The MS instrument setup for a mass range is dictated by the molecular mass of the intact peptide, e.g., 1261.8185 Da for ^{5Dab}HHC-36 in this protocol, as well as the molecular masses of the anticipated cleaved sequences in the enzymatic reaction.
9. The same experimental protocol may be used to investigate stability of the peptides against other serine and aspartate endopeptidases except that to measure the stability of the peptides to pepsin, a 2 % ammonium hydroxide solution should be used as the quenching cocktail.
10. The peptide's (plasma) half-life can be estimated using Eq. 1 (with slight modifications from http://www-users.med.cornell.edu/~spon/picu/calc/halfcalc.htm),

$$C_t = C_0 e^{\frac{-0.7t}{t_{1/2}}} \quad (1)$$

where C_0 and Ct are the RP-HPLC spectral peak areas before and after the incubation time t of a peptide with an enzyme, and $t_{1/2}$ is the half-life.

11. In the MS system, the total ion chromatogram (TIC) is created by scanning the mass range (e.g., from the m/z 50 to m/z 2000), and the intensities of all detected mass spectral peaks are summed after the acquisition time (~3–5 min). The mass spectral peaks are extracted from the TIC chromatogram using the MassHunter Workstation Software (Fig. 3). The observed mass may reflect the molecular masses of the peptides with a TFA and/or HOAc adduct ion, e.g., $[M + x\text{TFA} + y\text{HOAc} + n\text{H}]n^+$, where "$x$" and "$y$" are the number of TFA and HOAc adduct ions, respectively, and n represents multiple-charge ionization of peptides from protonation (Fig. 3). There are several online bioanalytical MS facilities, for instance, the software program "peptide mass calculator v3.2" http://rna.rega.kuleuven.ac.

be/masspec/pepcalc.htm, which may be used to confirm the observed mass spectral peaks and relate them to the primary structure of peptides (taking into consideration the molecular masses of the adduct ions). By comparison of the peptides' sequences before and after treatment with an enzyme, the enzymatic cleavage position is identified, and thus information on the mechanism of proteolysis is obtained (Fig. 3).

Acknowledgments

MB would thank the Iran National Science Foundation (reference No. 94012757), Iran's National Elites Foundation (reference No. 15/45362–1392), and University of Tehran for the financial supports. REWH would like to acknowledge research funding from the Canadian Institutes for Health Research, and he is the holder of a Canada Research Chair in Health and Genomics.

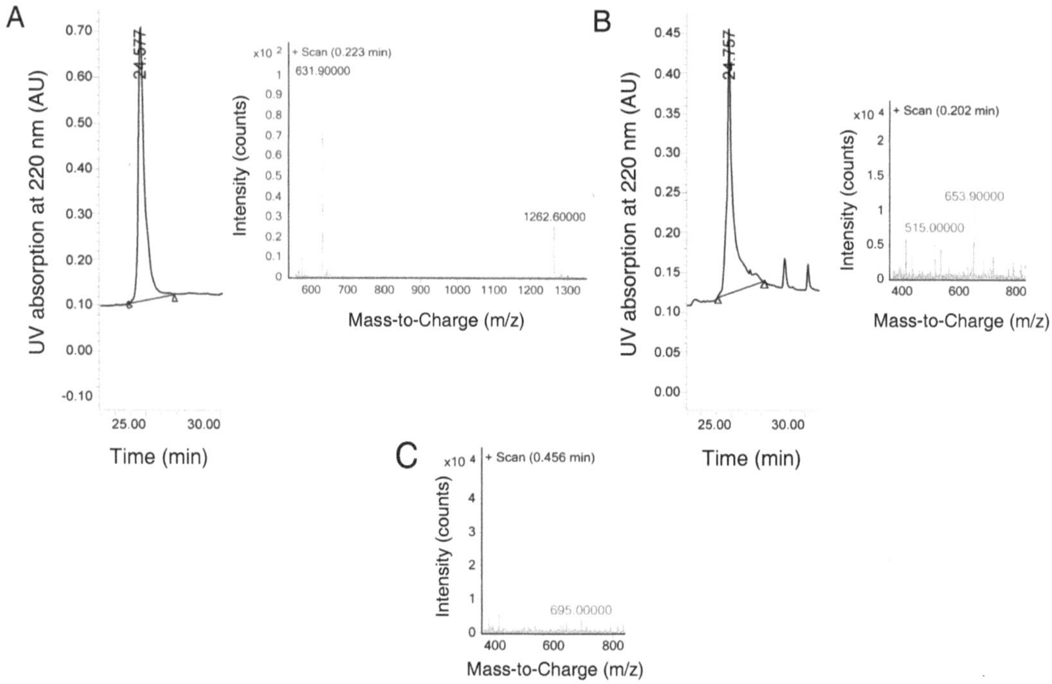

Fig. 3 Analytical RP-HPLC characterization and MS profiles. (**a**) intact ^{5Dab}HHC-36; $[M+1H]^{1+}=1262.8_{found}/1262.8263_{calculated}$, $[M+2H]^{2+}=631.9_{found}/631.9171_{calculated}$, (**b**) 2 h; $[M+4HOAc+4TFA+3H]^{+3}=653.9_{found}/653.6273_{calculated}$, and (**c**) 24 h; $[M+2TFA+2H]^{+2}=695.0_{found}/695.9_{calculated}$ after the peptide incubation with trypsin. The values are the monoisotopic masses in positive mode. The monoisotopic masses of TFA and HOAc are 113.99 Da and 60.02 Da, respectively. The values for the integrated peaks area in the RP-HPLC spectra for the untreated peptide and 2 h after the incubation with the enzyme are 24794329 and 18153357 (absorbance units × seconds), respectively. Thus, the peptide half-life is 4.49 h according to Eq. 1. The found positive ion modes correspond to Dab-Dab-Trp-Trp-Dab-Trp-Trp-Dab-Dab-NH_2 (for **a** and **b**) and Dab-Trp-Trp-Dab-Trp-Trp-Dab-Dab-NH_2 (for **c**) sequences. This indicates that ^{5Dab}HHC-36 is cleaved slowly from the peptide N-terminal by losing one Dab residue

References

1. Bagheri M (2015) Cationic antimicrobial peptides (AMPs): thermodynamic characterization of peptide-lipid interactions and biological efficacy of surface-tethered peptides. ChemistryOpen 4:389–393
2. Bagheri M, Keller S, Dathe M (2011) Interaction of W-substituted analogs of cyclo-RRRWFW with bacterial lipopolysaccharides: the role of the aromatic cluster in antimicrobial activity. Antimicrob Agents Chemother 55:788–797
3. Wenzel M, Chiriac AI, Otto A, Zweytick D, May C, Schumacher C, Gust R, Albada HB, Penkova M, Krämer U, Erdmann R, Metzler-Nolte N, Straus SK, Bremer E, Becher D, Brötz-Oesterhelt H, Sahl HG, Bandow JE (2014) Small cationic antimicrobial peptides delocalize peripheral membrane proteins. Proc Natl Acad Sci U S A 111:E1409–E1418
4. Park CB, Kim HS, Kim SC (1998) Mechanism of action of the antimicrobial peptide buforin II: buforin II kills microorganisms by penetrating the cell membrane and inhibiting cellular functions. Biochem Biophys Res Commun 244:253–257
5. Luque-Ortega JR, van't Hof W, Veerman EC, Saugar JM, Rivas L (2008) Human antimicrobial peptide histatin 5 is a cell-penetrating peptide targeting mitochondrial ATP synthesis in *Leishmania*. FASEB J 22:1817–1828
6. Overhage J, Campisano A, Bains M, Torfs EC, Rehm BH, Hancock REW (2008) Human host defense peptide LL-37 prevents bacterial biofilm formation. Infect Immun 76:4176–4182
7. Hilchie AL, Wuerth K, Hancock REW (2013) Immune modulation by multifaceted cationic host defense (antimicrobial) peptides. Nat Chem Biol 9:761–768
8. Fjell CD, Hiss JA, Hancock REW, Schneider G (2011) Designing antimicrobial peptides: form follows function. Nat Rev Drug Discov 11:37–51
9. Bagheri M, Arasteh S, Haney EF, Hancock REW (2016) Tryptic stability of synthetic bactenecin derivatives is determined by the side chain length of cationic residues and the peptide conformation. J Med Chem 59:3079–3086
10. Hancock REW, Sahl HG (2006) Antimicrobial and host-defense peptides as new anti-infective therapeutic strategies. Nat Biotechnol 24:1551–1557
11. Fosgerau K, Hoffmann T (2015) Peptide therapeutics: current status and future directions. Drug Discov Today 20:122–128
12. Bruno BJ, Miller GD, Lim CS (2013) Basics and recent advances in peptide and protein drug delivery. Ther Deliv 4:1443–1467
13. Carmona-Ribeiro AM, de Melo Carrasco LD (2014) Novel formulations for antimicrobial peptides. Int J Mol Sci 15:18040–18083
14. Chen Y, Mant CT, Farmer SW, Hancock REW, Vasil ML, Hodges RS (2005) Rational design of α-helical antimicrobial peptides with enhanced activities and specificity/therapeutic index. J Biol Chem 280:12316–12329
15. Torrent M, Valle J, Nogués MV, Boix E, Andreu D (2011) The generation of antimicrobial peptide activity: a trade-off between charge and aggregation? Angew Chem Int Ed Engl 150:10686–10689
16. Park IY, Cho JH, Kim KS, Kim YB, Kim MS, Kim SC (2004) Helix stability confers salt resistance upon helical antimicrobial peptides. J Biol Chem 279:13896–13901
17. Friedrich C, Scott MG, Karunaratne N, Yan H, Hancock REW (1999) Salt-resistant alpha-helical cationic antimicrobial peptides. Antimicrob Agents Chemother 43:1542–1548
18. Hedstrom L (2002) Serine protease mechanism and specificity. Chem Rev 102:4501–4524
19. Hudáky P, Kaslik G, Venekei I, Gráf L (1999) The differential specificity of chymotrypsin A and B is determined by amino acid 226. Eur J Biochem 259:528–533
20. Kageyama T (2002) Pepsinogens, progastricsins, and prochymosins: structure, function, evolution, and development. Cell Mol Life Sci 59:288–306
21. Zasloff M (2002) Antimicrobial peptides of multicellular organisms. Nature 415: 389–395
22. Tew GN, Scott RW, Klein ML, Degrado WF (2010) De novo design of antimicrobial polymers, foldamers, and small molecules: from discovery to practical applications. Acc Chem Res 43:30–39
23. Porter EA, Wang X, Lee HS, Weisblum B, Gellman SH (2000) Non-haemolytic β-amino-acid oligomers. Nature 404:565
24. Scott RW, DeGrado WF, Tew GN (2008) De novo designed synthetic mimics of antimicrobial peptides. Curr Opin Biotechnol 19:620–627
25. Cherkasov A, Hilpert K, Jenssen H, Fjell CD, Waldbrook M, Mullaly SC, Volkmer R, Hancock REW (2009) Use of artificial intelligence in the design of small peptide antibiotics effective against a broad spectrum of highly antibiotic-resistant superbugs. ACS Chem Biol 4:65–74
26. Schechter I, Berger A (1967) On the size of the active site in proteases. I. Papain. Biochem Biophys Res Commun 27:157–162

Chapter 6

Determination of Structure and Micellar Interactions of Small Antimicrobial Peptides by Solution-State NMR

Reinhard Wimmer and Lars Erik Uggerhøj

Abstract

NMR spectroscopy is a well-established technique to determine the structure of peptides and small proteins in solution, also when bound to detergent micelles or phospholipid bicelles. The structure of the peptide alone is, however, not conveying the full picture, if the peptide is bound to a micelle, since it does not tell anything about the orientation of the peptide in the micelle. This article describes how to obtain that information together with information on peptide structure.

Key words Antimicrobial peptides, Cell-penetrating peptides, Peptide-lipid interactions, Paramagnetic relaxation enhancement, NMR

1 Introduction

The central dogma in structural biology runs that knowledge of a biomolecule's three-dimensional structure is a prerequisite to understand its function. Cell-penetrating peptides (CPPs) and antimicrobial peptides (AMPs, also referred to as host-defense peptides, HDPs) are a structurally diverse group of peptides, many of which interact with bacterial cell membranes. A key aspect of the biological activity of some peptides is their interaction with biological membranes. Therefore, it is very relevant not only to study the structure of the peptides but also the details of their interaction with cell membranes. Cell membranes are complex, fragile entities that are difficult to work with. That also applies to high-resolution structural studies: it is extremely difficult to obtain the well-diffracting crystals of a protein-membrane complex required for X-ray crystallography, and solution-state NMR spectroscopy of protein-membrane complexes suffers from the inherently large particle sizes resulting in fast transverse relaxation. Solid-state NMR offers many advantages to studying peptide-lipid interactions; however, it is not the subject of this chapter, and the interested reader is referred to the review by Hong et al. [1].

Paul R. Hansen (ed.), *Antimicrobial Peptides: Methods and Protocols*, Methods in Molecular Biology, vol. 1548,
DOI 10.1007/978-1-4939-6737-7_6, © Springer Science+Business Media LLC 2017

Since solution NMR on intact peptide-membrane complexes is very difficult, suitable model systems are often applied instead of membrane systems. Most often, these are detergent (SDS or DPC) micelles, but also phospholipid bicelles [2] have been used with success. The different model systems used have been reviewed elsewhere [3, 4].

Still, determining the structure of a peptide in a micelle by standard methods only reveals the structure of the peptide; no information about the orientation of the peptide in the micelle or its penetration depth is obtained. Such information can be gained from paramagnetic relaxation enhancement (PRE). After excitation with radiofrequency radiation, nonequilibrium coherences are created, which are detected and form the basis of the observed NMR signal. Over time, the detectable signal vanishes, partly due to reestablishment of the equilibrium state (this process is referred to as T_1 relaxation), partly due to a loss of phase coherence (this process is referred to as T_2 relaxation.). Both processes can be enhanced by paramagnetic molecules (molecules with at least one unpaired electron). Paramagnetic relaxation enhancement (PRE) depends on the type of paramagnetic molecule used (its electron spin and electronic relaxation time), the molecular mobility of the nuclei subjected to the PRE, and the distance between the nucleus responsible for the observed NMR signal and the paramagnetic center. The PRE also depends on whether the nucleus and the paramagnetic center are bound to the same molecule or not and—if they are not—whether they are transiently interacting [5]. The basic idea of determining the orientation of a peptide in a micelle lies in adding paramagnetic centers to the aqueous bulk solution surrounding the micelle or to the micelle itself. Since the PRE is distance dependent, the NMR signals of individual peptide atoms will experience different PREs, depending on how far from the micellar surface they are located. Early studies exploited this principle by adding doxyl-labeled lipids to the micelle [6] or paramagnetic ions (Mn^{2+}, Ni^{2+}) to the water phase [7, 8]. These techniques, however, yield only qualitative or semiquantitative data, sufficient to elucidate the overall orientation of the peptide. Another problem associated with using unchelated metal ions is their ability to interact with proteins. Negatively charged groups in the protein will preferably interact with the metal ions. This will cause higher PREs in the vicinity of the metal ions than further away; thus the PRE will not be purely dependent on the distance of an amino acid from the micellar surface. This problem has been solved by using complexes of gadolinium that were developed for magnetic resonance imaging (MRI) and tested against interactions with proteins, lipids, and carbohydrates [9]. In the last decade, quantitative PRE-based methods for the determination of the orientation and insertion depth of peptides in micelles have been developed [10–12]. The different approaches and their advantages and applications have been reviewed [13].

The present chapter describes how to determine the position and orientation of a peptide in a micelle alongside with its structure. The chapter covers the necessary sample conditions, NMR experiments, and setup of calculations. The protocol applied here has been applied successfully for different α-helical AMPs [14, 15], chimera of peptides and peptoids ("peptomers") [16], and lantibiotics [17]. Figure 1 shows some illustrations of results obtained using this protocol.

2 Materials

NMR samples: all solutions are prepared in MilliQ H_2O with a suitable fraction of D_2O, depending on hardware requirements (we have good experience with 5 % D_2O.).

1. Perdeuterated dodecylphosphocholine (DPC-d_{38}).
2. D_2O.
3. 3-(Trimethylsilyl)propionic-2,2,3,3-d_4 acid sodium salt (TSP-d_4).
4. Gd(DTPA-BMA solution): 257.4 mM Gd(DTPA-BMA, known as gadodiamide or Omniscan®).
5. 10 mM phosphate buffer pH 6–7, containing 5–10 % of D_2O.

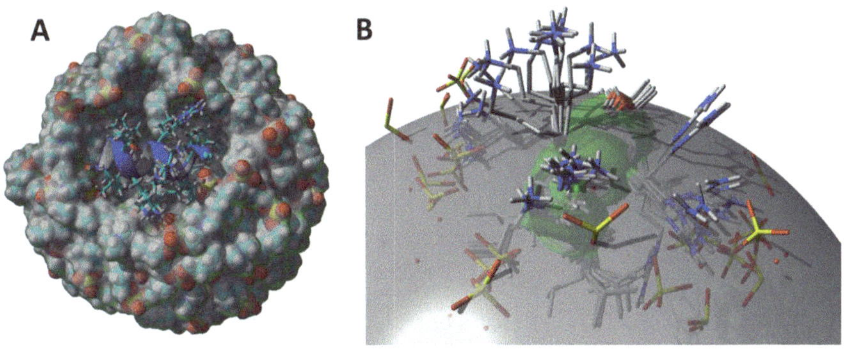

Fig. 1 Structure of the antimicrobial peptide anoplin (Gly-Leu-Leu-Lys-Arg-Ile-Lys-Thr-Leu-Leu-NH_2) in a DPC micelle: (**a**) anoplin embedded in the micelle shown with a transparent surface. (**b**) 20 superimposed models of anoplin (green helix and stick model) and the DPC micelle (gray sphere) with all phosphate groups (shown as yellow and red sticks) within 4 Å from a positively charged amino acid of anoplin, illustrating possible electrostatic interactions between the peptide and the micelle

3 Methods

3.1 NMR

1. Any modern NMR spectrometer can be used. At peptide concentrations of 2–3 mM, a 600 MHz NMR system without cryogenically cooled probe will yield good results. A whole week of NMR time must be calculated for obtaining all spectra for one peptide. When higher sensitivity is available, peptide concentrations or recording times can be reduced accordingly.
2. It is imperative that relaxation measurements are carried out at 310 K, since calibration of PREs is only possible at this temperature. An additional advantage of recording at 310 K is that the elevated temperature leads to enhanced molecular mobility, which in turn causes slower transverse relaxation and is thus beneficial for the linewidths of the NMR signals of large peptide/micelle systems (*see* **Note 1**). It is also necessary that the peptide is fully bound to the micelles, as the conversion of PREs to distances from the micelle center otherwise would fail (*see* **Note 2** for possible remedies for not fully bound peptide).
3. Peptides are dissolved to 2–3 mM in 10 mM phosphate buffer pH 6–7, containing 5–10% of D_2O.
4. Further, add 100 μM of TSP-d_4 as a chemical shift standard and 2 mM NaN_3 to prevent microbial growth and to improve samples' long-term stability. DPC-d_{38} was added to 150 mM (*see* **Note 3**). For an NMR spectrometer with a 5 mm probe, the total sample volume should be 525–600 μL, but different probe diameters could be used as well. The use of shaped tubes is not necessary since the zwitterionic DPC does not lead to RF heating.
5. Run a standard ^{1}H-NMR spectrum with water suppression, e.g., presaturation or excitation sculpting [18].
6. DOSY (diffusion ordered spectroscopy): in order to check whether the peptide binds fully to the micelle, diffusion measurements should be conducted. The apparent diffusion constant of the peptide should be the same as that of the DPC micelle (*see* **Note 4**).
7. Choose a convection-compensated double-stimulated echo experiment (e.g., pulse program dstebpgp3s) with $\Delta = 200$ ms, $\delta = 3$ ms, and 32–128 gradient strength steps ranging from 1 to 45 G/cm.
8. TOCSY (total correlation spectroscopy): any TOCSY sequence with water suppression is applicable. We have good results with the pulse program *mlevesgpph* using MLEV-17 mixing [19] and excitation sculpting [18]. We used a mixing time of 60 ms, a relaxation delay of 2.4 s, and $t_{1,max} \approx 50$ ms and $t_{2,max} \approx 300$ ms. On a 600 MHz NMR spectrometer with a room-temperature probe, NS = 32 yields a good result.

9. NOESY (nuclear Overhauser effect spectroscopy): use a standard three-pulse gradient-selected NOESY experiment, pulse program *noesyesgpph* with excitation sculpting water suppression, 60 ms mixing time, a relaxation delay of 2.4 s, and $t_{1,max} \approx 50$ ms and $t_{2,max} \approx 300$ ms. On a 600 MHz NMR spectrometer with a room-temperature probe, NS = 32 yields a good result.
10. DQF-COSY (double-quantum-filtered correlation spectroscopy): although not strictly required, a COSY spectrum sometimes helps distinguishing side-chain resonances. Use a gradient-selected double-quantum-filtered COSY with presaturation water suppression and pulse program *cosygpfmphpr*.
11. [^{1}H-^{13}C]-HSQC (heteronuclear single-quantum coherence spectroscopy): the natural abundance of ^{13}C is enough to obtain a ^{13}C-HSQC spectrum within reasonable time. ^{13}C-shift information is not necessary for structure determination, but it helps in two ways: firstly, C^α and C^β chemical shifts are valuable for obtaining torsion-angle constraints. Secondly, overlapping resonances in the ^{1}H spectra can often be resolved using the ^{13}C-HSQC.
12. Use standard HSQC pulse program *hsqcetgpsi* with a $t_{1,max} \approx 12$ ms and $t_{2,max} \approx 340$ ms with ^{13}C chemical shifts ranging from 5 to 75 ppm. On a 600 MHz NMR spectrometer with a room-temperature probe, NS = 32 yields a good result.
13. Relaxation measurements: if the structure of the peptide is sufficient and details about how the peptide is situated within the micelle (orientation and insertion depth) are not of interest, then this point is not necessary.
14. In order to determine the bulk solvent PRE of peptide H^α atoms, we have good experience using inversion-recovery weighted NOESY spectra [12, 14–16] (*see* **Note 5**). NOESY spectra are essentially set up as described above. We recorded NOESY spectra at six different relaxation delays and found it worthwhile to vary the relaxation time settings depending on Gd^{3+} concentrations. The faster T_1 relaxation caused by the paramagnetic agent allows for shorter relaxation delays and thus reduced recording times: at 0–1 mM Gd^{3+}, use 100 ms, 300 ms, 600 ms, 1.2 s, 2.6 s, and 4 s; at 2–4 mM Gd^{3+}, use 100 ms, 300 ms, 600 ms, 1.08 s, 1.8 s, and 3.6 s; at 5–8 mM Gd^{3+}, use 50 ms, 300 ms, 600 ms, 900 ms, 1.6 s, and 3.2 s; and above 8 mM Gd^{3+}, use 20 ms, 200 ms, 500 ms, 700 ms, 1.2 s, and 2.4 s.
15. The recycling delay can be set to the same value as the longest relaxation delay. The number of scans and the number of increments in the indirect ^{1}H dimension can be adjusted to meet sensitivity and resolution requirements and instrument time limitations.

16. The experiment is repeated at least at three, better four different concentrations of Gd^{3+}, starting with $[Gd^{3+}] = 0$ mM and not exceeding 8 mM. 0, 2, 5, and 8 mM is working fine (*see* **Note 6**).

Resonance assignment follows the standard "sequential walk" method for homonuclear spectra of peptides and proteins [20].

3.2 Determination of PREs and Structure Determination with CYANA

1. Fourier transform the inversion-recovery weighted NOESY spectrum along F3 and F2 (the direct and indirect ^{1}H dimensions) and apply phase and baseline corrections, as needed.
2. Pick the peaks you want to use for PRE determination—*see* **Note 7** on which peaks to choose. The chosen peaks have to be integrated; we used the NEASY module of CARA for integration; however, other software packages exist.
3. The Bruker Protein Dynamics Center offers a convenient way of integrating peaks in multiple planes of pseudo-3D spectra and fitting the obtained integrals (*see* **Note 8** on how to choose integration regions and **Note 9** on how PRE values are obtained from the peak integrals). The conversion of PREs to distance restraints is described in the following chapter.
4. Different software packages exist for the calculation of protein structures from NMR data. In this protocol, we will describe the use of CYANA. The determination of peptide structures from NMR data has been described extensively in the literature. For a basic protocol, not including PRE data, the reader is referred to Reference [21].
5. Many antimicrobial peptides feature an amidated C-terminus. Including this into the structure calculation with CYANA requires modifications in the CYANA library, which are explained in **Note 10**.
6. Preparation of input files for structure determination: torsion-angle constraints can be derived from chemical shifts by the TALOS, TALOS+, or TALOS-N software packages [22–24]. This is only recommended, when a near-complete assignment of C^α and C^β shifts has been achieved.
7. PRE-based constraints: PRE values are converted to distances as described in the literature (Eq. 3 of Reference [12]):

$$r = g + R - \left(\frac{\mathrm{PRE}}{k}\right)^{-1/3} \quad (1)$$

where r [Å] is the distance of the atom in question from the micelle center, g [Å] is the apparent average distance of the Gd complex from the micelle surface (8 ± 0.9 Å), R [Å] is the micelle radius of 22.55 Å (*see* **Note 11**), PRE [s^{-1} mM^{-1}] is the measured PRE value, and $k = 253.1 \pm 91.95$ [s^{-1} mM^{-1} $Å^3$] is the calibration constant. *See* **Note 12** for an example of how such constraints can be defined. Figure 2 illustrates the significance of the PRE-based restraints for the structure of the peptide/micelle complex.

8. CYANA sequence file containing the micelle center. The inclusion of PRE constraints needs the definition of the micelle for the structure calculation. This is most conveniently done by including the center of the micelle as a pseudo-atom. In CYANA, all atoms need to be organized in residues, and all residues need to be connected in one chain. This is achieved by starting with the peptide sequence and then connecting the micelle center to the peptide through a series of flexible linker residues (*see* **Note 10** for the CYANA library entries for the micelle and the sequence file to use for the definition of the peptide/micelle complex—both with and without C-terminal amide group).
9. Structure calculation: the structure calculation follows the standard protocol but using the modified sequence file and the extra PRE-derived constraints. Typically, 100 structures were calculated, and the 20 structures with lowest target functions were retained.

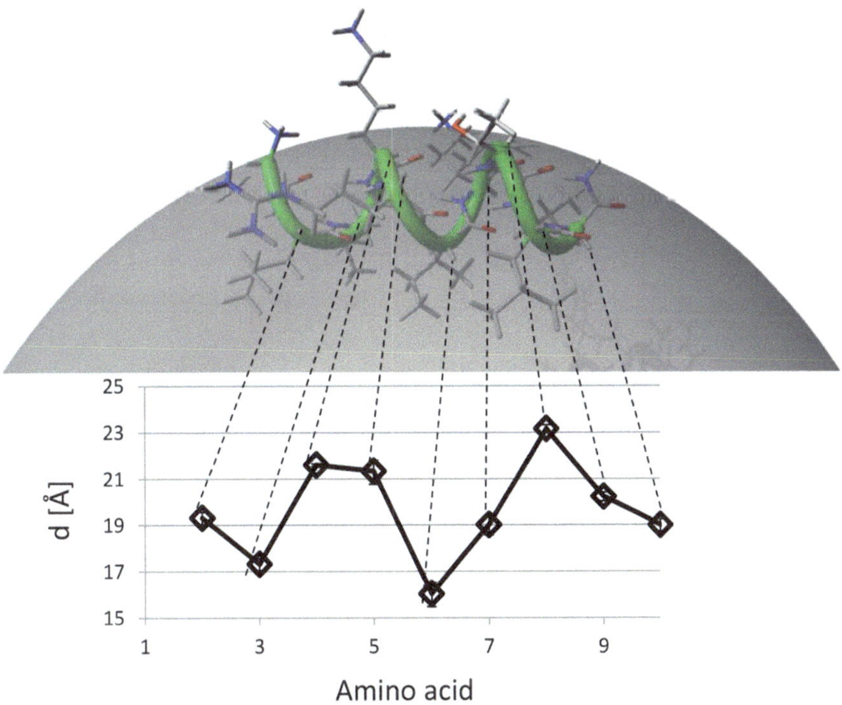

Fig. 2 Structure of anoplin (stick model) in a DPC micelle (transparent, gray sphere), with corresponding PRE-based distance restraints from H^{α} of each amino acid to the micelle center (lower graph)

10. Energy refinement: for energy refinement, many different software packages exist. We have good experience using YASARA [25]. Refinement was done in the following way: an energy-minimized DPC micelle was defined as described in [12]. Briefly, a preformed DPC micelle was obtained from the homepage of Professor Peter Tieleman at the University of Calgary (http://moose.bio.ucalgary.ca) [26], water molecules were defined surrounding the micelle, and a 3.3 ns MD simulation was applied. The PDB file of the micelle thus obtained is in the supplementary material.
11. Each of the 20 structures from CYANA is then placed at 20 random positions above the micellar surface; NOE, PRE, and dihedral angle constraints were added, and the system is energy minimized first in vacuo using the NOVA force field [27], then in water, using the YASARA force field [28], as described in [15]. Out of the 20 randomly placed peptide structures, the one with the lowest residual constraint violation energies was kept. In total, this yields 20 refined structures, which are merged to create the final bundle.

4 Notes

1. Prolonged NMR acquisition at 310 K causes non-negligible evaporation of water from the NMR sample. (We have experienced evaporation of up to 20%.) This leads to higher concentrations of paramagnetic agent than assumed and thus to an overestimation of PRE values. We therefore strongly recommend that the NMR sample is kept in an airtight tube at all times.
2. If the peptide is not fully bound to the micelle, it sometimes helps to increase the concentration of DPC and/or to add up to 300 mM NaCl. Be aware that the diffusion constant of the micelle changes and needs to be redetermined for each new sample condition.
3. If peptides do not aggregate, the number of peptides per micelle follows a Poisson distribution [29]:

$$P(n) = \frac{\lambda^n e^{-\lambda}}{n!} \quad (2)$$

where $P(n)$ is the probability of finding a micelle with n peptides bound and λ is the average number of peptide molecules per micelle. At 150 mM DPC, the concentration of DPC micelles is approximately 2.3 mM (the critical micelle concentration of DPC is approximately 1.1 mM. Therefore, at 150 mM DPC, 148.9 mM are to be found in micelles. At an aggregation number of approximately 65 DPC molecules per micelle, this gives 148.9/65 = 2.3 mM of micelles). Assuming fully bound peptide at a concentration of 2.3 mM, $\lambda = c_{\text{peptide}}/c_{\text{micelle}} \approx 1$. From Eq. 2 follows that 36.8 % of micelles will not contain any peptide, 36.8 % of micelles will contain exactly one peptide, 18.4 % of micelles will contain two peptide molecules, 6.1 % of peptides will contain three peptide molecules, and so forth. The micelle is much larger than the peptide; thus the accommodation of several peptide molecules in one micelle is not a problem for the analysis.

4. The apparent diffusion constant of DPC will be a concentration-weighted average of micelle and monomer diffusion constant. The size of the micelle and the viscosity of the solution depend on the DPC concentration, concentration of other salts, and temperature. Therefore, diffusion measurements of DPC micelles should always be done within the same sample as for the peptide. Since commercially available DPC is only ^{2}H labeled to approx. 98 %, there is sufficient ^{1}H sensitivity of DPC signals to measure DPC diffusion, too. Often, the apparent diffusion constant of the bound peptide will appear slightly lower than that of the DPC micelle, because the latter is a weighted average of monomer and micelle.

5. A pulse program *noesyesgpphir* based on Bruker's standard pulse program *noesyesgpph* is given in the Supplementary Material. Here, the experiment is set up in a pseudo-3D manner. This offers the advantage, that all planes are scaled equally during Fourier transformation, which would not automatically be the case if spectra are recorded as single 2D-NOESYs. In addition, when running the spectrum as a pseudo-3D, the acquisition order of the dimensions can be set such that all relaxation delays are cycled for each increment in the indirect frequency dimension, thus eliminating possible bias of obtained relaxation times by temporal changes in the sample or the spectrometer behavior.

   In principle, any 2D spectrum can be used. An inversion recovery element can, for instance, be introduced at the start of a TOCSY or HSQC pulse sequence. Other authors have reported the use of saturation-recovery weighted NOESY, which allows for faster repetition rates [10]. The NOESY spectrum offers the advantage that there often are several cross peaks for each H^{α} in the indirect dimension. In such a case, more than one PRE can be experimentally determined allowing for an estimate of precision in the data. At the same time, small peptides offer sufficient spectral resolution to distinguish most peaks in a 2D-NOESY. We have observed up to five useful cross peaks with the same H^{α} in the indirect dimension. The standard deviation for the distance constraints thus obtained was always below 1 Å. Hence, we allowed a deviation of ±1 Å around the mean distance constraint (derived from the mean PRE by Eq. 1) for the structure calculation. In addition, we weighted PRE-derived constraints with 10% compared to NOE or dihedral angle constraints in order to guarantee that PRE constraints do not distort the structure but merely serve to position the peptide correctly with respect to the micelle center.

6. At 10 mM Gd(III), detrimental line broadening can be detected, because not only T_1 but also T_2 relaxation is enhanced. Recently, the use of high-spin Fe(III) complexes instead of Gd(III) complexes has been advocated, because Fe(III) exhibits less line broadening than Gd(III) [30]. If Fe(III) were to be used instead of Gd(III), the conversion of PREs to distances would have to be recalibrated as described in [12].
7. Only PREs on H^{α} are recommended for use. The PRE depends on molecular mobility; thus side-chain atoms might yield deviating results due to internal mobility. Likewise, the use of H^{N} is strongly discouraged, since H^{N} atoms can exchange with bulk water that is experiencing a much stronger PRE. In principle, any peak featuring H^{α} in the indirect (!) dimension can be used (this is a consequence of inserting the inversion-recovery element at the start of the pulse sequence.). We experienced the best results with H^{N}/H^{α} cross peaks; in small helical peptides, there is usually more than one resolved cross peak for each H^{α}. Cross peaks from other atoms (e.g., H^{β}, H^{γ}, H^{δ}) to H^{α} can in principle also be used, but we found them more often overlapping with other peaks and weak. Intraresidual H^{β}/H^{α} cross peaks should only be used if they are free of zero-quantum artifacts.
8. It is important always to use identical integration regions in all planes of the pseudo-3D spectrum for a given Gd concentration. In NEASY/CARA, this can be achieved by drawing rectangles around the desired integration regions. These can then be written to the disk and loaded onto each single plane of the spectrum followed by integration of the rectangular region. The regions used for integration do not have to cover the whole cross peak, as long as they always cover an identical fraction of the cross peak. Different rectangular regions can be defined for each concentration of Gd. In case of overlapping peaks, define the region in a way that contributions from other nearby peaks are prevented. This becomes important at higher Gd concentrations, where signal overlap increases.
9. At first, longitudinal relaxation (T_1) times for each Gd titration step have to be determined: obtain the peak integral $I(\tau)$ for each peak p and for each relaxation delay τ at a given concentration of Gd. T_1 for a peak is obtained by fitting the integrals at different relaxation delays to Eq. 3:

$$I(\tau) = I(\tau = 0) \times \left(1 - 2e^{-\frac{\tau}{T_1}}\right) \quad (3)$$

Since it is not practicable to obtain a reliable data point for $\tau = 0$, the value $I(\tau = 0)$ should also be fitted together with T_1.

For each peak, PRE values are then obtained by a linear regression with $1/T_1$ [s^{-1}] values as y-values and Gd concentrations [mM] as x-values where the PRE [s^{-1} mM^{-1}] simply is the slope of the regression line.

If more than two PRE values are obtained for a given atom, the uncertainty of the obtained PRE can be estimated from the standard deviation of PREs.

10. The following modifications to the CYANA library need to be made:

amidated C-terminus:

The C-terminal amide group is easiest introduced as a separate residue. Below is the definition of the residue AMID, an amide group ending the peptide chain. If a micelle needs to be included into the calculations, the peptide chain in CYANA does not stop with the amide group, and the residue AML should be used. See below for examples of CYANA sequence files using these two residue types.

Add the following lines to the CYANA library file (the lines are available electronically in the supplementary file "residue definitions for CYANA library.docx"):

```
RESIDUE    AMID         1     6     3     6
   1 OMEGA   0   0   0.0000   2     1     3     4    0
   1 C       C_BYL     2     0.0000     0.0000
0.0000     0.0000     2     3     0     0     0
   2 O       O_BYL     2     0.0000    -1.1498
0.3122     0.1930     1     0     0     0     0
   3 N       N_AMI     2     0.0000     1.1461
0.9690     0.2222     1     4     5     0     0
   4 H1      H_AMI     0     0.0000     0.8883
1.8990     0.5404     3     0     0     0     6
   5 H2      H_AMI     0     0.0000     2.0843
0.6622     0.0457     3     0     0     0     6
   6 QH      PSEUD     0     0.0000     1.4863
1.2806     0.2931     0     0     0     0     0

RESIDUE    AML          1     9     3     8

    1 OMEGA       0     0     0.0000     2     1     3
4     0
   1 C       C_BYL     2     0.0000     0.0000
0.0000     0.0000     2     3     0     0     0
```

```
   2 O      O_BYL     2     0.0000    -1.1498
0.3122      0.1930     1     0     0     0     0
   3 N      N_AMI     2     0.0000     1.1461
0.9690      0.2222     1     4     5     0     0
   4 H1     H_AMI     0     0.0000     0.8883
1.8990      0.5404     3     0     0     0     6
   5 H2     H_AMI     0     0.0000     2.0843
0.6622      0.0457     3     0     0     0     6
   6 QH     PSEUD     0     0.0000     1.4863
1.2806      0.2931     0     0     0     0     0
   7 Q1     DUMMY     0     0.0000     0.0000
0.0000      0.0000     0     0     0     0     0
   8 Q2     DUMMY     0     0.0000     1.4142
0.0000     -1.4142     0     0     0     0     0
   9 Q3     DUMMY     0     0.0000     2.8284
0.0000      0.0000     0     0     0     0     0
```

Adding the micelle as a residue: simply add the following residue definition to the CYANA library (the lines are available electronically in the supplementary file "residue definitions for CYANA library.docx"). The residue is called DUM, and the micelle center is called X in order to comply with PDB naming conventions.

```
RESIDUE     DUM         1     7     3     6

   1 LB          0      0     0.0000     2     3     5
6     0
   1 Q1     DUMMY     0     0.0000     0.0000
0.0000      0.0000     0     0     0     0     0
   2 Q2     DUMMY     0     0.0000     1.4142
0.0000     -1.4142     0     0     0     0     0
   3 Q3     DUMMY     0     0.0000     2.8284
0.0000      0.0000     0     0     0     0     0
   4 X       P_ALI     0     0.0000     3.0000
0.0000      0.0000     0     0     0     0     0
   5 Q1     DUMMY     0     0.0000     4.2426
0.0000     -1.4142     0     0     0     0     0
   6 Q2     DUMMY     0     0.0000     2.8284
0.0000     -2.8284     0     0     0     0     0
   7 Q3     DUMMY     0     0.0000     4.2426
0.0000     -4.2426     0     0     0     0     0
```

The sequence file for a hypothetical peptide Gly-Leu-Ile with the flexible linkers and the micelle would look like this:

without amidated C-terminus		with amidated C-terminus	
GLY	1	GLY	1
LEU	2	LEU	2
ILE	3	ILE	3
PL	22	AML	4
LL2	23	PL	22
LL2	24	LL2	23
LL2	25	LL2	24
LL5	26	LL2	25
LL5	27	LL5	26
LL5	28	LL5	27
LL5	29	LL5	28
LL5	30	LL5	29
LL5	31	LL5	30
LL5	32	LL5	31
LL5	33	LL5	32
LL5	34	LL5	33
LL5	35	LL5	34
LL5	36	LL5	35
LL2	37	LL5	36
LL2	38	LL2	37
LL2	39	LL2	38
LL2	40	LL2	39
LL	41	LL2	40
LL	42	LL	41
DUM	43	LL	42
		DUM	43

11. Micelle size depends on the concentration of DPC and other parameters like temperature and ionic strength. Literature reports different values for the radius of a DPC micelle: 23.5 and 27.75 Å [31], 22.2 Å [32], 18.65 Å [33], 22.55 Å [12], 21.7–22.1 Å [34], and 23.3 ± 1.4 Å [35]. The PRE technique actually measures the distance of a given atom from the micellar surface; hence the obtained result is independent of the precise micellar radius assumed.

12. In the CYANA format, upper distance constraints and lower distance constraints have to be stored in separate files. In order to define a distance constraint between two atoms, e.g., H^{α} of residue Leu 2 and the micelle center (here defined as residue DUM 43), of 19.0 ± 1 Å, two constraints have to be given: a lower distance limit of 18.0 Å and an upper distance limit of 20.0 Å. The following two lines have to be included in two separate files:

upper distance limit: {filename}.upl

```
2   LEU   HA   43   DUM   X   20.000  0.10
```

lower distance limit: {filename}.lol

```
2   LEU   HA   43   DUM   X   18.000  0.10
```

The final "0.10" denotes the weight of the constraint in the CYANA target function.

The XPLOR format allows for a more compact definition of a 19.0 ± 1 Å distance constraint in one line, but the relative weighting of classes of constraints has to be written explicitly into the calculating script:

```
assi (resi   2 and name   HA) (resi   43 and name   X) 19.00 1.00 1.00
```

Acknowledgments

This work was supported by the Villum Kann Rasmussen Foundation (BioNET) and the Danish Research Councils for Free and for Strategic Research (DanCARD, project number 09-067075). The NMR laboratory at Aalborg University is supported by the Obel, SparNord, and Carlsberg Foundations. Gd(DTPA-BMA) was a gift from the local hospital.

Supplementary Material

Supplementary material is available electronically from the author's university homepage:

http://www.en.bio.aau.dk/research/biotechnology/nmr-spectroscopy/mmib-supplementary

References

1. Su Y, Li S, Hong M (2013) Cationic membrane peptides: atomic-level insight of structure-activity relationships from solid-state NMR. Amino Acids 44:821–833
2. Dittmer J, Thøgersen L, Underhaug J et al (2009) Incorporation of antimicrobial peptides into membranes: a combined liquid-state NMR and molecular dynamics study of alamethicin in DMPC/DHPC bicelles. J Phys Chem B 113:6928–6937
3. Raschle T, Hiller S, Etzkorn M et al (2010) Nonmicellar systems for solution NMR spectroscopy of membrane proteins. Curr Opin Struct Biol 20:471–479
4. Mäler L (2012) Solution NMR studies of peptide-lipid interactions in model membranes. Mol Membr Biol 29:155–176
5. Botta M (2000) Second coordination sphere water molecules and relaxivity of gadolinium(III) complexes: implications for MRI contrast agents. Eur J Inorg Chem 2000:399–407
6. Brown LR, Braun W, Kumar A et al (1982) High resolution nuclear magnetic resonance studies of the conformation and orientation of melittin bound to a lipid-water interface. Biophys J 37:319–328
7. Lindberg M, Biverståhl H, Gräslund A et al (2003) Structure and positioning comparison of two variants of penetratin in two different membrane mimicking systems by NMR. Eur J Biochem 270:3055–3063
8. Wimmer R, Andersen KK, Vad B et al (2006) Versatile interactions of the antimicrobial peptide novispirin with detergents and lipids. Biochemistry 45:481–497
9. Pintacuda G, Otting G (2002) Identification of protein surfaces by NMR measurements with a pramagnetic Gd(III) chelate. J Am Chem Soc 124:372–373
10. Respondek M, Madl T, Gobl C et al (2007) Mapping the orientation of helices in micelle-bound peptides by paramagnetic relaxation waves. J Am Chem Soc 129:5228–5234
11. Zangger K, Respondek M, Gobl C et al (2009) Positioning of micelle-bound peptides by paramagnetic relaxation enhancements. J Phys Chem B 113:4400–4406
12. Franzmann M, Otzen D, Wimmer R (2009) Quantitative use of paramagnetic relaxation enhancements for determining orientations and insertion depths of peptides in micelles. Chembiochem 10:2339–2347
13. Schrank E, Wagner GE, Zangger K (2013) Solution NMR studies on the orientation of membrane-bound peptides and proteins by paramagnetic probes. Molecules 18:7407–7435
14. Vad B, Thomsen LA, Bertelsen K et al (2010) Divorcing folding from function: how acylation affects the membrane-perturbing properties of an antimicrobial peptide. Biochim Biophys Acta 1804:806–820
15. Uggerhøj LE, Poulsen TJ, Munk JK et al (2015) Rational design of alpha-helical antimicrobial peptides: do's and don'ts. Chembiochem 16:242–253
16. Uggerhøj LE, Munk JK, Hansen PR et al (2014) Structural features of peptoid-peptide hybrids in lipid-water interfaces. FEBS Lett 588:3291–3297
17. Münch D, Müller A, Schneider T et al (2014) The Lantibiotic NAI-107 Binds to Bactoprenol-bound Cell Wall Precursors and Impairs Membrane Functions. J Biol Chem 289:12063–12076
18. Hwang TL, Shaka AJ (1995) Water suppression that works. Excitation sculpting using arbitrary wave-forms and pulsed-field gradients. J Mag Res A 112:275–279
19. Bax A, Davis DG (1985) MLEV 17 based two-dimensional homonuclear magnetization transfer spectroscopy. J Mag Res 65:355–360
20. Wüthrich K (1986) NMR of proteins and nucleic acids. Wiley, New York, NY
21. Güntert P (1997) Calculating protein structures from NMR data. Methods Mol Biol 60:157–194
22. Cornilescu G, Delaglio F, Bax A (1999) Protein backbone angle restraints from searching a database for chemical shift and sequence homology. J Biomol NMR 13:289–302
23. Shen Y, Delaglio F, Cornilescu G et al (2009) TALOS+: a hybrid method for predicting protein backbone torsion angles from NMR chemical shifts. J Biomol NMR 44:213–223
24. Shen Y, Bax A (2013) Protein backbone and sidechain torsion angles predicted from NMR chemical shifts using artificial neural networks. J Biomol NMR 56:227–241
25. Krieger E, Vriend G (2014) YASARA View - molecular graphics for all devices - from smartphones to workstations. Bioinformatics 30:2981–2982
26. Tieleman DP, van der Spoel D, Berendsen HJC (2000) Molecular dynamics simulations of dodecylphosphocholine micelles at three different aggregate sizes: micellar structure and chain relaxation. J Phys Chem B 104:6380–6388

27. Krieger E, Koraimann G, Vriend G (2002) Increasing the precision of comparative models with YASARA NOVA-a self-parameterizing force field. Proteins 47:393–402
28. Krieger E, Joo K, Lee J et al (2009) Improving physical realism, stereochemistry and side-chain accuracy in homology modeling: four approaches that performed well in CASP8. Proteins 77:114–122
29. Papoulis A (1991) Probability, random variables and stochastic processes. McGraw Hill, New York, NY
30. Oktaviani NA, Risør MW, Lee Y-H et al (2015) Optimized co-solute paramagnetic relaxation enhancement for the rapid NMR analysis of a highly fibrillogenic peptide. J Biomol NMR 62:129–142
31. Lauterwein J, Bösch C, Brown LR et al (1979) Physicochemical studies of the protein-lipid interactions in melittin-containing micelles. Biochim Biophys Acta 556:244–264
32. Gao X, Wong TC (1998) Studies of the binding and structure of adrenocorticotropin peptides in membrane mimics by NMR spectroscopy and pulsed-field gradient diffusion. Biophys J 74:1871–1888
33. Kallick DA, Tessmer MR, Watts CR et al (1995) The use of dodecylphosphocholine micelles in solution NMR. J Mag Res B 109:60–65
34. Abel S, Dupradeau F-Y, Marchi F (2012) Molecular dynamics simulations of a characteristic DPC micelle in water. J Chem Theory Comput 8:4610–4623
35. Göbl C, Dulle M, Hohlweg W et al (2010) Influence of phosphocholine alkyl chain length on peptide-micelle interactions and micellar size and shape. J Phys Chem B 114:4717–4724

Chapter 7

Mass Spectrometry Approaches for Determining the Structure of Antimicrobial Peptides

Jiongyu Liu and Jianping Jiang

Abstract

In the past decades, a great amount of antimicrobial peptides (AMPs) has been discovered, the structure identification of which relies heavily on de novo sequencing by Edman degradation or mass spectrometry. Here we outline the basic procedures for the exact mass measurement approaches that use off-line low-energy CID ESI Qq-TOF MS/MS in positive-ion mode, which is typically applied to de novo sequencing of peptides, to elucidate the structure of AMPs. Ambiguity I/L and partial sequence order were elucidated by Edman degradation or/and structural similarity analysis to known sequence. The approaches can determine the structure of peptides composed of as much as 38 amino acids in our practice.

Key words Antimicrobial peptides, Mass spectrometry, De novo, CID ESI Qq-TOF, MS/MS

1 Introduction

Antimicrobial peptides (AMPs) are important compounds that can kill microbes and may be an alternative to conventional antibiotics. The AMP field is advancing rapidly in response to the demand of the novel therapeutically valuable agents against microbes resistant to the traditional antibiotics. A large number of AMPs have been identified from almost all the groups of organisms, including virus, bacteria, fungi, plants, and animals [1, 2]. The Antimicrobial Peptide Database (APD3, http://aps.unmc.edu/AP/) has collected 2706 antimicrobial peptide and protein sequences with validated activities [3]. The Collection of Antimicrobial Peptides (CAMP$_{R3}$, http://www.camp.bicnirrh.res.in) holds 10,247 AMP antimicrobial peptide or protein sequences that include 757 antimicrobial structures, of which 4857 are experimentally validated and 5390 are predicted [4]. The majority of these sequences contain less than 100 amino acid residues, typically, less than 50 amino acid residues. The structure of AMPs can be determined by Edman degradation or mass spectrometry. An alternative approach to de novo sequencing became feasible after a type of mass spectrometer,

Paul R. Hansen (ed.), *Antimicrobial Peptides: Methods and Protocols*, Methods in Molecular Biology, vol. 1548, DOI 10.1007/978-1-4939-6737-7_7,

ESI Qq-TOF mass spectrometry, was introduced, where ESI refers to an electrospray ionization source, Q refers to a quadrupole, q refers to a collision cell or RF-only quadrupole, and TOF refers to a time-of-flight mass spectrometer. The peptide solution is directly introduced to the mass spectrometer. Ions are formed in the ESI source, which are guided through the hexapole to enter the quadrupole mass filter. In MS mode, the ions in the quadrupole (transmission mode, RF only) are transmitted to the TOF analyzer to generate the MS spectrum. In MS/MS mode, particular ions in the quadrupole (isolation mode, RF + DC) are selected to enter the collision cell for collision-induced dissociation (CID); the generated fragments are sent to and measured in the TOF analyzer to generate the MS/MS spectrum. Recently our group developed exact mass measurement approaches that use off-line low-energy CID ESI Qq-TOF MS/MS in positive-ion mode, which is typically applied to de novo sequencing of peptides, to elucidate the structure of AMPs from Asian frogs *Odorrana jingdongensis*. Ambiguity I/L and partial sequence order were elucidated by Edman degradation or/and structural similarity analysis to known sequence in other species of *Odorrana* genus. The approaches can determine the structure of peptides composed of as much as 38 amino acids in our practice [5].

2 Materials

All reagents are prepared and stored at room temperature unless indicated otherwise.

2.1 Disulfide Bond Reduction and Thiol Group Alkylation

Prepare all solutions using ultrapure water and analytical grade reagents.

1. Purified or partially purified peptide (*see* **Note 1**).
2. SmartSpec™ Plus Spectrophotometer.
3. Disulfide bond reduction solution (10×): 5 mM dithiothreitol (DTT), 500 mM NH_4HCO_3. The solution is stored in single-use aliquots (1 mL in microcentrifuge tube) at −20 °C (*see* **Note 2**).
4. Thiol group alkylation solution (100×): 110 mM iodoacetamide (IAM) solution. The solution is stored in single-use aliquots (100 μL in microcentrifuge tube covered with aluminum foil) at −20 °C (*see* **Note 3**).

2.2 Purification of Alkylated Peptide

Prepare all solutions using ultrapure water and HPLC grade reagents.

1. ÄKTA™ purifier PHC 10 chromatography system.
2. UNICORN control software (version 5.11).

3. 100-5C-18 HPLC reverse phase column (4.6 × 250 mm).
4. Solvent A: 0.1 % v/v trifluoroacetic acid (TFA)/water filtered through a 0.45 μm membrane.
5. Solvent B: 0.1 % v/v TFA/acetonitrile.
6. Microcentrifuge.
7. Ultrasonic bath.

2.3 Mass Spectrometry

Prepare all solutions using HPLC grade reagents and water.

1. ESI Qq-TOF mass spectrometry.
2. Peptide solvent: 50:50:1 (v/v/v) MeOH/H_2O/HCOOH.
3. Tuning Mix.
4. SpeedVac™ Vacuum Concentrator.

2.4 De Novo Sequencing and Analysis of Mass Spectrometry Data

1. DataAnalysis software (Bruker Daltonics, version 4.0 SP1).
2. Biotools software (Bruker Daltonics, version 3.2).

2.5 Edman Degradation

Prepare all solutions by exclusive reagents for PPSQ-31A instructed by manufacturer.

1. PPSQ-31A protein sequencer equipped with a Wakopack™ Wakosil-PTH II OSD column (4.6 × 250 mm mm).
2. PPSQ-30 analysis software (version 1.30).
3. Purified peptide solution (the purity should be above 95 %).
4. Polybrene.
5. GFD (Glass fiber disks, TFA treated).
6. PTH-amino acids mixture: The PTH-AA standard mixture is diluted by 37 % acetonitrile solution to 500 pmol/mL. The solution is stored in single-use aliquots (100 μL in microcentrifuge tube) at −20 °C.
7. Solutions equipped on the machine: PTH-amino acids mobile phase, 5 % phenyl isothiocyanate *n*-heptane.

3 Methods

Carry out all procedures at room temperature unless otherwise specified.

3.1 Disulfide Bond Reduction and Thiol Group Alkylation

1. Peptide is quantified by measuring the absorbance of their peptide bonds at 215 nm and 225 nm by using a spectrophotometer according to the relation: Concentration (μg/mL) = 144 × (A_{215} − A_{225}) (*see* **Note 4**). The peptide solution is

accuracy diluted till the measured value near 25 μg/mL, and then, the formally measures are performed at three 1:1 dilutions and to average three results. The Quantity (μg) = Concentration (μg/mL) × Volume (mL).

2. To break disulfide bridge, the peptide is reduced with 0.5 mM dithiothreitol (DTT) solution in 50 mM NH_4HCO_3: Add 10 μmol peptide (*see* **Note 5**) into a microcentrifuge tube and wrap the tube in aluminum foil.
3. Pick up the stored disulfide bond reduction solution (10×) from refrigerator; after thawing, transfer 100 μL of the solution and 900 μL water into the microcentrifuge tube containing the peptide.
4. Heat the tube for 30 min in a water bath at 50 °C.
5. To stabilize cysteine side chains of the break disulfide bridge, the peptide is alkylated with 11 mM iodoacetamide (IAM) solution in 50 mM NH_4HCO_3: Pick up the stored thiol group alkylation solution (100×) from refrigerator; after thawing, transfer 10 μL of the solution into the microcentrifuge tube containing the peptide (*see* Subheading 3.1, **step 2**).
6. Place the tube in the dark overnight.
7. The resulting solution is concentrated to dry in a SpeedVac Vacuum Concentrator.

3.2 Purification of Alkylated Peptide

The HPLC system such as ÄKTA™ purifier PHC 10 chromatography system with 100-5C-18 HPLC reverse phase column (4.6 × 250 mm) is used for purification of the alkylated peptide.

1. Sample preparation: Redissolve the alkylated peptide in a microcentrifuge tube by using 1 mL Solvent A. Centrifuge for 15 min at 14,000 × *g* to ensure clarity. Transfer the supernatant to a new tube.
2. Solvent preparation for HPLC: Solvent A and Solvent B are degassed with an ultrasonic bath.
3. Column preparation: The column is washed at a flow rate of 1.05 mL/min with 100% of Solvent B for 10 min, and then, the concentration of Solvent B is decreased from 100 to 15% over 10 min using a linear gradient (accordingly, the concentration of Solvent A is increased from 0 to 85%, the same as below). The column is equilibrated by using 15% of Solvent B for 15 min.
4. Purification of alkylated peptide: The sample is injected into a 1 mL loop. The operation software is programmed to perform chromatography under the following conditions at a flow rate of 1.05 mL/min: Increase concentration of Solvent B from 15 to 60% over 12 column volume (CV) by linear gradient elution.

5. The absorbance is monitored at 215 nm. The fractions corresponding to predominated absorbing peak are collected into collection tubes by hand (*see* **Notes 6** and 7).
6. Alkylated peptide in the fraction is quantified as **step 1** in Subheading 3.1 and then dried in a SpeedVac Vacuum Concentrator.

3.3 Mass Spectrometry

The ESI Qq-TOF mass spectrometry (micrOTOF-Q II™) is used throughout the experiments in positive-ion mode (*see* **Note 8**). The instrument is calibrated every hour by the Tuning Mix with external standard method for keeping the accuracy. The first-stage MS (in MS mode) provides data both on accurate monoisotopic molecular mass and on purity/homogeneity of the peptide, in which the mass accuracy should be better than ±5 ppm. The second-stage MS (in MS/MS mode) provides data on the spectrum of peptide fragmentation to elucidate the amino acid sequence of AMPs.

1. Peptide is redissolved in peptide solvent. The concentration is about 1 μM (*see* **Note 5**), and then, the sample is directly injected into the system and introduced to ESI source using a syringe pump at a flow rate of 180 μL/h.
2. In MS/MS mode, multiply positively charged peptide precursor ions, chosen based on the data of the first-stage MS, selected manually, and fragmented in the collision cell.
3. Nitrogen gas is used as nebulizer, curtain, and collision gases, and argon gas is used as trap cooling gas. Spectral acquisition rate is 1 Hz.
4. The optimized ESI source conditions are as follows: Capillary V, −4500 V; end-plate voltage, −4000 V; capillary exit voltage, 120 V; and dry gas temperature, 180 °C. Collision energy is raised from 10 to 45 eV till the peptide is completely fragmented.

3.4 Edman Degradation

If necessary, N-terminal sequences are further determined by automated Edman degradation by using a PPSQ-31A protein sequencer with the standard GFD protocols instructed by manufacturer. The normal parameters are as follows: Detected absorbance, 269 nm; Flow rate, 1 mL/min; Column oven temperature, 40 °C.

1. Analyzing the PTH-amino acid mixed standard: Pick up the PTH-amino acid mixture (500 pmol/mL) from refrigerator, thaw, and mix.
2. After the mixture recovered to room temperature, add at least 80 μL of mixture into the sample tube and perform the "PTH-AA" schedule in the PPSQ-30 analysis software. The

results are used to establish a standard amino acid profile and to calibrate the HPLC analysis system.

3. Apply 15 μL of the polybrene to the GFD and dry.
4. Add the purified peptide solution (*see* **Note 9**) to the same GFD and dry again, and then, this GFD is put into the reactor.
5. The "GFD" schedule in the PPSQ-30 analysis software is performed with the amount of amino acids being set up to detect. Each time the chromatogram analysis of the sequence sample is completed, the sequences are estimated automatically (*see* **Note 10**).

3.5 De Novo Sequencing and Analysis of Mass Spectrometry Data

1. The mass data are processed in DataAnalysis software as follows: choosing several continuous acquisition spectra (at least three) at identical energy where the peptide is fragmented appropriately to average (*see* **Note 11**), using Apex algorithm to find mass list, and then, deconvoluting the data. After adjusting the peaks by hand (*see* **Note 12**), the mass data is sent to Biotools software to assist in de novo sequencing (*see* **Note 13**).
2. The tolerance in de novo sequencing is relaxed to within 0.0180 Da and simultaneously within 20 ppm (*see* **Note 14**). The quasi-isobaric pair (*K* vs. *Q*, mass difference 0.0364 Da) is assigned preliminarily at this restrictive condition (±0.0180 Da, less than half of ±0.0364 Da) theoretically, which is further confirmed by the monoisotopic mass (*K* 128.0950 ± 0.0180 Da; *Q* 128.0586 ± 0.0180 Da) directly deduced from contiguous sequence ion ladders; at last, the sequence is analyzed by the substituting of *K*/*Q* one by one.
3. It is impossible to discriminate the isobaric I/L based on the data of the low-energy CID MS/MS [6]. Thus, it has been accepted to assign I/L indirectly based on the known sequences of the cDNA clones in the same species or on the known sequences of analogues in the same or others species [7–10]. Thus, the assignment between the isobaric pair (I vs. L) may be predicted based on the sequences of special analogues in other species of same genus, which is named "counterpart" which has identical sequence with the peptide needed to be identified except the ambiguity of I/L.
4. An example of sequencing procedure was established during the identification of the antimicrobial peptide Brevinin-2JD [5]. The deconvoluted low-energy CID MS/MS spectra of the precursor ions, corresponding to the pre-alkylated peptide component (of *m*/*z* 841.9586 $[M+4H]^{4+}$ at 25 eV) and the post-alkylated peptide component (of *m*/*z* 1160.9596 $[M+3H]^{3+}$ at 40 eV) of the peptide, were shown in Fig. 1a, b, respectively. Some ion series generally observed in the low-

energy CID MS/MS spectrum were *a*-series including *a*-17 (*a*-NH_3), *a*-18 (*a*-H_2O); *b*-series including *b*-17 (*b*-NH_3), *b*-18 (*b*-H_2O); and *y*-series including *y*-17 (*y*-NH_3), *y*-18 (*y*-H_2O). The *b* and *y* ions, denoted as the solid line with a solid arrowhead, were schematized above the spectra; meanwhile some prospective ion series specifically observed in these spectra were internal cleavage ions *y22b*-series (combination of *b* type and *y22* cleavage) and MH^+-28 ion in the pre-alkylated spectrum. The quasi-isobaric pair (*K* vs. *Q*, mass difference 0.0364 Da) can be discriminated within the tolerance (±0.0180 Da, less than half of ±0.0364 Da) theoretically. They were confirmed further by the monoisotopic mass (*K* 128.0950 ± 0.0180 Da; *Q* 128.0586 ± 0.0180 Da) deduced directly from the contiguous sequence ion ladders. K^{15} (−0.0005 Da from *y*-ion ladders) was deduced from the pre-alkylated spectrum; K^7 (−0.0031 Da from *b*-ion ladders, −0.0092 Da from *y*-ion ladders), K^{15} (0.0066 Da from *y*-ion ladders), K^{28} (0.0049 Da from *y*-ion ladders), and K^{31} (−0.0149 Da from *b*-ion ladders) were deduced from the post-alkylated spectrum. At last, they were analyzed by substituting with *Q* one by one. The substitution caused the correlative subsequence fragment ion series assignment failure within the tolerance. In addition, the monoisotopic molecular mass, measured by the first-stage MS, will exceed the mass accuracy obviously if any *K* was substituted by *Q*. Specifically, the mass accuracy of the monoisotopic molecular mass, which should be less than ±5 ppm, will change to 11.8 ppm for the pre-alkylation and 8.5 ppm for the post-alkylation. Therefore, the assignment of *K* was reliable. The mass difference values of *y33*-*y31* (170.1073 Da, error 0.0018 Da) shown in Fig. 1a and *y33*-*y31* (170.0966 Da, error −0.0089 Da) shown in Fig. 1b indicated that the amino acid composition in position 1–2 should be GI/L or AV. The result of the automated Edman degradation revealed that the first five amino acids are GLLDT. So far, the complementary information, collected from the pair of the mass spectra for the *b* and *y* ions of pre- and post-alkylated Brevinin-2JD, was sufficient to elucidate the sequence of the peptide except the ambiguity of partial I/L. The peptide sequence is GLLDTFKNI/LAI/LNAAKSAGVSVI/LNSI/LSCKI/LSKTC-OH. There are two kinds of counterparts with the difference only at the second amino acid (I^2 or L^2). One kind is Brevinin-2-OA17 [11] predicted from mRNA of *O. andersonii*. The other kind includes Brevinin-2GRa [12] identified from *O. graham*, Brevinin-2-OA18 and Brevinin-2-Omar [11], predicted from mRNAs of *O. andersonii* and *O. margaretae*, respectively. The former kind was excluded based on the result of the automated Edman degradation. Thus, the full sequence of Brevinin-2JD was suggested as GLLDTFKNLALNAAKSAGVSVLNSLSCKLSKTC-OH.

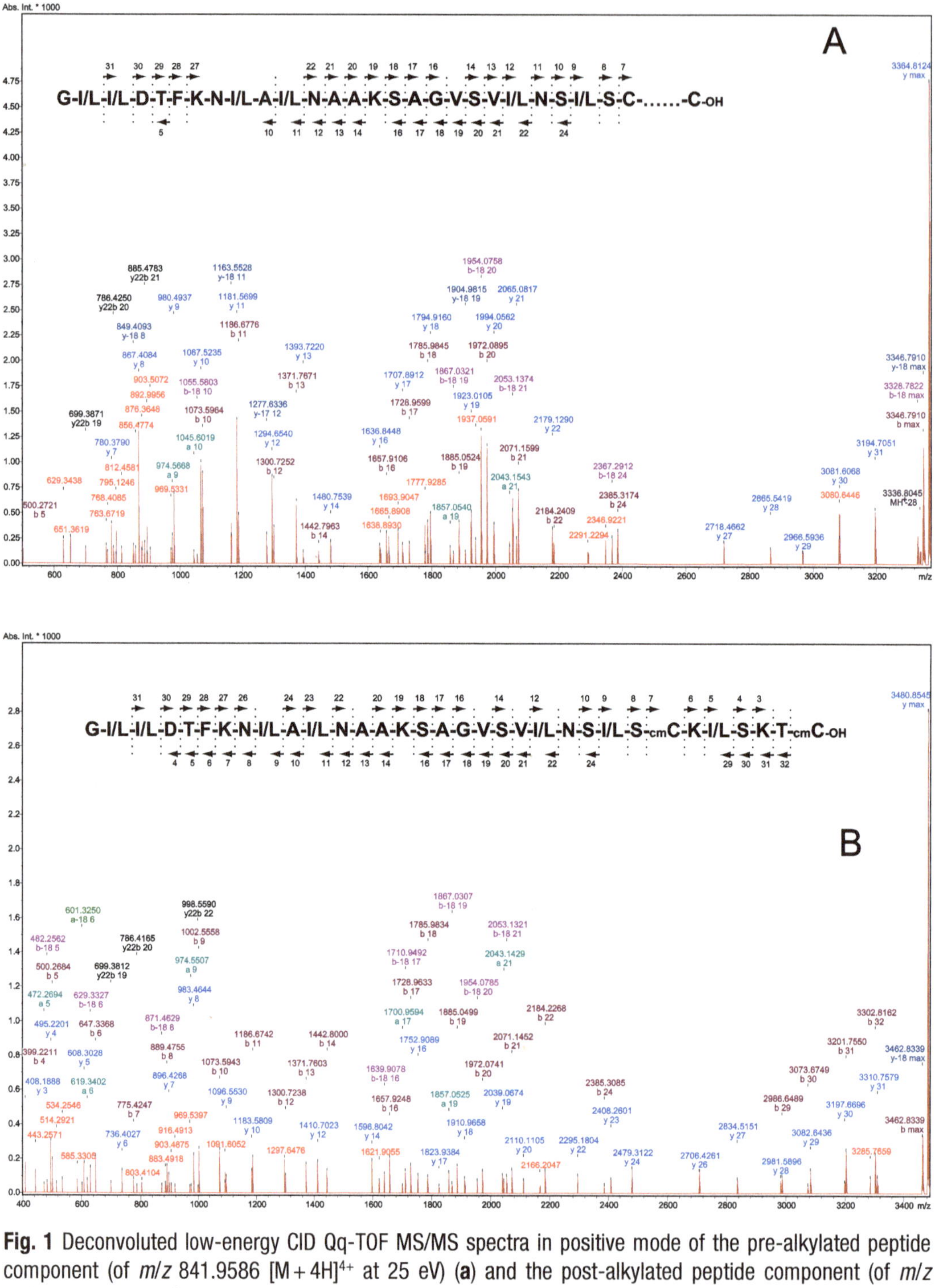

Fig. 1 Deconvoluted low-energy CID Qq-TOF MS/MS spectra in positive mode of the pre-alkylated peptide component (of *m/z* 841.9586 $[M+4H]^{4+}$ at 25 eV) (**a**) and the post-alkylated peptide component (of *m/z* 1160.9596 $[M+3H]^{3+}$ at 40 eV) (**b**) of the AMP Brevinin-2JD. The *b* and *y* ions, denoted as the *solid line* with a *solid arrowhead*, were schematized above the spectra (Reprinted from [5] with permission from Elsevier)

4 Notes

1. The purity of peptide should be higher than 30%; otherwise, the data quality of the MS/MS might decrease.
2. The solution may be stored at −20 °C for up to 6 months.
3. The solution may be stored at −20 °C within 3 days.
4. This formula gives near-linear results within narrow range of concentration (from 5 to 25 μg/mL) for ten measured proteins (coefficient of variation from 0.67 to 12.9%) [13].
5. Accuracy concentration is unnecessary here; thus, the estimated molar concentration can be used. Generally, the molecular mass of peptide is presumed to 2000 Da if molecular mass is unknown.
6. Collection peak by hand is preferable to by automatic fraction collector because automatic collecting often results in insufficiently pure peptide fractions.
7. In some cases, disulfide loop can reduce the difference of the amino acid composition in chromatographic behavior. Thus, if the post-alkylated peptide was separated more peaks than the pre-alkylated peptide with the same chromatograph, it indicates here might contain two or more peptides with similar sequence and equivalence monoisotopic molecular mass.
8. The instruments allow the acquisition of mass spectra with mass range: 20–40,000 m/z, mass resolution: 17,500 full width at half-maximum height (FWHM) for 922 peak, mass accuracy: 1–2 ppm root mean square (RMS) error by internal standard method, 5 ppm RMS error by external standard method, and sensitivity: 2.5 fmol fibrino peptide in the MS/MS mode will give a $S/N > 50$.
9. The amount of peptide needed dramatically increases along with the increasing number of the intendedly detected amino acids. For example, 50 pmol for five amino acids and 3000 pmol for 20 amino acids are necessary.
10. The cysteine is not detectable according to this method. Thus, if there is no PTH-AA peak on a cycle, it is generally considered that there is a cysteine residue in the position. However, the derivative of cysteine can be detected when the peptide is treated by pyridylethylation and run the "PE Analysis" schedule.
11. There is no formulary pattern to determine whether the peptide is fragmented appropriately. An appropriate deconvoluted MS/MS spectrum usually includes one to three predominated peaks with enough component peaks, the number of which is two to three times of the number of amino acids with the peak

intensity above 50. However, due to unusual cleavages, such as internal cleavages in the peptide bond and cleavages within the intramolecular disulfide loop, unusual dissociation, such as dissociation in the NQ residues within the intramolecular disulfide loop, or concomitance of peptides with equivalence theoretical monoisotopic molecular mass, the peak number might greatly increase [5].

12. The quality of the MS/MS spectrum is extremely important. Generally, for the deconvoluted peaks above 1800 Da, the first isotopic peak is lower than the second peak, even lower than the third peak when the molecular mass increases to above 3000 Da. In some case, due to miscellaneous peak interference or too low intensity, the picked isotopic peaks are not the first peak but second or third isotopic peaks by the software; on this condition, the picked isotopic peak should be adjusted to the first isotopic peak manually based on theoretical isotope pattern obtained by Compass Isotope Pattern tool. The interferential peaks should be deleted, and if the first isotopic peak could not be distinguished correctly, this ion peak should be shielded to avoid misallocation.
13. De novo peptide sequencing is the direct reading of the amino acid sequence from the MS/MS spectrum by considering the mass differences between adjacent labeled ions, which is analyzed for the presence of the *b* and *y* ions assistant with *b*-series ions including *b*-17 (*b*-NH_3) and *b*-18 (*b*-H_2O); *y*-series ions including *y*-17 (*y*-NH_3) and *y*-18 (*y*-H_2O); *a*-series ions including *a*, *a*-17 (*a*-NH_3), and *a*-18 (*a*-H_2O); internal cleavage ions *yb*-series (combination of *b* type and *y* cleavage at one site); etc. when employing this type of instrument [5]. It can be done manually with a great deal of experience. There are several published literatures, which should be consulted for further guidance, detailing the de novo sequencing of peptides based on tandem mass spectrometry data [14–16]. De novo sequencing can be time-consuming and sometimes may be impossible. The difficulty comes from low-quality data; on the other hand, partial fragment ions necessary for deriving a complete peptide sequence in single MS/MS spectrum are absent frequently. A function to produce sequence tags named "Find Tags…" under the "MSMS Analysis" tab in BioTools can enhance sequencing speed and flexibility.
14. The tolerance ±0.0180 Da is enough to distinguish *K/Q*; however, this limit is too relaxed to those ions in low mass when measured by ppm. Although the accuracy in MS/MS is not as good as in MS which should be better than ±5 ppm, the mass shift had very low probability to exceed 20 ppm in MS/MS. Thus, the ±20 ppm is another limit for the assignments to be more reliable in practice to ions in low mass [5].

Acknowledgment

We thank Zhijun Wu and Feng Xie. This work was supported by the National Natural Science Foundation of China (30800100), the Chinese Academy of Sciences (CIB-2007-LYQY-01), the Science and Technology Office of Guiyang (2012204), and the Science and Technology Department of Sichuan (2008JY0001).

References

1. Zasloff M (2002) Antimicrobial peptides of multicellular organisms. Nature 415:389–395
2. Sang Y, Blecha F (2008) Antimicrobial peptides and bacteriocins: alternatives to traditional antibiotics. Anim Health Res Rev 9:227–235
3. Wang G, Li X, Wang Z (2016) APD3: the antimicrobial peptide database as a tool for research and education. Nucleic Acids Res 44:1087–1093
4. Waghu FH, Barai RS, Gurung P, Idicula-Thomas S (2016) CAMP$_{R3}$: a database on sequences, structures and signatures of antimicrobial peptides. Nucleic Acids Res 44:1094–1097
5. Liu J, Jiang J, Wu Z, Xie F (2012) Antimicrobial peptides from the skin of the Asian frog, *Odorrana jingdongensis*: *de novo* sequencing and analysis of tandem mass spectrometry data. J Proteomics 75:5807–5821
6. Wang L, Evaristo G, Zhou M, Pinkse M, Wang M, Xu Y, Jiang XF, Chen TB, Rao PF, Verhaert P, Shaw C (2010) Nigrocin-2 peptides from Chinese Odorrana frogs - integration of UPLC/MS/MS with molecular cloning in amphibian skin peptidome analysis. FEBS J 277:1519–1531
7. Ueberheide BM, Fenyo D, Alewood PF, Chait BT (2009) Rapid sensitive analysis of cysteine rich peptide venom components. Proc Natl Acad Sci U S A 106:6910–6915
8. Thompson AH, Bjourson AJ, Orr DF, Shaw C, McClean S (2007) Amphibian skin secretomics: application of parallel quadrupole time-of-flight mass spectrometry and peptide precursor cDNA cloning to rapidly characterize the skin secretory peptidome of *Phyllomedusa hypochondrialis azurea*: discovery of a novel peptide family, the hyposins. J Proteome Res 6:3604–3613
9. Samgina TY, Artemenko KA, Gorshkov VA, Ogourtsov SV, Zubarev RA, Lebedev AT (2008) De novo sequencing of peptides secreted by the skin glands of the Caucasian Green Frog *Rana ridibunda*. Rapid Commun Mass Spectrosc 22:3517–3525
10. Samgina TY, Gorshkov VA, Vorontsov YA, Artemenko KA, Zubarev RA, Lebedev AT (2011) Mass spectrometric study of bradykinin-related peptides (BRPs) from the skin secretion of Russian ranid frogs. Rapid Commun Mass Spectrosc 25:933–940
11. Yang X, Lee WH, Zhang Y (2012) Extremely abundant antimicrobial peptides existed in the skins of nine kinds of Chinese odorous frogs. J Proteome Res 11:306–319
12. Conlon JM, Al-Ghaferi N, Abraham B, Hu JS, Cosette P, Leprince J, Jouenne T, Vaudry H (2006) Antimicrobial peptides from diverse families isolated from the skin of the Asian frog, *Rana grahami*. Peptides 27:2111–2117
13. Wolf P (1983) A critical reappraisal of Waddell's technique for ultraviolet spectrophotometric protein estimation. Anal Biochem 129:145–155
14. Kinter M, Sherman NE (2000) Protein sequencing and identification using tandem mass spectrometry. John Wiley & Sons Inc., New York, NY
15. Chapman J (2000) Mass spectrometry of proteins and peptides, vol 146. Humana Press, Totowa, NJ
16. Lipton M, Paša-Tolic L (2009) Mass spectrometry of proteins and peptides, vol 492. Humana Press, Totowa, NJ

Part II

Studying Interactions of AMPs with Model Membranes and Bacteria

Chapter 8

Molecular Dynamics Simulation and Analysis of the Antimicrobial Peptide–Lipid Bilayer Interactions

Shima Arasteh and Mojtaba Bagheri

Abstract

A great deal of research has been undertaken in order to discover antimicrobial peptides (AMPs) with unexploited mechanisms of action to counteract the health-threatening issues associated with bacterial resistance. The intrinsic effectiveness of AMPs is strongly influenced by their initial interactions with the bacterial cell membrane. Understanding these interactions in the atomistic details is important for the design of the less prone bacteria-resistant peptides. However, these studies always require labor-intensive and difficult steps. With this regard, modeling studies of the AMPs binding to simple lipid membrane systems, e.g., lipid bilayers, is of great advantage. In this chapter, we present an applicable step-by-step protocol to run the molecular dynamics (MD) simulation of the interaction between cyclo-RRWFWR (c-WFW) (a small cyclic AMP) and 1-palmitoyl-2-oleoyl-*sn*-glycero-3-phosphocholine (POPC) lipid bilayer using the Groningen machine for chemical simulations (GROMACS) package. The protocol as described here may simply be optimized for other peptide–lipid systems of interest.

Key words Cyclic antimicrobial peptides, Peptide structure, Lipid bilayers, Molecular dynamics simulation, Force fields

1 Introduction

Besides being involved in the regulation of many cellular processes such as apoptosis, cell signaling, and membrane trafficking [1, 2], protein (peptide)–lipid interactions are indeed considered as the central dogma in the dynamic coevolution relationships between the host and pathogens, in terms of the emergence of microbial resistance, as well as the models and strategies by which the antimicrobial host protein (peptide) targets the cells [3].

As compared with the electrically neutral phospholipids of the membrane of normal eukaryotic cells, the cationic AMPs tend to interact avidly with the overall negatively charged pathogens' cell components [4]. This follows by the modulation of the membrane physical dynamics (e.g., fluidity [5], thickness [6], curvature [7] and disruption of the lipid raft [8]) and thus, results in various

Paul R. Hansen (ed.), *Antimicrobial Peptides: Methods and Protocols*, Methods in Molecular Biology, vol. 1548, DOI 10.1007/978-1-4939-6737-7_8, © Springer Science+Business Media LLC 2017

mechanistic pathways to the AMP-induced cell death through, for instance, membrane lysis (and/or micellization) [9], lipid membrane demixing-induced formation of domains in the cell membrane [10] and membrane translocation [11].

Understanding of these initial events in the AMPs binding to the pathogens' cell membranes at the molecular levels with respect to the peptide secondary structure, as well as the nature of each amino acid residues will be beneficial in the design and generation of multifaceted peptide antibiotics with the reduced risk for the emergence of microbial resistance [12]. This, however, requires labor-intensive and time-consuming experimental approaches [12, 13], and becomes particularly more complicated for the small AMPs with ≥50 % of the hydrophobic amino acid residues where the structure cannot properly be resolved using techniques, such as nuclear magnetic resonance (NMR) [14], circular dichroism (CD) [15, 16] and X-ray crystallography [17] as a result of the peptides' low solubility, self-aggregation and difficulties in the peptide crystallization. Moreover, if the peptides have cyclic structure and contain aromatic residues, measurement interferences may arise and make light-based spectroscopic techniques used for the analysis of the peptide–lipid binding, i.e., fluorescence and CD spectroscopy, difficult to be interpreted [15, 16]. A particular example is cyclo-RRWFWR (c-WFW) which has shown excellent activities against bacteria with almost no hemolysis [18]. The cyclic conformation makes the peptide relatively stable against proteases and thus, suitable as a promising lead compound for the generation of peptide antibiotics against bacterial superbugs health-complicating issues.

The antibacterial activity of c-WFW depends highly on the hydrophobic nature of the aromatic residues in the sequence [18]. Whereas recent studies suggested that the peptide has high affinity for cardiolipin-rich lipid matrices and disrupts the bacterial metabolism and homeostasis by the cell lipid membrane demixing [19, 20], the molecular details of these interactions are still a matter of debates and remained to be elucidated. Unlike the easy synthesis of the peptide labeling with fluorescent probes, used for visualization of the peptide influences on the bacteria, the mechanistic analysis of c-WFW may be hampered due to the possible insolubility and the loss of activity associated with the peptide-tag conjugate [19].

To overcome the above-mentioned limitations, computational MD simulation is being used as a growing powerful tool to study in silico structural characterization of the AMPs and understand the peptides' antimicrobial modes of action and activity profiles at the molecular levels [21, 22]. This is done by investigating the interactions between the peptides and the simple lipid systems (e.g., lipid bilayers, micelles and liposomes) mimicking the physicochemical properties of the microorganisms' cell membrane in the absence of any conflicting experimental interpretation. In this chapter, therefore, a procedure for the MD simulation of the interaction between c-WFW [22] and the simple POPC lipid bilayer system is presented using the GROMACS package [23]. The protocol is simply described from the construction of the peptide–lipid system to its analysis and can be applied to a variety of AMPs and model lipid systems of different structures and phospholipid compositions.

2 Materials

1. Peptide model with atomic coordinate as a ".pdb" file.
2. Hyperchem v8.0.10 software http://www.hyper.com/.
3. Coordinate files for the lipid bilayers downloadable from Tieleman's group (http://people.ucalgary.ca/~tieleman/download.html) and Karttunen's group (http://www.softsimu.net/downloads.shtml), or constructed using the visual molecular dynamics (VMD) *membrane* plugin.
4. Chemistry at Harvard Macromolecular Mechanics (CHARMM) topology and parameter sets from MacKerell's lab (http://mackerell.umaryland.edu/) (*see* **Note 1**).
5. GROMACS simulation package v5.0 downloadable from http://www.gromacs.org/ (*see* **Note 2**).
6. UNIX operating systems.
7. VMD v1.9.2 as the visualization and analysis software, downloadable from http://www.ks.uiuc.edu/Research/vmd/ [24].
8. Graph plotting tools, e.g., gnuplot v5.0 (http://gnuplot.info/) and SigmaPlot v10.0 (http://www.sigmaplot.com/).
9. GridMAT-MD v2.0 downloadable from http://www.bevanlab.biochem.vt.edu/GridMAT-MD/ [25].

3 Methods

3.1 Setting Up MD Simulation of c-WFW in Water

3.1.1 Construction of the Cyclic Peptide Structure

1. Build the linear peptide acid model of RRWFWR (l-WFW) using HyperChem v8.0.10 software and save it as a ".pdb" file (*see* **Note 3**).
2. Run the following command on the ".pdb" saved structure (*see* **Note 4**):
```
$gmx pdb2gmx
```
3. Choose the desired number for the CHARMM force field by running this command:
```
$gmx pdb2gmx -f input_coordinates.pdb -o output.gro -water none
```
4. Generate a structural file (*topology*.top) with all parameters for the cyclic peptide structure by the modification of the initial linear conformation (*see* **Note 5**).

3.1.2 Solvation of the Cyclic Peptide and Neutralization

1. Place the cyclic peptide in a cubic water box filled with ~2500 TIP3P water molecules (*see* **Note 6**). Construct the box by running
```
$gmx editconf -f output.gro -o output_new-
box.gro -c -d 1.1 -bt cubic
```
2. Dissolve the peptide by running
$gmx solvate -cp output_newbox.gro -cs spc216.gro -o output_solv.gro -p topology.top
(*see* **Note 7**).

3.1.3 Neutralization and Minimization

1. To neutralize the system, counter ions are replaced with some water molecules. "grompp" is generally required for the neutralization and minimization (as well as the equilibration and simulation) of a system.
2. To run "grompp" a proper ".mdp" file (a text file) containing specified parameters should be generated. A sample ".mdp" file is available on the GROMACS online reference (http://manual.gromacs.org/online/mdp.html). However, one should make sure that the contents of the ".mdp" file are compatible with the selected force field used in the MD simulation.
3. Run "gmx grompp" to get a ".tpr" file.
```
$gmx grompp -f ions.mdp -c output_solv.gro
-p topology.top -o ions.tpr
```
4. Run "gmx genion" to neutralize the system. The system is rendered neutral by the addition of a sufficient number of sodium or chloride ions with respect to the total charge of the system.
```
$gmx genion -s ions.tpr -o output_solv_ions.
gro -p topology.top -nname CL -nn N -pname
NA -np M
```
where N and M are number of negative and positive ions, respectively.
5. Minimize the system by running the following commands (*see* **Note 8**):
```
$gmx grompp -f minim.mdp -c output_solv_
ions.gro -p topology.top -o minim.tpr
$mdrun -deffnm minim
```

3.1.4 Equilibration and MD Simulation

The minimized system is equilibrated under *NVT* and *NPT* conditions ($T = 310$ K, $p = 1$ atm) (*see* **Note 9**) until a root–mean–square deviation (RMSD) of atomic position values <0.4 nm is reached (*see* **Note 10**).

1. Run the following commands for the system equilibration under *NVT*:
```
$gmx grompp -f nvt.mdp -c minim.gro -p to-
pology.top -o nvt.tpr
```

```
$gmx mdrun -deffnm nvt
```

2. Run the following commands for the system equilibration under *NPT*:

```
$gmx grompp -f npt.mdp -c nvt.gro -p topol-
ogy.top -o npt.tpr
$gmx mdrun -deffnm npt
```

3. Run the MD simulation to generate the trajectory of the particles in the system (i.e., peptide, water and ion molecules) using the following commands (*see* **Note 11**):

```
$gmx grompp -f md.mdp -c npt.gro -t npt.cpt
-p topology.top -o md.tpr
$gmx mdrun md
```

3.1.5 Analysis

1. Run “gmx cluster” to generate clusters. The cluster with the high percentage of occurrence for the peptide is extracted by analyzing the trajectory and used to build up the models for the peptide structural analysis (*see* **Note 12**).

```
$gmx cluster -f md.trr -s md.tpr -cutoff
0.15 -skip 100 -cl cluster.pdb
```

2. Extract the Ramachandran plot by running “gmx rama” for the analysis of the peptide secondary structure (*see* **Note 13**).

```
$gmx rama -s md.tpr -f md.xtc -o rama.xvg
```

3. To extract Φ and Ψ angles over the C^{α} atoms for each amino acid residue along the trajectory, simple TCL commands in the VMD TK console can also work out:

```
set output [open "output" w]
setnf [molinfo top get numframes]
set selection [atomselect top "protein and
resid 1 and name CA"]
for {set i 0 } {$i<$nf } { incri } {
$selection frame $i
puts $output [$selection get {phi psi}]
}
close $output
```

The related torsion angles will be given in the output if the number of “resid” changes.

4. The peptide structure is visualized and analyzed using tools in the VMD and the SigmaPlot.

3.2 Setting Up MD Simulation of Cyclo-RRWFWR in POPC

The equilibrated lipid bilayer and the peptide structure (with the highest percentage of occurrence) in contact with water molecules are required for the MD simulation of the peptide–lipid system.

3.2.1 Preparation of POPC Lipid Bilayers

1. The coordinate files for the initial bilayer configuration is downloaded from Tieleman’s and Karttunen’s group, or constructed using the VMD *membrane* plugin.

2. Run the following command to generate the appropriate *topology*.top file (*see* **Note 14**):
```
$gmx pdb2gmx
```
3. Define a box with appropriate dimensions for the lipid system (*see* **Note 15**).
4. Lipid solvation (*see* **Note 16**).
5. Lipid neutralization (*see* **Note 17**).
6. Run *NVT* at 310 K.
7. Run *NPT* in 1 atm.
8. Run the MD simulation.

The commands for the **steps 4–8** are the same as previously described for the peptide in water.

3.2.2 Construction of the Peptide–Lipid System

1. Load POPC and c-WFW coordinate files (saved as ".pdb" or ".gro") using the VMD. The geometrical center point of the peptide is located ~5 Å far from the center point of the top lipid bilayer leaflet, such that the hydrophobic and hydrophilic amino acid residues are oriented toward the bilayer "x–y" plane and the bulk water, respectively (*see* **Note 18**).
2. Save separately the new coordinates for the lipid and the peptide as ".pdb" files.
3. Concatenate the peptide and the lipid bilayer ".pdb" files. This is done by cutting and pasting a ".pdb" file into another in a text editor.
4. Update the peptide's "*topology*.top" file by including the lipid parameters after the protein position restraint part as follows:
```
; Include POPC chain topology
#include "lipid.itp"
```
5. Solvate and neutralize the system (*see* **Note 16**).
6. Run the minimization.
7. Equilibrate the system. Restrain heavy and the phosphorus atoms in the peptide and POPC polar headgroups, respectively, as follow:

   (a) Make an index file containing atom numbers which are supposed to be restrained.
   ```
   $gmx make_ndx -f minimized_structure.gro
   -o index.ndx
   ```
   Then restrain atoms in "index.ndx" file.
   ```
   $gmx genrestr -f minimized_structure.
   gro -o posre_lipid.itp -fc 100000 100000
   100000 -n index.ndx
   ```

(b) Define the position restraint by adding the line "define = -DPOSRES_LIPID" to ".mdp" file.

(c) Modify the topology by adding a new "#ifdef" after lipid inclusion lines.

```
; Include lipid chain topology
#include "lipid.itp"
#ifdef POSRES_LIPID
#include "posre_lipid.itp"
#endif
```

8. Run *NVT* and *NPT* at 310 K and 1 atm (*see* **Note 19**).
9. Remove restraints by deleting "define = -DPOSRES_LIPID" in ".mdp" file.
10. Run the simulation.

3.2.3 Analysis

Selected the trajectory analyses are presented.

1. Run GridMAT by calling GridMAT perl script (*see* **Note 20**).
```
$perl GridMAT-MD.pl param
```
2. Use GridMAT-MD.pl to estimate the average area per lipids (APL) over time for the bilayer top and bottom leaflets.
3. Use GridMAT-MD.pl to estimate bilayer thickness.
4. Use "gmx order" to provide order parameters for the acyl chain carbons (*see* **Note 21**).
5. Make two index groups containing carbon atom names for each acyl chains, i.e., index1 and index2.
6. Next, run gmx order using the generated index files:
```
$gmx order -s md.tpr -f md.xtc -n index1.ndx
-d z -od deuter_index1.xvg
$gmx order -s md.tpr -f md.xtc -n index2.ndx
-d z -od deuter_index2.xvg
```
7. Run "gmx cluster".
8. Run "gmx rama" or TCL commands in the VMD TK console.
9. Run "gmx rms".
10. Run "gmx density" to extract the distribution of every components of the system through the box (*see* **Note 22**).
```
$gmx density -s md.tpr -f md.xtc -o density.
xvg -n index.ndx
```
11. Run "gmx mindist" to calculate the number of contacts between two selected groups of molecules in the peptide–lipid system, e.g., headgroups and peptide, peptide and waters, lipids and waters and etc (*see* **Note 23**).

```
$gmx mindist -s md.tpr -f md.xtc -on num-
cont.xvg -tu ns -n index.ndx
```

12. Run "gmx distance" to calculate the distance between positions of two selected groups.

```
$gmx distance -s md.tpr -f md.xtc -oall
dist.xvg -tu ns
```

13. If required, other analyses using available tools in the GROMACS, the VMD plugins, or one's own scripts are possible (*see* **Note 24**).

4 Notes

1. Depending upon the system of research interest, as well as parameters required for the simulation, all-atom, united-atom and coarse-grained force fields (e.g., GROMOS [26], CHARMM [27], OPLS [28], and AMBER [29]) may be utilized. CHARMM is an example for an all-atom force field that is commonly used for the MD simulation of lipid systems. This all-atom force field presents almost all the potential parameters for different type of atoms in the system.

   The SwissSidechain database (http://www.swisssidechain.ch/index.php), which provides the CHARMM compatible topology and parameters for the unnatural amino acid residues may be considered for use in the GROMACS where appropriate [30].
2. We recommend the GROMACS among other software packages for the MD simulation of peptide–lipid systems because it supports CHARMM36. Many parameters, such as surface tension and order parameters for lipid bilayers are in great agreement with the experimental values in the CHARMM36 [31].
3. The peptide initial NMR or crystal structures with the atomic coordinates are downloadable from the protein data bank (http://www.rcsb.org./pdb) as a ".pdb" file. When no experimental structural information is available, we recommend HyperChem databases to build a peptide structural conformation with respect to the sequence and chirality of each amino acid residue. An extended conformation is selected for the generation of the peptide structure and the output is saved as a ".pdb" file.
4. The commands are typed at the terminal and presented by "$*command*." By default, this command generates three different outputs from the initial ".pdb" file, which are in ".gro," ".top," and ".itp" formats. Whereas ".gro" file contains the atom coordinates, the remaining two files include all structural parameters.

Typing "-h" after each command gives a brief manual about that command, e.g., "gmx pdb2gmx -h".

5. The ".top" file of l-WFW is modified for c-WFW in a text editor. For this, an amide bond and other related parameters (i.e., angles, pairs and dihedrals) between the first residues at the C- and N-terminals of the linear sequence (Arg^1 and Arg^6) are generated by taking into the consideration the proper atom numbers matched to the ".pdb" file. Atom numbers are easily defined by labeling in the VMD. If required, the generated ". top" file is manually corrected for the missed bonds, dihedrals, pairs and angles.
6. A macroscopic system can be simulated using a limited number of particles (e.g., peptide, lipid and water molecules (here, TIP3P as an appropriate water model molecule with reduced cost of simulation), provided the periodic boundary conditions are considered. A periodic large system is produced by repeating a unit cell in all directions of space so that if a virtual particle goes out of the unit cell, a modulator particle comes in from the other side of the cell. Therefore, the number of particles in the unit cell is remained unchanged.

   Cubic, hexagonal, and dodecahedron are some examples for unit cells used in the MD simulations.
7. The calculation of trajectories for the solvent molecules is one of the most time-consuming parts in the MD simulations. The peptide solvation in water is considered as a simple system and its equilibration with significant error reductions is a prerequisite toward the simulation of a peptide–lipid system.

   Implicit and explicit solvents are two approximations commonly used in the solvation step in the MD simulation. The molecules of solvent are less fractional and behave as a continuous medium with fewer degrees of freedom for the simulation in the implicit solvent and thus, it is less difficult to get a detailed view on the atom motions. In contrast, the number of solvent molecules in the explicit approximation is discrete. This apparently increases the cost of simulation; however, more detailed results are obtained. Taken together, we recommend the latter for the MD simulation of a peptide–lipid system.

   A defined concentration of salt, e.g., NaCl, may be added to the box to reach a better equilibrated system.
8. In the minimization of the solvated peptide, all improper atom–atom overlaps and steric clashes are removed and relaxed. In this step, the cyclic structure of c-WFW is energetically minimized with respect to the topology file.
9. According to the statistical mechanics, *NVT* is a canonical ensemble where "*V*", volume; "*T*", temperature of a system;

and "*N*", number of particles are constant but the energy of the system is exchanged with the environment. Also, *NPT* is an isothermal–isobaric ensemble where "*T*", and "*P*", pressure are constant in a system with "*N*".

The particle motion in a system is dependent upon its mass and the temperature and described by the Boltzmann–Maxwell (BM) distribution. To model a real biological system, the MD simulations are done in *NVT* conditions. For this, the temperature is coupled to different thermal baths, such as Berendsen [32], Nosé–Hoover [33] and V-rescale [34]. In this protocol, V-rescale thermostat is used to keep the temperature at 310 K with a coupling time of 0.1 ps.

Also, the pressure is coupled to either Berendsen [32] or Parrinello–Rahman [35] barostats. We use the former to keep the pressure at 1 atm in the MD simulation. Vectors of the unit cell change along the simulation. Berendsen algorithm controls the pressure by adjusting the box vectors and coordinates in every step.

10. An RMSD value represents the structural changes of each individual particle that exists in a system (e.g., peptide, lipid bilayers or the whole peptide–lipid mixture), compared to the reference structure over time. RMSD values less than 0.4 nm indicate the existence of stable structures. To extract the RMSD values, run "gmx rms" in the GROMACS as follows:

```
$gmx rms -s md.tpr -f md.xtc -o rmsd.xvg
-tu ns
```

11. The optimum time required for the MD simulation of a system depends upon the RMSD values of the existing particles in there. An MD simulation can be stopped if RMSD < 0.4 nm. The time step of the simulations is set to 2 fs in the ".mdp" file.

12. Cluster analysis is done over the whole MD trajectory with cutoff 0.15 nm on the C^{α} atoms, which is a good value for the small peptides and those with flexible structures. At least one hundred configurations are selected from the simulation and subjected to a cluster analysis. For this, the time intervals for clustering should be specified depending upon the time steps and the frame coordinates recorded in an MD simulation (Fig. 1).

13. As a result of mismatching atom names between those in the "gmx rama" script with the real atom names in the CHARMM, some information might be missed if the Ramachandran plot is obtained using the tool in the GROMACS (Fig. 1). "gmx rama" is obviously designed for the GROMOS force field in the GROMACS.

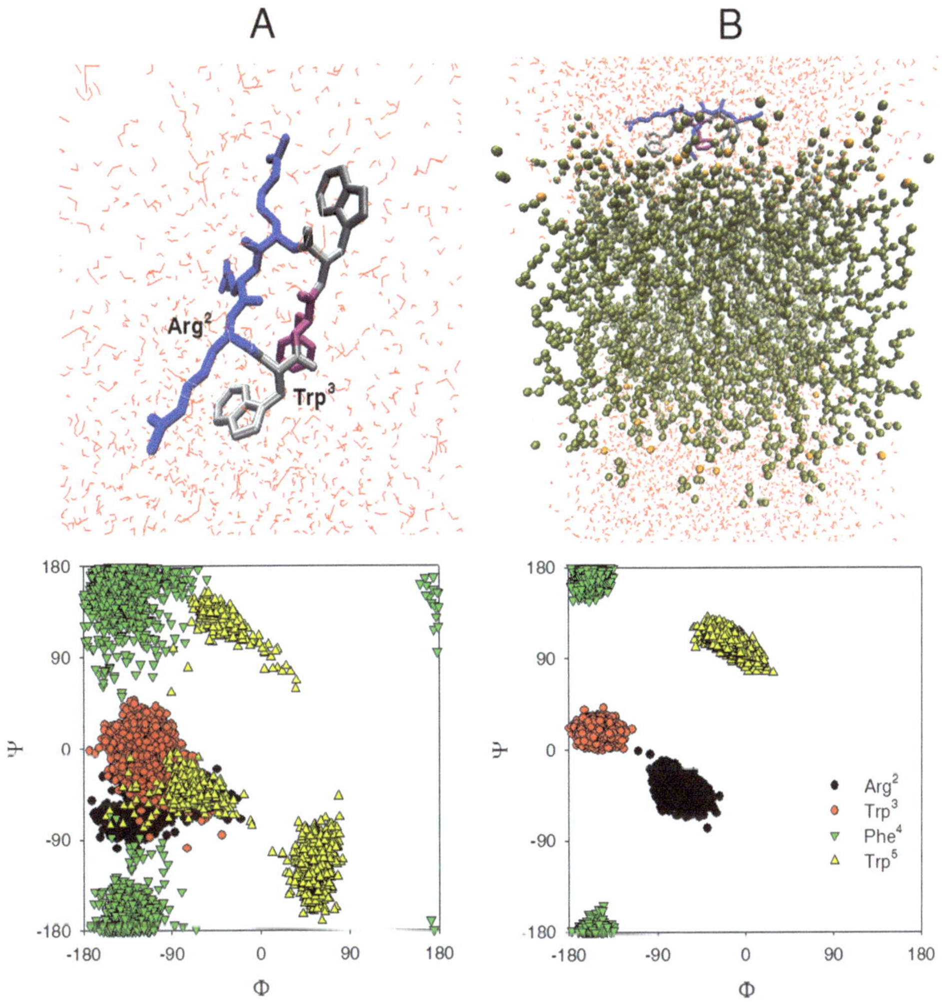

Fig. 1 The representative structures and the Ramachandran plots for cyclo-RRWFWR generated from the cluster analysis of the peptide MD simulation in water (box dimension, 80.51 nm^3) (**a**) and bound to POPC (box dimension, 250.00 nm^3) (**b**). The peptide is located ~6 Å of the POPC bilayer plane. The length of simulation is 105 ns and 165 ns for the peptide in water and bound to POPC, respectively. The percentage of occurrence of the peptide structures is ≥%50. The peptide in water and bound to POPC adapts a secondary structure close to a type I β-turn (*i* + 1, Arg^2; *i* + 2, Trp^3) with the latter being more restrained. The colors for the peptide and lipid structures define as follow: Trp (*silver*), Arg (*blue*), Phe (*pink*), lipid acyl chain (*tan*), and phosphate headgroups (*orange*). Water molecules are shown in *red*

14. A POPC lipid bilayer consisting of 72 lipid molecules has selected in this protocol. When the "*lipid*.itp" file is available, the lipid topology file may manually be generated as follow:

    Assuming that the CHARMM36.ff is located in current directory, add this line representing the force field parameters to the beginning of the "*lipid*.itp" file.

```
#include "./charmm36.ff/forcefield.itp"
```

Add also these lines to the end of the "*lipid*.itp" file.

```
; Include Position restraint file
#include "lipid.itp"
#ifdef POSRES
#include "lipid.itp"
#endif
; Include water topology
#include "./charmm36.ff/tip3p.itp"
#ifdef POSRES_WATER
; Position restraint for each water oxygen
[ position_restraints ]
; i funct fcx  fcy  fcz
 1 1 1000 1000 1000
#endif
; Include topology for ions
#include "./charmm36.ff/ions.itp"
[ system ]
; Name
lipid in water
[ molecules ]
; Compound  #mols
lipid  72
```

15. Type $*measure minmax [atomselect top "all"]* (a TCL command in the VMD) to estimate the system dimensions and define the box in the GROMACS. For example, a cubic box for POPC with 72 lipid molecules is defined by running

```
$gmx editconf -f lipid.pdb -o lipid_newbox.
gro -box 5.0 5.0 10.0
```

16. After the lipid (or later for the peptide–lipid) solvation, open the output in the VMD to make sure that no water molecule exists in the acyl chain region. Otherwise, label the molecule serial number and remove it from the ".pdb" file in a text editor.

17. If working with the uncharged lipid bilayers, e.g., POPC, this step can be skipped.

18. AMPs are directed toward the lipid bilayer using the nonspecific electrostatic interactions between the positively charge peptide and the negative components of the phospholipids. This is followed by the peptide hydrophobic insertion into the bilayer. In order to perform an all-atom MD simulation for a peptide–lipid system, it is recommended to put the peptide close to the lipid bilayer "x–y" plane (≤10 Å), such that the hydrophobic amino acid residues are in close contact with the phospholipid acyl chain. This will reduce the cost of the simulation and allows the system to reach an equilibration state within a time-saving process.

19. Semi-isotropic pressure coupling, where the isotropic scaling in the *x–y* direction is independent of the *z* direction, is recommended for the simulation of systems composed of bilayers.
20. GridMAT-MD.pl is a script written in "perl" and uses "param.txt" file which contains required parameters to calculate APL or bilayer thickness. The example of parameters in "param.txt" that should be cleared are the presence of peptide/protein, types of lipid molecules, phosphorus atoms name in head groups, solvent molecules, types of ions, box size and output resolution.
21. The APL, local thickness of lipid bilayer leaflets and lipid order parameters are some aspects of a lipid system to measure the impact of peptides upon the balanced stabilizing forces in the membrane (Fig. 2).
22. This analysis is useful if the peptide penetration in the lipid bilayer wants to be studied (Fig. 3). Generate any group consisting of the favourite atoms by running "gmx make_ndx" as appropriate. For instance, if a group of phosphorus atoms in polar headgroups is required, an "index.ndx" file is generated by choosing matched phosphorus atoms name.

```
$gmx make_ndx -f md.tpr -o index.ndx
...
>a P
>q
```

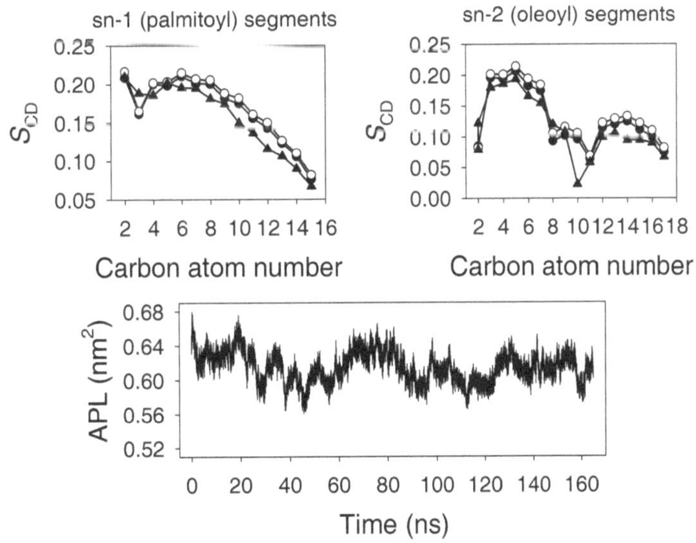

Fig. 2 The order parameter S_{CD} values for the sn-1 and sn-2 acyl chains of POPC alone (*filled triangle*) and in interaction with the peptides over $t = 0–35$ nm (*filled circle*) and $t = 140–160$ nm (*open circle*), and the average APL fluctuations for the top and bottom bilayer leaflets of the peptide-bound POPC over the course of simulation. The APL experimentally measured value for the pure POPC is 0.68 ± 0.01 nm^2 [36]

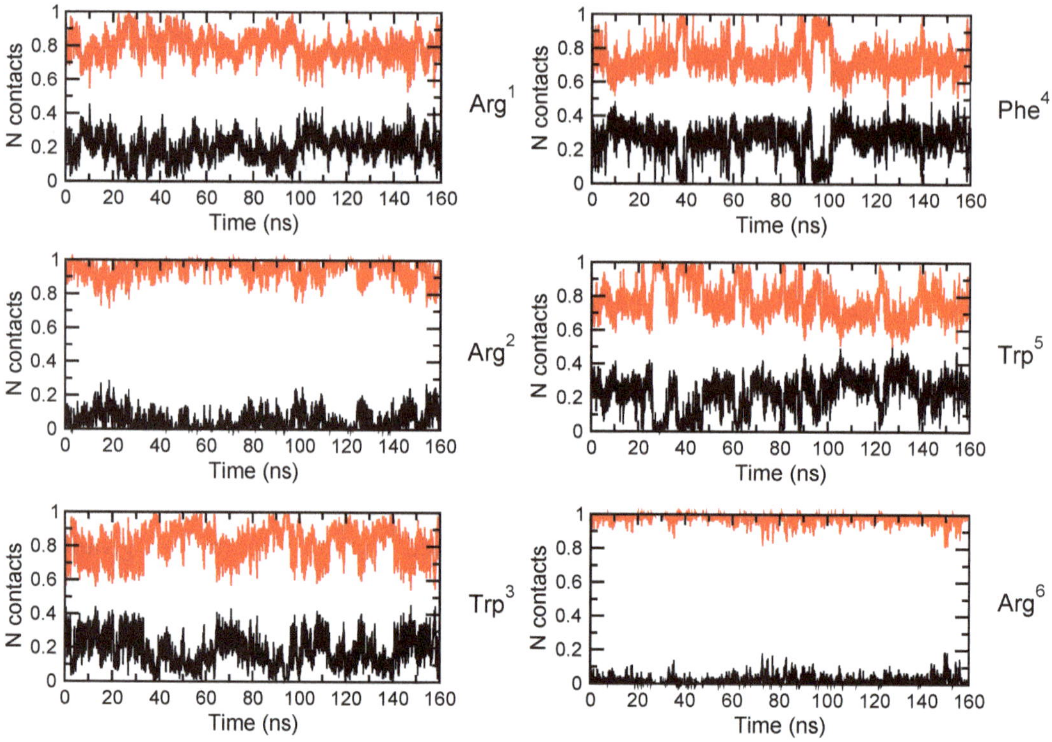

Fig. 3 Normalized contact number values for each amino acid residue in cyclo-RRWFWR. *Black* and *red lines* point to the values for the peptide contact with POPC and water, respectively. As seen, the normalized average contact number of the peptide atoms with POPC is higher for the hydrophobic amino acid residues, i.e., Trp and Phe (except Arg[1]). The peptide is ~6 Å far from the POPC bilayer plane

where "*P*" is the atom name for the phosphorus atoms in lipid molecules, e.g., POPC.

By adding "-n index.ndx" to the command, the defined groups in "index.ndx" appear. For other lipid molecules (e.g., phosphatidylglycerol, phosphatidylethanolamine and etc), the phosphorus atoms name is figured out by labeling them in the VMD.

23. After running the "gmx mindist" and "gmx distance" one will prompt to select two groups by entering the desired group's numbers.

24. These include cluster analysis, extraction of Ramachandran plot and other analyses such as, dipole moment, moment of inertia, electrostatic and van der Waals interaction energies between the peptide, lipid, and other groups in the peptide–lipid system, number of hydrogen bonding between peptide residues and polar headgroups, etc.

Acknowledgments

The financial supports to MB from the Iran National Science Foundation (reference No. 94012757), the Iran's National Elites Foundation (reference No. 15/45362-1392) and University of Tehran are gratefully acknowledged.

References

1. Cho W, Stahelin RV (2005) Membrane-protein interactions in cell signaling and membrane trafficking. Annu Rev Biophys Biomol Struct 34:119–151
2. Volinsky R, Kinnunen PK (2013) Oxidized phosphatidylcholines in membrane-level cellular signaling: from biophysics to physiology and molecular pathology. FEBS J 280:2806–2816
3. Peschel A, Sahl HG (2006) The co-evolution of host cationic antimicrobial peptides and microbial resistance. Nat Rev Microbiol 4:529–536
4. Zasloff M (2002) Antimicrobial peptides of multicellular organisms. Nature 415:389–395
5. Willumeit R, Kumpugdee M, Funari SS, Lohner K, Navas BP, Brandenburg K, Linser S, Andrä J (2005) Structural rearrangement of model membranes by the peptide antibiotic NK-2. Biochim Biophys Acta 1669:125–134
6. Pabst G, Grage SL, Danner-Pongratz S, Jing W, Ulrich AS, Watts A, Lohner K, Hickel A (2008) Membrane thickening by the antimicrobial peptide PGLa. Biophys J 95:5779–5788
7. Schmidt NW, Wong GC (2013) Antimicrobial peptides and induced membrane curvature: geometry, coordination chemistry, and molecular engineering. Curr Opin Solid State Mater Sci 17:151–163
8. McHenry AJ, Sciacca MF, Brender JR, Ramamoorthy A (2012) Does cholesterol suppress the antimicrobial peptide induced disruption of lipid raft containing membranes? Biochim Biophys Acta 1818:3019–3024
9. Bagheri M (2015) Cationic antimicrobial peptides (AMPs): thermodynamic characterization of peptide-lipid interactions and biological efficacy of surface-tethered peptides. ChemistryOpen 4:389–393
10. Arouri A, Dathe M, Blume A (2009) Peptide induced demixing in PG/PE lipid mixtures: a mechanism for the specificity of antimicrobial peptides towards bacterial membranes? Biochim Biophys Acta 1788:650–659
11. Kobayashi S, Chikushi A, Tougu S, Imura Y, Nishida M, Yano Y, Matsuzaki K (2004) Membrane translocation mechanism of the antimicrobial peptide buforin 2. Biochemistry 43:15610–15616
12. Nascimento JM, Oliveira MD, Franco OL, Migliolo L, de Melo CP, Andrade CA (2014) Elucidation of mechanisms of interaction of a multifunctional peptide *Pa*-MAP with lipid membranes. Biochim Biophys Acta 1838:2899–2909
13. Legrand B, Laurencin M, Sarkis J, Duval E, Mouret L, Hubert JF, Collen M, Vié V, Zatylny-Gaudin C, Henry J, Baudy-Floc'h M, Bondon A (2011) Structure and mechanism of action of a *de novo* antimicrobial detergent-like peptide. Biochim Biophys Acta 1808:106–116
14. Bechinger B (1999) The structure, dynamics and orientation of antimicrobial peptides in membranes by multidimensional solid-state NMR spectroscopy. Biochim Biophys Acta 1462:157–183
15. Nichols M, Kuljanin M, Nategholeslam M, Hoang T, Vafaei S, Tomberli B, Gray CG, DeBruin L, Jelokhani-Niaraki M (2013) Dynamic turn conformation of a short tryptophan-rich cationic antimicrobial peptide and its interaction with phospholipid membranes. J Phys Chem B 117:14697–14708
16. Bagheri M, Arasteh S, Haney EF, Hancock REW (2016) Tryptic stability of synthetic bactenecin derivatives is determined by the side chain length of cationic residues and the peptide conformation. J Med Chem 59:3079–3086
17. Nauli S, Farr S, Lee YJ, Kim HY, Faham S, Bowie JU (2007) Polymer-driven crystallization. Protein Sci 16:2542–2551
18. Bagheri M, Keller S, Dathe M (2011) Interaction of W-substituted analogs of cyclo-RRWFWR with bacterial lipopolysaccharides: the role of the aromatic cluster in antimicrobial activity. Antimicrob Agents Chemother 55:788–797
19. Scheinpflug K, Krylova O, Nikolenko H, Thurm C, Dathe M (2015) Evidence for a novel mechanism of antimicrobial action of a cyclic R-, W-rich hexapeptide. PLoS One 10:e0125056
20. Finger S, Kerth A, Dathe M, Blume A (2015) The efficacy of trivalent cyclic hexapeptides to

induce lipid clustering in PG/PE membranes correlates with their antimicrobial activity. Biochim Biophys Acta 1848:2998–3006
21. Tieleman DP, Sansom MSP (2001) Molecular dynamics simulations of antimicrobial peptides: from membrane binding to trans-membrane channels. Int J Quantum Chem 83:166–179
22. Arasteh S, Bagheri M, Goliaei B (2014) Membrane selectivity of small cyclic antimicrobial hexapeptides studied by molecular dynamics simulations. J Pept Sci 20:S279–S280
23. Pronk S, Páll S, Schulz R, Larsson P, Bjelkmar P, Apostolov R, Shirts MR, Smith JC, Kasson PM, van der Spoel D, Hess B, Lindahl E (2013) GROMACS 4.5: a high-throughput and highly parallel open source molecular simulation toolkit. Bioinformatics 29:845–854
24. Humphrey W, Dalke A, Schulten K (1996) VMD: visual molecular dynamics. Graphics 14:33–38
25. Allen WJ, Lemkul JA, Bevan DR (2009) GridMAT-MD: a grid-based membrane analysis tool for use with molecular dynamics. J Comput Chem 30:1952–1958
26. Oostenbrink C, Villa A, Mark AE, van Gunsteren WF (2004) A biomolecular force field based on the free enthalpy of hydration and solvation: the GROMOS force-field parameter sets 53A5 and 53A6. J Comput Chem 25:1656–1676
27. Brooks BR, Bruccoleri RE, Olafson BD, States DJ, Swaminathan S, Karplus M (1983) CHARMM: a program for macromolecular energy, minimization, and dynamics calculations. J Comput Chem 4:187–217
28. Jorgensen WL, Maxwell DS, Tirado-Rives J (1996) Development and testing of the OPLS all-atom force field on conformational energetics and properties of organic liquids. J Am Chem Soc 118:11225–11236
29. Pearlman DA, Case DA, Caldwell JW, Ross WS, Cheatham TE III, DeBolt S, Ferguson D, Seibele G, Kollman P (1995) AMBER, a package of computer programs for applying molecular mechanics, normal mode analysis, molecular dynamics and free energy calculations to simulate the structural and energetic properties of molecules. Comput Phys Commun 91:1–41
30. Gfeller D, Michielin O, Zoete V (2013) SwissSidechain: a molecular and structural database of non-natural sidechains. Nucleic Acids Res 41:D327–D332
31. Klauda JB, Venable RM, Freites JA, O'Connor JW, Tobias DJ, Mondragon-Ramirez C, Vorobyov I, MacKerell AD Jr, Pastor RW (2010) Update of the CHARMM all-atom additive force field for lipids: validation on six lipid types. J Phys Chem B 114:7830–7843
32. Berendsen HJC, Postma JPM, van Gunsteren WF, DiNola A, Haak JR (1984) Molecular dynamics with coupling to an external bath. J Chem Phys 81:3684–3690
33. Cheng A, Merz KM Jr (1996) Application of the Nosé–Hoover chain algorithm to the study of protein dynamics. J Phys Chem 100:1927–1937
34. Bussi G, Donadio D, Parrinello M (2007) Canonical sampling through velocity rescaling. J Chem Phys 126:014101
35. Parrinello M, Rahman A (1981) Polymorphic transitions in single crystals: a new molecular dynamics method. J Appl Phys 52:7182–7190
36. Kucerka N, Tristram-Nagle S, Nagle JF (2005) Structure of fully hydrated fluid phase lipid bilayers with monounsaturated chains. J Membr Biol 208:193–202

Chapter 9

Calorimetry Methods to Study Membrane Interactions and Perturbations Induced by Antimicrobial Host Defense Peptides

Mauricio Arias, Elmar J. Prenner, and Hans J. Vogel

Abstract

Biological membranes play an important role in determining the activity and selectivity of antimicrobial host defense peptides (AMPs). Several biophysical methods have been developed to study the interactions of AMPs with biological membranes. Isothermal titration calorimetry and differential scanning calorimetry (ITC and DSC, respectively) are powerful techniques as they provide a unique label-free approach. ITC allows for a complete thermodynamic characterization of the interactions between AMPs and membranes. DSC allows one to study the effects of peptide binding on the packing of the phospholipids in the membrane. Used in combination with mimetic models of biological membranes, such as phospholipid vesicles, the role of different phospholipid headgroups and distinct acyl chains can be characterized. In these protocols the use of ITC and DSC methods for the study of peptide–membrane interactions will be presented, highlighting the importance of membrane model systems selected to represent bacterial and mammalian cells. These studies provide valuable insights into the mechanisms involved in the membrane binding and perturbation properties of AMPs.

Key words Antimicrobial peptides, Calorimetry, Isothermal titration calorimetry, Differential scanning calorimetry, Peptide–membrane interactions

1 Introduction

Antimicrobial host defense peptides (AMPs) are a large family of peptides that are ubiquitously distributed throughout all forms of life, from bacteria to mammals. In higher organisms these peptides constitute an important part of the host defense system that can contribute as a first and direct response to fight bacterial infections or indirectly by activating the immune system. Consequently, some AMPs are also referred to as host defense peptides. AMPs have been extensively studied as potential candidates for the generation of new bactericidal drugs, which could counteract the ever-growing number of pathogenic bacterial strains that are becoming resistant to conventional antibiotics. Several studies aimed at understanding

Paul R. Hansen (ed.), *Antimicrobial Peptides: Methods and Protocols*, Methods in Molecular Biology, vol. 1548,
DOI 10.1007/978-1-4939-6737-7_9, © Springer Science+Business Media LLC 2017

the mechanism of action of AMPs have identified the bacterial membrane as the main target for most cationic peptides [1–5]. In many cases the perturbation and/or permeabilization of the lipid bilayer is directly related to the killing efficiency of AMPs. However, for a smaller group of AMPs, the membrane has been identified as an initial physical barrier that needs to be overcome in order for the peptide to reach intracellular targets, such as essential proteins/enzymes and/or nucleic acids [6]. Clearly, the initial interaction of AMPs with bacterial membranes constitutes a crucial step for their mechanism of action.

A very important property of AMPs is their ability to preferentially "attack" prokaryotic (bacterial) rather than eukaryotic (host) cells [7, 8]. The difference in membrane composition between prokaryotic and eukaryotic cells can explain the selective activity exhibited by cationic AMPs. Bacterial membranes are characterized by an overall negative charge, provided by a high content of phospholipids with negatively charged headgroups, such as phosphatidylglycerol (PG). In contrast, the outer leaflet of eukaryotic membranes is often neutral, due to the presence of phospholipids with zwitterionic headgroups, such as phosphatidylcholine (PC) [9]. These distinct characteristics of these two membranes, taken together with the net positive charge of most AMPs, highlight the favorable electrostatic interactions that take place between AMPs and bacterial membranes.

Since the peptide–membrane interactions play such a key role in the selectivity and activity of AMPs, these interactions are frequently studied by different biophysical techniques. Calorimetry methods, such as differential scanning calorimetry (DSC) and isothermal titration calorimetry (ITC), represent a valuable approach, because the experiments are done in a non-perturbing and label-free manner. In the case of optical techniques such as fluorescence spectroscopy, when intrinsic Trp residues are not present in AMPs, bulky fluorescent dyes are often attached to the peptides. Unfortunately, these fluorescent dyes can disturb the structure and membrane binding properties of the AMPs. Since no labels are required for calorimetric measurements, they provide a distinct advantage.

Comparisons of results obtained with well-defined membrane model systems that are comprised of different lipid components facilitate an understanding of the role of the lipid composition of biological membranes in the antibacterial mechanisms of AMPs. In most cases multilamellar, large, and small unilamellar vesicles are used to comprehensively study the influence of different lipid compositions on the peptide–membrane interactions. Several calorimetric studies have addressed the effects of the composition of the lipid bilayer by changing the phospholipid headgroup and the acyl chain length and/or their saturation [10, 11]. Furthermore, the use of vesicles with a defined and varying composition allows for comparison of different types of cell membranes. For example, membranes from bacterial and cancer cells can be represented through the use of phospholipids with negatively charged headgroups (phosphatidylglycerol and phosphatidylserine, respectively) when preparing the vesicles. At the same time, vesicles with zwitterionic phosphatidylcholine headgroups are frequently used to represent the outer leaflet of the membranes of normal eukaryotic cells.

In differential scanning calorimetry (DSC), a sample and an inert reference (which exhibits no thermotropic events within the temperature range of interest) are both heated at a predetermined rate. The instrument then measures the differential flow of heat that is required in order to maintain this temperature difference. When a thermotropic event occurs in the sample, additional power is required to maintain this difference in temperature. The additional power (joules/second) is registered as a function of the temperature [12]. By measuring changes in heat (dQ) as a function of the changes in temperature (dT), the DSC instrument establishes the specific heat capacity at constant pressure (Cp) of the system (of mass m) at each temperature, according to Eq. 1.

$$dQ = m \cdot C_p \cdot dT \tag{1}$$

The heat for thermally induced transitions is then calculated as the area under the curve and represents the change in enthalpy of the transition (ΔH_{cal}, Eq. 2).

$$\Delta H_{cal} = m \int_{T_1}^{T_2} C_p \cdot dT \tag{2}$$

For a deeper understanding, the reader is advised to consult several review papers for a detailed description of the DSC instrumentation and theory [13–18].

Because of the ability to measure changes in the sample due to increasing temperatures, DSC is often used to study the thermal stability of biomolecules [19]. However, in the context of peptide–membrane interactions, DSC is frequently used to study the thermotropic behavior of lipid bilayers and the perturbations induced by their interactions with peptides. During a typical DSC experiment, a simplified model membrane, such as a multilamellar lipid vesicle (MLV), is normally used. These model systems are composed of individual or combinations of phospholipids, and they are characterized by specific thermotropic phase transitions (Figs. 1 and 2). Commonly used phospholipids for the study of peptide–membrane interactions show a main phase transition (highly cooperative) between the gel (Lβ) and the liquid crystalline (Lα) phases. This transition is identified by its transition temperature (T_m) and enthalpy (ΔH). For some lipid bilayers, an additional weak and broader (less cooperative) pre-transition (T_p), between the gel (Lβ)

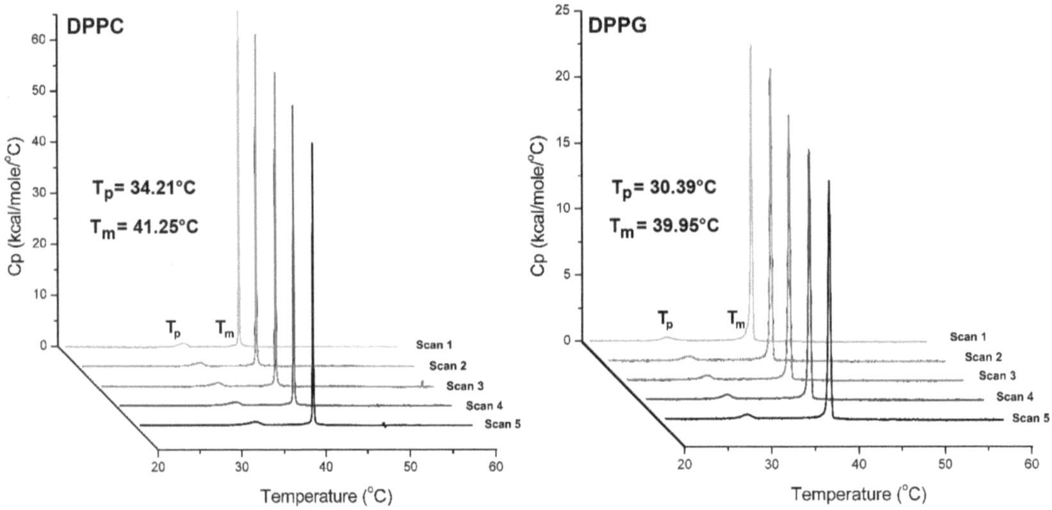

Fig. 1 DSC thermograms for pure phospholipid MLVs, 1,2-dipalmitoyl-sn-glycero-3-phosphocholine (DPPC), and 1,2-dipalmitoyl-sn-glycerol-3-phospho-(1′-rac-glycerol) (DPPG). An increasing gray color scale depicts the five consecutive scans. The main transition (T_m) and pre-transition (T_p) were determined for the last scan and are depicted in the figure. The changes in enthalpy for the pre-transition ($\Delta H_{DPPC} = 1110 \pm 8.5$ kcal/mol and $\Delta H_{DPPG} = 527 \pm 4.8$ kcal/mol) and main transition ($\Delta H_{DPPC} = 7377 \pm 21.3$ kcal/mol and $\Delta H_{DPPG} = 5986 \pm 11.9$ kcal/mol) were calculated for the last scan using the fitting model *Non-2-State with zero DCp*. Both phospholipids were prepared at 0.5 mg/ml concentration in DSC buffer (Tris-HCl 10 mM pH 7.4, NaCl 150 mM, EDTA 1 mM)

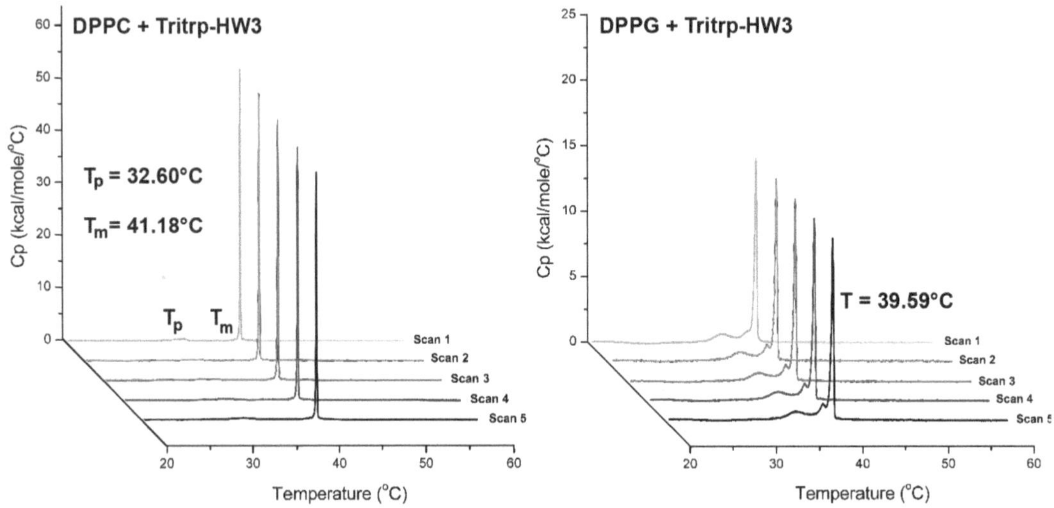

Fig. 2 DSC thermograms for phospholipid MLVs in the presence of the antimicrobial peptide Tritrp-HW3 (VRRFPWW[5OHW]PFLRR-NH_2). DPPC (*left*) and DPPG (*right*) were exposed to 1:10 (peptide/lipid) molar ratio of peptide after the formation of the MLVs. An increasing gray color scale depicts the five consecutive scans. The main transition (T_m) and pre-transition (T_p) were determined for the last scan and are depicted in the figure. The changes in enthalpy for the pre-transition ($\Delta H_{DPPC+TritrpHW3} = 722.8 \pm 5.7$ kcal/mol) and main transition ($\Delta H_{DPPC+TritrpHW3} = 6267 \pm 42.1$ kcal/mol) were only calculated for the DPPC and Tritrp-HW3 using the last scan the fitting model *Non-2-State with zero DCp*. Both phospholipids were prepared at 0.5 mg/ml concentration in DSC buffer (Tris 10 mM pH 7.4, NaCl 150 mM, EDTA 1 mM)

and ripple (Pβ) phases, can be observed at lower temperatures [20, 21]. An example of the thermotropic phase transitions observed for DPPC and DPPG MLVs is shown in Fig. 1. When a peptide is bound to the lipid bilayer, it will disturb the thermotropic behavior of the phospholipids, giving rise to modified phase transitions (Fig. 2). Therefore, DSC can be successfully used to determine the ability of an AMP to interact with a specific lipid bilayer and disturb its packing organization.

Isothermal titration calorimetry (ITC) allows for in-depth characterization of biomolecular interactions. This experimental technique permits one to determine all the thermodynamic parameters of an interaction reaction in a single experiment [22–24]. These parameters include the binding constant (K_a), the free energy ($\Delta G°$), enthalpy ($\Delta H°$), and entropy ($\Delta S°$) for the reaction, for which the relationships are shown by Eqs. 3 and 4.

$$\Delta G^o = -RT \cdot \ln(K_a) \quad (3)$$

$$\Delta G^o = \Delta H^o - T \cdot \Delta S^o \quad (4)$$

In several cases the dissociation constant (K_d), which corresponds to the inverse of the K_a (Eq. 5), is used to describe the affinity of an interaction.

$$K_a = 1/K_d \quad (5)$$

The ITC method is based on a measurement of the heat that is generated or absorbed when two binding partners interact. This is achieved by separating both binding partners (normally referred to as "ligand" and "macromolecule") in two different compartments before they are combined in controlled manner. The instrument contains a sample cell holding a diluted solution of the "macromolecule," while the highly concentrated "ligand" is located in an injection/titration syringe. It is important to note that the "ligand" and "macromolecule" terms only make reference to their location and not to any other intrinsic characteristics of the molecules. In the specific case of peptide–membrane interactions, solubility limitations and the aggregation tendencies of some AMPs limit the use of these peptides to the sample cell (as "macromolecules"), where their concentration is relatively lower. In this setup, the model membranes are normally used in the titration syringe (as "ligands"), as they can be prepared as a stable solution with a high concentration of phospholipids. Large unilamellar lipid vesicles (LUVs) or small unilamellar vesicles (SUVs) can be used as model membranes for ITC experiments (Fig. 3, *upper*). Upon titration the LUVs will bind to the peptides and these interactions will generate or absorb heat, which is detected by the ITC instrument. Following repeated injections of LUVs, fewer peptides will be available for interaction; consequently, smaller heat signals will be produced. Eventually, the system will reach saturation, where all peptides will already be bound to LUVs. Further injections of LUVs will then result in small heat signals corresponding to the heats of dilution of the LUV suspension. The area under each heat signal/peak in the raw data corresponds to the heat generated by each injection. Analysis and fitting of all the heats per injection during the titration experiment to a proper binding model generate the thermodynamic parameters of the binding event (Fig. 3, *lower*). A "one set of sites" model is often used, which assumes that the peptide–membrane interactions take place in one or several identical and independent binding sites. The interested reader should consult the tutorial guide for ITC data analysis in the Origin® manual for a detailed description of the analysis and interpretation of ITC data. Similarly, a detailed explanation of the ITC instrumentation and theory can be found in several comprehensive reviews [22, 24].

It is important to consider that the interaction between a negatively charged vesicle surface and positively charged peptides can be described by a two-step model. An initial electrostatic attraction will drive the peptide to the membrane surface, effectively increasing the concentration of the peptide in the vicinity of the membrane. Then a second process, involving hydrophobic interactions, will allow the peptide to partially penetrate into the bilayer. However,

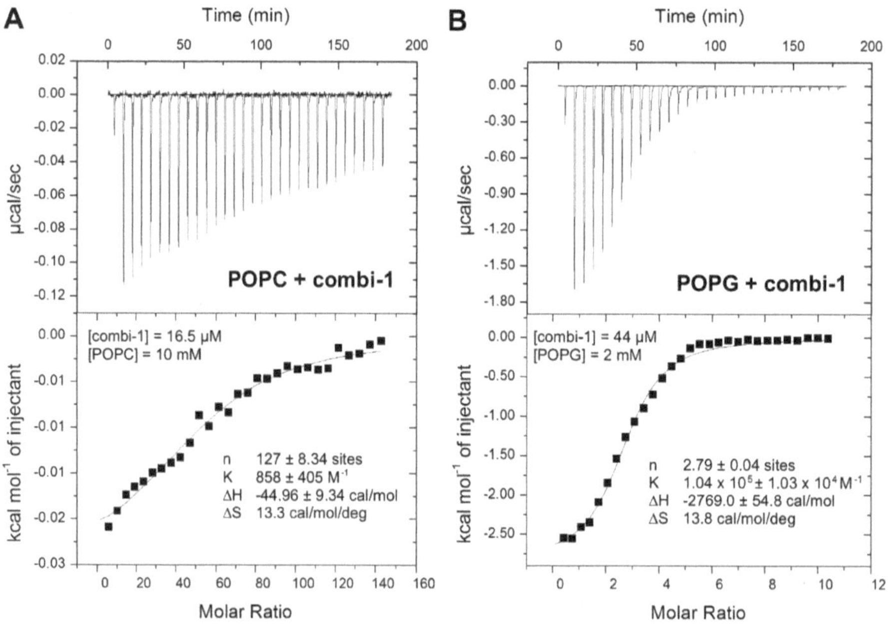

Fig. 3 ITC experiment raw data (*upper*) and integrated heat for each injection (*lower*) for the interaction of the antimicrobial peptide combi-1 (Ac-RRWWRF-NH_2) with POPC (**a**) or POPG (**b**) LUVs in ITC buffer (HEPES 10 mM pH 7.0, NaCl 100 mM). The concentrations of peptide and LUVs for each experiment are depicted in the corresponding figure. The thermodynamic parameters (K, ΔH, ΔS) were determined by fitting the integrated heat using the model *One Set of Sites*, after subtraction of the heats of dilution for the corresponding LUV system

the binding of peptides to the membrane will decrease the surface potential. Therefore, the electrostatic attraction between the peptides in the bulk solution and the membrane surface will be reduced. Similarly, the initial interaction of a positively charged peptide with a neutrally charged membrane would generate a positive charge in the surface of the membrane, resulting in increased repulsion of other peptides in the bulk solution. Therefore, in most cases the binding constant (K_a) derived from a direct analysis of the ITC-binding isotherm between AMPs and lipid vesicles will describe the so-called apparent binding constant (K_{app}). This K_{app} does not consider the nonspecific electrostatic effects directly. However, Seelig and colleagues have described a detailed treatment of ITC data in order to obtain the intrinsic binding constant, which takes into account these electrostatic effects [25–27]. Unfortunately, the interpretation of the ITC results is not always straightforward as the heat signals can be related to processes that are different from the direct peptide–membrane interactions. Phenomena such as oligomerization, pore formation, and/or changes in peptide conformation upon membrane binding, among others, can also generate heat signals that can complicate the interpretation of the ITC data. Thus ITC is usually most successfully used when a series of related AMPs are studied and the results can be directly compared. Be that as it may, in combination with other biophysical techniques, the deconvolution of the heat signals involved in such processes has been successful in specific cases [28–31].

The following protocols will describe the use of DSC and ITC in the study of peptide–membrane interactions as applied to the AMPs Tritrp-HW3 and Combi-1, as representative examples. Specific membrane models to emulate the negatively charged and neutral membrane surfaces of prokaryotic and eukaryotic cells, respectively, were used in order to glean information about the selectivity of these two peptides.

2 Materials

2.1 Peptides

1. Peptides are synthesized as described in Chapter 3 or 4 or obtained as a lyophilized powder from commercial suppliers. The purity (>95%) and molecular weight are confirmed by HPLC and mass spectrometry.
2. Peptide stock solution for DSC experiments: ~1 mg of peptide in ultrapure (Milli-Q) water and stored at −20 °C.
3. Peptide stock solution for ITC experiments: ~1 mg of peptide in ITC buffer (*see* Subheading 2.3, **item 1**) and stored at −20 °C (*see* **Note 1**).

2.2 Differential Scanning Calorimetry

1. DSC buffer: 10 mM Tris–HCl pH 7.4, 150 mM NaCl, 1 mM EDTA (*see* **Note 2**).
2. Gastight syringe (Hamilton 2.5 ml, 20 gauge, 4.375″ long needle, point style 3).
3. MicroCal VP-DSC instrument and MicroCal ThermoVac accessory for sample degassing and cell cleaning.
4. VPViewer™ and Origin® software for DSC instrument setup and data analysis, respectively.
5. Phospholipids: 1,2-dipalmitoyl-sn-glycero-3-phosphocholine (DPPC) and 1,2-dipalmitoyl-sn-glycero-3-phospho-(1′-rac-glycerol) sodium salt (DPPG) (*see* **Note 3**). These lipids are supplied as dry or chloroform solutions. If dry lipids are used, a stock solution (10 mg/ml) is prepared by dissolving the lipids in chloroform/methanol (2:1). This solution is stored at −20 °C until use.

2.3 Isothermal Titration Calorimetry

1. ITC buffers: 10 mM HEPES pH 7.0, 100 mM NaCl (*see* **Note 4**).
2. Gastight syringe (Hamilton 2.5 ml, 18 gauge, 8″ long needle, point style 3).
3. MicroCal VP-ITC instrument and MicroCal ThermoVac accessory for sample degassing and cell cleaning.
4. VPViewer™ and Origin® software for ITC instrument setup and data analysis, respectively.
5. Phospholipids: 1-palmitoyl-2-oleoyl-sn-glycero-3-phosphocholine (POPC) and 1-palmitoyl-2-oleoyl-3-phosphoglycerol (POPG). Lipids can be obtained in chloroform solutions or as dry lipids. If dry lipids are used, a stock solution (12.5 mg/ml) is prepared by dissolving the lipids in chloroform/methanol (2:1). This solution is stored at −20 °C until use.
6. Avanti® Mini-Extruder.
7. Two gastight syringes (Hamilton 1000 μl, 0.75″ long needle, point style 3).
8. Polycarbonate membrane filters (0.1 μm pore size) and filter supports.
9. Freeze-thawing setup: liquid nitrogen and water bath (35–40 °C).

2.4 Phosphate Assay (Phospholipid Concentration Determination)

1. Reagent A: 10% $MgNO_3$ in ethanol.
2. Reagent B: mix one part (i.e., 1.5 ml) of 10% ascorbic acid with six parts (e.g., 9 ml) of 0.42% ammonium molybdate tetrahydrate $(NH_4)_6Mo_7O_{24}\cdot 4H_2O$ in 1 N H_2SO_4.
3. HCl solution: aqueous solution of 0.5 N HCl.
4. Water bath (water boiling temperature)

3 Methods

3.1 Determination of Peptide Concentration

1. The peptide (*see* Subheading 2.1) concentration is established by measuring the UV absorbance for Trp at 280 nm and using the extinction coefficient provided by the software tool ProtParam from ExPASy: SIB bioinformatics resource portal [32].

3.2 Differential Scanning Calorimetry

This section describes the protocol used to study the effect of antimicrobial peptides on the thermotropic phase transitions of multilamellar lipid vesicles composed of single phospholipids.

3.2.1 Multilamellar Lipid Vesicle (MLV) Preparation With or Without AMPs

1. Take the appropriate amount of the phospholipid stock (in chloroform or chloroform/methanol) in order to have 0.5 mg of lipids in a clean glass vial. The final sample contains 1 ml of the phospholipid solution at a concentration of 0.5 mg/ml (*see* **Note 5**).
2. Evaporate the solvent (chloroform/methanol) inside a fume hood under a soft stream of nitrogen gas, rotating the vial in order to create a thin lipid film over the glass walls.
3. Remove the remaining solvent by leaving the open vial under vacuum overnight. The dried lipid films can be stored at −20 °C until use. In addition overlay with an inert gas in case of long-term storage.
4. Sample containing MLVs without AMPs: the lipid films (*see* **step 3**) are warmed to room temperature, while the DSC buffer (~2 ml) is heated above the melting temperature (T_m) of the phospholipids for 10 min (*see* **Note 6**). Then hydrate the lipid films with 1 ml of the preheated DSC buffer, and vortex vigorously in order to bring all lipids in solution. In some cases several cycles of heating and vortexing are required to solubilize all the lipids from the vial walls. At this point the solution contains mainly MLVs.
5. Sample containing MLVs with AMPs: add the required amount of peptide stock (*see* Subheading 2.1, **item 2**) to the MLV suspension without AMPs (*see* **step 4**). The peptide-to-lipid (P:L) molar ratio for our experiment is setup as 1:10 (*see* **Note 7**). The solution is then mixed by extensive vortexing (~1 min), allowing for the peptide to bind to the lipid bilayers.

3.2.2 Sample Degassing and Loading

As mentioned above the DSC instrument contains two cells, one for the actual sample and the second for the reference containing the DSC buffer. Three separate experiments should be performed; first, a buffer-buffer experiment is required to establish the baseline. Second, the buffer-MLV experiment will determine the thermotropic phase transitions (unaltered) of the pure phospholipid bilayers (Fig. 1). Third, buffer-MLV/AMP experiment (Fig. 2) will highlight the changes induced by the peptide on the phase transitions of the phospholipid bilayers. The following steps describe the setup for the third experiment (buffer-MLV/AMP). However, the same steps should be followed for the other two experiments.

1. *Sample and DSC buffer degassing*: Place both uncapped glass vials containing the MLVs/AMP suspension (*see* Subheading 3.2.1, **step 5**) and DSC buffer (~2 ml) in the tube holder of the MicroCal ThermoVac accessory. The instrument is then set up to create vacuum for 10 min keeping the temperature at 20 °C. Use the vacuum cap to cover the tube holder in order to start the degassing process (*see* **Note 8**).
2. Load the degassed MLVs/AMP sample and DSC buffer into the sample and reference cells, respectively, following the instructions in the DSC.

   Sample loading: Rinse the sample and reference cells with DSC buffer at least two times. Using the gastight Hamilton syringe, take 1 ml of degassed buffer or MLV sample from the glass vial. Then place the filling funnel in the corresponding instruments cells, and insert the syringe needle through the center of the funnel. At this point the top of the needle should be located ~2 mm from the bottom of the cell. Slowly start delivering most of the content of the syringe into the cell until the solution can be seen coming out from the filling funnel. In order to remove trapped air bubbles, sharply pump the solution (~0.3 ml) in and out from the cell. At this point air bubbles can be observed coming out of the cell. This procedure should be repeated several times (>5 times) until no air bubbles are observed. The excess volume in the sample cell is then removed by placing the delivery syringe needle on the inner ledge of the cell and drawing out the excess solution until you see air being drawn into the syringe. Proper sample loading is crucial in order to prevent the presence of air bubbles in the DSC sample and reference cells (*see* **Note 8**).

3.2.3 Instrument Preparation and Data Collection

1. After loading the sample and reference cells, close the DSC cell with the DSC cap. First secure the lower metal portion of the cap against the DSC cells port, and then start rotating the upper plastic (white) part of the cap. At this point the pressure in the cells will start increasing (as indicated in the VPviewer window). Stop tightening the upper cap when the pressure stops increasing. Be careful not to over-tighten the cap, as it can break. At this point the pressure in the cells should be around 27–28 psi, for the conditions and MicroCal instrument used in our laboratory (*see* **Note 9**).

2. Set up the following experimental parameter in the DSC instrument, using the VPviewer window (*left panel*) under the *DSC controls* tab.

   Number of scans, 5. The thermal history of the samples before the DSC experiment can influence the thermotropic behavior of a lipid sample. Therefore, it is recommended to run three or more scans in order to ensure similarity between consecutive scans. This would indicate that equilibrium has been reached (Fig. 1). Similarly, in the case of MLV samples where peptide is externally added, the peptides could take time to reach the target membranes, which would be reflected in changes over time in the thermotropic behavior of the sample (Fig. 2). In all cases the number of scans should allow for the sample to reach equilibrium conditions. Frequently, the reported thermograms correspond to the last scan.

   Post-cycle thermostat, 60 °C. This is the temperature that the instrument would reach after the last scan is performed.

3. Set up the following scan parameter in the DSC instrument, using the VPviewer window (*right panel*) under the *DSC controls* tab.

   Starting temperature, 20 °C. The starting temperature should allow for a stable baseline before the phospholipid phase transitions.

   Final temperature, 60 °C. The final temperature should allow for a stable baseline after the phospholipid phase transitions.

   In our case the T_m and T_p for DPPC and DPPG are around 40 and 30 °C. Therefore, ±10 °C allows for stable baselines before the pre-transition and after the main phase transitions (*see* **Note 10**).

   Scan rate, 10°/h. High scan rates are necessary for broad peaks and could have large effects on the DSC data for both heat-conduction or power-compensation calorimeters [13]. However, single lipids normally exhibit very sharp peaks; therefore, it is recommended to use lower scan rates in such cases.

   Prescan thermostat, 15 min. This time allows for a proper equilibration of the thermal core of the instrument, which is important for the repeatability and shape of the baseline.

   Postscan thermostat, 0 min. This represents the time to thermostat at the final temperature before starting the next scan, and it is normally used to incubate the DSC cells at high temperature for cleaning purposes. Therefore, during the DSC experiments, it is set to 0 min.

   Filtering period, 4 s. This represents the time period in which the data samples are averaged and a data point is generated and stored. For the fast and sharp lipid transitions, short filter periods are normally used (1–5 s).

Feedback (FB) mode/gain: High. In the case of fast and sharp transitions for lipid samples, a fast and high compensation response is required.

Scan edit mode: Identical scans. This option will set all the scan parameters for all the scans in an experiment to be identical.

4. Start the data collection.

3.2.4 Data Analysis

The data analysis was performed following the instructions of the tutorial guide: DSC Data Analysis in Origin®, provided by the manufacturer of the instrument.

1. The final thermograms, corresponding to the MLV alone (buffer-MLV experiment) and in the presence of peptide (buffer-MLV/AMP experiment), are prepared by subtracting the last DSC buffer-buffer scan. In our laboratory the Origin 7.0 software from MicroCal Inc. is used for data analysis. The option *Subtract Reference* is found in the DSC Main Control section of the RawDSC window.
2. Normalization and fitting of the data will allow the determination of the transition temperature (T_m and T_p) as well as the calorimetric heat change (ΔH). Briefly, after subtraction of the reference data, the normalization is performed by inputting the lipid concentration value (mM). Once normalized a baseline needs to be created and is then subtracted from the normalized data. At this point the data is ready for the fitting process. In our laboratory model number 2, *Non-2-State with zero DCp* is normally used to fit the thermograms involving lipids and peptides. Regularly, the fitting process is easier if the T_m is initially setup (as close as possible to the actual T_m) as a non-variable, while the ΔH is left to vary during the iterations. Once a reasonable ΔH is achieved, the T_m can be included in the iteration process as well. Alternatively, the ΔH can also be determined by manually integrating the area under the peak, which is also available as an option in the Origin 7.0 software.
3. In Fig. 1 representative thermograms for neutral (DPPC) and negatively charged (DPPG) MLVs are presented. Characteristic main and pre-transition temperatures for both lipid systems are depicted as well as the corresponding changes in enthalpy. In Fig. 2 representative results for an AMP Tritrp-HW3 (VRRFPWW[5OHW]PFLRR-NH_2) [33], interacting with DPPC and DPPG vesicles, are presented. The main DPPC phase transition is essentially not altered in the presence of peptide (Tritrp-HW3). The disappearance of the pre-transition can be observed. In combination, these data indicate that the peptide is only able to superficially interact with the DPPC lipid bilayer, leaving the packing of the acyl chains unchanged.

In contrast, large changes in both the pre- and main phase transitions of DPPG are observed in the presence of the peptide (Fig. 2). This suggests that considerable alterations in the acyl chain packing in the DPPG lipid bilayer occurred.

3.3 Isothermal Titration Calorimetry

This protocol describes the steps to be followed in order to study the interaction of AMPs with large unilamellar vesicles (LUVs) composed of single or combined phospholipids. It is important to note that the specific setup of an ITC experiment depends markedly on the expected binding constant (K_a) for the system. The protocol described here should only be used as a reference when similar K_a values are suspected or as an initial attempt when the K_a value is completely unknown.

3.3.1 Large Unilamellar Vesicle (LUV) Preparation

This procedure describes the preparation of LUVs for ITC experiments (also see the manual for liposome preparation from supplier).

1. Combine the appropriate amounts of phospholipid stocks (in chloroform/methanol) in a clean glass vial in order to have ~12 mg of lipids (*see* **Note 11**).
2. Evaporate the solvent (chloroform/methanol) from the lipid solution as previously described (*see* Subheading 3.2.1, **steps 2** and **3**).
3. Hydrate the lipid films with 1 ml of ITC buffer and vortex vigorously to bring all lipids in suspension. At this point the solution is composed of multilamellar lipid vesicles (MLVs).
4. Freeze-thaw the suspended MLVs by submerging the glass flask in liquid nitrogen for a couple of minutes and then thawing the suspension in a warm water bath (35–40 °C). Repeat this procedure at least five times.
5. Assemble the lipid extruder according to the manufacturer's instructions. The pore size of the polycarbonate filter membrane will determine the average size of the resulting LUVs. In our laboratory, LUVs with a 100 nm diameter are normally used (filter pore size 0.1 μm).
6. Load ~1 ml of leakage buffer into one syringe, and pass the solution through the extruder, in order to check for any leakages on the extrusion device and the proper attachment of the second syringe. Discard all liquids from the second syringe once all buffers have passed through the device.
7. Load the MLV suspension into one of the syringes and pass the suspension through the extruder device. The suspension will become more translucent indicating the formation of LUVs. Once the second syringe is loaded, return the suspension through the extruder to the initial syringe. Repeat this procedure

at least 15 times. Make sure that the final extrusion ends with the LUV suspension in the syringe that is opposite to the one you started with. At this point the suspension is composed of LUVs with an average diameter of 100 nm (*see* **Note 12**). Transfer this LUV suspension to a glass vial.

3.3.2 Lipid Concentration Determination

This procedure is a modified version of the phosphorus assays described by Ames et al. [34]:

1. Pipette 5 μl of the LUV suspension into phosphate-free Pyrex test tube. Additionally, pipette 5 μl of ITC buffer into an additional test tube as control. These samples should be prepared in triplicate.
2. Add 30 μl of Reagent A to each tube, and carefully ash the samples over a hot flame (*see* **Note 13**). Once all ethanol is evaporated, leave the tubes over the flame until there is no more smoke coming from the tubes and a white/gray precipitate (ash) is observed. Let the tubes cool down.
3. Add 300 μl of 0.5 N HCl into each tube. Vortex the tubes until the gray precipitate is resuspended. During this step the $MgNO_3$ and HCl hydrolyze the lipids releasing free PO_4^{3-} groups.
4. Cap the tubes with aluminum foil and place them in boiling water for 15 min. Then remove the tubes from the water bath and allow cooling.
5. Add 700 μl of Reagent B to each tube, cap them, and incubate for 1 h at 37 °C. During this step the samples acquire a light blue color, indicating the presence of phosphorous.
6. Measure the OD_{820} for each tube, and calculate the concentration of phosphate according to the following ratio: 0.01 μmol will correspond to $OD_{820}=0.240$ [34]. Alternatively, the phosphorus concentration can be calculated by preparation of a standard curve using a phosphorus standard solution. Divide the number of moles by 5 μl in order to establish the final phosphate concentration in the samples. Each lipid molecule in the LUVs contains one phosphate group; therefore, the lipid concentration corresponds directly to the phosphate concentration.
7. Dilute the LUV suspension with ITC buffer in order to have ≥10 mM lipid concentration. At least 650 μl of this suspension is required for one ITC experiment (*see* **Note 14**).

3.3.3 Peptides

1. The peptide stock prepared for the ITC experiments (*see* Subheading 2.1, **item 3**) is diluted in ITC buffer in order to have a final concentration of 10–50 μM. At least 2.0 ml of this peptide solution is necessary for one ITC experiment. The actual concentration would depend on the specific peptide and lipids being used and their expected affinity (*see* **Note 14**).

3.3.4 Sample Degassing and Loading

1. *Samples and ITC buffer degassing*: Place the peptide (*see* Subheading 3.3.3, **step 1**) and LUV samples (*see* Subheading 3.3.2, **step 7**), as well as the buffer (~2 ml) into separate glass or plastic vials. Place the vials in the tube holder of the MicroCal ThermoVac accessory, and start the degassing process as described before (*see* Subheading 3.2.2, **step 1**).
2. Load the reference cell (~1500 μl) with the degassed ITC buffer and the sample cell (1415 μl) with the degassed peptide solution, following the VP-ITC instructions manual, briefly described next (*see* **Note 15**).

   Sample loading: Rinse the sample and reference cells with ITC buffer at least two times. Then using the gastight Hamilton syringe take ~2 ml of degassed buffer or peptide sample from the vials. Insert the syringe needle into the corresponding cell. Raise the end of the needle ~2 mm from the cell bottom, and slowly start delivering the content of the syringe into the cell until the solution can be seen coming out. Remove all trapped air bubbles and remove excess volume in the sample cell (*see* Subheading 3.2.2, **step 2**). Cover the ITC chamber in order to prevent contamination of the sample and reference cells.
3. Load the ITC titration syringe with 300 μl of the degassed LUV suspension following the instructions described in the VP-ITC instruction manual and briefly described next (*see* **Note 15**).

   Titration syringe/auto-pipette loading: Rinse the MicroCal auto-pipette syringe needle with ITC buffer (~5 ml) before loading the LUV suspension. Then submerge the auto-pipette syringe needle into the vial containing the LUV suspension (650 μl), and slowly draw in the suspension, making sure that the auto-pipette is full and no air bubbles are introduced (*see* **Note 16**). The volume of the auto-pipette is ~300 μl; therefore, some LUV suspension (>200 μl) should remain in the vial. Using the option "purge and refill" from the ITC control window (VPviewer), mix the content of the auto-pipette with the remaining LUV suspension in the vial, ensuring that there are no air bubbles in the pipette. Dry the outside of the auto-pipette syringe needle with a Kimwipe. Be careful not to touch the bottom of the needle as the Kimwipe can take some of the solution from the syringe.
4. Remove the cover from the ITC chamber, and insert the auto-pipette into the sample cell. Make sure the pipette is tightly adjusted to the ITC chamber by pushing the auto-pipette firmly downward.

3.3.5 Instrument Preparation and Data Collection

1. Set up the following experimental parameter in the ITC instrument, using the VPviewer window (*left panel*) under the *ITC controls* tab.

   Total number of injections, 20. It is recommended that a large number of low-volume injections are used for systems with a strong heat signal. Instead, if a system exhibits weak heat signals, a smaller number of high-volume injections would be advantageous (*see* **Note 17**).

   Cell temperature, 30 °C. It is recommended to thermostat the instrument at a slightly lower temperature (−1.0 °C) while preparing the experiment in order to reduce the waiting time for the initial equilibration. This can be done by adjusting the temperature set point under the *Thermostat/Calib* tab in the VPviewer.

   Reference power, 10 μCal/s. This value will approximately determine the baseline value of the system at equilibrium. For experiments with strong exothermic reactions, a large reference power (~30 μCal/s) will be required. For strong endothermic reactions, a low reference power (~2 μCal/s) will work better.

   Initial delay, 5 min. This delay is required to establish the baseline before the first injection.

   Syringe concentration (*mM*): Use the value determined during the LUV sample preparation (*see* **Note 18**).

   Cell concentration (*mM*): Use the value determined during peptide sample preparation (*see* **Note 19**).

   Stirring speed, 300 rpm. High stirring speed is normally required if the suspension tends to precipitate or if an extremely high affinity is expected.

   Feedback mode/gain: High. Most reactions will require a high gain, which provides the fastest response time.

   ITC equilibration options: Auto. The auto option allows the instrument to proceed automatically through the equilibration.

2. Set up the following injection parameters in the ITC instrument, using the VPviewer window (*right panel*) under the *ITC control* tab.

   Volume (μl), 10 μl. The injections are normally 3–15 μl, which allows for proper temperature equilibration of the ligand solution. Considering the sensitivity limitations of the VP-ITC instrument (~0.1 μCal), the injection volume or concentration of the ligand should be increased if the heat signals are too small. Due to diffusion of the ligand into the sample cell during the initial delay and equilibration time, the first injection is usually set up for a lower volume (3 μl), and this data point is normally excluded from the final analysis.

Duration(s), 20 s. By default the VPviewer selects twice the number set up in the volume box. This will correspond to a ligand delivery rate of 0.5 μl/s.

Spacing(s), 360 s. The spacing time should always be large enough to allow the heat signal to return to the baseline before the next injection.

Filter period(s), 2 s. For fast reactions (as in this case) a period of 2 s is normally sufficient to represent the heat peaks accurately.

3. Start the data collection.

3.3.6 Control Experiments

1. An additional ITC experiment is required to measure the heat of dilution during injection of the ligand (LUVs) into the sample cell. In this experiment the LUV suspension (*see* Subheading 3.3.2, **step 7**) is titrated into ITC buffer (sample cell). The experimental and injection parameters should be the same (*see* Subheading 3.3.5).

3.3.7 Data Analysis

The data analysis is performed following the instructions of the tutorial guide ITC Data Analysis in Origin®, provided by MicroCal.

1. In our laboratory the software Origin 7.0 is used for data analysis. The software automatically transforms the raw data (μcal/s vs time) into heat changes of injectant (ΔH, kcal/mol) versus the molar ratio of ligand/binding partner, by integrating the injection peaks (Fig. 3, *lower*).
2. Before analysis of the AMP-LUV experiment results, the heats of dilution for the LUVs into ITC buffer (*see* Subheading 3.3.6) are subtracted. The option *Subtract Reference Data* is found in the Data Control section of the DeltaH window.
3. The final data, after reference subtraction, is fitted by using the model *one set of sites* (*see* **Note 20**).
4. In Fig. 3 representative results for an AMP interacting with different LUV systems are presented. The interaction of the peptide combi-1 (Ac-RRWWRF-NH_2) [35], with negatively charged LUVs composed of POPG (Fig. 3b), was characterized by a strong binding constant ($K_a = 1.04 \times 10^5$ M^{-1}). However, the interaction of the peptide with zwitterionic LUVs composed of POPC (Fig. 3a) was considerably weaker ($K_a = 858$ M^{-1}). In both cases the peptide–membrane interactions were characterized by the release of heat, indicating that the process was exothermic. When comparing the ΔH and ΔS values, it is clear that the interaction of combi-1 with POPG is mainly driven by the change in enthalpy ($\Delta H = -2769$ cal/mol).

4 Notes

1. Consistency in the buffer composition and pH for all the solutions used in isothermal titration calorimetry is a requirement to obtain high-quality data. Poorly matched buffers between the peptide and lipid solutions can lead to undesired heat signals during the titration process that are related to dilution and/or ionization effects [22, 36].
2. The pH of the buffer selected should have a low temperature dependence, as the DSC experiment encompasses a broad range of temperature. Tris buffer was used in this protocol as the ionization state of the peptides and the phospholipids was not expected to be affected by changes in pH inside the range of temperatures used. For Tris buffer the ΔpK/pH ~ 1 when the temperature changes from 20 to 60 °C (pK 7.2098 at 333.15 K and pK 8.221 at 293.15 K) [37].
3. In our laboratory DPPC and DPPG are normally used because the zwitterionic headgroup phosphatidylcholine (PC) is mainly found in the outer leaflet of mammalian membranes, while the negatively charged phosphatidylglycerol (PG) is a main component of the bacterial cytoplasmic membranes [9, 18]. Similarly, palmitate (16:0) constitutes one of the most prominent fatty acid chains in *E. coli* and human red blood cells [38, 39]. In addition, the transition temperatures of these phospholipids are inside the temperature range (20–80 °C) used in our experiments.
4. Buffers with low enthalpy of ionization should be selected to minimize the appearance of artificial heat signals. Buffer systems such as phosphate, citrate, and acetate that have low enthalpy of ionization are often recommended for ITC experiments [36, 37]. The HEPES buffer used in this protocol has been successfully used in our laboratory for several ITC experiments.
5. The volume of the sample cells in the DSC instrument is approximately 520 μl; however, in our laboratory 1 ml of sample is normally used in order to ensure proper loading of the solution into the sample cell (*see* Subheading 3.2.2).
6. Heating of the DSC buffer should allow for complete hydration and proper formation of the lipid particles. The temperature requirements may vary according to the specific phospholipid being used. We have found that for DPPC and DPPG (T_m ~ 41 °C), a temperature of 60 °C can be used for the formation of the MLVs.

7. A MLV suspension of DPPG (MW_t 744.95) of 0.5 mg/ml will correspond to 0.671 mM. Therefore, in order to have a peptide/lipid ratio of 1:10, the final concentration of the peptide should be 0.0671 mM. If a peptide stock solution of 2 mM is used, 33.6 μl of solution should be added to 1 ml of MLV (DPPG) suspension. In our laboratory we consider that adding the AMPs after the formation of the MLVs resembles a biologically relevant scenario. However, in some DSC studies, the authors have chosen to form the MLVs in the presence of peptides. It is important to note that the 1:10 ratio can be seen as the higher peptide ratio to be used in this type of interaction studies. Lower ratios such as 1:50 or 1:100 are recommended.
8. Air bubbles constitute a problem during the DSC experiments, as the bubbles displace liquids, changing the heat capacity of the sample. In addition, the bubbles can be dissolved in solution over time inducing an increase in heat capacity. In order to make sure the samples are properly degassed, check that the vacuum cap is adjusted over the vacuum sealing O-ring. The sound of the vacuum pump will change pitch indicating that the vacuum cap has been sealed to the O-ring. Additionally, if resistance is felt when trying to remove the cap from the tube holder, this means that the vacuum is being produced, and the cap is properly adjusted.
9. A different final pressure can indicate the presence of air bubbles. In this case remove the DSC cap and reload the sample and/or reference cells. Additionally, the final solution volume in the sample and reference cells can modify the final pressure values; ensure that no excess solution is present in either cell.
10. If the phospholipids to be used for the DSC studies are not well known, a broader range of temperature could be advantageous; ±15–20 °C from the suspected T_m should be considered for the start and final temperatures.
11. The composition of the LUVs is highly customizable, depending on the known characteristics of the biological membranes that one wants to study.
12. A larger number of extrusion cycles should generate a more homogenous sample; therefore, no less than 15 cycles are suggested. Finalizing the extrusion procedure in the opposite syringe will prevent the final sample from having large lipid aggregates that did not pass through the first filter. The average size of the resulting LUVs can be confirmed by dynamic light scattering (DLS) experiments.
13. The evaporation of the ethanol needs to be done carefully, as it can occur violently if the samples are left directly over an open flame.

14. The final concentration of the ligand (LUVs) to be used in the auto-pipette should be ~20–50 times the concentration of the binding partner in the sample cell. The actual value would depend on the expected binding affinity (K_a). Higher concentrations might be required for low affinities, while lower concentration can be used when the interaction exhibits high affinities. The parameter c, where $c = K_a \cdot [M]$, is normally used to establish the appropriate concentration of the binding partner ($[M]$). This parameter should be around 10–100 [22]. Ideally, the concentration of the ligand should allow for the system (binding partner and ligand) to reach saturation between the first third or half of the experiment.
15. Loading of the ITC sample cell can also be viewed in a JoVE publication [23].
16. In our experience there is often a small air bubble at the top of the auto-pipette; however, this does not affect the experiment results.
17. The total number of injection as well as the volume of each injection should be carefully selected. Ideally, the number and volume of the injections should allow for the ligand to reach a 1:1 ratio in the sample cells by the middle of the ITC experiments (normally at the 10–15 injection) if the reaction stoichiometry is expected to be one to one.
18. The accurate determination of the concentration of the ligand in the titration syringe is essential in order to obtain reliable thermodynamics parameters for the interaction [40]. In some cases it is recommended to measure the concentration of the ligand using the remaining ligand solution left in the vial after loading and purge/refill of the titration syringe.
19. Determination of the concentration of the ligand partner in the sample cell is important to obtain an accurate value of the reaction stoichiometry (n) [40]. However, it has been noted that during the ITC study of peptide and LUV interaction, the n value cannot be assigned to the stoichiometry of the reaction. The peptides are often used as the ligand partner (sample cell) due to solubility limitations, as formation of aggregates/oligomers has been observed at high peptide concentrations. In addition, the cost of the synthetic peptides is normally higher than the cost of the purified or synthetic phospholipids. Therefore, it is more cost-effective to have the peptides in the sample cell.
20. The *one set of sites* model is used on the assumption that all the binding sites between the peptide and the vesicles are equal (same K and ΔH).

References

1. Yeaman MR, Yount NY (2003) Mechanisms of antimicrobial peptide action and resistance. Pharmacol Rev 55:27–55
2. Shai Y (2002) Mode of action of membrane active antimicrobial peptides. Biopolymers 66:236–248
3. Nguyen LT, Haney EF, Vogel HJ (2011) The expanding scope of antimicrobial peptide structures and their modes of action. Trends Biotechnol 29:464–472
4. Jenssen H, Hamill P, Hancock REW (2006) Peptide antimicrobial agents. Clin Microbiol Rev 19:491–511
5. Epand RM, Vogel HJ (1999) Diversity of antimicrobial peptides and their mechanisms of action. Biochim Biophys Acta 1462:11–28
6. Nicolas P (2009) Multifunctional host defense peptides: intracellular-targeting antimicrobial peptides. FEBS J 276:6483–6496
7. Matsuzaki K, Sugishita K, Fujii N et al (1995) Molecular basis for membrane selectivity of an antimicrobial peptide, magainin 2. Biochemistry 34:3423–3429
8. Matsuzaki K (2009) Control of cell selectivity of antimicrobial peptides. Biochim Biophys Acta 1788:1687–1692
9. Lohner K, Sevcsik E, Pabst G (2008) Liposome-based biomembrane mimetic systems: implications for lipid-peptide interactions. In: Advances in planar lipid bilayers and liposomes. Elsevier, Amsterdam, pp 103–132
10. Jing WG, Prenner EJ, Vogel HJ et al (2005) Headgroup structure and fatty acid chain length of the acidic phospholipids modulate the interaction of membrane mimetic vesicles with the antimicrobial peptide protegrin-1. J Pept Sci 11:735–743
11. Bozelli JC, Sasahara ET, Pinto MRS et al (2012) Effect of head group and curvature on binding of the antimicrobial peptide tritrpticin to lipid membranes. Chem Phys Lipids 165:365–373
12. McElhaney RN (1982) The use of differential scanning calorimetry and differential thermal analysis in studies of model nad biological membranes. Chem Phys Lipids 30:229–259
13. Lewis RNAH, Mannock DA, Mcelhaney RN (2007) Differential scanning calorimetry in the study of lipid phase. Practical considerations. Methods Mol Biol 400:171–195
14. Mavromoustakos TM (2007) The use of differential scanning calorimetry to study drug – membrane interactions. Methods Mol Biol 400:587–600
15. Cañadas O, Casals C (2013) Differential scanning calorimetry of protein-lipid interactions. In: Kleinschmidt JH (ed) Lipid-protein interaction: methods and protocols. Humana Press, New York, NY, pp 55–71
16. Demetzos C (2008) Differential scanning calorimetry (DSC): a tool to study the thermal behavior of lipid bilayers and liposomal stability. J Liposome Res 18:159–173
17. Spink CH (2008) Differential scanning calorimetry. Methods Cell Biol 84:115–141
18. Lohner K, Prenner EJ (1999) Differential scanning calorimetry and X-ray diffraction studies of the specificity of the interaction of antimicrobial peptides with membrane-mimetic systems. Biochim Biophys Acta 1462:141–156
19. Chiu MH, Prenner EJ (2011) Differential scanning calorimetry: an invaluable tool for a detailed thermodynamic characterization of macromolecules and their interactions. J Pharm Bioallied Sci 3:39–59
20. Riske KA, Barroso RP, Vequi-suplicy CC et al (2009) Lipid bilayer pre-transition as the beginning of the melting process. Biochim Biophys Acta 1788:954–963
21. Heimburg T (2000) A model for the lipid pretransition: coupling of ripple formation with the chain-melting transition. Biophys J 78:1154–1165
22. Freyer MW, Lewis EA (2008) Isothermal titration calorimetry: experimental design, data analysis, and probing macromolecule/ligand binding and kinetic interactions. Methods Cell Biol 84:79–113
23. Duff Jr MR, Grubbs J, Howell EE (2011) Isothermal titration calorimetry for measuring macromolecule-ligand affinity. J Vis Exp (55): 2796
24. Lewis EA, Murphy KP (2005) Isothermal titration calorimetry. In: Nienhaus GU (ed) Protein-ligands interactions methods and applications. Humana Press, Totowa, NJ, pp 1–15
25. Bagheri M (2013) Synthesis and thermodynamic characterization of small cyclic antimicrobial arginine and tryptophan-rich peptides with selectivity for Gram-negative bacteria. In: Giuliani A, Rinaldi AC (eds) Antimicrobial peptides methods and protocols. Humana Press, New York, NY, pp 87–109
26. Seelig J (2004) Thermodynamics of lipid – peptide interactions. Biochim Biophys Acta 1666:40–50
27. Wieprecht T, Seelig J (2002) Peptide-lipid interactions. Curr Top Membr 52:31–56

28. Wenk MR, Seelig J (1998) Magainin 2 amide interaction with lipid membranes: calorimetric detection of peptide binding and pore formation. Biochemistry 37:3909–3916
29. Henriksen JR, Andresen TL (2011) Thermodynamic profiling of peptide membrane interactions by isothermal titration calorimetry: a search for pores and micelles. Biophys J 101:100–109
30. Klocek G, Schulthess T, Shai Y et al (2009) Thermodynamics of melittin binding to lipid bilayers. Aggregation and pore formation. Biochemistry 48:2586–2596
31. Wieprecht T, Apostolov O, Beyermann M et al (1999) Thermodynamics of the alpha-helix-coil transition of amphipathic peptides in a membrane environment: implications for the peptide-membrane binding equilibrium. J Mol Biol 294:785–794
32. Artimo P, Jonnalagedda M, Arnold K et al (2012) ExPASy: SIB bioinformatics resource portal. Nucleic Acids Res 40:W597–W603
33. Arias M, Jensen KV, Nguyen LT et al (2014) Hydroxy-tryptophan containing derivatives of tritrpticin: modification of antimicrobial activity and membrane interactions. Biochim Biophys Acta 1848:277–288
34. Ames BN, Neufeld EF, Ginsberg V (1966) Assay of inorganic phosphate, total phosphate and phosphatases. Methods Enzymol 8: 115–118
35. Jing W, Hunter HN, Hagel J et al (2003) The structure of the antimicrobial peptide micelles Ac-RRWWRF-NH2 bound to micelles and its interactions with phospholipid bilayers. J Pept Res 61:219–229
36. Fukada H, Takahashi K (1998) Enthalpy and heat capacity changes for the proton dissociation of various buffer components in 0.1 M potassium chloride. Proteins Struct Funct Genet 33:159–166
37. Goldberg RN (2002) Thermodynamic quantities for the ionization reactions of buffers. J Phys Chem Ref Data 31:231
38. Phillips GB, Dodge JT (1967) Composition of phospholipids and of phospholipid fatty acids in human red cells. J Lipid Res 8:667–675
39. Marr AG, Ingraham JL (1962) Effect of temperature on the composition of fatty acids in *Escherichia coli*. J Bacteriol 84:1260–1267
40. Tellinghuisen J, Chodera JD (2011) Systematic errors in isothermal titration calorimetry: concentrations and baselines. Anal Biochem 414:297–299

Chapter 10

Fluorescence and Absorbance Spectroscopy Methods to Study Membrane Perturbations by Antimicrobial Host Defense Peptides

Mauricio Arias and Hans J. Vogel

Abstract

Antimicrobial peptides (AMPs) are currently intensely studied because of their potential as new bactericidal and bacteriostatic agents. The mechanism of action of numerous AMPs involves the permeabilization of bacterial membranes. Several methods have been developed to study peptide–membrane interactions; in particular optical spectroscopy methods are widely used. The intrinsic fluorescence properties of the Trp indole ring in Trp-containing AMPs can be exploited by measuring the fluorescence blue shift and acrylamide-induced fluorescence quenching. One important aspect of such studies is the use of distinct models of the bacterial membrane, in most cases large unilamellar vesicles (LUVs) with different, yet well-defined, phospholipid compositions. Deploying LUVs that are preloaded with fluorescent dyes, such as calcein, also allows for the study of vesicle permeabilization by AMPs. In addition, experiments using genetically engineered live *Escherichia coli* cells can be used to distinguish between the effects of AMPs on the outer and inner membranes of gram-negative bacteria. In combination, these methods can provide a detailed insight into the mode of action of AMPs.

Key words Fluorescence spectroscopy, Absorbance spectroscopy, Antimicrobial peptides, Tryptophan, *Escherichia coli*, Membrane permeabilization, Large unilamellar vesicles

1 Introduction

The mechanism of action of most antimicrobial peptides (AMPs) can be divided into two main steps. First, the peptides bind to the bacterial membrane through electrostatic and hydrophobic attractions. The cationic nature of most AMPs, in conjunction with the characteristic negatively charged membrane surfaces of bacterial cells, provides the basis for initial electrostatic attractions [1–4]. The presence of hydrophobic residues (~50%) in the amino acid sequence of most AMPs and their contribution to forming an amphipathic peptide structure [3], together with the highly hydrophobic nature of the core of the bacterial membrane, drive the subsequent formation of hydrophobic interactions. Once bound

Paul R. Hansen (ed.), *Antimicrobial Peptides: Methods and Protocols*, Methods in Molecular Biology, vol. 1548,
DOI 10.1007/978-1-4939-6737-7_10, © Springer Science+Business Media LLC 2017

to the membrane, a second step usually takes place. The accumulation of the membrane-bound peptides disturbs the organization of the lipid bilayer [3, 5], thereby leading to an increase in membrane permeability and simultaneously disrupting the vital electrochemical gradients across the bacterial membrane [4, 6]. Several mechanisms have been proposed for the bacterial membrane permeabilization process [1, 5]; however, a complete molecular level understanding of this process has remained elusive.

Given the important role that peptide–membrane interactions play in the mechanism of action of AMPs, considerable attention has been focused on studying this process. A powerful approach to study these interactions is the use of phospholipid vesicles/liposomes as simplified model membranes. These defined systems allow us to study the effects of different lipid headgroups, fatty acid chain length, and fatty acid (un)saturation on the interactions of membranes with AMPs. Large unilamellar vesicles (LUVs), normally 100 nm in diameter, with different lipid compositions are among the most commonly used membrane mimetics for studies with AMPs [7]. Phospholipids with the negatively charged phosphatidylglycerol (PG) headgroups and the neutrally charged phosphatidylethanolamine (PE) headgroups are found in high concentration in the membranes of prokaryotic cells and are frequently used in LUVs representing bacterial membranes [7]. In contrast, in the outer leaflet of the membranes of eukaryotic cells, phospholipids with neutrally charged phosphatidylcholine (PC) headgroups and cholesterol are normally observed, and these constitute the main components of LUVs simulating this type of cells [7]. Additionally, phospholipids with the negatively charged phosphatidylserine (PS) headgroups can be used as one of the components of LUVs mimicking the outer leaflet of cancer cell membranes. Several cancer cell lines have been shown to lose their cytoplasmic membrane leaflet asymmetry, thereby exposing PS to the extracellular environment [8].

Several biophysical techniques (i.e., isothermal titration calorimetry (ITC), dynamic light scattering (DSC), circular dichroism (CD), etc.) can be used for the study of peptide–LUV interactions [9–14]. However, fluorescence spectroscopy is particularly useful due to its high sensitivity. When Trp residues are present in the peptide, the intrinsic fluorescence of the indole side chains can be followed to directly monitor the binding of the peptides to the membrane. In aqueous solution Trp exhibits an absorption maximum at 280 nm and a maximal emission wavelength around 350 nm [15]. However, upon interacting with a lipid bilayer, the Trp residue(s) experiences a change in the polarity of the environment, from a hydrophilic (polar) to a more hydrophobic (apolar) surrounding. This is accompanied by a reduction in the maximal emission wavelength, a phenomenon known as blue shift [15].

Another complementary method that is widely used to study the interaction of Trp-containing AMPs and model membranes is the quenching of the Trp fluorescence by specific small soluble molecules (quenchers) that can drastically reduce the fluorescence intensity of Trp. The accessibility of the quenchers to the Trp residues in a peptide determines the amount of fluorescence quenching observed. Different mechanisms, such as collisional (dynamic) quenching, static quenching, quenching by energy transfer, or charge transfer reactions, can all give rise to quenching effects [16]. However, in this protocol only collisional quenching will be considered and discussed. In collisional quenching direct molecular contact between the quencher and the fluorophore (in this case Trp) is required. During this contact the fluorophore returns from the excited to the ground state without the emission of a photon. Several substances can induce collisional quenching, among the most widely used are molecular oxygen (O_2), heavy atoms such as iodide (I^-) and cesium (Cs^+), and acrylamide [16, 17]. For studies of peptide–membrane interactions, charged quenchers such as I^- and Cs^+ should be used with extreme caution, as electrostatic attraction/repulsion can influence the accessibility of these reagents [17, 18]. Consequently, the neutral quencher acrylamide is frequently used. In aqueous solution, the Trp residues in an AMP will be fully exposed to the solvent, and addition of acrylamide would result in a large quenching of the Trp fluorescence. In contrast, when bound to model membranes, the Trp sidechains are likely less exposed to the aqueous environment, thereby reducing the accessibility for the acrylamide and giving rise to less efficient quenching. In this protocol utilization of these intrinsic Trp fluorescence methods will be highlighted to study the binding of AMPs to LUVs.

Fluorescence spectroscopy in combination with LUVs can also be used to assess the ability of AMPs to permeabilize the lipid bilayer. Leakage of fluorescence dyes from preloaded LUVs induced by AMPs indicates that membrane permeabilization might be part of the mechanism of action of a peptide. Several fluorescent molecules (8-aminonaphthalene-1,3,6-trisulfonic acid, disodium salt (ANTS)/p-Xylene-bis-pyridinium bromide (DPX), calcein, carboxyfluorescein, FITC-labeled dextrans, among others) can be used as probes in these experiments [19–23]. In this protocol we will only discuss the use of calcein, as it is the most simple and widely used method. LUVs can be preloaded with high concentrations of calcein (e.g., 70 mM), in which case the fluorescence of the calcein is self-quenched. Upon permeabilization of the model membrane, the calcein is released into the extra-vesicular media, where it is diluted and a large increase in the fluorescence is observed. The flexibility of preparing LUVs with different compositions in combination with the ability to mutate specific residues in the peptide sequence allows one to pinpoint the key elements contributing to membrane permeabilization.

Of course, model membranes cannot account for the possible roles that transmembrane proteins and other bacterial membrane components play during membrane permeabilization in living cells by AMPs. Consequently, the results obtained from experiments with LUVs should always be interpreted with care. Fortunately, some complementary methods have been developed to study the effects of AMPs on real bacterial membranes [24, 25]. Such studies were originally developed by Lehrer et al. [24], and some minor modifications were introduced in recent years [26, 27]. In this approach a genetically modified strain of *E. coli* (*E. coli ML35p*) is used in combination with easily hydrolysable enzyme substrates that function as permeabilization probes (Fig. 1). The *E. coli ML35p* strain constitutively expresses the cytoplasmic enzyme β-galactosidase, as well as the plasmid-encoded β-lactamase protein, which is located in the periplasmic space. In addition, this *E. coli* strain lacks the lactose permease, making the membrane impermeable to β-galactosides [24, 28]. Ortho-nitrophenyl-β-galactoside (ONPG) is then used as a membrane impermeable substrate for the cytoplasmic β-galactosidase. Only after permeabilization of the inner membrane by an AMP can the enzyme hydrolyze the ONPG. Similarly, nitrocefin (NCF) is a specific substrate for the periplasmic enzyme β-lactamase, and this enzyme can only act after NCF passes through the outer membrane. The hydrolysis of these two substrates can be followed by absorbance spectroscopy, by following the increase in absorbance at 420 nm and 490 nm for ONPG and NCF, respectively [26]. These methods provide a powerful approach for the study of the permeabilization induced by AMPs in a representative gram-negative bacterium, where the outer and inner membranes can be affected differently.

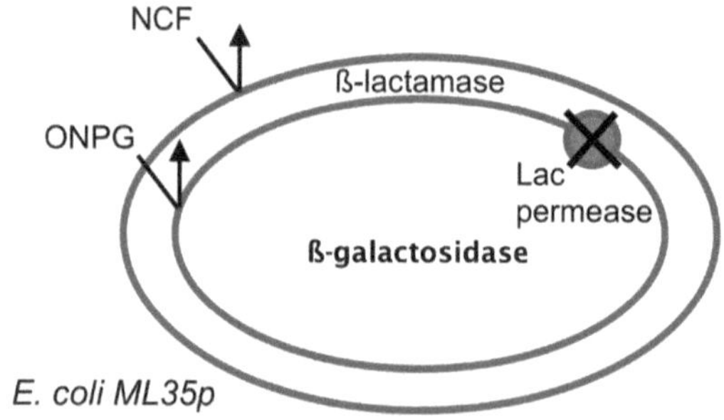

Fig. 1 Schematic representation of the bacteria *E. coli ML35p* strain. The enzymes β-galactosidase and β-lactamase are expressed by the bacteria but have different locations inside the cells. β-galactosidase is located in the cytoplasm, while β-lactamase is found in the periplasmic space. The substrates for these enzymes ONPG (for β-galactosidase) and NCF (for β-lactamase) are impermeable to the inner and outer membranes

2 Materials

2.1 Peptides

1. Peptides are synthesized as described in Chapter 3 or 4 or obtained as a lyophilized powder from commercial suppliers. The purity (>95 %) and molecular weight are confirmed by HPLC and mass spectrometry (*see* **Note 1**).
2. Peptide stock solution: ~1 mg of peptide in ultrapure (Milli-Q) water and stored at −20 °C.
3. 0.1 mM peptide stock: 0.1 mM peptide in leakage buffer (*see* Subheading 2.2, **item 2**).
4. 0.01 mM peptide stock: 0.01 mM peptide in leakage buffer.

2.2 Large Unilamellar Vesicles (LUVs)

1. Phospholipids: Chicken egg-derived l-α-phosphatidylcholine (ePC) and chicken egg-derived l-α-phosphatidylglycerol (ePG) in chloroform.
2. Leakage buffer: 10 mM Tris–HCl pH 7.4, 150 mM NaCl, and 1 mM EDTA. For calcein leakage assays, the leakage buffer is supplemented with 70 mM calcein (*see* **Note 2**).
3. Avanti® Mini-Extruder and gastight syringes (short needle Hamilton syringe 1 ml).
4. Filter supports.
5. Polycarbonate membrane filters 0.1 μm pore sizes.
6. Freeze–thawing setup: Liquid nitrogen and water bath (35–40 °C).
7. Gel filtration chromatography column media: Sephadex G-50 resin hydrated in water.

2.3 Phosphate Assay, Calcein Leakage, and Tryptophan Fluorescence Quenching

1. Reagent A: 10 % $Mg(NO_3)_2$ in ethanol.
2. Reagent B: mix one part (i.e., 1.5 ml) of 10 % ascorbic acid with six parts (e.g., 9 ml) of 0.42 % ammonium molybdate tetrahydrate $(NH_4)_6Mo_7O_{24} \cdot 4H_2O$ in 1 N H_2SO_4.
3. HCl solution: aqueous solution of 0.5 N HCl.
4. Fluorescence quartz cuvette (1.5 ml for magnetic stirrers) and magnetic stirring bar for cuvette.
5. Triton X-100: aqueous solution of 10 % (v/v) of the detergent Triton X-100.
6. Acrylamide stock solution: 4 M acrylamide in leakage buffer.

2.4 Outer and Inner Membrane Permeabilization in E. coli ML35p

1. Autoclaved LB Lennox broth media prepared according to the manufacturer's instructions.
2. The bacterial strain *E. coli ML35p* is kept at −80 °C as a glycerol stock. Before each experiment an LB–agar plate with 100 μg/ml ampicillin is streaked with the bacteria using a sterile

inoculation loop. The agar plate is incubated upside down overnight at 37 °C. After incubation individual colonies should be observed.

3. Bacterial assay buffer: 10 mM Na^+–phosphate pH 7.4, 100 mM NaCl, and 300 μg/ml LB.
4. Ortho-nitrophenyl-β-galactoside (ONPG) 2 mM solution: ONPG (~10 mg) is dissolved in leakage buffer, and the solution has to be protected from light, as ONPG in solution is light sensitive. The solution is prepared the same day of the experiments and is stored at 4 °C until use.
5. Nitrocefin (NCF) 120 μM solution: NCF (~1 mg) is first dissolved in DMSO (1.0 ml) and then diluted in assay buffer to prepare a stock solution of 120 μM. This solution is protected from direct light, and it is prepared the same day of the experiments. Store at room temperature until use.

3 Methods

3.1 Determination of Peptide Concentration

1. The peptide concentration is established by measuring the UV absorbance for Trp at 280 nm and using the extinction coefficient provided by the software tool ProtParam from ExPASy: SIB bioinformatics resource portal [29].

3.2 Large Unilamellar Vesicles (LUVs)

This procedure describes the preparation of LUVs for either calcein leakage or acrylamide quenching experiments. The preparation method for LUVs is based on the manual for liposome preparation [30], with some minor modifications.

3.2.1 Multilamellar Vesicle (MLV) Preparation

1. Combine the appropriate amounts of the desired phospholipid stocks (in chloroform) in a clean glass vial, in order to reach the desired molar ratio and a combined mass of ~1.0 mg (*see* **Note 3**).
2. Evaporate the solvent (chloroform) inside a fume hood under a soft stream of nitrogen gas, rotating the vial in order to create a thin lipid film over the glass walls.
3. Remove the remaining chloroform by leaving the vial under vacuum overnight (*see* **Note 4**).
4. Hydrate the lipid films with 500 μl of the leakage buffer (for *acrylamide quenching*) or leakage buffer supplemented with calcein (for *calcein leakage*). In order to get the lipids in solution, vigorous vortexing is normally required. At this point the solution is composed of multilamellar vesicles (MLVs).
5. Freeze-thaw the suspended MLVs by submerging the glass flask in liquid nitrogen and then thawing the suspension in a warm water bath (35–40 °C). Repeat this procedure at least five times.

3.2.2 LUV Extrusion

1. Assemble the lipid extruder according to the manufacturer's instructions. The size of the polycarbonate filter membrane determines the average size of the resulting LUVs. In our laboratory, LUVs with a 100 nm diameter are normally used (filter pore size 0.1 μm).
2. Load ~1 ml of leakage buffer into a Hamilton syringe, and pass the solution through the extruder, in order to check for any leakage and to test the proper attachment of the second syringe. Discard all liquids from the second syringe once all buffer passed through the device.
3. Load the MLV suspension into one of the Hamilton syringes, and pass the suspension through the extruder device. The suspension will become more translucent indicating the formation of LUVs. Once the second syringe is loaded, return the suspension through the extruder to the initial syringe. Repeat this procedure 15 times. Make sure that the final extrusion ends when the suspension of the LUVs is in the opposite syringe than the one that you started with. At this point the suspension is composed of LUVs with an average diameter of 100 nm (*see* **Note 5**). The average size of the resulting LUVs can be confirmed by dynamic light scattering [31].

3.2.3 Removal of Free Calcein

In order to run the calcein leakage experiments, the free calcein that is present in the LUV suspension has to be removed by gel filtration chromatography.

1. Prepare a Sephadex G-50 column by decanting hydrated Sephadex resin into a 1 cm diameter glass column. The resin is left to settle, forming a column with a height of ~10 cm. The column is then equilibrated with at least two column volumes of leakage buffer.
2. Load the calcein-loaded LUV suspension into the equilibrated column. Run the suspension through the column with leakage buffer while observing the formation of two bands. The first band, which contains the LUVs, will be light yellow (slightly turbid), while the second band, containing the free calcein, will have a strong orange color. Start collecting fractions (0.5–1.0 ml) once the first band is about to come out of the column.
3. Measure the turbidity of the samples at OD_{450}, and select and mix the fractions with higher turbidity. Proceed to measure the lipid concentration.

3.2.4 Lipid Concentration Determination

This procedure is a modified version of the phosphorus assays described by Ames et al. [32]. The lipid concentration is estimated for LUV suspensions for calcein leakage and acrylamide quenching experiments.

1. Pipette 5 μl of the LUV suspension into a phosphate-free Pyrex test tube. Additionally, pipette 5 μl of the leakage buffer into an additional test tube as control. These samples should be prepared in triplicate.
2. Add 30 μl of Reagent A to each tube, and carefully ash the samples over a hot flame (*see* **Note 6**). Once all ethanol is evaporated, leave the tubes over the flame until there is no more smoke coming from the tubes and a gray precipitate (ashes) is observed. Let the tubes cool down.
3. Add 300 μl of 0.5 N HCl solution into each tube. Vortex the tubes until the gray precipitate is resuspended. During this step the $MgNO_3$ and HCl destroy the phospholipids releasing the PO_4^{3-} groups.
4. Cap the tubes with aluminum foil and place them in boiling water for 15 min. Then remove the tubes from the water bath and allow cooling.
5. Add 700 μl of Reagent B to each tube, cap them, and incubate them for 1 h at 37 °C. During this step the samples acquire a light blue color, indicating the presence of phosphorus.
6. Measure the OD_{820} for each tube and calculate the concentration of phosphate (*see* **Note 7**).
7. Dilute the regular and calcein-loaded LUV suspensions in leakage buffer in order to have a 2 mM (regular LUVs) and 1 mM (calcein-loaded LUVs) stock solution.

3.3 Tryptophan Fluorescence Blue Shift

1. Set up the fluorimeter to scan the fluorescence intensity over the emission wavelength range 300–500 nm (*scan function* in a Cary Eclipse fluorimeter), with excitation wavelength at $\lambda_{exc} = 280$ nm (slit width 5 nm) (*see* **Note 8**). The experiments can be run at room temperature or at 37 °C, with constant stirring. Additionally, the final scan is setup to be the average of ten consecutive scans.
2. Add 940 μl of leakage buffer to the fluorescence quartz cuvette with a magnetic stirring bar, and record the first fluorescence scan. This scan corresponds to the background fluorescence.
3. Add 10 μl of the peptide stock (0.1 mM) in order to have ~1 μM peptide in solution and record the second fluorescence scan. Record the emission wavelength for the maximum fluorescence intensity (Fig. 2).
4. Add 50 μl of LUV suspension (2 mM) in order to have 100 μM lipids in the sample, which corresponds to a peptide-to-lipid molar ratio of 1:100, and record the third fluorescence scan (*see* **Note 9**). Register the emission wavelength for the maximum fluorescence intensity (Fig. 2).

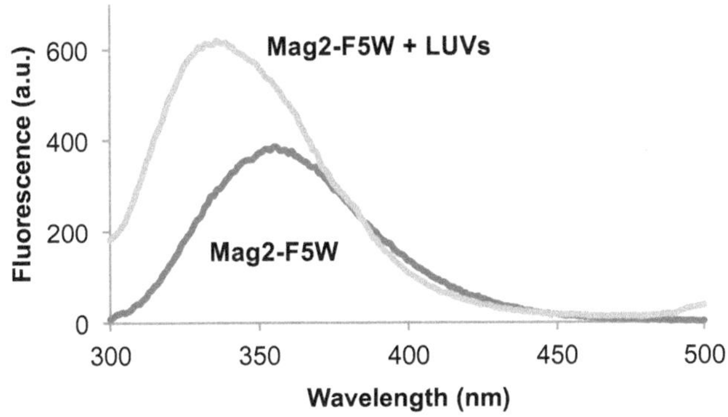

Fig. 2 Trp fluorescence blue shift induced by binding of the antimicrobial peptide magainin2-F5W (Mag2-F5W) with negatively charged LUVs (ePC/ePG ratio 1:1). The peptide (1 μM) was initially incubated in leakage buffer and the fluorescence spectrum was recorded. Later LUVs (100 μM) were added to the peptide solution and a new fluorescence spectrum was recorded. The fluorescence spectra were recorded at λ_{exc} 280 nm (slit width 5 nm) and λ_{em} 300–500 nm (slit width 5 nm). The leakage buffer fluorescence spectrum (background) was subtracted from each spectrum

5. The blue shift ($\Delta\lambda$) corresponds to the difference between the wavelength maxima recorded for the peptide in leakage buffer and the peptide in the presence of LUVs (P/L ratio 1:100).
6. Figure 2 presents typical results obtained for the Trp fluorescence blue shift of AMPs in the absence and presence of negatively charged vesicles. For this experiment the peptide magainin2-F5W (Mag2-F5W) and ePC/ePG LUVs were used.

3.4 Tryptophan Fluorescence Quenching

To measure the change in the accessibility of the Trp sidechain when the peptides bind to the LUVs, two independent fluorescence quenching experiments need to be performed. The first experiment measures the accessibility (K_{sv}) for the peptide in leakage buffer (*peptide only*) (Fig. 3a), while the second experiment will register the K_{sv} for the peptide in the presence of LUVs (*peptide+LUVs*) (Fig. 3b).

1. Set up the fluorimeter to scan the fluorescence intensity over the emission wavelength range 300–500 nm (*scan function* in a Cary Eclipse fluorimeter), with excitation wavelength at λ_{exc} = 295 nm (slit width 5 nm) (*see* **Note 10**). The experiments can be run at room temperature or at 37 °C, with constant stirring.
2. For *peptide only*: Add 990 μl of leakage buffer to the fluorescence quartz cuvette containing a magnetic stirring bar, and record the first fluorescence scan.

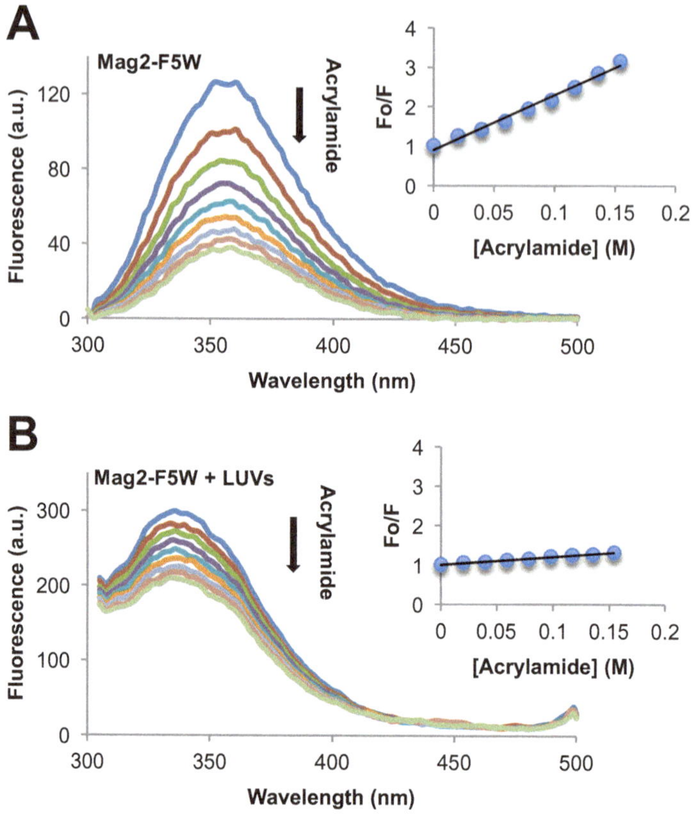

Fig. 3 Quenching of the Trp fluorescence as induced by adding increasing concentrations of acrylamide for the peptide magainin2-F5W (Mag2-F5W) in the absence or presence of negatively charged LUVs (ePC/ePG). The peptide (1 μM) (**a**) or the peptide (1 μM) in the presence of LUVs (100 μM) (**b**) was incubated in leakage buffer, and the fluorescence spectrum was registered (*top-blue curves*). Then eight sequential additions of 5 μl acrylamide (4 M) were made and the fluorescence spectrum was collected after each addition. The fluorescence spectra were recorded at λ_{exc} 295 nm (slit width 5 nm) and λ_{em} 300–500 nm (slit width 5 nm). The fluorescence spectrum of the leakage buffer (background) was subtracted from each spectrum

For *peptide+ LUVs*: Add 940 μl of leakage buffer to the fluorescence quartz cuvette containing a magnetic stirring bar, and record the first fluorescence scan.

3. For *peptide only*: Add 10 μl of the peptide stock (0.1 mM) in order to have ~1 μM peptide in solution and record the second fluorescence scan. Register the maximum emission wavelength and the maximum fluorescence intensity.

   For *peptide+ LUVs*: Add 10 μl of peptide stock (0.1 mM) and 50 μl of LUV stock (2 mM). Record the second fluorescence scan. Register the maximum emission wavelength and its maximum fluorescence intensity (*F*).

4. For both experiments: Add 5 μl acrylamide solution (4 M) and record the next fluorescence scan. Determine the fluorescence intensity for the maximum emission wavelength found in **step 3**.
5. Continue sequentially adding 5 μl of the acrylamide stock and recording the fluorescence spectra (*see* **step 4**), until 40 μl acrylamide has been added to the sample, corresponding to ~0.15 M.
6. The Stern–Volmer quenching constant (K_{sv}) is calculated with the Eq. 1:

$$\frac{F_o}{F} = 1 + K_{sv}[Q] \quad (1)$$

where F_o and I represent the fluorescence intensities without quencher (Q) and with each concentration of the quencher $[Q]$ (*see* **Notes 11–13**).
7. Figure 3 presents typical results obtained for the Trp fluorescence quenching induced by acrylamide in AMPs in the absence and presence of negatively charged vesicles. In these experiments the peptide magainin2-F5W (Mag2-F5W) and ePC/ePG LUVs were used.

3.5 Calcein Leakage

1. Set up the fluorimeter to measure the fluorescence intensity (*kinetics* function in a Cary Eclipse fluorimeter), with excitation and emission wavelengths at $\lambda_{exc} = 490$ nm and $\lambda_{em} = 520$ nm, respectively. The excitation and emission slit widths are normally set at 5 nm. The experiments are run at room temperature or 37 °C, with constant stirring.
2. Add 1 ml of leakage buffer to the fluorescence quartz cuvette containing a magnetic stirring bar, and start recording the fluorescence intensity.
3. After 1 min, pause the fluorescence recording and add 10 μl of the calcein-loaded LUV suspension (1 mM) in order to have a final lipid concentration of 10 μM. Continue recording for 1 min. The background level of fluorescence after 1 min will represent the 0% fluorescence (Fig. 4).
4. Pause the fluorescence reading and add the required amount of the peptide stock solution (0.1 mM) in order to obtain a peptide-to-lipid (P/L) molar ratio of 1:10–1:200 (*see* **Note 14**). Continue recording the fluorescence for 10 min or until the fluorescence signal is stable. If the peptide induces membrane permeabilization, the fluorescence intensity would increase during this time. The fluorescence intensity after this 10 min will represent the leakage induced by the peptide (Fig. 4).
5. Pause the fluorescence reading and add Triton X-100 (10 μl of 0.1% stock) in order to lyse all remaining LUVs. Continue

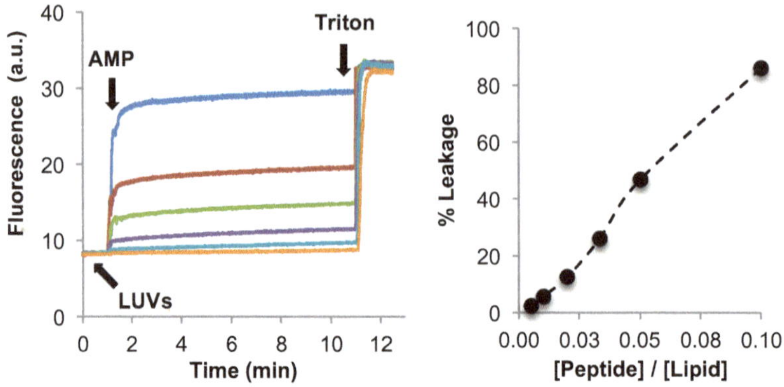

Fig. 4 Calcein leakage from negatively charged LUVs (ePC/ePG) as induced by the antimicrobial peptide Tritrpticin-1 (*left*). LUVs (100 μM) were incubated at different peptide-to-lipid ratios, and the fluorescence of the released calcein was registered at λ_{exc} 490 nm (slit 5 nm) and λ_{em} 520 nm (slit 5 nm). The detergent Triton X-100 was used to induce 100 % leakage (*left*). The percentage of leakage (*right*) was calculated as indicated in Subheading 3.5, **step 6**

recording for 1 min. The final value of the fluorescence intensity corresponds to 100 % fluorescence (Fig. 4).

6. Calculate the percentage of calcein leakage according to the following Eq. 2:

$$\%\text{Leakage} = \frac{F_{\text{Peptide}} - F_o}{F_{\text{Triton}} - F_o} \quad (2)$$

where F_o, F_{Peptide}, and F_{Triton} represent the fluorescence intensities after the addition of LUVs, peptide, and Triton X-100, respectively.

7. Figure 4 presents typical results obtained for calcein leakage from vesicles with negatively charged phospholipid headgroups induced by a membrano-lytic AMP. In this experiment the peptide Tritrpticin-1 and ePC/ePG LUVs were used.

3.6 Outer and Inner Membrane Permeabilization of *E. coli* ML 35p

These procedures are based on the work of Lehrer et al. [24, 33] and Epand et al. [26]. The experiment is set up in a 96-well plate allowing for the simultaneous analysis of several twofold dilutions of up to 12 different peptides.

3.6.1 Peptides

1. Stock solutions are prepared in bacterial assay buffer at 16× the maximum concentration required. In our laboratory the maximum concentration used is normally 2× the MIC of the peptide. Determination of MIC is described in Chapter 22 Subheading 3.2.

3.6.2 Bacterial Cells and Plate Setup

1. Prepare two *E. coli ML35p* cultures by incubating single bacterial colonies in two culture tubes containing 5 ml of LB broth

media supplemented with ampicillin (100 μg/ml). Grow the cells at 250 rpm and 37 °C until $OD_{600} \geq 0.6$.

2. Collect the bacteria by centrifugation at 3220×*g*, 6 min, and resuspend the bacterial pellet in 2–5 ml of the bacterial assay buffer. Repeat this washing step two more times.
3. Measure the OD_{600} and dilute the final bacterial suspension, with bacterial assay buffer, in order to reach $OD_{600} = 0.6$. If OD_{600} is less than 0.6, the bacterial suspension can be spun down one more time and resuspended in a lower volume. In order to set up a 96-well plate, 5 ml of the final bacterial suspension is required. This suspension should be used in the next 20 min in order to maintain the OD_{600}.
4. Set up the 96-well plate with 25 μl of the peptide dilutions in each well and 25 μl of assay buffer as control for each peptide. The peptides can be set up in several ways, but we have found that the following procedure works well for twofold dilutions of up to 12 different peptides per plate: add 25 μl of assay buffer to all wells, then add 25 μl of peptide stock solution (16×) to the first row (row A), and mix up and down by pipette. Take 25 μl from the first row and mix into the second row (row B). Continue the twofold dilutions until the penultimate row (row G) and discard 25 μl from row G. At this point the plate will contain 25 μl of twofold diluted peptides at 8× concentration.
5. Add respective enzyme substrate:

   For *inner membrane permeabilization*, add 25 μl of ONPG (2.0 mM).

   For *outer membrane permeabilization*, add 25 μl of NCF (120 μM).
6. Add 50 μl *E. coli ML35p* suspension ($OD_{600} = 0.6$) in bacterial assay buffer. Make sure that there are no bubbles in any well, and transfer the 96-well plate to the plate reader.
7. Set up the plate reader for absorbance readings every 2 min for 1 h at 37 °C, and shake the sample for 10 s between readings. Set absorbance wavelength at 420 nm or 490 nm for *inner* and *outer membrane permeabilization*, respectively (Fig. 5).
8. Figure 5 presents typical results obtained for permeabilization of the inner and outer membrane of *E. coli* ML35p induced by the membrano-lytic peptide Tritrpticin-1.

4 Notes

1. Commercially synthesized peptides are normally cleaved and purified in the presence of trifluoroacetic acid (TFA). Therefore many peptides contain TFA (CF_3COOH) as a counterion. In some instances replacement of TFA is necessary, as it can interfere

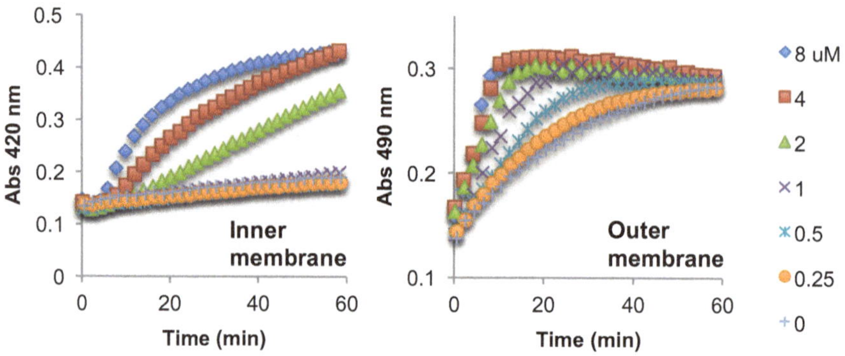

Fig. 5 Permeabilization of the inner and outer membranes of *E. coli ML35p* as induced by addition of the antimicrobial peptide Tritrpticin-1. The bacteria (OD_{600} 0.3) were incubated in the twofold dilutions of the peptide (0–8 μM). The permeabilization of the inner (*left*) and outer (*right*) membrane was registered by following the absorbance at 420 nm and 490 nm, respectively. The MIC of Tritrpticin-1 against *E. coli* was considered 4 μM

with some biophysical and biological assays. Several techniques for TFA removal have been developed; however, one of the most efficient methods involves the lyophilization of the peptides in the presence of HCl [34].

2. During preparation of the calcein solution, the pH rapidly drops, limiting the further dissolution of the calcein. Saturated aqueous solution of NaOH can be added drop wise until the calcein is dissolved. If the final pH is higher than 7.4, concentrated HCl can be used to compensate.
3. In our laboratory, initial assays are normally performed with an ePC/ePG or ePE/ePG ratio of 1:1, intended to simulate the negatively charged membranes of bacteria. A common mix would contain ePC 25 μl (20 mg/ml chloroform stock) and ePG 50 μl (10 mg/ml chloroform stock).
4. In case the LUVs would be prepared the same day, at least 4 h is required in order to remove all solvent.
5. According to the manufacturer, a larger number of extrusion cycles would generate a more homogenous sample; therefore, no less than five cycles are suggested. Finalizing the extrusion procedure in the opposite syringe will prevent the final sample from having large lipid aggregates that did not pass through the filter.
6. The evaporation of the ethanol needs to be done very carefully, as the evaporation can occur violently if the samples are placed directly over the open flame.
7. 0.01 μmol of organic phosphate will correspond to $OD_{820} = 0.240$ [32]. Divide the number of moles by 5 μl in order to establish the final phosphate concentration in the samples. Every phospholipid molecule in the LUVs contains one phosphate group; therefore, the lipid concentration will

correspond to the phosphate concentration. Alternatively, the phosphorus concentration can be calculated by preparation of a calibration curve using a commercially available phosphorus standard solutions [14] or aqueous solutions of monobasic sodium phosphate [25].

8. In the case of peptides containing Tyr in addition to Trp residues, the λ_{exc} for the fluorescence experiment should be set to 295 nm (rather than 280 nm). This would selectively excite the Trp and not the Tyr residues.
9. Sequential additions of lower peptide-to-lipid ratios can be performed to register the sequential change in fluorescence intensity and the maximum wavelength.
10. $\lambda_{exc} = 295$ nm is selected instead of 280 nm in order to reduce the absorbance by acrylamide ($\varepsilon^{280} = 4.3$ M^{-1} cm^{-1} and $\varepsilon^{295} = 0.24$ M^{-1} cm^{-1}) [35].
11. The quenching data is normally presented as Stern–Volmer plots, where I_o/I is plotted versus $[Q]$. It is normally expected that these plots are linear, with the line intercepting the y-axis at one and the slope representing the K_{sv}. However, linear Stern–Volmer plots can only be obtained if collisional quenching is solely responsible for the fluorescence quenching, in addition to having only a single class of fluorophores with the same accessibility to the quencher [16]. The presence of static quenching and/or two different classes of fluorophores, where one class is not accessible to the quencher, will give rise to deviations of the Stern–Volmer plots from linearity. The presence of both static and collisional quenching would induce an upward curvature, while the presence of a second class of fluorophores would deviate the curve toward the x-axis [16].
12. An upward deviation from linearity has been observed for the acrylamide-induced fluorescence quenching of NATA (C- and N-terminal blocked l-Trp analog), indole, or the Trp residue of the protein monellin [16, 17]. This deviation was interpreted as an apparent static quenching, which was not related with the formation of a ground-state complex. Instead, this “static” quenching contribution is the result of the quencher being located close (inside the active volume) to the fluorophore at the moment of excitation, resulting in an instantaneous loss of the fluorophore’s excited state [17]. In this case the modified Stern–Volmer considers the contribution of the static quenching in the form of the term exp($V[Q]$) [17]:

$$\frac{I_o}{I} = \left(1 + K_{sv}\left[Q\right]\right)e^{V[Q]}$$

where V represents the static quenching constant. However, in most of our studies with AMPs and LUVs, the Stern–Volmer plots are very close to linearity and the original Stern–Volmer equation was used.

13. When plotting I_o/I versus $[Q]$, the dilution effect should be considered for fluorescence intensity (I) values after each quencher addition.
14. In our laboratory the peptide-to-lipid ratios normally assessed are 1:10, 1:20, 1:30, 1:50, 1:100, and 1:200, which correspond to 10 μl, 5 μl, 3.3 μl, and 2 μl of 0.1 mM or 0.01 mM peptide stocks.

Acknowledgments

Research on antimicrobial peptides has been funded by a grant from the "Novel alternatives to antibiotics" program from the Canadian Institutes of Health Research.

E. coli ML35p was kindly provided by Dr. Robert Lehrer from the David Geffen School of Medicine at UCLA, Los Angeles, CA.

References

1. Nguyen LT, Haney EF, Vogel HJ (2011) The expanding scope of antimicrobial peptide structures and their modes of action. Trends Biotechnol 29:464–472
2. Matsuzaki K (2009) Control of cell selectivity of antimicrobial peptides. Biochim Biophys Acta 1788:1687–1692
3. Yeaman MR, Yount NY (2003) Mechanisms of antimicrobial peptide action and resistance. Pharmacol Rev 55:27–55
4. Epand RM, Vogel HJ (1999) Diversity of antimicrobial peptides and their mechanisms of action. Biochim Biophys Acta 1462:11–28
5. Jenssen H, Hamill P, Hancock REW (2006) Peptide antimicrobial agents. Clin Microbiol Rev 19:491–511
6. Shai Y (2002) Mode of action of membrane active antimicrobial peptides. Biopolymers 66:236–248
7. Lohner K, Sevcsik E, Pabst G (2008) Liposome-based biomembrane mimetic systems: implications for lipid-peptide interactions. In: Advances in planar lipid bilayers and liposomes. Elsevier, Amsterdam, pp 103–132
8. Riedl S, Rinner B, Asslaber M et al (2011) In search of a novel target - phosphatidylserine exposed by non-apoptotic tumor cells and metastases of malignancies with poor treatment efficacy. Biochim Biophys Acta 1808:2638–2645
9. Abraham T, Lewis RN, Hodges RS et al (2005) Isothermal titration calorimetry studies of the binding of the antimicrobial peptide gramicidin S to phospholipid bilayer membranes. Biochemistry 44:11279–11285
10. Lohner K, Prenner EJ (1999) Differential scanning calorimetry and X-ray diffraction studies of the specificity of the interaction of antimicrobial peptides with membrane-mimetic systems. Biochim Biophys Acta 1462:141–156
11. Andrushchenko VV, Aarabi MH, Nguyen LT et al (2008) Thermodynamics of the interactions of tryptophan-rich cathelicidin antimicrobial peptides with model and natural membranes. Biochim Biophys Acta 1778:1004–1014
12. Andrushchenko VV, Vogel HJ, Prenner EJ (2006) Solvent-dependent structure of two tryptophan-rich antimicrobial peptides and their analogs studied by FTIR and CD spectroscopy. Biochim Biophys Acta 1758:1596–1608
13. Andrushchenko VV, Vogel HJ, Prenner EJ (2007) Interactions of tryptophan-rich cathelicidin antimicrobial peptides with model membranes studied by differential scanning calorimetry. Biochim Biophys Acta 1768:2447–2458
14. Pate M, Blazyk J (2008) Methods for assessing the structure and function of cationic antimicrobial peptides. In: Champney WS (ed) New antibiotics targets. Humana Press, Totowa, NJ, pp 155–173

15. Lakowicz JR (2006) Protein fluorescence. In: Lakowicz JR (ed) Principles of fluorescence spectroscopy. Springer, New York, NY, pp 529–575
16. Lakowicz JR (2006) Quenching of fluorescence. In: Lakowicz JR (ed) Principles of fluorescence spectroscopy. Springer, New York, NY, pp 277–330
17. Eftink MR, Ghiron CA (1981) Fluorescence quenching studies with proteins. Anal Biochem 114:199–227
18. Shashidhara KS, Gaikwad SM (2007) Fluorescence quenching and time-resolved fluorescence studies of alpha-mannosidase from Aspergillus fischeri (NCIM 508). J Fluoresc 17:599–605
19. Hristova K, Selsted ME, White SH (1997) Critical role of lipid composition in membrane permeabilization by rabbit neutrophil defensins. J Biol Chem 272:24224–24233
20. Wimley WC, Selsted ME, White SH (1994) Interactions between human defensins and lipid bilayers: evidence for formation of multimeric pores. Protein Sci 3:1362–1373
21. Stella L, Mazzuca C, Venanzi M et al (2004) Aggregation and water-membrane partition as major determinants of the activity of the antibiotic peptide trichogin GA IV. Biophys J 86:936–945
22. Ambroggio EE, Separovic F, Bowie JH et al (2005) Direct visualization of membrane leakage induced by the antibiotic peptides: maculatin, citropin, and aurein. Biophys J 89:1874–1881
23. Belokoneva OS, Satake H, Mal'tseva EL et al (2004) Pore formation of phospholipid membranes by the action of two hemolytic arachnid peptides of different size. Biochim Biophys Acta 1664:182–188
24. Lehrer RI, Barton A, Ganz T (1988) Concurrent assessment of inner and outer membrane permeabilization and bacteriolysis in *E. coli* by multiple-wavelength spectrophotometry. J Immunol Methods 108:153–158
25. Chongsiriwatana NP, Barron AE (2010) Comparing bacterial membrane interactions of antimicrobial peptides. In: Giuliani A, Rinaldi AC (eds) Antimicrobial peptides methods and protocols. Humana Press, Totowa, NJ, pp 171–182
26. Epand RF, Pollard JE, Wright JO et al (2010) Depolarization, bacterial membrane composition, and the antimicrobial action of ceragenins. Antimicrob Agents Chemother 54:3708–3713
27. Eriksson M, Nielsen PE, Good L (2002) Cell permeabilization and uptake of antisense peptide-peptide nucleic acid (PNA) into *Escherichia coli*. J Biol Chem 277:7144–7147
28. Kaback HR, Sahin-tóth M, Weinglass AB (2001) The kamikaze approach to membrane transport. Nat Rev Mol Cell Biol 2:610–620
29. Artimo P, Jonnalagedda M, Arnold K et al (2012) ExPASy: SIB bioinformatics resource portal. Nucleic Acids Res 40:W597–W603
30. Avanti Polar Lipids Inc. (2015) Preparation of liposomes
31. Mayer LD, Hope MJ, Cullis PR (1986) Vesicles of variable sizes produced by a rapid extrusion procedure. Biochim Biophys Acta 858:161–168
32. Ames BN, Neufeld EF, Ginsberg V (1966) Assay of inorganic phosphate, total phosphate and phosphatases. Methods Enzymol 8:115–118
33. Lehrer R, Barton A, Daher KA et al (1989) Interaction of human defensins with Escherichia coli. Mechanism of bactericidal activity. J Clin Invest 84:553–561
34. Andrushchenko VV, Vogel HJ, Prenner EJ (2007) Optimization of the hydrochloric acid concentration used for trifluoroacetate removal from synthetic peptides. J Pept Sci 13:37–43
35. De Kroon AIPM, Soekarjo MW, De Gier J et al (1990) The role of charge and hydrophobicity in peptide-lipid interaction: a comparative study based on tryptophan fluorescence measurements combined with the use of aqueous and hydrophobic quenchers. Biochemistry 29:8229–8240

Chapter 11

Applying Fluorescence Correlation Spectroscopy to Investigate Peptide-Induced Membrane Disruption

Kasper Kristensen, Jonas R. Henriksen, and Thomas L. Andresen

Abstract

There is considerable interest in understanding the interactions of antimicrobial peptides with phospholipid membranes. Fluorescence correlation spectroscopy (FCS) is a powerful experimental technique that can be used to gain insight into these interactions. Specifically, FCS can be used to quantify leakage of fluorescent molecules of different sizes from large unilamellar lipid vesicles, thereby providing a tool for estimating the size of peptide-induced membrane disruptions. If fluorescently labeled lipids are incorporated into the membranes of the vesicles, FCS can also be used to obtain information about whether leakage occurs due to localized membrane perturbations or global membrane destabilization. Here, we outline a detailed step-by-step protocol on how to optimally implement an FCS-based leakage assay. To make the protocol easily accessible to other researchers, it has been supplemented with a number of practical tips and tricks.

Key words Fluorescence correlation spectroscopy (FCS), Antimicrobial peptides, Membrane-active peptides, Peptide-lipid membrane interactions, Lipid vesicles, Vesicle leakage assay, Membrane disruption, Pore formation

1 Introduction

It is of fundamental scientific interest to understand the mechanisms by which antimicrobial peptides interact with and disrupt phospholipid membranes [1, 2]. One common approach to study the impact of the peptides on membrane integrity is based on the release of quenched fluorophores from large unilamellar lipid vesicles (LUVs) [3–5]. In this approach, the LUVs are initially loaded with high self-quenching concentrations of fluorophores. Subsequently, the LUVs are incubated with a given peptide. If the peptide permeabilizes the LUVs, the entrapped fluorophores will leak from the LUVs to become dequenched. Thus, the extent and kinetics of the leakage process can be gauged by evaluating the increase in fluorescence emission

The original version of this chapter was revised. The erratum to this chapter is available at: DOI 10.1007/978-1-4939-6737-7_32

Paul R. Hansen (ed.), *Antimicrobial Peptides: Methods and Protocols*, Methods in Molecular Biology, vol. 1548, DOI 10.1007/978-1-4939-6737-7_11, © Springer Science+Business Media LLC 2017

intensity. Other aspects of the leakage process can also be investigated by different variations of the quenching-based assay [6–8].

Fluorescence correlation spectroscopy (FCS) is a powerful experimental technique that can distinguish fluorescent species of different diffusion coefficients, thereby offering another approach to study leakage from LUVs [9–15]. Importantly, we have recently shown that FCS can be used to quantify leakage of fluorescent molecules of different sizes from LUVs, providing information about the size of disruptions in the membranes [14]. It has previously also been shown that FCS can be used to evaluate whether leakage is associated with aggregation, fusion, or micellization of the LUVs. Thus, FCS represents a highly useful alternative to the existing quenching-based leakage assays.

Here, we describe how to optimally use FCS for studying peptide-induced leakage from LUVs. Thus, after a brief outline of the underlying FCS theory, we describe a detailed step-by-step protocol on how to perform such FCS leakage experiments. As we go through the protocol, we highlight a number of practical tips and tricks to make the protocol easily accessible to other researchers.

1.1 Basic Introduction to FCS

In FCS, the emission intensity of fluorescent particles diffusing through a tiny focal detection volume is measured (Fig. 1a). Because the fluorescent particles are subject to Brownian motion, the number of particles in the detection volume will fluctuate in time. Consequently, the fluorescence emission intensity from the volume will fluctuate in time (Fig. 1b). The fluorescent particles may also be subject to certain chemical reactions or photophysical processes that cause their emission intensity to fluctuate in time; these reactions/processes may also contribute to the temporal intensity fluctuations from the detection volume [16, 17].

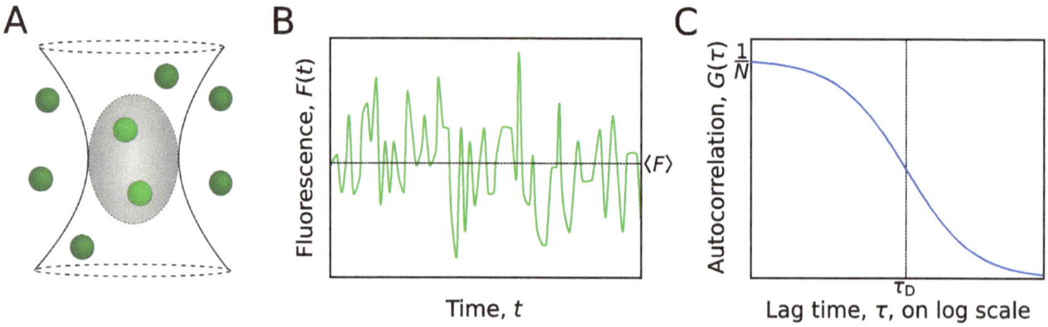

Fig. 1 Schematic illustration of the basic principle of FCS. (**a**) Fluorescent particles diffuse through a tiny focal detection volume. (**b**) The fluorescence emission intensity from the detection volume fluctuates as a function of time. (**c**) The temporal intensity fluctuations are statistically analyzed by calculating an autocorrelation function. The autocorrelation function contains information about the dynamic and kinetic properties of the fluorescent particles. The plot shows an idealized single-component diffusion autocorrelation function (Eq. 2): the amplitude of the function is inversely proportional to the mean number of fluorescent particles in the detection volume, and the decay of the function is related to the characteristic diffusion time, which reflects the typical time that it takes for the particles to diffuse across the detection volume

The intensity fluctuations from the detection volume can be statistically analyzed by calculating the autocorrelation function, $G(\tau)$ (Fig. 1c) [16, 17]:

$$G(\tau) = \frac{\langle \delta F(t) \cdot \delta F(t+\tau) \rangle}{\langle F \rangle^2} \quad (1)$$

where $F(t)$ is the fluorescence emission intensity recorded from the detection volume at time t, τ is the lag time, the angular brackets denote time averaging, and $\delta F(t) = F(t) - \langle F \rangle$ is the deviation of the fluorescence emission intensity at time t from the average fluorescence emission intensity. Basically, the autocorrelation function provides a measure of the self-similarity of the fluorescence emission intensity after a given lag time. Generally, the autocorrelation function can be fitted with appropriate theoretical models to deduce information about the dynamic and kinetic properties of the fluorescent particles in a given sample.

In the simple but common experimental case in which a single species of fluorescent particles undergoes free three-dimensional translational diffusion through a Gaussian-shaped detection volume, the autocorrelation function is modeled by [16, 17] (*see* **Note 1**)

$$G(\tau) = \frac{1}{N} g(\tau) = \frac{1}{N}\left(1 + \frac{\tau}{\tau_D}\right)^{-1}\left(1 + \frac{\tau}{S^2 \tau_D}\right)^{-\frac{1}{2}} \quad (2)$$

where N is the mean number of fluorescent particles in the detection volume, S is the ratio of the axial to radial dimensions of the detection volume, and τ_D is the characteristic diffusion time of the fluorescent particles. Figure 1c shows an idealized example of such a single-component diffusion autocorrelation function. In case there are multiple species of fluorescent particles, the autocorrelation function is a linear combination of the autocorrelation functions of the individual species [16]:

$$G(\tau) = \sum_i A_i g_i(\tau) \quad (3)$$

where i is the index of the individual species of fluorescent particles, A_i is the amplitude associated with the ith species of fluorescent particles, and $g_i(\tau)$ is the normalized diffusion autocorrelation function (defined in Eq. 2) of the ith species of fluorescent particles.

1.2 Studying Leakage from LUVs by FCS

To understand how FCS can be used to study leakage from LUVs, consider a solution of LUVs encapsulating fluorescent molecules. As long as the fluorescent molecules are inside the LUVs, they will be restricted to diffuse together with the LUVs. However, if a leakage-inducing peptide is added to the LUV solution, some of the fluorescent molecules will leak from the LUVs to diffuse freely in the solution. Consequently, there will be two diffusing fluorescent species: free fluorescent molecules (index f) and LUVs encapsulating fluorescent molecules (index v). According to Eq. 3, the autocorrelation function will then be given by [14]

$$G(\tau) = A_f \mathcal{g}_f(\tau) + A_v \mathcal{g}_v(\tau). \tag{4}$$

The magnitude of A_f will increase, and the overall amplitude of the autocorrelation function will decrease as the fluorescent molecules leak from the LUVs (*see* **Note 2**). Figure 2a schematically illustrates how the autocorrelation function will change as a function of

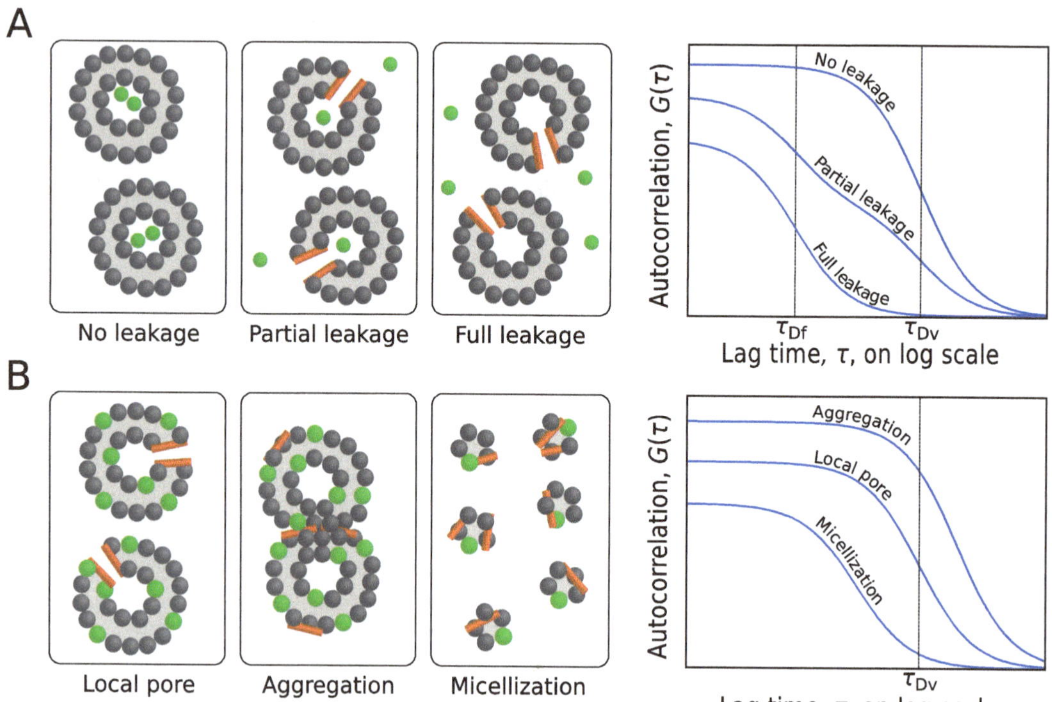

Fig. 2 Schematic illustration of how FCS can be used to study the effect of peptides on LUVs. (**a**) If fluorescent molecules (*green spheres*) are encapsulated in LUVs, they will effectively be restricted to diffuse with the LUVs; consequently, the characteristic diffusion time of the autocorrelation function will be equal to the diffusion time of the LUVs. If a given peptide (*red bars*) induces leakage of the fluorescent molecules from the LUVs, the autocorrelation function will shift toward the diffusion time of free fluorescent molecules, and the amplitude of the autocorrelation function will decrease. (**b**) If fluorescently labeled lipids (*green headgroups*) are incorporated into the membranes of the LUVs, it is possible to obtain information about whether leakage is associated with aggregation, fusion, or micellization of the LUVs. If the peptide induces leakage by localized membrane perturbations, such as the formation of aqueous pores, the autocorrelation function will be similar to one recorded from a solution of unperturbed LUVs. If the peptide induces the LUVs to assemble into large structures, for example, by aggregation or fusion, then the characteristic diffusion time and the amplitude of the autocorrelation function will increase. If the peptide induces formation of small mixed peptide-lipid micelles, then the characteristic diffusion time and the amplitude of the autocorrelation function will decrease

leakage. The fraction, L, of fluorescent molecules that has leaked from the LUVs upon addition of the peptide can be calculated by solving the following equation [14] (*see* **Note 3**):

$$A_f(1-k)^2 L^2 + \left(2A_f k(1-k) - A_{f100}\right)L + A_f k^2 = 0. \qquad (5)$$

Here, A_{f100} is the value of A_f of an LUV solution in which all of the LUVs have released all of their contents, i.e., it is the value of A_f for $L = 1$. The parameter k is the brightness ratio between encapsulated and free fluorescent molecules (*see* **Note 4**).

To understand how FCS can be used to deduce whether leakage is associated with aggregation, fusion, or micellization of the LUVs, consider a solution of LUVs with fluorescently labeled lipids incorporated in their membranes [9, 14]. In this solution, the labeled LUVs will be the only diffusing fluorescent species; accordingly, the autocorrelation function will be given by the single-component model (Eq. 2). Consider then the case in which a leakage-inducing peptide is added to the LUV solution. The peptide may exert various effects on the LUVs. For example, the peptide may induce localized membrane perturbations. In that case, the LUVs as such will remain intact and, therefore, the autocorrelation function will be similar to that of a solution of unperturbed LUVs [15]. The peptide may also induce LUV aggregation and/or fusion [7], causing the individual LUVs to assemble into fewer but bigger particles. In that case, the mean particle number, N, will decrease, and the diffusion time, τ_D, will increase as compared to the case of an unperturbed LUV solution [15]. The peptide may alternatively cause solubilization/micellization of the LUVs, leading to the formation of small mixed peptide-lipid micelles [18]. In that case, the fluorescently labeled lipids are expected to incorporate into the micelles. Consequently, the mean particle number, N, will increase, and the diffusion time, τ_D, will decrease as compared to the case of an unperturbed LUV solution [9]. Figure 2b schematically illustrates how the autocorrelation function will change in the three different situations just described.

2 Materials

The following section lists the materials that are used in the experimental protocol. The manufacturers of some specific reagents and specialist equipment are listed, whereas the manufacturers of more standard routine reagents and equipment are not.

2.1 Sample Preparation

1. 1-Palmitoyl-2-oleoyl-*sn*-glycero-3-phosphocholine (POPC) in powder form (Avanti Polar Lipids, Alabaster, AL).
2. 1-Palmitoyl-2-oleoyl-*sn*-glycero-3-[phospho-rac-(1-glycerol)], sodium salt (POPG) in powder form (Avanti Polar Lipids).

3. 1-Palmitoyl-2-[11-(dipyrrometheneboron difluoride) undecanoyl]-*sn*-glycero-3-phosphocholine (TopFluor PC) in powder form (Avanti Polar Lipids).
4. Chloroform.
5. Methanol.
6. Glass sample vials.
7. *N*-(2-hydroxyethyl)piperazine-*N'*-(2-ethanesulfonic acid) (HEPES) buffer: 10 mM HEPES, 100 mM NaCl, pH 7.4, prepared using ultrapure water with a resistivity of 18.2 MΩ · cm produced by a Milli-Q system.
8. Alexa Fluor 488 hydrazide, sodium salt (Alexa488) (Molecular Probes, Thermo Fisher Scientific, Waltham, MA).
9. Alexa Fluor 488 dextran, 3000 MW (Alexa488-3kMW) (Molecular Probes, Thermo Fisher Scientific).
10. Alexa Fluor 488 dextran, 10,000 MW (Alexa488-10kMW) (Molecular Probes, Thermo Fisher Scientific).
11. UV-VIS spectrophotometer.
12. Vacuum pump system.
13. Isopropanol.
14. Dry ice.
15. Warm water bath.
16. Mini-extruder with 250 μL syringes (Avanti Polar Lipids).
17. Nuclepore track-etched polycarbonate filters with 100 nm pore diameter (Whatman, GE Healthcare, Little Chalfont, UK).
18. Sepharose CL-4B slurry.
19. Glass chromatography column with dimensions 1.5 × 20 cm connected to a peristaltic pump.
20. Mastoparan X (*see* **Note 5**).
21. Protein LoBind tubes (Eppendorf, Hamburg, Germany).

2.2 FCS Instrumentation

DCS-120 confocal scanning FLIM system (Becker & Hickl, Berlin, Germany):

1. The excitation source of the system is a 473 nm picosecond diode laser (BDL-473-SMC) operated at a pulse repetition rate of 50 MHz.
2. The excitation light is coupled into an Axio Observer Z1 inverted microscope equipped with a C-apochromat 40×/1.2 W Corr UV-VIS-IR water immersion objective (Carl Zeiss, Jena, Germany).
3. The emission light passes through a confocal pinhole (1.2 airy units) and a 485 nm longpass filter (HQ485LP).
4. The fluorescence emission is detected by an HPM-100-40 hybrid detector connected to an SPC-150 module.

5. Lifetime gating is used to partially suppress background noise.
6. The autocorrelation curves are calculated and shown in the SPCM software.

The following experimental protocol is generally also applicable to other commercial FCS setups.

2.3 FCS Measurements and Data Analysis

1. Uncoated μ-slide 8 wells (ibidi, Martinsried, Germany).
2. Optical power meter.
3. MATLAB.

3 Methods

The following protocol describes how we in our lab use FCS to study the effect of the antimicrobial peptide mastoparan X on POPC/POPG (3:1) LUVs [14]. However, the protocol can easily be used to study the effect of other peptides on LUVs composed of other lipids.

3.1 LUV and Peptide Stock Preparation

3.1.1 LUV Stock Preparation

1. Dissolve 25 μmol POPC/POPG (3:1) in 1 mL chloroform/methanol (9:1) in a glass sample vial to obtain a homogenous lipid mixture (*see* **Note 6**). Add 25 nmol TopFluor PC (corresponding to 0.1 mol%) if fluorescently labeled lipids are to be incorporated in the membranes of the LUVs to be prepared.
2. Remove the solvent using a gentle nitrogen flow. A lipid film then appears at the bottom of the sample vial.
3. Remove residual solvent from the lipid film by keeping the sample in vacuum (~0.5 mbar) overnight.
4. Add 500 μL HEPES buffer to the lipid film. This gives a suspension with 50 mM lipid. If fluorescent molecules are to be entrapped in the LUVs to be prepared, the buffer should contain 4 μM Alexa488, 5 μM Alexa488-3kMW, or 2.5 μM Alexa488-10kMW (*see* **Notes 7** and **8**). The concentration of Alexa488 in buffer can be determined from the absorbance at 493 nm by using that the extinction coefficient of Alexa488 at this wavelength is 71,000 cm^{-1} M^{-1}. (This extinction coefficient was stated by the manufacturer.) To determine the concentration of the Alexa488-labeled dextrans, it is also necessary to take the degree of labeling (number of fluorophores per dextran) into account.
5. Regularly vortex the sample every 5 min for a total period of 30 min to completely suspend the lipids (*see* **Note 9**).
6. Subject the sample to five freeze-thaw cycles by alternately placing the sample vial in an isopropanol/dry ice bath and a warm water bath of ~50 °C. The sample should be completely

frozen before being transferred from the isopropanol/dry ice bath to the warm water bath and completely melted before being transferred from the warm water bath to the isopropanol/dry ice bath (*see* **Note 10**). This treatment is done to equilibrate and reduce the lamellarity of the lipid vesicles.

7. Extrude the lipid suspension 21 times through a 100 nm polycarbonate filter (*see* **Notes 9** and **11**). This will result in the formation of LUVs (*see* **Note 12**).
8. For LUVs prepared with fluorescent molecules in the buffer: Separate the LUVs encapsulating fluorescent molecules from free fluorescent molecules by size exclusion chromatography using a Sepharose CL-4B column equilibrated in HEPES buffer. Use a flow rate of 1 mL/min and collect 1 min fractions. The LUVs will elute ~10 min after they have been added to the column. By visual inspection, it will be possible to identify the fractions that contain the LUVs because these fractions will appear opaque. Pool the 2–3 fractions with the highest opacity to a final LUV stock solution (*see* **Note 13**).
9. Determine the lipid concentration of the final LUV stock solution using the colorimetric assay of Rouser et al. [19] (*see* **Note 14**).

3.1.2 Peptide Stock Preparation

1. Dissolve an appropriate amount of mastoparan X in HEPES buffer in a Protein LoBind tube to a concentration of 100 μM (*see* **Note 15**).
2. Confirm the concentration of mastoparan X by measuring the absorbance at 220 nm. The extinction coefficient of mastoparan X at this wavelength is 40,100 cm^{-1} M^{-1} [14] (*see* **Notes 16** and **17**).

3.2 Configuration of FCS Setup

It is important that the FCS setup is properly configured [20–22].

3.2.1 Adjustment and Calibration Experiment

Adjust the FCS setup and perform a calibration experiment by the following protocol:

1. Prepare a 10 nM Alexa488 solution in HEPES buffer. The concentration of Alexa488 in more concentrated stock solutions can be determined from the absorbance at 493 nm by using that the extinction coefficient of Alexa488 at this wavelength is 71,000 cm^{-1} M^{-1}. (This extinction coefficient was stated by the manufacturer.)
2. Transfer 100 μL of the 10 nM Alexa488 solution to a μ-slide 8 well.
3. Add a drop of water to the objective, and place the μ-slide 8 well on the microscope stage.
4. Set the excitation power, as measured at the objective back aperture, to 11 μW (*see* **Note 18**).

5. Position the laser focus ~50 μm above the top of the cover slip.
6. Adjust the correction collar of the objective to maximize the brightness of Alexa488 (*see* **Note 4**). Maximal brightness is obtained when the product of the photon count rate and the amplitude of the autocorrelation curve is maximal. This adjustment is done to correct for spherical aberrations introduced by the cover slip [20] (*see* **Notes 19** and **20**). After the setup has been properly adjusted, we typically get a photon count rate of ~25 kHz from the 10 nM Alexa488 solution, corresponding to an Alexa488 brightness of ~4 kHz.
7. Perform 10 × 30 s FCS measurements on the 10 nM Alexa488 solution.
8. Export the ten experimental autocorrelation curves for handling and fitting in MATLAB.

3.2.2 Data Analysis

Analyze the autocorrelation curves of the calibration experiment by the following protocol for the purpose of determining the structural *S*-parameter of the detection volume:

1. Fit each of the ten experimental autocorrelation curves with the single-component autocorrelation function (Eq. 2) by use of weighted least squares fitting with N, τ_D, and S as free parameters (*see* **Notes 21** and **22**).
2. Determine the *S*-parameter by taking the average value of the 10 fits (*see* **Note 23**). In our setup, we typically determine the *S*-parameter to be between 7 and 10 (*see* **Notes 24** and **25**).

3.3 Measurement of Peptide-Induced Leakage by FCS

3.3.1 Sample Preparation

Prepare the mastoparan X-LUV samples for a given leakage experiment by the following protocol:

1. Prepare a number of mastoparan X-LUV samples by mixing appropriate volumes of (i) HEPES buffer; (ii) a stock solution of POPC/POPG (3:1) LUVs encapsulating Alexa488, Alexa488-3kMW, or Alexa488-10kMW; and (iii) a stock solution of mastoparan X. The samples should be mixed in Protein LoBind tubes (*see* **Note 15**). The peptide should be added to the tubes as the last component. The final lipid concentration of the samples should be 1 mM (*see* **Notes 26** and **27**). For samples with LUVs containing Alexa488, the final mastoparan X concentration should vary between 0 and 35 μM. For samples with LUVs containing Alexa488-3kMW, the final mastoparan X concentration should vary between 0 and 50 μM. For samples with LUVs containing Alexa488-10kMW, the final mastoparan X concentration should vary between 0 and 70 μM (*see* **Note 28**).
2. Immediately after mastoparan X has been added to a given sample, vortex the sample for a few seconds (*see* **Note 29**).

3. Incubate each sample for 1 h (*see* **Note 30**).

3.3.2 FCS Experiments

After a given sample has been incubated for 1 h, investigate it by FCS using the following protocol:

1. Transfer 100 μL of the sample to a μ-slide 8 well.
2. Add a drop of water to the objective, and place the μ-slide 8 well on the microscope stage.
3. Set the excitation power, as measured at the objective back aperture, to 11 μW (*see* **Note 18**).
4. Position the laser focus ~50 μm above the top of the cover slip.
5. Perform a 5 min FCS measurement (*see* **Note 31**). We typically get a photon count rate of ~20–25 kHz.
6. Export the experimental autocorrelation curve for handling and fitting in MATLAB (*see* **Note 32**).

3.3.3 Data Analysis

Analyze the autocorrelation curves of a given leakage experiment by the following protocol for the purpose of determining the leakage levels of each of the samples investigated in the experiment [14]. Data fitting of the autocorrelation curves should be done by use of non-weighted least squares fitting with S kept fixed to the value determined in the calibration experiment (*see* **Notes 33** and **34**).

1. Fit an experimental autocorrelation curve acquired from a sample with 0% leakage (an LUV solution to which no peptide has been added) with the single-component autocorrelation function (Eq. 2) with N and τ_D as free parameters. The value of τ_D is equal to the LUV diffusion time, τ_{Dv}, which is a parameter to be used in the subsequent data analysis. We typically get that τ_{Dv} is equal to ~4700 μs [14] (*see* **Note 35**).
2. Fit an experimental autocorrelation curve acquired from a sample with 100% leakage (an LUV solution with complete leakage of the fluorescent molecules) with the single-component autocorrelation function (Eq. 2) with N and τ_D as free parameters (*see* **Note 36**). The value of τ_D is equal to the diffusion time of free fluorescent molecules, τ_{Df}, which is a parameter to be used in the subsequent data analysis. We typically get that τ_{Df} is equal to ~43 μs for Alexa488, ~100 μs for Alexa488-3kMW, and ~210 μs for Alexa488-10kMW [14] (*see* **Note 35**).
3. Fit the experimental autocorrelation curve acquired from the sample with 0% leakage (the sample used in **step 1**) with the equation

$$G(\tau) = A_{v0}\left(z_0 g_f(\tau) + g_v(\tau)\right) \qquad (6)$$

with A_{v0} and z_0 as free parameters and with τ_{Dv} and τ_{Df} kept fixed to the values determined in **steps 1** and **2**, respectively.

The parameter z_0 is a correction parameter, which is going to be used in the subsequent data analysis (*see* **Note 37**). We typically get that z_0 is equal to ~0.033 for LUVs containing Alexa488, ~0.028 for LUVs containing Alexa488-3kMW, and ~0.025 for LUVs encapsulating Alexa488-10kMW [14].

4. Fit all of the autocorrelation curves of the given leakage experiment with a modified two-component autocorrelation function:

$$G(\tau) = (A_f + z_0 A_v) g_f(\tau) + A_v g_v(\tau) \quad (7)$$

with A_f and A_v as free parameters and with τ_{Dv} and τ_{Df} kept fixed to the values determined in **steps 1** and **2**, respectively, and z_0 kept fixed to the value determined in **step 3**. The values of A_f are going to be used in the subsequent data analysis. The value of A_f acquired from the sample with 100% leakage (the sample used in **step 2**) gives A_{f100}, which also is a parameter to be used in the subsequent data analysis.

5. Determine the brightness ratio, k, by use of the equation

$$k = \frac{\langle F_0 \rangle}{\langle F_{100} \rangle} \quad (8)$$

where $\langle F_0 \rangle$ and $\langle F_{100} \rangle$ are the photon count rates of the samples with 0 and 100% leakage, respectively (the samples used in **steps 1** and **2**, respectively). We typically get that k is equal to ~0.85 for samples with LUVs containing Alexa488 and ~0.95 for samples with LUVs containing Alexa488-3kMW or Alexa488-10kMW [14] (*see* **Notes 38** and **39**).

6. Solve Eq. 5 for L using each of the values of A_f (determined in **step 4**) together with the value of A_{f100} (determined in **step 4**) and the value of k (determined in **step 5**). This gives the leakage level, L, for each of the investigated samples.

7. Plot L as a function of the peptide-to-lipid ratio.

3.3.4 Example of Results

Figure 3a shows autocorrelation curves acquired from an experiment in which varying amounts of mastoparan X were incubated for 1 h with 1 mM POPC/POPG (3:1) LUV containing Alexa488. As the peptide-to-lipid ratio increases, the curves shift toward shorter lag times, and the amplitude of the curves decreases. These changes of the autocorrelation curves occur because mastoparan X induces leakage of Alexa488 from the LUVs.

Figure 3b shows the leakage levels measured in an experiment in which varying amounts of mastoparan X were incubated for 1 h with 1 mM POPC/POPG (3:1) LUV containing Alexa488, Alexa488-3kMW, or Alexa488-10kMW. Generally, leakage increases as the peptide-to-lipid ratio increases. Also, leakage is

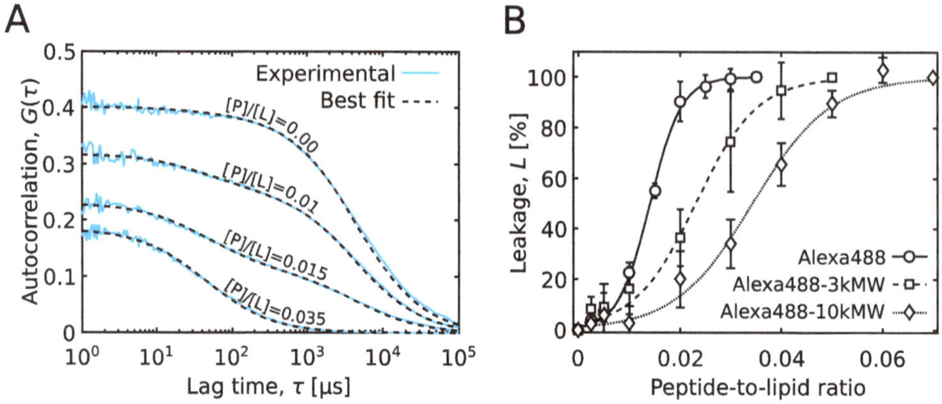

Fig. 3 Example of results of an FCS leakage experiment in which varying amounts of mastoparan X was incubated for 1 h with 1 mM POPC/POPG (3:1) LUV containing 4 μM Alexa488, 5 μM Alexa488-3kMW, or 2.5 μM Alexa488-10kMW. (**a**) Example of experimental autocorrelation curves of an experiment in which the LUVs contained Alexa488. The curves shift toward shorter lag times and the amplitude of the curves decreases as the peptide-to-lipid ratio increases. The dashed black lines show the best fit of the two-component model (Eq. 7) to the experimental curves. (**b**) Leakage as a function of the peptide-to-lipid ratio. The data are the average of three separate experiments. The error bars represent the standard deviations. The lowest and highest peptide-to-lipid ratios in each curve correspond to no leakage and complete leakage, respectively; therefore, they are by definition set to 0 and 100 %, respectively. Sigmoidal trend lines have been added to guide the eye. It is observed that leakage increases as the peptide-to-lipid ratio increases. Moreover, it is observed that leakage depends on the size of the encapsulated fluorescent molecules. Panel (**b**) of the figure has been adapted from [14] with permission from Elsevier

dependent on the size of the encapsulated fluorescent molecules: Alexa488, with a hydrodynamic diameter of ~1.1 nm, is more effectively released than Alexa488-3kMW, with a hydrodynamic diameter of ~2.5 nm, and Alexa488-10kMW, with a hydrodynamic diameter of ~5.3 nm. This observation suggests that the size of the membrane disruptions is between ~1.1 and ~5.3 nm, at least for peptide-to-lipid ratios below 0.02 (*see* **Note 40**).

3.4 Evaluation of the Mechanism of Peptide-Induced Leakage by FCS

3.4.1 Sample Preparation

Prepare the mastoparan X-LUV samples by the following protocol (*see* **Note 41**):

1. Prepare a number of mastoparan X-LUV samples by mixing appropriate volumes of (i) HEPES buffer, (ii) a stock solution of POPC/POPG (3:1) LUVs labeled with TopFluor PC, and (iii) a stock solution of mastoparan X. The samples should be mixed in Protein LoBind tubes (*see* **Note 15**). The peptide should be added to the tubes as the last component. The final lipid concentration of the samples should be 1 mM. The final mastoparan X concentration of the samples should vary between 0 and 70 μM (*see* **Note 42**).
2. Immediately after mastoparan X has been added to a given sample, vortex the sample for a few seconds (*see* **Note 29**).
3. Incubate each sample for 1 h (*see* **Note 42**).

3.4.2 FCS Experiments

After a given sample has been incubated for 1 h, investigate it by FCS using the following protocol:

1. Transfer 100 μL of the sample to a μ-slide 8 well.
2. Add a drop of water to the objective and place the μ-slide 8 well on the microscope stage.
3. Set the excitation power, as measured at the objective back aperture, to 2.5 μW (*see* **Note 18**).
4. Position the laser focus ~50 μm above the top of the cover slip.
5. Perform a 5 min FCS measurement (*see* **Note 31**). We typically get a total photon count rate of ~50 kHz.
6. Export the experimental autocorrelation curve for handling and fitting in MATLAB (*see* **Note 32**).

3.4.3 Data Analysis

Analyze the experimental autocorrelation curves by the following protocol [14].

1. Fit all of the experimental autocorrelation curves with the single-component autocorrelation function (Eq. 2) by use of use of non-weighted least squares fitting with N and τ_D as free parameters and S kept fixed to the value determined in the calibration experiment (*see* **Notes 33** and **34**).
2. Plot the mean particle number, N, and the particle diffusion time, τ_D, as a function of the peptide-to-lipid ratio.

3.4.4 Example of Results

Figure 4 shows the results of an experiment in which varying amounts of mastoparan X was incubated for 1 h with 1 mM POPC/POPG (3:1) LUV labeled with TopFluor PC. Specifically,

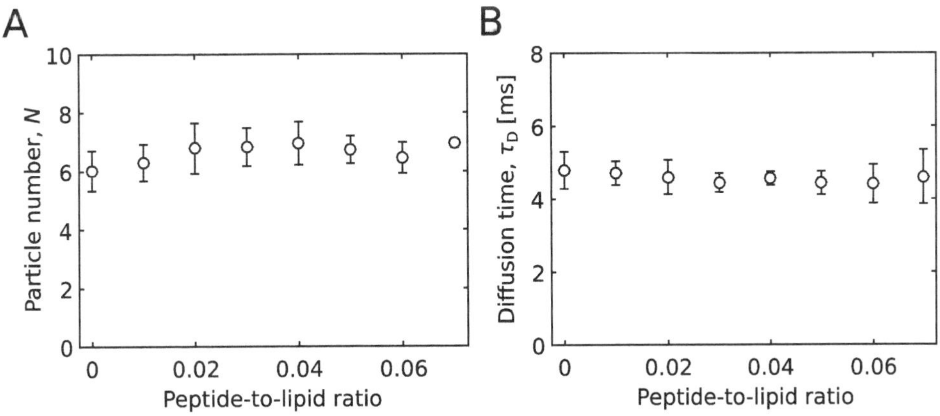

Fig. 4 Example of results of an FCS experiment in which varying amounts of mastoparan X was incubated for 1 h with 1 mM POPC/POPG (3:1) LUV labeled with 0.1 % TopFluor PC. (**a**) The mean number of fluorescent particles in the detection volume as a function of the peptide-to-lipid ratio. (**b**) The particle diffusion time as a function of the peptide-to-lipid ratio. The data in both panels are the average of three separate experiments. The error bars represent the standard deviations. Both the mean particle number and the diffusion time are largely unaffected by the peptide-to-lipid ratio, indicating that leakage occurs via localized perturbations in the membrane. The figure has been adapted from [14] with permission from Elsevier

Fig. 4 shows how the number of fluorescent particles in the detection volume, N, and their diffusion time, τ_D, depends on the peptide-to-lipid ratio. Both N and τ_D are largely unaffected by the peptide-to-lipid ratio, indicating that the LUVs as such remain intact and that leakage occurs via localized peptide-induced perturbations in the lipid membranes.

4 Notes

1. In case the fluorescent particles are subject to significant triplet blinking, the single-component autocorrelation function is given by [16]

$$G(\tau) = \frac{1}{N} g(\tau) = \frac{1}{N}\left(1 + \frac{T}{1-T} e^{-\frac{\tau}{\tau_T}}\right)\left(1 + \frac{\tau}{\tau_D}\right)^{-1}\left(1 + \frac{\tau}{S^2 \tau_D}\right)^{-\frac{1}{2}} \quad (9)$$

where T is the fraction of the fluorescent particles in the triplet state and τ_T is the characteristic triplet state lifetime.

2. To understand why the amplitude of the autocorrelation function commonly decreases as a function of leakage, consider that the LUVs in a typical leakage experiment initially contain multiple fluorescent molecules. Yet, each of the LUVs will only be counted as one single particle. However, as the leakage process proceeds, the fluorescent molecules will be released from the LUVs to diffuse freely and independently. Consequently, the number of independent particles will increase during the leakage process, and, thus, the amplitude of the autocorrelation function will decrease. For completeness, it should be mentioned that in the special case in which the average number of fluorescent molecules per LUV is much lower than one, so that practically no LUVs contain more than one molecule, then the number of independent particles will not change during the leakage process, and, therefore, the amplitude of the autocorrelation function will remain constant.

3. Equation 5 has been derived without any specific assumptions about how the brightness of the LUVs is affected by the leakage process. Consequently, Eq. 5 can be used to quantify leakage, regardless of whether leakage occurs due to all-or-none leakage or graded leakage.

4. The brightness, B, of a given type of fluorescent particle is the photon count rate per particle:

$$B = \frac{\langle F \rangle}{N} \quad (10)$$

where $\langle F \rangle$ is the photon count rate of the particles in the detection volume and N is the number of particles in the volume.

5. In our experience, commercial peptides may not always have the purity stated by the manufacturer. Accordingly, we always check the quality of commercial peptides by analytical HPLC and MALDI-TOF.
6. We generally keep the chloroform/methanol solutions in glass vessels. Plastic vessels may not be chemically resistant to this solvent.
7. We use Alexa488, Alexa488-3kMW, and Alexa488-10kMW due to the following considerations: First, Alexa488 and conjugates of this dye are generally a popular choice for FCS, inter alia, due to the bright and photostable properties of the dye [16]. Second, Alexa488, Alexa488-3kMW, and Alexa488-10kMW do not bind to the membrane of the POPC/POPG (3:1) LUVs. In this context, it is relevant to mention that we found that Rhodamine 6G, which is another dye that has been used in FCS, binds to the POPC/POPG (3:1) LUVs [14]; therefore, this dye is not suitable for our experiments. Third, Alexa488, Alexa488-3kMW, and Alexa488-10kMW do not spontaneously leak from the POPC/POPG (3:1) LUVs. Fourth, Alexa488, Alexa488-3kMW, and Alexa488-10kMW represent three distinct sizes, allowing us to gain insight into how peptide-induced leakage depends on the size of the entrapped molecules.
8. The number of fluorescent molecules encapsulated in the LUVs should be sufficiently high that a reasonable photon count rate is measured in the FCS leakage experiments. On the other hand, it is a basic requirement of the FCS leakage experiments that the number of fluorescent molecules encapsulated per LUV is kept low [14]. We choose the concentrations of fluorescent molecules in the lipid hydration buffer to balance these two requirements. Specifically, with the given concentrations of fluorescent molecules in the buffer, each LUV will on average contain ~1.5–3 molecules. It may be possible to encapsulate a slightly higher number of fluorescent molecules per LUV without greatly compromising the accuracy of the FCS leakage measurements. Further information on how to evaluate the accuracy of the leakage measurements can be found in [14].
9. During the hydration and extrusion process, it is necessary to keep the temperature of the samples above the transition temperature of the lipids. The transition temperature of both POPC and POPG is −2 °C; consequently, it is acceptable to hydrate and extrude samples containing these lipids at room temperature. However, some saturated lipids have a transition temperature above room temperature; therefore, it is necessary to heat samples containing such lipids during the hydration and extrusion process.

10. Before a given sample is transferred from the warm water bath in the freeze-thaw process, the temperature of the sample should be higher than the transition temperature of the lipids in the sample. For a sample containing POPC and POPG, which both have a transition temperature of −2 °C, this criterion is achieved once the sample has melted. However, if a given sample contains saturated lipids with a high transition temperature, it may be necessary to keep the sample in the water bath for longer periods of time and/or to raise the temperature of the water bath to more than 50 °C to ensure that the sample is heated to an appropriate temperature.
11. It varies from lab to lab how many times the lipid suspensions are extruded through a filter during the LUV preparation process. Thus, the 21 extrusions that we suggest in this text should not be strictly regarded as the only valid number of extrusions to prepare LUVs. However, lipid suspensions should always be extruded an odd number of times so that the final LUV sample is in the opposite syringe than the initial lipid suspension. This minimizes the presence of large lipid aggregates and other particles in the final LUV sample.
12. The extrusion procedure gives a batch of monodisperse lipid vesicles. This is crucial since the analysis of the experimental autocorrelation curves acquired in the FCS experiments is based on the assumption that all of the LUVs in a given sample under investigation have similar diffusion properties; that is, it is assumed that the LUVs can be modeled by use of only one diffusion time.
13. It is essential that the LUVs encapsulating fluorescent molecules are completely purified from the free fluorescent molecules before conducting the FCS leakage experiments. We have found the Sepharose CL-4B column to be effective in performing this purification. However, other size exclusion chromatography columns or other separation methods, such as centrifugation or dialysis, may also be applicable.
14. The lipid concentration can also be determined by various other methods, for example, by use of the colorimetric assays of Bartlett [23] or Stewart [24] or by use of ICP-OES or ICP-MS.
15. We recently showed that cationic membrane-active peptides to a great extent may be lost from samples kept in common glass and plastic sample vessels due to adsorption of the peptides to the walls of the vessels [25]. Because of this observation, we generally handle peptide solutions in Protein LoBind tubes, and we keep the concentrations of peptide stock solutions high at ~100 μM.

16. We determined the extinction coefficient of mastoparan X at 220 nm by correlating the peptide concentration of a given sample determined by a chemiluminescent nitrogen detector to the absorbance at 220 nm of the same sample [14].
17. The concentration of mastoparan X can also be determined from the absorbance at 280 nm by using that the peptide contains a tryptophan residue; that is, to a good approximation, mastoparan X has an extinction coefficient of 5690 cm^{-1} M^{-1} at 280 nm [26].
18. The optimal choice of laser power may be different for other FCS setups and fluorophores. Generally, the excitation power should be sufficiently high that an adequately high brightness is measured in the FCS experiments. On the other hand, the excitation power should still be sufficiently low that there is no significant photobleaching and/or optical saturation in the system [20, 21]; that is, the excitation power should be sufficiently low that the shape and amplitude of the experimental autocorrelation curves do not depend on the excitation power.
19. The correction collar should be adjusted using the same types of cover slips to be used in the intended FCS experiments.
20. In other confocal FCS setups, additional settings may be available to optimize the configuration of the setup. For example, the position of the confocal pinhole is adjustable in some setups [20].
21. We perform the weighted least squares fitting of a given theoretical autocorrelation function, $G(\tau)$, to an experimental autocorrelation curve by minimizing the following expression:

$$R^2 = \sum_i \left(\frac{G_{ex}(\tau_i) - G(\tau_i)}{\sigma(\tau_i)} \right)^2. \tag{11}$$

Here, $G_{ex}(\tau_i)$ is a data point of the experimental autocorrelation curve, i is the index of the data points along the curve, and $\sigma(\tau_i)$ is the standard deviation of $G_{ex}(\tau_i)$. The value of $\sigma(\tau_i)$ can be calculated in multiple ways [27]. In the calibration experiment, we calculate $\sigma(\tau_i)$ by use of the 10 experimental autocorrelation curves acquired from the 10 nM Alexa488 solution:

$$\sigma(\tau_i) = \sqrt{\frac{1}{9} \sum_{\ell=1}^{10} \left(G_{ex,\ell}(\tau_i) - \bar{G}_{ex}(\tau_i) \right)^2}. \tag{12}$$

Here, ℓ is the index of the experimental autocorrelation curves, and $\bar{G}_{ex}(\tau_i)$ is the mean experimental autocorrelation curve:

$$\bar{G}_{ex}(\tau_i) = \frac{1}{10} \sum_{\ell=1}^{10} G_{ex,\ell}(\tau_i). \tag{13}$$

22. In our experience, it is necessary to use weighted least squares fitting to make the *S*-parameter converge to a reasonable value.
23. For some autocorrelation curves, the *S*-parameter may not converge to a reasonable value, even when weighted least squares fitting is used. We generally discard such outliers from the estimation of the *S*-parameter.
24. The value of the *S*-parameter is often reported to range between 3 and 8 for other confocal FCS setups [20, 28, 29].
25. From the *N*-values determined in the calibration experiment, it is possible to estimate the effective size of the focal detection volume, V_{det}, by use of the following equation [30]:

$$V_{det} = \frac{N}{N_A C} \tag{14}$$

where N_A is Avogadro's number and *C* is the molar concentration of Alexa488. For our setup, the effective size of the detection volume is typically determined to be ~1 fL. It should be mentioned that there also are other approaches to determine the effective size of the focal detection volume [30].

26. It is also feasible to work with lipid concentrations lower than 1 mM. However, in that case, it may be necessary to entrap more fluorescent molecules in the LUVs to ensure that the experimental photon count rate is significantly higher than the background count rate. On the other hand, it is important that the number of fluorescent molecules encapsulated per LUV is still kept low (*see* **Note 8**). This requirement sets a lower limit to the applicable lipid concentration.
27. Equation 5 was derived under the assumption that the overall concentration of fluorescent molecules does not change during the leakage process [14]. Consequently, the concentration of fluorescent molecules—and thus the lipid concentration—of all the samples in a given leakage experiment should be constant.
28. The leakage levels of the samples after incubation for the entire incubation time should range between 0 and 100%. The peptide concentration range should be chosen according to this criterion. Consequently, the relevant peptide concentration range to investigate may be different for experiments on LUVs of other lipid compositions or other peptides than mastoparan X. If a given peptide under investigation, even at high concentrations, does not lead to 100% leakage, it may also be possible to use detergents, such as Triton X-100, to prepare a sample with 100% leakage.

29. In our experience, the vortexing intensity used in the sample preparation will not affect the eventual results of the FCS experiments.
30. The FCS experiments can also be conducted with shorter incubation times than 1 h to obtain more information about the time kinetics of the leakage process. However, the time resolution of the FCS method is limited to ~1 min (*see* **Note 31**), causing detailed information about the time kinetics of the leakage process to be inaccessible by FCS.
31. The acquisition time could be shortened to 1 min, albeit this will lead to slightly higher uncertainties in the estimated leakage values. However, it is not feasible to obtain robust FCS results on LUV-containing samples using an acquisition time that is much shorter than 1 min. Accordingly, FCS cannot be used to deduce information about the time kinetics of the leakage process on a time scale faster than ~1 min [14].
32. In some instances, we observe that single bright events completely dominate the experimental autocorrelation curves. However, given that such bright events are rare, we do not ascribe them any significant importance for the interpretation of the leakage experiments. Consequently, autocorrelation curves dominated by single bright events are discarded from the subsequent data analysis [14, 15].
33. We perform the non-weighted least squares fitting of a given theoretical autocorrelation function, $G(\tau)$, to an experimental autocorrelation curve by minimizing the following expression:

$$R^2 = \sum_i \left(G_{ex}(\tau_i) - G(\tau_i)\right)^2. \quad (15)$$

Here, $G_{ex}(\tau_i)$ is a data point of the experimental autocorrelation curve, and i is the index of the data points along the curve.
34. We find that in cases in which the value of the S-parameter is kept fixed, we can fit our experimental autocorrelation curves reasonably well by use of non-weighted least squares fitting.
35. The diffusion time, τ_D, can be related to the diffusion coefficient, D, of the given type of fluorescent particles under investigation by solving the following equation [16]:

$$w_{xy}^2 = 4\tau_D D \quad (16)$$

where w_{xy} is the lateral radius of the detection volume. The value of w_{xy} can be estimated in multiple ways. If the effective size of the detection volume is known (*see* **Note 25**), then w_{xy} can be determined by solving the relation [16]

$$V_{\text{det}} = \pi^{3/2} S w_{xy}^3. \tag{17}$$

In our setup, w_{xy} is typically equal to ~270 nm. From the diffusion coefficient, D, the hydrodynamic diameter, d_H, of the fluorescent particles can be estimated by use of the Stokes-Einstein relation:

$$d_H = \frac{k_B T}{3\pi\eta D} \tag{18}$$

where k_B is Boltzmann's constant, T is the temperature, and η is the viscosity of the aqueous buffer. Typically, η can be assumed to be equal to the viscosity of water.

36. Complete leakage can be confirmed by comparing the measured value of τ_D to the diffusion time of the fluorescent molecules in a solution without any LUVs.
37. The correction parameter, z_0, is introduced to enforce that $L = 0$ in cases where there has been no leakage from the LUVs [14]. If z_0 is not included in the protocol, then we typically find that A_f, and thus L, is equal to a small but nonzero value in cases where there has been no leakage from the LUVs.
38. In order for Eq. 5 to be applicable to quantify leakage, k should be close to 1 [14]. In this context, it should be mentioned that the value of k may depend on the excitation power used in the FCS experiments. Specifically, the value of k may decrease when the excitation power is increased [14]. Consequently, k cannot be determined by use of a conventional fluorometer; it is necessary to measure k in the FCS setup.
39. To get more robust estimates of k, we typically average the photon count rates of multiple experiments.
40. Dextran molecules are not perfect spheres; rather dextran molecules are prolate ellipsoids with a short and long axis [31]. Consequently, the size of membrane disruptions estimated by use of the hydrodynamic diameter of dextran molecules should be regarded with caution [14]. Other types of fluorescently labeled molecules may be more appropriate for estimating the size of the disruptions in the membrane.
41. If the LUVs are labeled with a fluorescently labeled lipid that is spectrally separated from the encapsulated fluorescent molecules, it would in principle also be possible to measure leakage and evaluate the leakage mechanism in one single experiment instead of in two different experiments.
42. To allow for comparison between the experiments, the lipid and peptide concentrations and the incubation time used in the experiments to evaluate the leakage mechanism (the experiments described in the protocol in Subheading 3.4)

should be the same as those used in the experiments in which leakage is measured (the experiments described in the protocol in Subheading 3.3).

Acknowledgments

We thank Jannik B. Larsen and Anncatrine L. Petersen for critical reading of the manuscript and valuable comments. Financial support for this work was kindly provided by the NanoMorph consortium—funded by the Danish Council for Technology and Innovation—and the Lundbeck Foundation Research Initiative on Brain Barriers and Drug Delivery.

References

1. Wimley WC, Hristova K (2011) Antimicrobial peptides: successes, challenges and unanswered questions. J Membr Biol 239:27–34
2. Almeida PF (2014) Membrane-active peptides: binding, translocation, and flux in lipid vesicles. Biochim Biophys Acta 1838:2216–2227
3. Allen TM, Cleland LG (1980) Serum-induced leakage of liposome contents. Biochim Biophys Acta 597:418–426
4. Weinstein JN, Yoshikami S, Henkart P et al (1977) Liposome-cell interaction: transfer and intracellular release of a trapped fluorescent marker. Science 195:489–492
5. Matsuzaki K, Murase O, Miyajima K (1995) Kinetics of pore formation by an antimicrobial peptide, magainin 2, in phospholipid bilayers. Biochemistry 34:12553–12559
6. Ladokhin AS, Wimley WC, White SH (1995) Leakage of membrane vesicle contents: determination of mechanism using fluorescence requenching. Biophys J 69:1964–1971
7. Wimley WC (2015) Determining the effects of membrane-interacting peptides on membrane integrity. Methods Mol Biol 1324:89–106
8. Patel H, Tscheka C, Heerklotz H (2009) Characterizing vesicle leakage by fluorescence lifetime measurements. Soft Matter 5:2849–2851
9. Pramanik A, Thyberg P, Rigler R (2000) Molecular interactions of peptides with phospholipid vesicle membranes as studied by fluorescence correlation spectroscopy. Chem Phys Lipids 104:35–47
10. Magzoub M, Oglęcka K, Pramanik A et al (2005) Membrane perturbation effects of peptides derived from the N-termini of unprocessed prion proteins. Biochim Biophys Acta 1716:126–136
11. Yu L, Guo L, Ding JL et al (2009) Interaction of an artificial antimicrobial peptide with lipid membranes. Biochim Biophys Acta 1788:333–344
12. Blicher A, Wodzinska K, Fidorra M et al (2009) The temperature dependence of lipid membrane permeability, its quantized nature, and the influence of anesthetics. Biophys J 96:4581–4591
13. Macháň R, Jurkiewicz P, Olżyńska A et al (2014) Peripheral and integral membrane binding of peptides characterized by time-dependent fluorescence shifts: focus on antimicrobial peptide LAH_4. Langmuir 30:6171–6179
14. Kristensen K, Henriksen JR, Andresen TL (2014) Quantification of leakage from large unilamellar lipid vesicles by fluorescence correlation spectroscopy. Biochim Biophys Acta 1838:2994–3002
15. Kristensen K, Ehrlich N, Henriksen JR, Andresen TL (2015) Single-vesicle detection and analysis of peptide-induced membrane permeabilization. Langmuir 31:2472–2483
16. Krichevsky O, Bonnet G (2002) Fluorescence correlation spectroscopy: the technique and its applications. Rep Prog Phys 65:251–297
17. Ries J, Schwille P (2012) Fluorescence correlation spectroscopy. Bioessays 34:361–368
18. Henriksen JR, Andresen TL (2011) Thermodynamic profiling of peptide membrane interactions by isothermal titration calorimetry: a search for pores and micelles. Biophys J 101:100–109

19. Rouser G, Siakotos AN, Fleischer S (1966) Quantitative analysis of phospholipids by thin-layer chromatography and phosphorus analysis of spots. Lipids 1:85–86
20. Kim SA, Heinze KG, Schwille P (2007) Fluorescence correlation spectroscopy in living cells. Nat Methods 4:963–973
21. Enderlein J, Gregor I, Patra D et al (2005) Performance of fluorescence correlation spectroscopy for measuring diffusion and concentration. ChemPhysChem 6:2324–2326
22. Hess ST, Webb WW (2002) Focal volume optics and experimental artifacts in confocal fluorescence correlation spectroscopy. Biophys J 83:2300–2317
23. Bartlett GR (1959) Phosphorus assay in column chromatography. J Biol Chem 234:466–468
24. Stewart JC (1980) Colorimetric determination of phospholipids with ammonium ferrothiocyanate. Anal Biochem 104:10–14
25. Kristensen K, Henriksen JR, Andresen TL (2015) Adsorption of cationic peptides to solid surfaces of glass and plastic. PLoS One 10:e0122419
26. Gill SC, von Hippel PH (1989) Calculation of protein extinction coefficients from amino acid sequence data. Anal Biochem 182:319–326
27. Wohland T, Rigler R, Vogel H (2001) The standard deviation in fluorescence correlation spectroscopy. Biophys J 80:2987–2999
28. Melo AM, Prieto M, Coutinho A (2014) Quantifying lipid-protein interaction by fluorescence correlation spectroscopy (FCS). Methods Mol Biol 1076:575–595
29. Sahoo B, Drombosky KW, Wetzel R (2016) Fluorescence correlation spectroscopy: a tool to study protein oligomerization and aggregation in vitro and in vivo. Methods Mol Biol 1345:67–87
30. Rüttinger S, Buschmann V, Krämer B et al (2008) Comparison and accuracy of methods to determine the confocal volume for quantitative fluorescence correlation spectroscopy. J Microsc 232:343–352
31. Bohrer MP, Deen WM, Robertson CR et al (1979) Influence of molecular configuration on the passage of macromolecules across glomerular capillary wall. J Gen Physiol 74:583–593

Chapter 12

Macromolecule Biosynthesis Assay and Fluorescence Spectroscopy Methods to Explore Antimicrobial Peptide Mode(s) of Action

Bimal Jana, Kristin Renee Baker, and Luca Guardabassi

Abstract

Antimicrobial peptides (AMPs) are viable alternatives to the currently available antimicrobials, and numerous studies have investigated their possible use as therapeutic agents for specific clinical applications. AMPs are a diverse class of antimicrobials that often act upon the bacterial cell membrane but may exhibit additional modes of action. Identification of the multiple modes of action requires a comprehensive study at subinhibitory concentrations and careful data analysis since additional modes of action can be eclipsed by AMP action on the cell membrane.

Techniques that measure the biosynthesis rate of macromolecules (e.g., DNA, RNA, protein, and cell wall) and the cytoplasmic membrane proton motive force (PMF) energy can help to unravel the diverse modes of action of AMPs. Here, we present an overview of macromolecule biosynthesis rate measurement and fluorescence spectroscopy methods to identify AMP mode(s) of action. Detailed protocols designed to measure inhibition of DNA, RNA, protein, and cell wall synthesis or membrane de-energization are presented and discussed for optimal application of these two techniques as well as to enable accurate interpretation of the experimental findings

Key words Antimicrobial peptide, Mode(s) of action, Macromolecule biosynthesis, Fluorescence spectroscopy, Proton motive force, Membrane de-energization/depolarization

1 Introduction

The increasing prevalence of bacterial resistance to clinically important antimicrobials has resulted in an urgent need for alternative antimicrobial drugs [1]. Antimicrobial peptides (AMPs) are an integral part of most host defense systems, and have favorable features as antimicrobial drugs [2]. In contrast to conventional antibiotics, development of resistance to AMPs can be more difficult and costly for the cell to maintain as a consequence of membrane permeabilization and multi-target antimicrobial activity [2–4].

Paul R. Hansen (ed.), *Antimicrobial Peptides: Methods and Protocols*, Methods in Molecular Biology, vol. 1548,
DOI 10.1007/978-1-4939-6737-7_12,

Accurate elucidation of AMP modes of action requires a combination of several experimental approaches. The most commonly reported AMP mode of action, membrane permeabilization, is often caused by pore formation and can be identified by tracing the leakage of material using small molecule probes in lipid vesicles, fluorescence microscopy, and electron microscopy methods [5, 6]. However, identification of additional AMP modes of action using inhibitory concentrations is often eclipsed by membrane depolarization, which causes secondary effects on DNA, RNA, protein, and cell wall synthesis [7]. Instead, macromolecule biosynthesis experiments to assess AMP inhibition on DNA, RNA, protein, or cell wall synthesis should be conducted at subinhibitory concentrations, identified from growth curve analysis. Because partial membrane permeabilization can lower macromolecule biosynthesis rates, additional experiments to measure relative proton motive force (PMF) are required to determine if macromolecule synthesis inhibition is due to targeted AMP action or is a consequence of membrane permeabilization. Measurement of macromolecule biosynthesis rates coupled with analysis of AMP impact on PMF, using probe-based fluorescence spectroscopy, is an accurate approach to explore the existence of multiple modes of action.

Macromolecule biosynthesis assays measure the synthesis rate of four main cellular components: (DNA, RNA, protein, and cell wall) by determining the incorporation rate of corresponding radiolabeled precursors [8]. These assays estimate the percent inhibition of precursor incorporation in the presence of subinhibitory AMP concentrations with respect to a negative (unexposed) control. A reduced incorporation rate upon AMP exposure indicates possible AMP interference in a particular cellular process. However, this interference can be indirectly caused by cellular PMF energy dissipation at the cytoplasmic membrane [9], a common effect of the membrane permeabilization activity of many AMPs. Hence, an additional probe-based fluorescence spectroscopy approach is needed to measure relative PMF, allowing evaluation of the AMP effects on membrane potential.

Fluorescence spectroscopy measures relative change of PMF of the cytoplasmic membrane based on the fluorescence of a fluorophore probe, which fluoresces proportionally to dissipation of membrane potential [10]. Under normal conditions the fluorophore concentrates on the cytoplasmic membrane. Upon dissipation of PMF, the fluorophore molecule dissociates from the cytoplasmic membrane and is released into the assay medium, increasing fluorescence emission. Exposure of bacteria to subinhibitory concentrations of AMP could be sufficient to partially permeabilize the membrane, allowing diffusion of ions and weakening PMF without significant leakage of cellular contents. Concentration-dependent effects of AMPs on PMF can be evaluated by this probe-based fluorescence spectroscopy method.

2 Materials

2.1 Macromolecule Biosynthesis Assay

1. Cation-adjusted Mueller-Hinton Broth II (MHB II): 0.3% (w/v) beef extract, 1.75% (w/v) acid hydrolysate of casein, 0.15% (w/v) starch in distilled water, and supplement with 20–25 mg/l of calcium and 10–12.5 mg/l magnesium, pH 7.3–7.4 (*see* **Note 1**).
2. Spectrophotometer.
3. Centrifuge.
4. ThermoMixer.
5. Radionuclide: DNA, RNA, protein, and cell wall precursors are tritiated (^{3}H)-thymidine, (^{3}H)-uridine, (^{3}H)-isoleucine, and (^{3}H)-glucosamine hydrochloride, respectively (*see* **Note 2**).
6. 30% trichloroacetic acid (TCA) (w/v) solution in water.
7. Vacuum manifold including chimney weights and vacuum pump.
8. Blotting filter paper.
9. Cellulose membrane filter.
10. Scintillation vial.
11. Scintillation liquid.
12. Scintillation counter.

2.2 Fluorescence Spectroscopy

1. 1 mM 3, 3′-Dipropylthiadicarbocyanine iodide [DiSC3(5)] solution in DMSO and stored at 4 °C with a cover of aluminum foil to protect it from direct exposure of light.
2. Spectrofluorometer.
3. Transparent 2 ml plastic disposable cuvette.

3 Methods

3.1 Macromolecule Biosynthesis Assay

Effect of subinhibitory or sub-bactericidal concentrations of AMP on the macromolecule biosynthesis rate can identify AMP inhibition of cellular processes. Following this method, the rate of any cellular process can be measured using radiolabeled precursors specific to that particular process. In general, four major macromolecule biosynthesis (DNA, RNA, protein, and cell wall) rate studies can provide initial hints regarding the intracellular targets of AMPs. Nisin, a lantibiotic, known to possess multiple modes of action, can be used in this method as a positive control in Gram-positive species [11]. The protocol below has been optimized for staphylococci.

1. Bacteria are inoculated in 5 ml MBH II and incubated at 37 °C overnight in the incubator with shaking 150 rpm (*see* **Note 1**).

2. Overnight cultures are subcultured at 1:100 dilution in 25 ml MHB II media and incubated at 37 °C with shaking at 150 rpm. Cultures are grown up to optical density (OD) ~ 0.2 at 600 nm. The OD of the culture needs to be monitored by checking the OD of 1 ml culture in a 30 min interval using a spectrophotometer.
3. Pellet down the entire volume of freshly grown cells by centrifugation at speed 1500 × *g* for 10 min at room temperature (RT), and suspend the pellet in 25 ml MHB II media.
4. Distribute 2 ml of the cell suspension to four 2.2 ml microfuge tubes (*see* **Note 3**).
5. Add the four radiolabeled precursors, to a final concentration of 0.1 mCi/ml, individually to the four 2.2 ml tubes (*see* **Note 4**). Vortex and immediately transfer 0.5 ml volumes to four new 1.5 ml tubes to create a duplicate set with a technical replicate (two for control and two for the test compound). To one pair of tubes, add a subinhibitory concentration of AMP, and to the control pair, add an equal volume of solvent. This arrangement shall be repeated for each precursor (*see* **Note 5**).
6. Pulse-label the samples by incubation for 20 min at 37 °C (*see* **Note 6**).
7. Add an equal volume of 30 % ice-cold TCA to precipitate down the radiolabeled cells. Keep in ice for 1–2 h.
8. Place cellulose membrane filters on vacuum manifold, add the chimney weights, and switch on the vacuum pump (*see* **Note 7**).
9. Transfer 1 ml cell precipitates to the cellulose membrane filters, and subsequently wash three times with 3 ml ice-cold 15 % TCA and then two times with 3 ml ice-cold water (five washes in total).
10. Turn off the pump and move the cellulose membrane filters to blotting paper to air-dry overnight.
11. Transfer the membrane to 10 ml scintillation vials with forceps and add 3 ml scintillation fluid to each vial to soak the membrane in liquid for a minimum of 1 h.
12. Finally, transfer the scintillation vials, containing membrane and scintillation fluid, to the holding rack of the scintillation counter and place inside the counter.
13. Select the flag that corresponds to 1 min ^{3}H count program and start the run of counting.
14. Radioactive count of a control without antimicrobial treatment is considered as 100 % precursor incorporation, and accordingly, percentage rate of precursor incorporation for other samples shall be calculated relative to the untreated control. Examples of DNA and cell well synthesis rates are shown in Fig. 1.

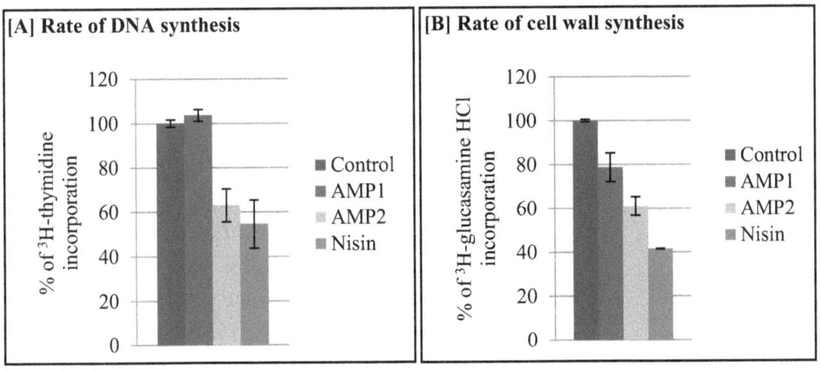

Fig. 1 Rate of radiolabel precursor incorporation for (**a**) ^{3}H-thymidine (DNA synthesis) and (**b**) ^{3}H-Glucosamine hydrochloride (cell wall synthesis). Initial rates of radiolabeled precursor incorporation were determined relative to the untreated sample (100 %)

3.2 Fluorescence Spectroscopy

Effect of sub-lysis concentrations of AMP on bacterial PMF can be measured by fluorescence spectroscopy. Comparison of the degree of PMF dissipation to the reduced rate of radiolabeled precursor incorporation can determine if inhibition of a cellular process is specifically targeted in that synthetic pathway or if inhibition is due to de-energization of the cytoplasmic membrane (i.e., PMF dissipation). Nisin can be used in this method as a control to compare the effect of antimicrobial peptide on membrane PMF and macromolecule biosynthesis. This protocol has been optimized for staphylococci.

1. Bacteria are inoculated in 5 ml MBH II and incubated at 37 °C overnight in the incubator with shaking 150 rpm.
2. Overnight cultures are subcultured in 1:100 dilution in 25 ml MHB II media with shaking at 150 rpm. Cultures are grown up to optical density (OD) ~ 0.2 at 600 nm. OD of the culture is monitored by checking the OD of 1 ml culture in 30 min interval using a spectrophotometer.
3. Dilute the freshly grown cells to OD ~ 0.1 at 600 nm with equal volume of MHB II and aliquot 1 ml of diluted culture to a 2 ml plastic disposable cuvette.
4. Label the cells with 1 μM DiSC3(5) and incubate at RT for 2–3 min covered by aluminum foil to avoid direct exposure to light (*see* **Note 8**).
5. Set the fluorescence spectrometer with excitation (Ex) and emission (Em) wavelengths 546 nm and 573 nm, respectively, in time drive continuous mode with Ex/Em slits 5 nm/5 nm (*see* **Note 9**).
6. Place the cuvette of DiSC3(5)-labelled cells in the holder of the spectrofluorometer, and start to take the continuous emission spectra over time for 2–3 min. The emission spectra will stabilize over time.

7. When emission spectra has stabilized, AMP or nisin or solvent control can be added to the labeled cell suspension with the continuous recording of emission spectra (*see* **Note 5**). For every individual exposure, a new cuvette with DiSC3(5) freshly labeled cells needs to be used (*see* **Note 10**).
8. Transfer DiSC3(5) emission spectra to excel (or similar plotting software), and plot the emission spectra over time. The change of DiSC3(5) fluorescence upon AMP or nisin addition ($FU_{after\ treatment} - FU_{before\ treatment}$) can be plotted as column plot. (FU = arbitrary fluorescence unit). An example of each plot is shown in Fig. 2. Interpretation of AMP mode of action is achieved by comparison of fluorescence spectroscopy and macromolecule biosynthesis assay results (*see* **Note 11**).

4 Notes

1. The final pH of the media needs to be adjusted to 7.3–7.4 with 1 normal (N) sodium hydroxide (NaOH). Dispense media aliquots in glass bottles, autoclave them at 120 °C, 15 psi for 20 min, and store the sterilized media at room temperature. MBH II media is also available from commercial suppliers. In general, minimal growth media is the best choice to grow bacteria for macromolecule biosynthesis rate measurement because the low abundance of natural macromolecule precursors in this media increases cellular utilization of radiolabeled precursors. Unfortunately, AMP activity is salt concentration dependent, and that limits the use of standard minimal media for AMP mode of action studies. Therefore, to increase radiolabeled precursor incorporation in MHB II media, which increases assay sensitivity, a higher concentra-

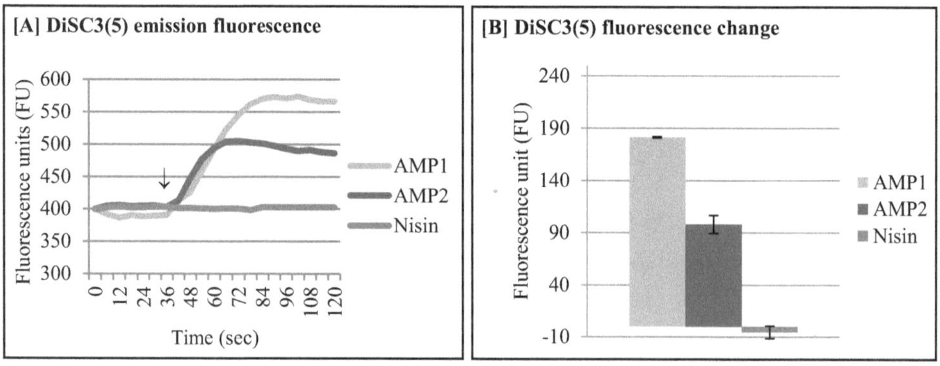

Fig. 2 Emission of DiSC3(5) fluorescence. (**a**) Time-spectra plot of DiSC3(5) emission over time. AMPs or nisin was added at 36 s, indicated by *arrow*. (**b**) Maximum change in DiSC3(5) fluorescence after AMP or nisin addition

tion of labeled precursor probe can be considered. The concentration given in this protocol (0.1 mCi/ml) should be sufficient for most applications.

2. Although the decay rate of ^3H is very slow, radionuclide-handling guidelines must be followed.
3. Four tubes will be needed to measure the rate of incorporation of radiolabeled DNA, RNA, protein, and cell wall precursors of one AMP sample and one unexposed control sample. For additional samples, increase the cell aliquot volume and precursor such that it can be easily divided. For example, three samples and one control need a 4 ml aliquot of cell suspension and shall be divided to equal 0.5 ml aliquots after precursor addition.
4. For slow-acting AMPs, bacteria can be exposed to AMP prior to radiolabeled precursor addition. However, cell numbers between control and exposed samples need to be equalized by measuring and adjusting the OD before the addition of labeled precursor. Additional precautions need be taken for slow-acting AMPs because prolonged inhibition of one macromolecule synthesis pathway may collaterally reduce the synthesis rate of others. Consequently, if inhibition in all synthesis pathways is detected, a shorter incubation time is recommended.
5. AMP concentrations need to be carefully considered and determined by growth curve analysis in growth conditions identical to the desired assay. It may be necessary to choose several different subinhibitory concentrations to achieve a full view of relative macromolecule synthesis inhibition and PMF dissipation. Carbonyl cyanide 3-chlorophenylhydrazone (CCCP), a known protonophore, can be used to study or standardize the effect of PMF dissipation on cellular growth and its consequence on macromolecule biosynthesis rate [12]. In addition, the bactericidal effect can be measured by plating aliquots of the culture during growth curve analysis and by colony counting after plate incubation.
6. Do not increase incubation time to increase radioactive counts in the sample. Instead, if higher counts are desired, use a higher mCi/ml concentration to increase counts. An increased incubation time may decrease assay sensitivity to discriminate between inhibited and uninhibited samples, as explained below. Longer incubations will decrease the concentration of the radiolabeled precursor in the media (and thus the rate of precursor incorporation) over time in a non-inhibited sample. In contrast, in an inhibited sample, the concentration of the radiolabeled precursor and the rate of precursor incorporation will remain almost constant over time. Thus long incubations may equalize the counts of inhibited and uninhibited samples

due to the limitation of precursor. Hence, short incubations keep the radiolabeled precursor concentration constant in the medium for both control and AMP-exposed samples, allowing accurate measurement of the relative initial rate of radiolabeled precursor incorporation.

7. To miniaturize the assay and perform the experiment in a high-throughput manner, the macromolecule synthesis assay can be performed in a 96-well format using multiscreen filter plates. In such case, after the wash and dry steps, the scintillation counting of samples from individual wells of the plate can be taken using a Microplate Scintillation Counter. Volumes of cell suspension, cellular precipitates, and washes should be reduced when using this approach.
8. For aerobic bacteria, DiSC3(5)-labelled cells cannot be stored for a long time (>15 min) because cellular PMF dissipates under anaerobic conditions.
9. The spectrofluorometer needs to be turned on 30 min in advance, to avoid the fluctuation of light source. Slits may need to be optimized depending on the model or light source of the spectrofluorometer used and the fluorescence property of the fluorophore to get clear signal with respect to background noise.
10. Membrane potential of bacteria is dependent on the cellular energy state that relies on the respiration of aerobic bacteria, which necessitates continuous aeration. Since the lack of aeration may affect the PMF of the cell, maximum 2–3 samples should be handled at a time for PMF measurement. Subculturing of cells in 20–30 min intervals can enable continuous assay of large numbers of samples.
11. Comparison of fluorescence spectroscopy and macromolecular biosynthesis experiments results is required to determine if inhibition of a cellular process, detected by a decreased rate of macromolecule biosynthesis, is due to the direct AMP action or is a consequence of AMP action on the membrane resulting in PMF dissipation. Figures 1 and 2 are used to illustrate the results of this analysis. Nisin does not have any effect on PMF (Fig. 2), indicating that the cell wall and DNA synthesis inhibition by nisin (Fig. 1) are not due to a secondary effect of membrane permeabilization. However, it should be noted that DNA synthesis inhibition may be a consequence of cell wall synthesis inhibition. Interpretation of the results is more difficult for the two AMP compounds (AMP1 and AMP2). Both AMP1 and AMP2 decreased the rate of incorporation of cell wall synthesis precursors (Fig. 1). Since the relative PMF dissipation mediated by the AMPs (Fig. 2) correlates similarly to the degree of inhibition in macromolecule biosynthesis, inhibition could be indirectly caused by AMP action on the membrane and subsequent PMF dissipation. Although AMP1

inhibited 20% cell wall synthesis, it did not hamper the synthesis rate of DNA, suggesting that either DNA synthesis is less sensitive to partial PMF dissipation or that this AMP can indeed inhibit cell wall synthesis. Although AMP2 dissipated PMF to a lesser degree than AMP1, it had a greater degree of DNA synthesis inhibition, suggesting a potential intracellular mode of action of AMP2. Additional experiments with these peptides at varying concentrations are needed to clearly establish if PMF dissipation is directly correlated to a corresponding degree of DNA and cell wall synthesis inhibition. Comparison of macromolecule synthesis inhibition and PMF dissipation of CCCP-treated control samples may further aid interpretation. Additional studies, e.g., measurement of accumulated precursors of inhibited cellular pathways by mass spectrometry of cellular content, can be performed to further detail the mode of action [8].

Acknowledgments

The work reported here has been supported by grants from the University of Copenhagen Centre for Control of Antimicrobial Resistance (UC-Care) and Zoetis.

References

1. European Centre for Disease Prevention and Control (2015) Antimicrobial resistance surveillance in Europe 2014. Annual report of the European Antimicrobial Resistance Surveillance Network (EARS-Net). ECDC, Stockholm. doi:10.2900/23549
2. Tavares LS, Silva CSF, de Souza VC, da Silva VL, Diniz CG, Santos MO (2013) Strategies and molecular tools to fight antimicrobial resistance: resistome, transcriptome, and antimicrobial peptides. Front Microbiol 4:1–11
3. Nizet V (2006) Antimicrobial peptide resistance mechanisms of human bacterial pathogens. Curr Issues Mol Biol 8:11–26
4. Guilhelmelli F, Vilela N, Albuquerque P, Derengowski L d S, Silva-Pereira I, Kyaw CM (2013) Antibiotic development challenges: the various mechanisms of action of antimicrobial peptides and of bacterial resistance. Front Microbiol 4:1–12
5. Marcellini L, Giammatteo M, Aimola P, Mangoni ML (2010) Fluorescence and electron microscopy methods for exploring antimicrobial peptides mode(s) of action. Methods Mol Biol 618:249–266
6. Wimley WC (2015) Determining the effects of membrane-interacting peptides on membrane integrity. In: Cell-penetrating peptides: methods and protocols. Humana Press, New York, NY, pp 89–106
7. Hancock REW, Rozek A (2002) Role of membranes in the activities of antimicrobial cationic peptides. FEMS Microbiol Lett 206:143–149
8. Ling LL, Schneider T, Peoples AJ, Spoering AL, Engels I, Conlon BP, Mueller A, Hughes DE, Epstein S, Jones M, Lazarides L, Steadman V, Cohen DR, Felix CR, Fetterman KA, Millett WP, Nitti AG, Zullo AM, Chen C, Lewis K (2015) A new antibiotic kills pathogens without detectable resistance. Nature 517:455–459
9. Weigel PH, Englund PT (1976) Inhibition of DNA replication in *Escherichia coli* by dibromophenol and other uncouplers. J Biol Chem 252:1148–1155
10. Farha MA, Verschoor CP, Bowdish D, Brown ED (2013) Collapsing the proton motive force to identify synergistic combinations against *Staphylococcus aureus*. Chem Biol 20: 1168–1178

11. Münch D, Müller A, Schneider T, Kohl B, Wenzel M, Bandow JE, Maffioli S, Sosio M, Donadio S, Wimmer R, Sahl HG (2014) The lantibiotic NAI-107 binds to bactoprenol-bound cell wall precursors and impairs membrane functions. J Biol Chem 289:12063–12076
12. Bulathsinghala CM, Jana B, Baker KR, Postle K (2013) ExbB cytoplasmic loop deletions cause immediate, proton motive force-independent growth arrest. J Bacteriol 195:4580–4591

Chapter 13

Using Confocal Microscopy and Computational Modeling to Investigate the Cell-Penetrating Properties of Antimicrobial Peptides

Gabriel Del Rio, Edda Klipp, and Andreas Herrmann

Abstract

Antimicrobial peptides (AMPs) may display the ability to penetrate cells, which may be relevant for their antibiotic activity. To investigate the relevance of the penetrating activity for the antibiotic activity of AMPs, here we describe a method based on the combined use of confocal microscopy and computational modeling coupled with cell death kinetics.

Key words Antimicrobial peptides, Cell-penetrating peptides, Mathematical modeling, Confocal microscopy, Multifunctional peptides, Cell death kinetics

1 Introduction

The widespread use of AMPs among living forms [1] and their potential use in pharmaceutical industry [2] constitute a driving force to characterize the activity of these peptides. Different studies have shown that several AMPs display other activities such as cell-penetrating activity [3], antiviral activity [4], or immunomodulating activity [5]. Of particular interest is the relationship between AMPs and cell-penetrating peptides (CPPs) since AMP and CPP activities may be just the same activity, as has been early noted [6]. For instance, many AMPs and CPPs are cationic and amphipathic, features that may be relevant for membrane translocation; in agreement with this idea, it has been noted that at high CPP concentrations these peptides may permeabilize cell membranes acting as pore-forming AMPs [7], while AMPs may reach intracellular targets before membrane permeabilization thus acting as CPPs [8]. Hence, to understand the mechanism of action of AMPs, it is relevant to establish the significance of the CPP activity for the antimicrobial activity.

Paul R. Hansen (ed.), *Antimicrobial Peptides: Methods and Protocols*, Methods in Molecular Biology, vol. 1548,
DOI 10.1007/978-1-4939-6737-7_13,

AMP activity is commonly determined by following the optical density of cells that have been exposed to AMPs as a way to interfere with cell growth [9]. Such effect may be the consequence of cell growth arrest or cell death, and further experiments are required to discriminate between these two possibilities, such as determining the number of colony-forming units (CFUs) and the use of fluorescent markers of cell metabolic activity to distinguish the living from the death cells [10]. For any AMP the mechanism of action involves the peptide interaction with the cell membrane. As a result of this interaction, the peptide may disrupt the membrane integrity through different mechanisms [11], e.g., pore-forming mechanism and carpet-like mechanism. Alternatively, the peptide may penetrate the cell membrane without any apparent damage to the membrane integrity to find an intracellular target (e.g., DNA [12]) and exerts its antimicrobial activity. The peptide internalization may as well be accomplished by different mechanisms [13], including those involving energy-dependent mechanisms (e.g., receptor-mediated internalization, endocytosis) or energy-independent translocation of the peptide.

Attaching a fluorogenic compound to a peptide may be used to detect the peptide in the interior of a cell using confocal microscopy; this is commonly used to monitor CPP activity [14]. An important limitation of this approach is that the labeled peptide may be internalized into cells independently of the peptide's antimicrobial activity. To test whether the CPP activity is relevant for the AMP activity, it is desirable to correlate the observed internalization rate of the peptide with the AMP activity. For instance, it is possible to compare the kinetics of the cell death induced by an AMP with the cell internalization rate by mathematical modeling; this modeling in turn may be used to predict possible mechanisms of internalization at play during the AMP activity.

In this work we describe a procedure aimed to test for the relevance of CPP activity for AMP activity [15]. The basic idea of our work is described in Fig. 1. Briefly, live cells are exposed to an AMP known to kill them [16]; in parallel, using confocal microscopy, the intracellular presence of a fluorescent version of this peptide is observed. Once both activities are observed, the rate of cell death may be quantitatively estimated to fit the parameters of two distinct internalization mathematical models: one energy-dependent and another energy-independent. The observed fitting of the models to the experimental data may provide evidence supporting the relevance of the internalization for the observed phenotype (e.g., cell death), and the best model fitting the experimental data may predict the mechanism of internalization. This method has been applied to the study of the CPP activity observed in the AMP referred here as Iztli peptide 1, IP-1 [15]. Our results showed that the best model fitting the experimental data was the energy-independent one. We show by genetic studies that indeed endocytosis does not participate in the AMP activity of IP-1, yet the Ste2 receptor was required for this AMP activity [16].

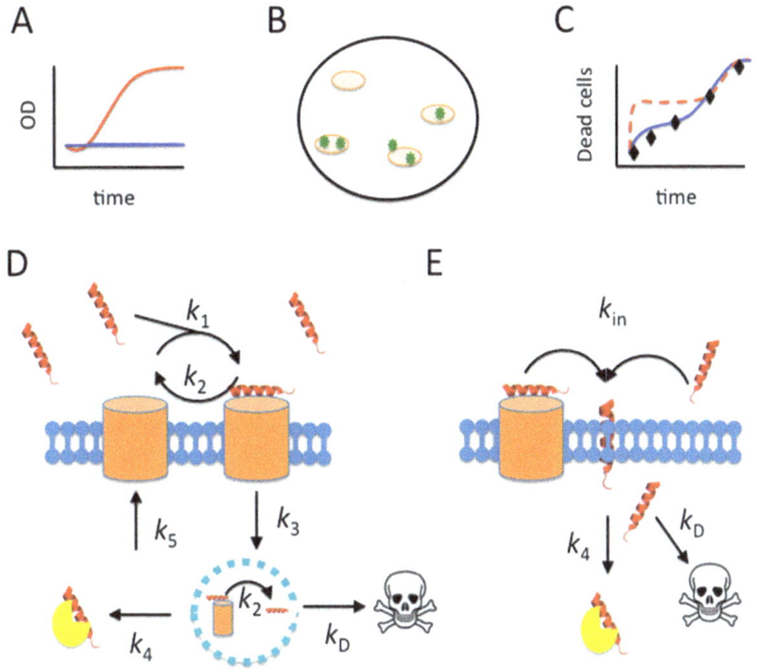

Fig. 1 Method summary. (**A**) Broth microdilution antimicrobial assay of peptides the red curve corresponds with bacterial cells growing in rich media and the blue curve corresponds to those bacteria exposed to an AMP. (**B**) Confocal microscopy of live cells (*orange ovals*) in the presence of a fluorescently labeled peptide (*green stars*). (**C**) Comparing two models fitted to the measured number of dead cells; in this example, the *blue* model fits best the experimental data represented by *black rhombs*. (**D**) Model assuming an energy-dependent internalization of an AMP (in this example, a receptor-mediated internalization). (**E**) Model assuming internalization of an AMP where the peptide may cross the membrane independent of receptor-mediated endocytosis

2 Materials

2.1 Peptides and Fluorophores

1. 5-(and-6)-carboxytetramethylrhodamine succinimidyl ester (5(6)-TAMRA SE), mixed isomers.
2. HiLyte Fluor™ 488 succinimidyl ester.
3. Peptides (>95% purity) may be synthesized as described in Chaps. 3 and 4 or obtained commercially. *See* Table 1 for full sequences.
4. IP-1 (KFLNRFWHWLQLKPGQPMY).
5. 70 μM TAMRA-IP-1.
6. 90 μM HiLyte Fluor™-IP-1.

2.2 Microbial Growth

1. Yeast synthetic complete media: yeast nitrogen base without amino acids 0.67%, potassium phosphate monobasic 0.1%, glucose 2%, and amino acids 0.079%.
2. *S. cerevisiae* strain BY4741.

Table 1
Fluorescent peptides

Name	Fluorophore	Peptide sequence	Modification point
IP-1	None	KFLNRFWHWLQLKPGQPMY	None
TAMRA-IP-1	TAMRA	KFLNRFWHWLQLKPGQPMY	(K1)
HiLyte Fluor-IP-1	HiLyte Fluor™ 488	KFLNRFWHWLQLKPGQPMY	(K13)

The table specifies the name of the peptide used in this work, the fluorophore name, the peptide sequence, and the modification in the peptide sequence where the fluorophore was linked. In parenthesis, the amino acid position is indicated, e.g., the first lysine is indicated by K1. Amino acids are indicated by one-letter code

3. Petri dishes (disposable 100 × 15 mm).
4. Solid YPD: dissolve 1.0 g yeast extract, 2.0 g Bacto™ peptone, 2.0 g glucose, 1.5 g agar in 100 mL of distilled water.
5. Sterilized media is poured into the petri dish when the media is still liquid.

2.3 Computational Modeling

1. Cell death kinetic data (number of dead cells vs. time) is used to fit two different internalization models (see below).
2. This fitting is done using the LSODA solver and an evolutionary optimization algorithm.
3. Here, we use the implementations within the modeling software COPASI (http://copasi.org).
4. For the statistical analysis of the data generated with the mathematical models and the experimental data, we used the R package (https://www.r-project.org) (*see* **Note 1**).

3 Methods

3.1 Labeling of Peptides

1. Peptides are essentially labeled with HiLyte Fluor™ 488 or TAMRA (*see* **Note 2**) as described in Chapter 22, Subheading 3.1. IP-1 is both N- and C-terminus-free (*see* **Note 3**).
2. All the fluorescent forms of IP-1 must be tested for MIC activity (*see* **Note 4**) as described in Chapter 22, Subheading 3.2.

3.2 Sample Preparation for Microscopy

1. Yeast cells are grown overnight on synthetic complete media (*see* **Note 5**).
2. Cells are diluted to a final optical density of 0.4 at 600 nm.
3. MATa cells are exposed to TAMRA-IP-1 (70 μM) or HiLyte Fluor-IP-1 (90 μM) and observed under the microscope after 15, 30, 45, and 60 min (*see* **Note 6**). It is recommended to use

a confocal microscope with at least 60× amplification for this analysis using yeast cells.

3.3 Microbial Growth

1. To determine the number of viable or death cells exposed to IP-1 over time, *S. cerevisiae* strain BY4741 is grown in YPD at 0.03 optical density (A600 nm).
2. Cells are exposed to IP-1 for 0, 30, 60, 120, and 180 min.
3. As a control, BY4741 cells are grown in the absence of IP-1 (*see* **Note 7**).
4. The number of live cells for both cells treated or not with IP-1 is estimated by counting at each time the number of living cells derived from a typical colony-forming units (CFU) assay.
5. Briefly, cells (treated or not with IP-1) are diluted to reach concentrations that are between 10 and 10^2 cells per sample (as a guide, 1.0 OD_{600} has on average 10^7 yeast cells).
6. The difference at each time between the non-treated cells with the treated cells corresponds with the number of cells killed by IP-1. This data may be used to fit the internalization models as described below.

3.4 Mathematical Modeling

Once the internalization and cell death kinetics have been measured, the kinetics data are fitted to two internalization models. For convenience, we have made available the models implemented for COPASI at http://bis.ifc.unam.mx:8090/CPP-AMP_Models

3.4.1 Energy-Dependent Model

In this model the internalization depends on a receptor (R) that recognizes the AMP. The following set of ordinary differential equations (square brackets denote concentrations, and the indices *e*, *m*, *v*, and *c* denote locations: extracellular, membrane, vacuole, and cytosol, respectively) summarizes this model (*see* **Note 8**):

$$\frac{d[R_m]}{dt} = k_2[R_m AMP_m] - k_1[R_m][AMP_e] + k_5 \quad (1)$$

$$\frac{d[R_m AMP_e]}{dt} = k_1[R_m][AMP_m] - (k_2 + k_3)[R_m AMP_m] \quad (2)$$

$$\frac{d[R_v AMP_v]}{dt} = k_3[R_m AMP_m] - (k_4 + k_5)[R_v AMP_v] \quad (3)$$

$$\frac{d[R_v]}{dt} = k_E[R_v AMP_v] - k_4[R_v] \quad (4)$$

$$\frac{d[AMP_c]}{dt} = k_E[R_v AMP_v] - k_4[AMP_c] \quad (5)$$

$$n_{\mathrm{D}} = n_{\mathrm{tot}} \frac{\mathrm{AMP_c}}{k_{\mathrm{D}}} \tag{6}$$

k_1 and k_2 are the association and dissociation constants for the receptor-AMP complex. The rates of endocytosis, degradation, and synthesis for the receptor (R) are represented by k_3, k_4, and k_5, respectively. Constants k_1 through k_5 are all derived from the literature in the case where R is Ste2p from *S. cerevisiae*. The last two constants, k_{D} and k_{E}, are obtained by adjusting the cell death kinetics to the model; k_{E} is the rate of endosome escape for the receptor, and k_{D} represents the intracellular concentration of the AMP required to kill a cell. The number of dead cells (n_{D}; the number of initial cells is n_{tot}) is evaluated in each step of the ordinary differential equations and compared with that observed in experiments.

3.4.2 Energy-Independent Model

In this model, the internalization is independent of endocytosis, yet it is assumed that there is an interaction between the receptor and the AMP; thus Eqs. 1 through 5 are ignored, and a new equation is proposed to model this internalization process:

$$\frac{\mathrm{d}[\mathrm{AMP_c}]}{\mathrm{d}t} = k_{\mathrm{in}}[AMP_{\mathrm{e}}] - k_4[AMP_{\mathrm{c}}] \tag{7}$$

Here, k_{in} represents the rate of internalization, and this is the parameter that is fitted from the experimental data. This fitting is achieved by using Eqs. 6 and 7 (*see* **Note 9**).

3.5 Computational Modeling

1. LSODA is accessible through the Time Course widget within the task menu of COPASI (see COPASI tutorial for a more detailed description: http://copasi.org/Support/User_Manual/Tasks/Time_Course_Simulation/).
2. Briefly, LSODA is used to calculate time courses in a deterministic fashion, while the Gibson and Bruck method is used to calculate time courses in a stochastic fashion. To run these simulations, it is necessary first to fit the missing parameters for each model described above. To do so, COPASI has a menu named Parameter Estimation under the Multiple Task menu (see COPASI tutorial for a more detailed description:http://copasi.org/Support/User_Manual/Tasks/Parameter_Estimation/).
3. After running the deterministic or stochastic time course simulations, the predicted number of dead cells is compared with the experimental values and the F statistical test performed to determine the best fitting model. For that goal the R package is used (for further details, visit the R package web site: https://cran.r-project.org/manuals.html).

4 Notes

1. Note that COPASI and R packages are two freeware available, but other packages may as well be used for similar purposes. For instance, for statistical analysis there is a wiki page that lists several packages for this end (https://en.wikipedia.org/wiki/List_of_statistical_packages). The web site http://systems-biology.org maintains a list of software that may be used for similar purposes than COPASI.
2. HiLyte Fluor™ 488 fluorescence is stable and independent of pH from pH 4–10, which is a major improvement over other dyes such as fluorescein or fluorescein isothiocyanate (FITC); TAMRA is a hydrophobic dye, facilitating the study of proteins or peptides associated to membranes, and its fluorescence is independent of pH between 4 and 10. Peptides modified with TAMRA while they may look soluble, sometimes form aggregates due to the hydrophobic nature of TAMRA. These aggregates may be observed under the microscope. In that case, it is recommended to sonicate the solution with the labeled peptide before this can be used for the internalization assay.
3. Modification of the C-terminus with a NH_2 group is commonly used to stabilize the helical structure of peptides. Yet, modification of the N- or C-terminus of peptides may affect positively or negatively the peptide activity. Thus, it is recommended to test the activity of peptides with and without modification of their N- or C-terminus.
4. The modification of peptides by attaching a fluorogenic dye sometimes renders slightly less active or inactive peptides. Thus, it is recommended to: (1) try different dyes and/or (2) try different sites where to attach the dye.
5. The media must be selected depending on the cell type or strain to be used. It is recommended that the same media used for cell growth is the one used for internalization assays. Since many AMPs are cationic, using a media with ions sometimes reduces the AMP activity and/or the penetrability of peptides.
6. The concentration of labeled peptides has to be determined in each case. Since AMP may kill cells, it is important to consider whether to use a sublethal or lethal concentration for the internalization assays. It is recommended to use sublethal concentrations of the labeled peptide to prevent the staining of dead cells that would be a false positive for the internalization assay. Note that using sublethal concentrations of the labeled peptide may induce longer times for the internalization to be observed; this is particularly the case for those peptides that are internalized independent of any receptor.

7. To expose cells for a specific time to the action of the peptide, cells are commonly centrifuged and washed to remove the peptide from the media. This process may reduce cell viability or cell number (or both) and hence, it is recommended that this procedure must be applied to cells not treated with peptide (control). Depending on the cell type or strain, these may be more or less sensitive to this washing procedure and may as well limit the accuracy to determine the number of living cells present at different times.
8. The differential equations presented may be used for any receptor-mediated endocytosis phenomena, since these equations represent the basic aspect of such type of endocytosis.
9. Note that the binding to the receptor in this model only affects the availability of the AMP when the receptor is degraded (k_4 in Eq. 7).

Acknowledgment

We would like to thank Dr. Maria Teresa Lara Ortiz for her technical assistance in the writing of this work. Grants from the Alexander von Humboldt Foundation (FKZ: 3.4-IP-DEU/113263) and PAPIIT (IN205911) to GDR supported this work.

References

1. Wang G, Li X, Wang Z (2009) APD2: the updated antimicrobial peptide database and its application in peptide design. Nucleic Acids Res 37:D933–D937
2. Seo MD, Won HS, Kim JH, Mishig-Ochir T, Lee BJ (2012) Antimicrobial peptides for therapeutic applications: a review. Molecules 17:12276–12286
3. Splith K, Neundorf I (2011) Antimicrobial peptides with cell-penetrating peptide properties and vice versa. Eur Biophys J 40:387–397
4. Falco A, Ortega-Villaizan M, Chico V, Brocal I, Perez L, Coll JM, Estepa A (2009) Antimicrobial peptides as model molecules for the development of novel antiviral agents in aquaculture. Mini Rev Med Chem 9:1159–1164
5. Haney EF, Hancock RE (2013) Peptide design for antimicrobial and immunomodulatory applications. Biopolymers 100:572–583
6. Henriques ST, Melo MN, Castanho MA (2006) Cell-penetrating peptides and antimicrobial peptides: how different are they? Biochem J 399:1–7
7. Palm C, Netzereab S, Hällbrink M (2006) Quantitatively determined uptake of cell-penetrating peptides in non-mammalian cells with an evaluation of degradation and antimicrobial effects. Peptides 27:1710–1716
8. Matsuzaki K, Murase O, Fujii N, Miyajima K (1995) Translocation of a channel-forming antimicrobial peptide, magainin 2, across lipid bilayers by forming a pore. Biochemistry 34:6521–6526
9. Otvos L, Cudic M (2007) Broth microdilution antibacterial assay of peptides. Methods Mol Biol 386:309–320
10. Jorgensen JH, Ferraro MJ (2009) Antimicrobial susceptibility testing: a review of general principles and contemporary practices. Clin Infect Dis 49:1749–1755
11. Gee ML, Burton M, Grevis-James A, Hossain MA, McArthur S, Palombo EA, Wade JD, Clayton AH (2013) Imaging the action of antimicrobial peptides on living bacterial cells. Sci Rep 3:1557
12. Marchand C, Krajewski K, Lee HF, Antony S, Johnson AA, Amin R, Roller P, Kvaratskhelia

M, Pommier Y (2006) Covalent binding of the natural antimicrobial peptide indolicidin to DNA abasic sites. Nucleic Acids Res 34:5157–5165

13. Shai Y (2002) Mode of action of membrane active antimicrobial peptides. Biopolymers 66:236–248
14. Bechara C, Sagan S (2013) Cell-penetrating peptides: 20 years later, where do we stand? FEBS Lett 587:1693–1702
15. Rodriguez Plaza JG, Morales-Nava R, Diener C, Schreiber G, Gonzalez ZD, Lara Ortiz MT, Ortega Blake I, Pantoja O, Volkmer R, Klipp E, Herrmann A, Del Rio G (2014) Cell penetrating peptides and cationic antibacterial peptides: two sides of the same coin. J Biol Chem 289:14448–14457
16. Rodríguez Plaza JG, Villalón Rojas A, Herrera S, Garza-Ramos G, Torres Larios A, Amero C, Zarraga Granados G, Gutiérrez Aguilar M, Lara Ortiz MT, Polanco Gonzalez C, Uribe Carvajal S, Coria R, Peña Díaz A, Bredesen DE, Castro-Obregon S, del Rio G (2012) Moonlighting peptides with emerging function. PLoS One 7:e40125

Chapter 14

Atomic Force Microscopy Study of the Interactions of Indolicidin with Model Membranes and DNA

Peter Fojan and Leonid Gurevich

Abstract

The cell membrane is the first barrier and quite often the primary target that antimicrobial peptides (AMPs) have to destroy or penetrate to fulfill their mission. Upon penetrating through the membrane, the peptides can further attack intracellular targets, in particular DNA. Studying the interaction of an antimicrobial peptide with a cell membrane and DNA holds keys to understanding its killing mechanisms. Commonly, these interactions are studied by using optical or scanning electron microscopy and appropriately labeled peptides. However, labeling can significantly affect the hydrophobicity, conformation, and size of the peptide, hence altering the interaction significantly. Here, we describe the use of atomic force microscopy (AFM) for a label-free study of the interactions of peptides with model membranes under physiological conditions and DNA as a possible intracellular target.

Key words Atomic force microscopy (AFM), Antimicrobial peptides (AMPs), Indolicidin, Supported planar bilayers (SPBs), Peptide–DNA interaction

1 Introduction

Atomic force microscopy (AFM) has proven to be a useful technique to elucidate surface structures of biological materials with unprecedented resolution (*see*, e.g., [1] for review). In particular, AFM imaging in liquid has seen significant improvement, allowing operation under physiological conditions. This has made it possible to monitor important biological processes in real time at high resolution (for review *see* [2] and references therein). Studies [3–5] have shown that in situ AFM, using supported planar bilayers (SPBs) as membrane mimics, is a very well-suited technique for studying peptide–membrane interactions. Specifically, AFM has been used to directly visualize how indolicidin induces pore formation in the gel state of negatively charged SPBs [6]. Furthermore, it has been found that indolicidin–membrane interactions are dependent on the composition of the model membranes as well as the peptide concentration [7, 8]. In the case of phospholipid model

Paul R. Hansen (ed.), *Antimicrobial Peptides: Methods and Protocols*, Methods in Molecular Biology, vol. 1548,
DOI 10.1007/978-1-4939-6737-7_14, © Springer Science+Business Media LLC 2017

membranes composed of phase-segregated phosphatidylcholine (PC) and phosphatidylglycerol (PG), indolicidin has been shown to interact with the membrane in a concentration-dependent manner. At low concentrations (<5 μM), indolicidin forms an amorphous layer localized at the fluid domains, in the case they contain anionic lipids (i.e., PG). At high peptide concentrations (>5 μM), indolicidin appears to initiate the melting of gel-phase domains, leading to a reduction in height, which is independent of the presence of an anionic lipid [9]. Another study, involving simultaneous AFM and confocal microscopy imaging, revealed that indolicidin is preferentially inserted into the fluid phase of the PC domains [10]. These data suggest that the indolicidin–membrane association is influenced greatly by specific electrostatic interactions, lipid fluidity, and peptide concentration [10]. This leads to the suggestion that, similar to the preferred activation mechanism of *Candida rugosa* lipase [11], the edges of the initial structural lipid defects are the preferred interaction sites for indolicidin with phospholipid bilayers composed of PC [9, 10]. Previously, we have described that the mechanism of action of indolicidin could be best seen and studied at the boundary of the gel–fluid domains when imaged by in situ AFM [9]. These studies showed that the physical properties of the lipids play a crucial role as the specificity of the peptide is dependent upon the composition of the membrane. The different phases of lipid membranes observed by AFM suggest that the process of indolicidin-induced membrane thinning requires phase segregation, where indolicidin acts at the boundary of the fluid and gel domains. It has also been suggested that indolicidin induces a gel–fluid phase transition in phospholipid bilayers [9]. Furthermore, as has been found in the above studies as well as in earlier works [8, 12], indolicidin does not lead to lysis of the cell membrane at physiologically relevant concentrations, which implies the presence of secondary mechanisms of antimicrobial actions. Several studies [13–15] have demonstrated a strong affinity between indolicidin and DNA, which leads to indolicidin–DNA complex formation and inhibits DNA replication, transcription, and repair mechanisms. Although the actual interaction mechanisms are still under debate, interaction with both membrane and DNA holds the key to understanding of the broad activity spectrum of indolicidin and eventually can pave the way to novel AMP-inspired antibiotics. With appropriate knowledge about the target organism, it should be possible to engineer a peptide with charge properties complementary to that of the membrane—optimal to bind but not too strong to prevent aggregation—and with hydrophobicity best suited to permeabilize the membrane while maintaining its DNA-binding properties.

In this chapter we show how AFM can be applied to study the interactions between indolicidin and differently charged model membranes on one side and indolicidin and DNA on the other side. In situ AFM imaging of the peptide–membrane interactions was carried out using SPBs of a one-component phospholipid, being either zwitterionic, cationic, or anionic (results with zwitterionic PC membranes were presented in our previous study [9]). To further approach an actual bacterial membrane, the results on a lipid mixture (20% PG and 80% PC), corresponding to a typical membrane composition of a bacterium [16], are shown. We further demonstrate the formation of indolicidin–DNA complexes, observed by AFM as a significant height increase of DNA molecules.

2 Materials

2.1 Reagents

1. Chloroform 99.8%.
2. Acrylamide solution 40% for electrophoresis.
3. l-α-Phosphatidylcholine (PC) from egg yolk (60% TLC purity).
4. (3-Aminopropyl)-trimethoxysilane (APTMS).
5. Ammonium acetate (99.99%).
6. Water (DNAse free).
7. Anhydrous toluene.
8. HEPES sodium salt ≥99.0%.
9. 1,2-Distearoyl-sn-glycero-3-phospho-(1′-rac-glycerol) (sodium salt) (DSPG).
10. 1,2-Dioleoyl-sn-glycero-3-phospho-(1′-rac-glycerol) (sodium salt) (DOPG).
11. 1,2-Distearoyl-sn-glycero-3-ethylphosphocholine (chloride salt) (ESPC).
12. 1,2-Dioleoyl-sn-glycero-3-ethylphosphocholine (chloride salt) (EOPC).

All the above chemicals were used as is without prior purification.

All buffer solutions were prepared using DI water (18 MΩ) at 25 °C and pH adjusted accordingly.

2.2 Peptide Synthesis

1. 9-Flurenylmethyloxycarbonyl (Fmoc)-protected amino acids.
2. Fmoc-protected Rink–MBHA resins.
3. 2-(1H-Benzotriazol-1-yl)-1,1,3,3-tetramethyluronium hexafluorophosphate (HBTU).
4. 1-Hydroxybenzoetriazole (HOBt).
5. Trifluoroacetic acid (TFA).
6. Triisopropylsilane (TIS).

7. Peptide cleavage mixture: (95 % TFA, 2.5 % TIS, and 2.5 % water).
8. 4-(2-Hydroxyethyl)-1-piperazineethanesulfonic acid (HEPES) buffer: 10 mM HEPES adjusted to pH 7.0.

2.3 HPLC

1. Semi-preparative C-18 reversed phase column.
2. Buffer A: H_2O with 0.1 % TFA.
3. Buffer B: Acetonitrile.

2.4 Vesicle Preparation for AFM and Supported Planar Lipid Bilayer Formation

1. Lipid solutions: Either PC, ESPC/EOPC (EPC) 4:1 w/w %, DSPG/DOPG (PG) 4:1 w/w %, or PC/PG (PC/PG) 4:1 w/w % at a concentration of 3 mg/ml in chloroform.
2. Round mica sheets (grade V1 muscovite mica, 9.5 mm diam.).
3. Mica surface solution: 5 mM $MgCl_2$.
4. Teflon disks, glued on metal chucks (specimen metal disks for AFM, 15 mm diameter).

2.5 AFM Imaging of DNA and DNA–Indolicidin Complexes

1. APTMS in anhydrous toluene (1:3 v/v).
2. Triangular silicon nitride cantilever (NP-S or NP series).
3. Small linearized pUC19 plasmid (2686 bps).
4. Ammonium acetate solution: 20 mM (AmAc).
5. Sephadex column: Illustra Nap-5 Sephadex G-25 gravity column equilibrated with 20 mM ammonium acetate.

2.6 Apparatus

1. Automatic peptide synthesizer.
2. Freeze dryer.
3. Preparative HPLC.
4. Analytical HPLC.
5. UV–Vis spectrophotometer.
6. LiposoFast extruder.
7. Nanoscope IIIa atomic force microscope.
8. All AFM images were processed using the WSxM software package [17].

3 Methods

3.1 Indolicidin Synthesis and Stock Solution Preparation

1. Indolicidin is synthesized by standard solid-phase peptide synthesis using Fmoc chemistry as described in Chapters 3 and 4. The obtained resin peptide is dried in a freeze dryer (*see* **Note 1**).
2. After synthesis, 3 ml of cleavage mix (*see* **Note 2**) is added to the entire amount of dried resin peptide and mixed vigorously for 60 min using a manual synthesizer.

3. The 3 ml cleavage mix is removed from the resin and collected in a 15 ml PP-tube. The resin is washed once with 3 ml pure TFA, and the liquid is collected into the same 15 ml PP-tube.
4. From the resulting 6 ml solution, the peptide is precipitated with 10 ml of ice-cold diethyl ether and centrifuged for 10 min.
5. The supernatant is decanted and the white precipitate is washed with 12 ml of ice-cold diethyl ether by centrifugation for 10 min. The procedure is repeated twice.
6. The white precipitate is dried in a freeze dryer overnight.
7. Indolicidin is purified by HPLC on a C-18 reversed phase column. For this, 2 ml of the dissolved peptide in 0.1% TFA in water is injected onto a semi-preparative C-18 reversed phase column, running a 10–80% acetonitrile gradient with an initial step of 10% acetonitrile.
8. The collected indolicidin is freeze-dried overnight.
9. Indolicidin stock solution is prepared by dissolving 5 mg of the freeze-dried powder in 10 ml of a 10 mM HEPES pH 7.0.
10. Indolicidin concentration is determined by absorbance measurement at 280 nm on a UV–Vis spectrophotometer. The peptide concentration is calculated using a theoretical molar extinction coefficient of 28,450 M^{-1} cm^{-1} [8].
11. The obtained solution is diluted to a final stock concentration of 25.0 μM, and 1 ml aliquots are stored at −21 °C (*see* **Note 3**).

3.2 Vesicle Preparation for Atomic Force Microscopy Experiments

1. Lipid solutions are prepared (*see* Subheading 2.4, **step 1**), and the solvent is removed by rotary evaporation overnight to form a dry lipid film on the inside of the round-bottom flask.
2. The lipids are rehydrated in 5 ml of 10 mM HEPES buffer (pH 7.0) in the rotary evaporator at 80 rpm and temperature above the phase transition temperature for the respective lipids used (50 °C for PC, 60 °C for ESPC/EOPC, and 60 °C for DSPG/DOPG are used in the experiment; *see* **Note 4**).
3. The vesicles are extruded 21 times using a LiposoFast extruder equipped with a polycarbonate filter with a pore diameter of 100 nm (*see* **Note 5**).

3.3 Supported Planar Lipid Bilayer Formation

1. Round mica sheets are glued with epoxy glue onto a 15 mm diam. Teflon disks, glued on metal chucks (*see* **Notes 6** and **7**).
2. The mica disks are freshly cleaved with adhesive tape right before use.
3. 50 μl of a respective hydrated vesicle solution is applied to the mica surface and incubated, in the case of PC, ESPC/EOPC, and DSPG/DOPG, for 20 min at room temperature. In the case of PG and PC/PG, the vesicle solution is mixed with

5 mM $MgCl_2$, applied to a freshly cleaved mica surface and incubated for 30 min (*see* **Note 8**).

4. The surface is rinsed with centrifuged 10 mM HEPES buffer at pH 7.0 (15 °C, 15 min, *see* **Note 9**).
5. All samples are used for AFM measurements immediately after preparation.

3.4 Atomic Force Microscopy Imaging of SPBs

1. The samples are prepared on mica substrates as described above.
2. A Nanoscope IIIa, equipped with a J-scanner and liquid cell cantilever holder, is used throughout the experiment. A 115 μm long triangular silicon nitride cantilever with a nominal spring constant of 0.58 N/m is used for imaging. The liquid cell is filled with centrifuged 10 mM HEPES buffer, pH 7.0, through the inlet of the liquid cell using a 2 ml syringe.
3. After filling the cell, the syringe is carefully removed avoiding formation of an air bubble between the cantilever and the cell (*see* **Note 10**).
4. AFM images are acquired using tapping mode at approximately 7 kHz and are captured as 512 × 512 pixel images. The scan rate is adjusted for each sample and is between 0.4 and 1 Hz.

3.5 Imaging of the Interaction Between Indolicidin and SPBs

1. The contact/tapping mode liquid cell placed over the Teflon sample holder with the attached sample is equipped with inlet tubes to allow injections of fluid into the chamber between measurements without removing the liquid cell. Reference images of liquid-equilibrated SPBs are acquired at this point (Fig. 1), prior to injection of indolicidin, to ensure surface stability and optimal imaging conditions (*see* **Note 11**).
2. Once a stable SPB had been formed as determined by in situ AFM imaging, 100 μl of the 25 μM indolicidin stock solution is injected into the liquid cell. Starting from this point, the effect of indolicidin on the SPBs is continuously monitored by in situ AFM in real time. The experiments are performed in identical manner on EPC (Fig. 2), PG (Fig. 3), and PG/PC (Fig. 4) planar bilayers (*see* **Note 12**).
3. To avoid evaporation of the liquid and formation of air bubbles in the liquid cell, the injections of 100 μl indolicidin stock solution are repeated at time intervals of approximately 2 h during the experiment.

3.6 AFM Imaging of DNA and DNA–Indolicidin Complexes

1. Mica sheets are attached to 15 mm metal chucks using double-sided adhesive and placed into a desiccator (*see* **Note 13**).
2. The desiccator is evacuated using a solvent-resistant membrane pump down to approximately 100 mTorr and flushed with

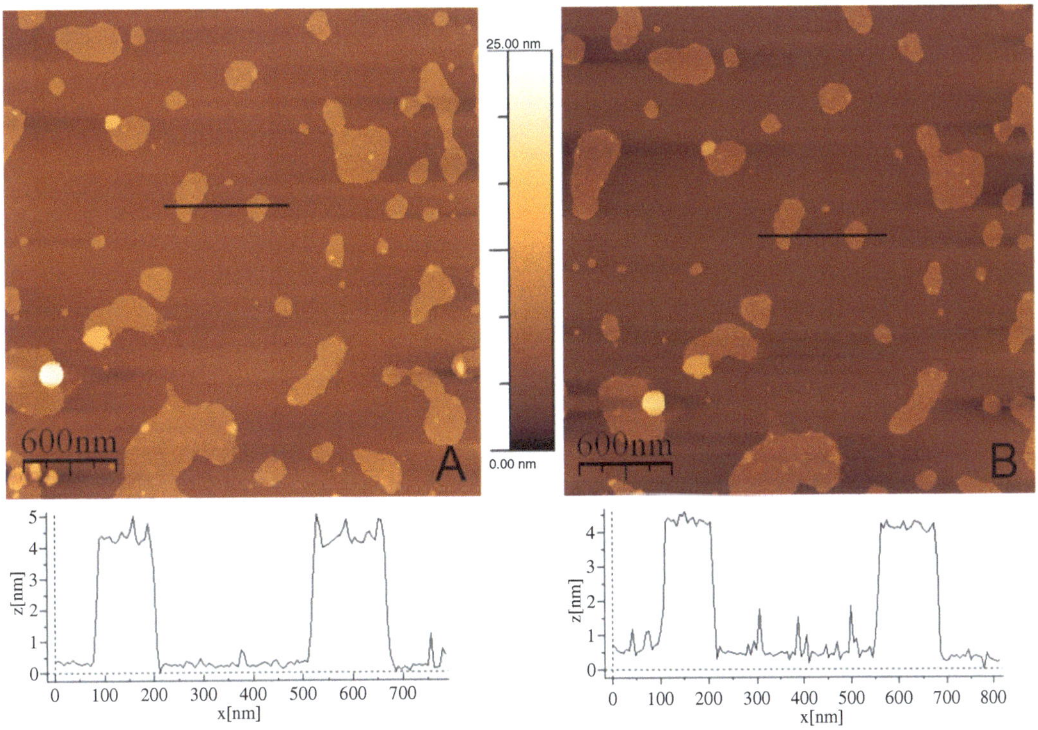

Fig. 1 AFM images of SPB formed from a 3 mg/ml EPC vesicle preparation (**a**) immediately after the addition of 10 mM HEPES buffer pH 7.0 and (**b**) after 2 h of continuous scanning on the same surface. Both images are 3 μm × 3 μm, scale bar is 600 nm, and height bar is 25 nm. Both images have been equalized

argon. The procedure is repeated at least three times to ensure that the atmospheric humidity is largely removed. In the end of the step, the desiccator is filled with argon up to atmospheric pressure (*see* **Note 14**).

3. A test tube containing 1 ml of APTMS in anhydrous toluene (1:3 v/v) is gently placed into the bottom of the desiccator, which is immediately evacuated down to 100 mTorr, flushed three times with argon, and left under argon at a pressure slightly below atmospheric pressure for 30 min. The resulting modified surface is positively charged and extremely flat and can be stored under argon for weeks (*see* **Note 15**).
4. For imaging of the DNA–indolicidin complexes, SmaI-linearized pUC19 plasmid is used (*see* **Note 16**). The pUC19 solution is purified using an Illustra Nap-5 Sephadex G-25 gravity column equilibrated with 20 mM ammonium acetate. Obtained aliquots are checked using a UV–Vis spectrometer, and those exhibiting absorption at 250 nm are retained and used as a stock solution of DNA.
5. 3 μl of the stock solution is diluted with 15 μl of 20 mM AmAc and deposited on the substrate (*see* **Note 17**).

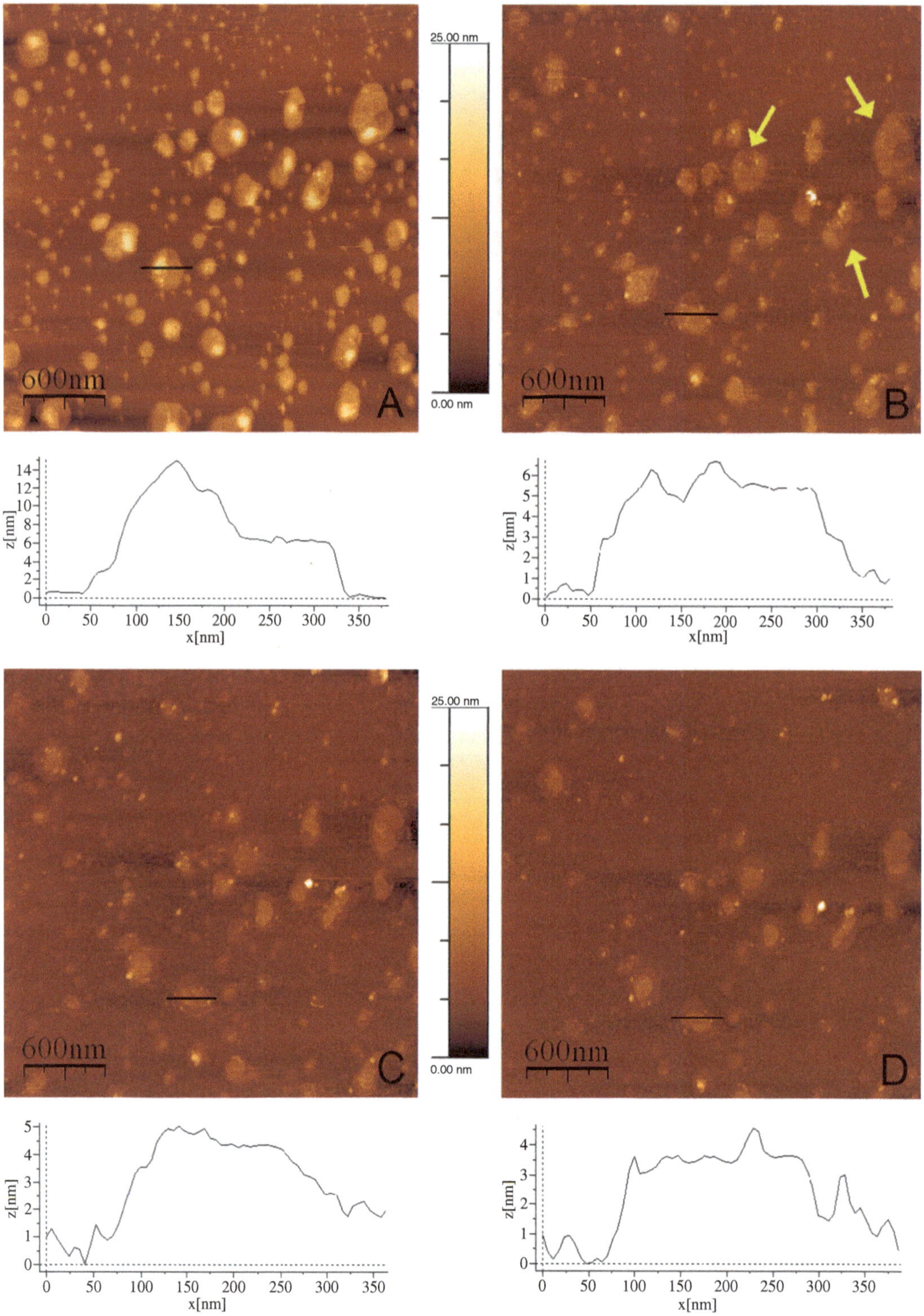

Fig. 2 AFM images of SPB formed from 3 mg/ml EPC vesicle preparation, equilibrated in 10 mM HEPES buffer pH 7.0: (**a**) prior to the addition of indolicidin and (**b**) after the addition of 25 μM indolicidin (approximately 25 min). Images (**c**) and (**d**) were acquired after approximately 90 min and 150 min of continuous scanning of the same surface, respectively. All images are 3 μm × 3 μm, scale bar is 600 nm, and height bar is 25 nm. 100 μl of 25 μM indolicidin was injected again after 1 h to keep the liquid level in the cell constant

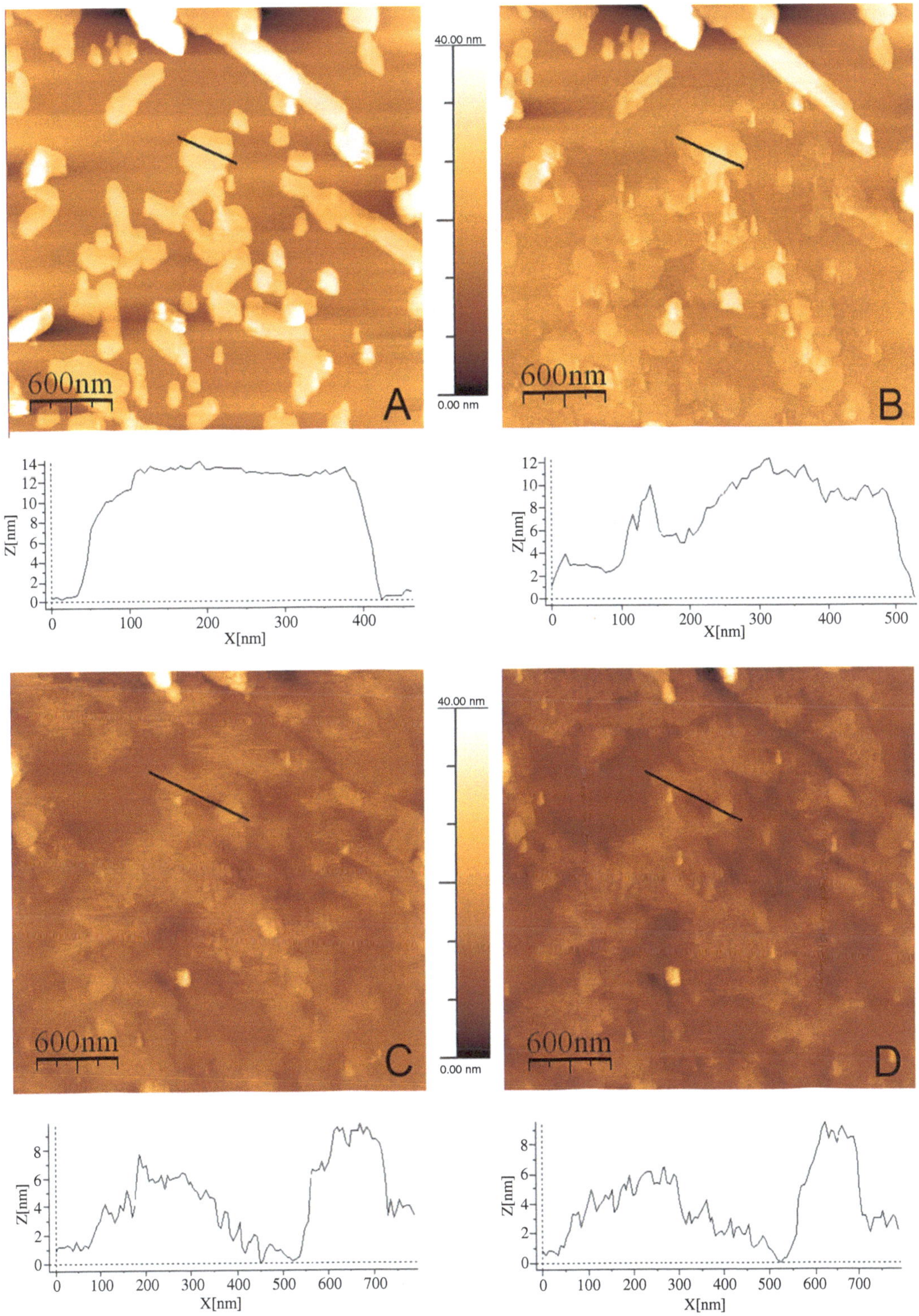

Fig. 3 AFM images of SPB formed from 3 mg/ml PG/PC vesicle preparation, equilibrated in 10 mM HEPES buffer pH 7.0: (**a**) prior to the addition of indolicidin and (**b**) immediately after the addition of 25 μM indolicidin (approximately 25 min). (**c**) and (**d**) correspond to approximately 110 and 180 min of continuous scanning of the same surface. All images are 3 μm × 3 μm, scale bar is 600 nm, and height bar is 40 nm

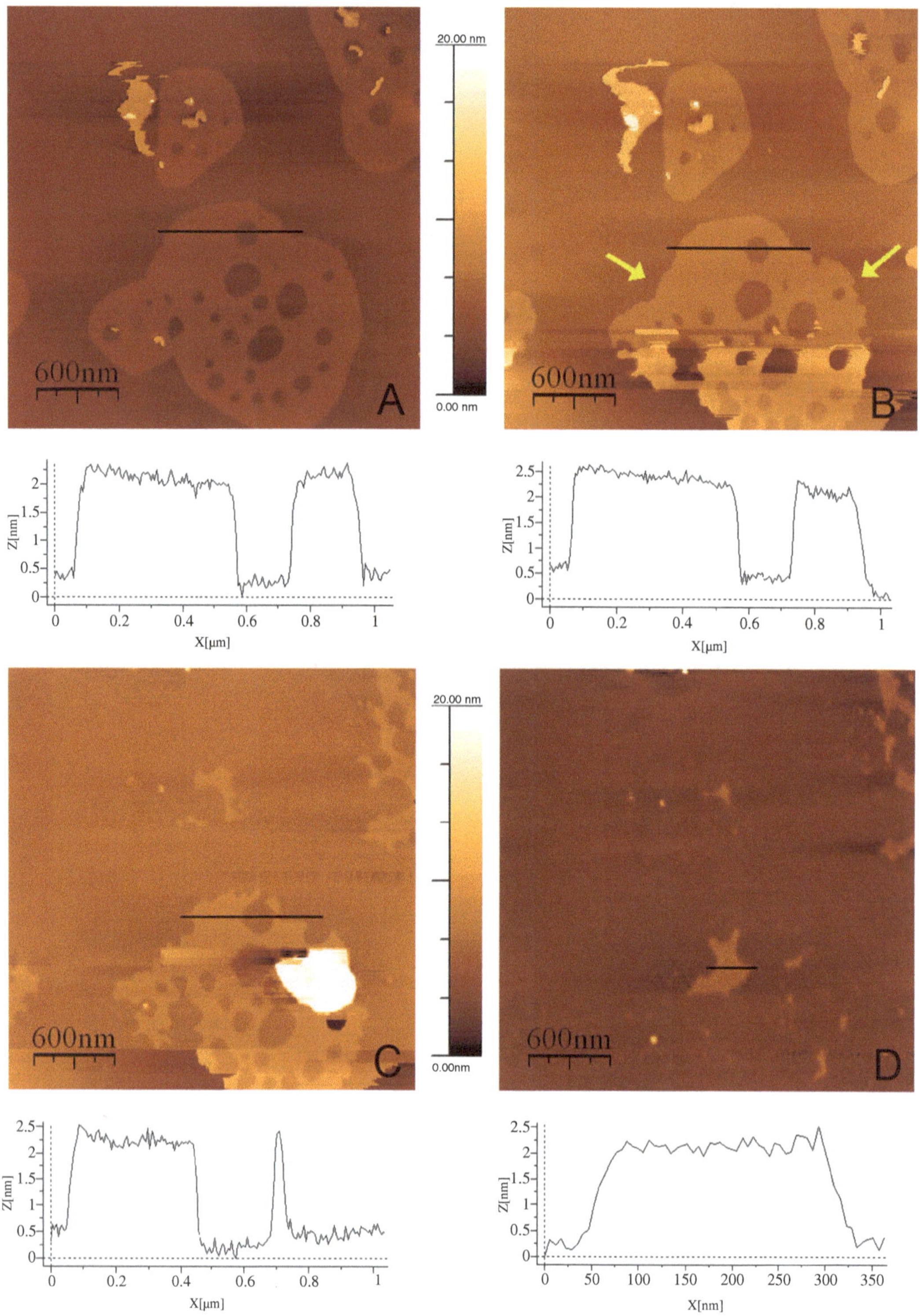

Fig. 4 AFM images of SPB formed from 3 mg/ml PG/PC vesicle preparation equilibrated in 10 mM HEPES buffer pH 7.0: (**a**) prior to indolicidin injection and (**b**) immediately after the addition of 25 μM indolicidin (approximately 14 min). Images (**c**) and (**d**) were acquired after approximately 28 and 42 min of continuous scanning on the same surface. All images are 3 μm × 3 μm, scale bar is 600 nm, and height bar is 20 nm. Height bar is 10 nm in image (**d**)

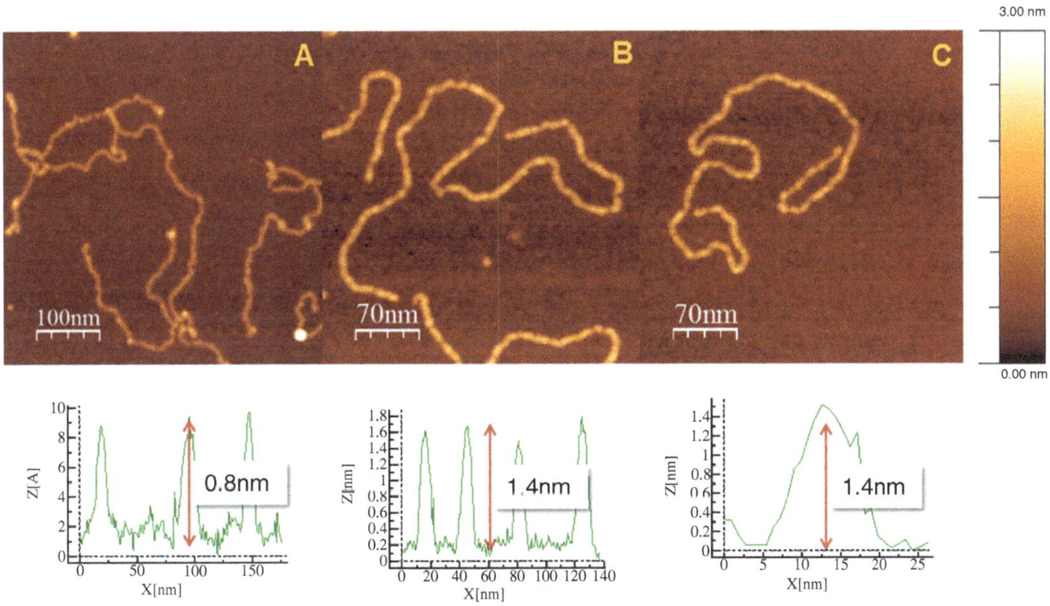

Fig. 5 AFM images of (**a**) pUC19 DNA and the respective indolicidin–DNA complexes obtained by (**b**) on-surface reaction of indolicidin with pre-adsorbed DNA and (**c**) by in-solution pre-formed indolicidin–DNA complexes deposited on the surface. The *lower panel* shows the respective heights of the DNA molecules in the panels above

6. After 5–6 min incubation, the droplet of DNA solution is removed from the surface by blowing it off with a stream of nitrogen.
7. A 10 μl droplet of indolicidin stock solution diluted with DNAse-free water (1:1 v/v) is deposited on the surface.
8. After 5–10 min of incubation, the surface is rinsed with 1–2 ml of DNAse-free water and the substrate is blown dry with nitrogen.
9. The obtained samples are scanned using a Nanoscope IIIa AFM in air at ambient conditions in tapping mode using soft noncontact cantilevers (nominal spring constant 1.8 N/m, resonant frequency ~70 kHz). Figure 5 shows a comparison of bare DNA molecules and DNA–indolicidin complexes obtained in two different ways (*see* **Note 18**).

4 Notes

1. Drying of the resin peptide can be also accomplished by placing the reactor tube into a desiccator evacuated with a solvent-resistant membrane vacuum pump until dry.
2. The cleavage mix has to be prepared fresh right before use.
3. Indolicidin stock solutions with an indolicidin concentration higher than 25 μM have been shown to lose their activity

within very short times, presumably, due to indolicidin self-aggregation. Therefore, it is important to immediately adjust the initial high indolicidin concentration to 25 μM in order to preserve the stock solution over a prolonged period of time. The 1 ml volume for the aliquots is chosen because it fits very well with the need for the AFM measurements. This is to avoid freeze thawing cycles of indolicidin stock solutions.

4. The formation of lipid vesicles from a dried lipid film through film rehydration is most efficient when the lipids are in a molten form, which is achieved by heating the system above the glass transition temperature of the respective lipids. The formed vesicles are multi-lamellar vesicles, having onion-like structure which consists of many layers on top of each other. To convert these vesicles into unilamellar vesicles, they need to be extruded.
5. The vesicle extrusion through the LiposoFast membrane is done 21 times. The odd number of extrusions is to avoid having the resulting unilamellar vesicles on the same side of the filter as the initial material. This provides a cleaner vesicle preparation.
6. The round shape of the mica disks helps to stabilize the droplet and avoids sharp corners that might break the surface tension leading to leakage. In general manufacturers tend to use a denser mica for the production of round disks which are less prone to delamination and water penetration between the mica layers, hence contributing to more stable imaging conditions.
7. The Teflon disk glued between the standard metal sample holder and the mica disk acts as a hydrophobic barrier to prevent liquid from leaking into the piezo tube scanner of the AFM. This setup stabilizes the liquid meniscus at the edge of the mica disk, allowing for an open cell scanning. An important advantage of this approach compared to the use of the standard closed cell is that the silicon o-ring between the cantilever cell holder and the sample surface is no longer required, hence eliminating drift and other imaging artifacts caused by friction between the sample scanner and the cantilever holder.
8. The 5 mM $MgCl_2$ is introduced to compensate for the negative charge on the vesicle surface and allow them to fuse with the mica surface.
9. HEPES buffer is centrifuged to avoid contaminating particles on top of the membrane layer. Over time, buffer solutions are known to form small nanocrystals due to temperature changes and aging phenomena. A large number of particles are found on the samples washed with uncentrifuged buffers. Typically, only the top third of the solution after centrifugation is used in the experiments.

10. Filling the liquid cell requires a careful and steady flow; therefore, a syringe with a total volume of 2 ml gave the best results. It is critical to flush out all air bubbles from the liquid cell, in particular those trapped between the cantilever and the cell top, as they make laser alignment impossible due to scattering and refraction. Furthermore, air bubbles in the vicinity of the cantilever will also be collected during scanning and eventually disturb imaging. If an air bubble is visible in the video microscope, careful cycling the liquid in and out can remove it.
11. Imaging of SPBs in HEPES buffer over a prolonged period of time provides an essential reference and also allows the system to equilibrate. In general, scanning of very soft materials is challenging as special care should be taken not to introduce scanning artifacts due to cantilever–surface interactions over a prolonged period of time. To ensure surface integrity and stability (stability refers to cessation of vesicle fusion and surface topography remaining unaffected by continuous scanning), the surface layer of the SPBs is continuously imaged over a period of 3 h prior to the addition of AMPs. The liquid cell is constantly monitored to prevent it from drying out. It is essential to ensure that the surface topology does not change during this prolonged imaging period [9].
12. The indolicidin injections are carried out by inserting a flat-tip syringe needle into the end of the inlet tube to ensure minimal pressure disturbance in the system. As the total volume of the liquid cell is just below 100 μl, an injection of 100 μl of a 25 μM indolicidin solution in HEPES buffer will exchange the buffer in the cell; however, small amounts of the remaining buffer inside the cell will lead to a dilution of indolicidin upon the first injection.
13. Due to the fact that mica acquires a negative surface charge in aqueous solution, direct attachment of negatively charged DNA is impossible. To achieve DNA deposition, either divalent metal cations, such as Mg^{2+}, bridging DNA to the surface, or covalent surface modification rendering it positively charged is necessary [18].
14. In general, the procedure can also be run with nitrogen; however, argon has some advantages over nitrogen. First of all, it is heavier than air, so it stays in the bottom of the desiccator and better protects the sample from humidity and atmospheric oxygen. Furthermore, nitrogen has a higher surface reactivity than argon which can be critical for surface modification purposes.
15. Samples prepared according to the described procedure exhibit positively charged surface at normal pH and a contact angle of about 60° and can be stored without losing their activity for

several hours at ambient conditions or for several days or even weeks in a desiccator under argon. Positively charged mica surface binds DNA molecules while preventing adsorption of cationic indolicidin. Another important advantage of this procedure, besides its simplicity, is that it essentially preserves atomic flatness of mica by forming a uniform (sub)monolayer of APTMS. Alternatively, good results are also obtained with the procedure involving silatrane described by Shlyakhtenko et al. [19]. This procedure leads to a water-resistant silane, which is preferable for surface modification in aqueous environments. However, the silatrane preparation and surface modification are somewhat more time consuming.

16. Linearized pUC19 has a length convenient for AFM measurements (ca. 700 nm in length). Moreover, SmaI produces blunt ends upon cutting and therefore the DNA is less prone to form networks upon surface deposition.
17. AmAc is a convenient and very clean, even on the nanoscale, salt which is used to adjust the ionic strength in solutions accordingly. An additional advantage of AmAc over classical used salts to regulate the ionic strength is its relative volatility, so it can evaporate from the substrate without leaving residual crystals.
18. The AFM images (Fig. 5) reveal that an indolicidin-coated DNA exhibits a different morphology and is significantly higher as compared to the native DNA. Alternatively, it is also possible to form the DNA–indolicidin complex in solution first, separate the complexes with gravity columns, and deposit them on an APTMS-treated mica surface. This method however produces the DNA–indolicidin complexes of the same morphology as the ones described above (cf. Fig. 5b, c), but has a significantly lower yield and is more time consuming.

Acknowledgments

This work was generously supported by grants from the Spar Nord Foundation and the Det Obelske Familiefond. The authors also acknowledge fruitful discussions with Patrick Markus.

References

1. Casuso I, Rico F, Scheuring S (2011) Biological AFM: where we come from–where we are–where we may go. J Mol Recognit 24:406–413
2. Lyubchenko YL (2013) AFM visualization of protein–DNA interactions. In: Oberhauser AF (ed) Single-molecule studies of proteins. Springer, New York, NY, pp 97–117
3. Yip CM, McLaurin J (2001) Amyloid-β peptide assembly: a critical step in fibrillogenesis and membrane disruption. Biophys J 80: 1359–1371

4. Slade A, Luh J, Ho S, Yip CM (2002) Single molecule imaging of supported planar lipid bilayer—reconstituted human insulin receptors by in situ scanning probe microscopy. J Struct Biol 137:283–291
5. Epand RF, Martinou J-C, Montessuit S, Epand RM, Yip CM (2002) Direct evidence for membrane pore formation by the apoptotic protein Bax. Biochem Biophys Res Commun 298:744–749
6. Ha TH, Kim CH, Park JS, Kim K (2000) Interaction of indolicidin with model lipid bilayer: quartz crystal microbalance and atomic force microscopy study. Langmuir 16:871–875
7. Zhao H, Mattila J-P, Holopainen JM, Kinnunen PK (2001) Comparison of the membrane association of two antimicrobial peptides, magainin 2 and indolicidin. Biophys J 81:2979–2991
8. Falla TJ, Karunaratne DN, Hancock RE (1996) Mode of action of the antimicrobial peptide indolicidin. J Biol Chem 271: 19298–19303
9. Askou HJ, Jakobsen RN, Fojan P (2008) An atomic force microscopy study of the interactions between indolicidin and supported planar bilayers. J Nanosci Nanotechnol 8:4360–4369
10. Shaw JE, Alattia J-R, Verity JE, Privé GG, Yip CM (2006) Mechanisms of antimicrobial peptide action: studies of indolicidin assembly at model membrane interfaces by in situ atomic force microscopy. J Struct Biol 154:42–58
11. Prim N, Iversen L, Diaz P, Bjørnholm T (2006) Atomic force microscope studies on the interactions of Candida rugosa lipase and supported lipidic bilayers. Colloids Surf B Biointerfaces 52:138–142
12. Subbalakshmi C, Sitaram N (1998) Mechanism of antimicrobial action of indolicidin. FEMS Microbiol Lett 160:91–96
13. Hsu C-H, Chen C, Jou M-L, Lee AY-L, Lin Y-C, Yu Y-P, Huang W-T, Wu S-H (2005) Structural and DNA-binding studies on the bovine antimicrobial peptide, indolicidin: evidence for multiple conformations involved in binding to membranes and DNA. Nucleic Acids Res 33:4053–4064
14. Marchand C, Krajewski K, Lee H-F, Antony S, Johnson AA, Amin R, Roller P, Kvaratskhelia M, Pommier Y (2006) Covalent binding of the natural antimicrobial peptide indolicidin to DNA abasic sites. Nucleic Acids Res 34:5157–5165
15. Ghosh A, Kar RK, Jana J, Saha A, Jana B, Krishnamoorthy J, Kumar D, Ghosh S, Chatterjee S, Bhunia A (2014) Indolicidin targets duplex DNA: structural and mechanistic insight through a combination of spectroscopy and microscopy. ChemMedChem 9:2052–2058
16. Som A, Tew GN (2008) Influence of lipid composition on membrane activity of antimicrobial phenylene ethynylene oligomers. J Phys Chem B 112:3495–3502
17. Horcas I, Fernández R, Gómez-Rodríguez JM, Colchero J, Gómez-Herrero J, Baro AM (2007) WSXM: a software for scanning probe microscopy and a tool for nanotechnology. Rev Sci Instrum 78:013705
18. Shlyakhtenko LS, Gall AA, Lyubchenko YL (2013) Mica functionalization for imaging of DNA and protein-DNA complexes with atomic force microscopy. In: Cell imaging techniques. Springer, New York, NY, pp 295–312
19. Shlyakhtenko LS, Gall AA, Filonov A, Cerovac Z, Lushnikov A, Lyubchenko YL (2003) Silatrane-based surface chemistry for immobilization of DNA, protein-DNA complexes and other biological materials. Ultramicroscopy 97:279–287

Chapter 15

Protocols for Studying the Interaction of MSI-78 with the Membranes of Whole Gram-Positive and Gram-Negative Bacteria by NMR

Nury P. Santisteban, Michael R. Morrow, and Valerie Booth

Abstract

Antimicrobial peptides (AMPs) may interact with a variety of target cell components, including the lipid bilayer, non-lipidic cell envelope components, and/or intracellular targets. However, most biophysical experiments aimed at elucidating the detailed mechanism of AMPs are limited to simple model membrane systems and neglect potentially functional interactions between AMPs and non-lipidic cell components. One of the biophysical techniques commonly used to study how AMPs interact with lipid bilayers is solid-state deuterium NMR. In this chapter we provide protocols to prepare deuterium-labeled intact Gram-negative and Gram-positive bacteria and to observe these samples using solid-state deuterium NMR. Such experiments have the potential to provide important information about how non-lipidic cell envelope components modulate AMP interactions with the cytoplasmic membrane of bacteria.

Key words AMP, Antimicrobial peptide, Solid-state NMR, Deuterium NMR, Whole-cell NMR

1 Introduction

The detailed mechanisms by which antimicrobial peptides (AMPs) interact with and disrupt lipid bilayers are often studied via "biophysical" methods such as solid-state NMR, fluorescence, or molecular dynamics simulation. The norm for such methods is to apply them solely to simplified model membrane systems, i.e., lipid bilayers with simple compositions. While these biophysical approaches provide an excellent starting place for understanding the details of AMP/pathogen interactions, they fail to capture all of the myriad of potentially important interactions that AMPs may have with their target microbes. For instance, before reaching the cytoplasmic membrane of bacteria, AMPs must traverse the thick peptidoglycan layer of Gram-positive bacteria or, in the case of Gram-negative bacteria, a thin peptidoglycan layer and an outer membrane containing lipopolysaccharide (Fig. 1). Other potential

Paul R. Hansen (ed.), *Antimicrobial Peptides: Methods and Protocols*, Methods in Molecular Biology, vol. 1548,
DOI 10.1007/978-1-4939-6737-7_15, © Springer Science+Business Media LLC 2017

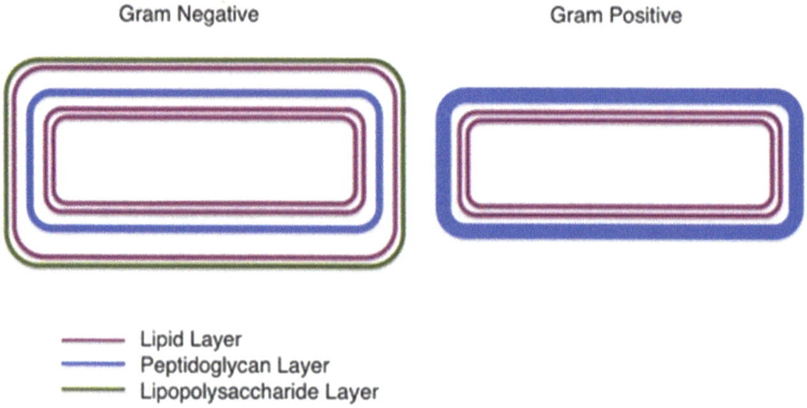

Fig. 1 Schematic view of Gram-negative and Gram-positive bacterial cell envelopes

Table 1
Summary of bacteria and labeling strategies employed

	Strain	Type	Function of mutation(s)	^{2}H-labeling pattern
Mutated bacteria	K1060 (CGSC#: 5040)	*Gram negative*	Cannot synthesize or degrade unsaturated fatty acids	Unsaturated acyl chains
	LA8 [1]	*Gram negative*	No fatty acid synthesis or metabolism	Saturated acyl chains
Unmutated bacteria	JM109 (CGSC#: 8267)	*Gram negative*	–	Saturated lipid acyl chains
	B. subtilis (ATCC#: E6051)	*Gram positive*	–	Saturated lipid acyl chains

modifiers of AMP action on bacteria are the presence of membrane proteins, lipid domains, bilayer asymmetry, and other complexities of cell envelope composition and organization not reflected by the one- or two-component lipid systems typically used in biophysical studies. It is important to consider AMP interactions with non-lipidic cell components, both from the point of view obtaining a more comprehensive understanding of how AMPs function and for the optimization of effective drugs based on AMPs.

In order to assess the effect that non-lipidic cell components have on AMP-membrane interactions in bacteria, our group and others have developed methods to apply deuterium solid-state NMR methods to intact, deuterium-labeled bacteria [1–4]. We have introduced deuterated acyl chains into the cell envelope lipids via two main strategies, and, as of now, we have applied each strategy to two strains of bacteria (summarized in Table 1). The first strategy was pioneered in our group [1, 5] and employs mutated bacteria grown in media with deuterated palmitic acid. The mutations affect fatty acid synthesis and/or metabolism and are employed with a view to increasing the fraction of lipids in the cell membrane that are deuterated, while preventing deuteration of non-lipid cell components. One mutated strain, referred to as LA8, cannot synthesize (*accB* mutation) or metabolize (*fadE* mutation) fatty acids [1]. The other mutated strain, K1060, cannot synthesize or degrade fatty acids that are unsaturated [6]. The second strategy we use is a modified version of the protocol established by the Marcotte group [2]. It employs nonmutated bacteria and achieves the desired deuteron labeling by incorporating deuterated palmitic acid into dodecylphosphocholine (DPC) micelles and adding these to the bacterial growth media. The use of DPC is expected to increase the uptake of the deuterated palmitic acid by the bacteria. In this work, the main alteration to the protocol described by Tardy-Laporte and coworkers [2] is the addition of oleic acid to the growth media in our protocol.

2 Materials

2.1 Bacterial Strains and Peptide

All bacterial strains are stored at −80 °C in 1 mL aliquots of 50:50 glycerol: growth medium. *E. coli* LA8, starting with the L8 strain of bacteria (GCSC #5637), *E. coli* K1060 bacteria (CGSC #5040).

1. *E. coli JM109* bacteria (CGSG #8267).
2. *Bacillus subtilis* (*B. subtilis*) bacteria (ATCC # E6051).
3. Antimicrobial peptide MSI-78.

2.2 Stocks for Making Up Media

1. 5 mg/mL ferrous sulfate ($FeSO_4 \cdot 7H_2O$) stock: 100 mg of $FeSO_4 \cdot 7H_2O$ is dissolved in 20 mL of 95 % ethanol in a 50 mL Falcon tube, mix until the solution is clear, and store at −20 °C.
2. 200 mg/mL casamino acid stock: Dissolve 9 g of casamino acids in 45 mL of deionized water in a 50 mL Falcon tube; mix until the solution is clear, using a 20 mL syringe; pass the solution through a 0.2 μm filter unit in a new 50 mL Falcon tube; and store at −20 °C.
3. 7.25 mg/mL deuterated palmitic acid (PA-d_{31}) stock: Dissolve 0.3 g of hexadecanoic-d_{31} acid in 41.4 mL of 95 % ethanol in a 50 mL Falcon tube, mix, and store as described above.
4. 63.85 mg/mL oleic acid stock: Dissolve 2 g of oleic acid in 31.32 mL of 95 % ethanol in a 50 mL Falcon tube, mix, and store as described above.
5. 7.14 mg/mL deuterated oleic acid (OA-d_{34}) stock: Dissolve 250 mg of deuterated oleic acid in 35 mL of 95 % ethanol in a 50 mL Falcon tube, mix, and store as explained above.

6. 56 mg/mL *n*-dodecylphosphocholine (DPC) stock: Dissolve 560 mg of DPC in 10 mL of 95% ethanol in a 50 mL Falcon tube, mix, and store as explained above.
7. 1 mg/mL thiamine stock: 20 mg of thiamine is dissolved in 20 mL of deionized water in a 50 mL Falcon tube, mix, filter, and store as explained above.
8. 50 mg/mL Brij-58 stock: Dissolve 2.25 g of Brij-58 in 45 mL of deionized water in a 50 mL Falcon tube, mix, filter, and store as explained above.
9. 30 mg/mL kanamycin stock: Dissolve 600 mg of kanamycin in 20 mL deionized water in a 50 mL Falcon tube, mix, filter, and store as explained above.

2.3 Culture Media: Medium 63

1. Medium 63 – 13.6 g/L KH_2PO_4, 2.0 g/L $(NH_4)_2SO_4$, 0.2 g/L $MgSO_4 \cdot 7H_2O$, 0.5 mg/L $FeSO_4 \cdot 7H_2O$, 4.0 g/L glycerol, pH 7.0, prepared as follows: 13.6 g of potassium phosphate monobasic (KH_2PO_4), 2 g of ammonium sulfate $(NH_4)_2SO_4$, 200 mg of magnesium sulfate heptahydrate ($MgSO_4 \cdot 7H_2O$), 4 g of glycerol, and 100 μL of the 5 mg/mL $FeSO_4 \cdot 7H_2O$ stock are dissolved in 800 mL of deionized water, pH 7.0
2. pH is adjusted to 7.0 using NaOH (*see* **Note 1**).
3. Adjust volume to 1 L.
4. 400 mL of the mixture (800 mL total) is placed in each of two 4 L flasks (one for each of two experiments). Each flask is closed with a sponge plug and foil. The remaining volume of the mixture is placed in a 500 mL media storage bottle.
5. The medium is autoclaved and stored at room temperature. The two 4 L flasks are each used for one experiment, and the remaining volume is used for growth of overnight culture and for the washing of cells.

2.4 LB Medium

1. LB medium – 10 g/L tryptone, 5 g/L NaCl, and 5 g/L yeast extract prepared as follows: 10 g of tryptone, 5 g of yeast extract, and 5 g of NaCl are dissolved in 900 mL of distilled water.
2. Adjust volume to 1 L.
3. Repeat Subheading 2.3, **items 4** and **5**.

2.5 2XYT Medium

1. 2XYT medium – 16 g/L tryptone, 5 g/L NaCl, and 10 g/L yeast extract prepared as follows: 16 g of tryptone, 10 g of yeast extract, and 5 g of NaCl are dissolved in 900 mL of distilled water.
2. Adjust volume to 1 L.
3. Repeat Subheading 2.3, **items 4** and **5**.

3 Methods

3.1 *E. coli* Mutant K1060 Labeled with Deuterated Unsaturated Chains

1. 100 mL of medium 63 contained in a 500 mL Erlenmeyer flask is enriched with 100 μL of the 1 mg/mL thiamine stock, 1.5 mL of the 200 mg/mL casamino acid stock, 100 μL of the 30 mg/mL kanamycin stock, 2 mL of the 50 mg/mL Brij-58 stock, and 700 μL of the 7.14 mg/mL deuterated oleic acid stock.
2. The flask with the medium is placed in a 37 °C shaking incubator.
3. When the medium equilibrates to the incubator's temperature, a 1 mL aliquot of 50:50 glycerol: growth medium containing K1060 is added to the medium and the shaking incubator is set at 150 rpm.
4. Check the culture's absorbance at 600 nm (A_{600}) every 45 min using the same media for the blank until $A_{600} \approx 2.0$ is reached.
5. To harvest, the cell culture is transferred into two 50 mL Falcon tubes and centrifuged at 2590 × *g*, 4 °C for 10 min. Then, the cells are washed twice with 20 mL of cold medium 63 containing 0.1 % Brij-58 and centrifuged as before.
6. Finally, the pellet is removed from the centrifuge vessel using a clean spatula and placed in a 400 μL, 8 mm diameter, NMR sample tube which is closed with a Teflon stopper. The NMR tube top is wrapped with Teflon tape to avoid leakage.

3.2 *E. coli* Mutant LA8 Labeled with Deuterated Saturated Chains

1. In a 50 mL Falcon tube, 10 mL of medium 63 at room temperature is enriched with 10 μL of the 1 mg/mL thiamine stock, 150 μL of the 200 mg/mL casamino acids stock, and 10 μL of the 30 mg/mL kanamycin stock.
2. Then, the enriched medium is inoculated with 50 μL of *E. coli* LA8 and incubated overnight.
3. All overnight cultures are incubated at 30 °C with a shaking speed of 150 rpm. The end of this step is indicated by clouding of the solution.
4. 400 mL of medium 63 contained in a 4 L flask is enriched with 400 μL of the 1 mg/mL thiamine stock, 6 mL of the 200 mg/mL casamino acids stock, 400 μL of the 30 mg/mL kanamycin stock, 8 mL of the 50 mg/mL Brij-58 stock, 2.78 mL of the 7.14 mg/mL deuterated palmitic acid stock, and 313 μL of the 63.85 mg/mL oleic acid stock.
5. The flask with the medium is placed in a 37 °C incubator and once equilibrated, 4 mL of the overnight culture is added, and the media is incubated at 37 °C with a shaking speed of 150 rpm.

6. Check the culture's absorbance at 600 nm (A_{600}) every ~45 min using the same media for the blank until A_{600} = 0.6–1.0.
7. To harvest, the cells are centrifuged at 4 °C for 10 min at 4100 × *g*. The resulting pellet should have a whitish color and be adhered to the centrifuge vessel wall.
8. Add 60 mL of cold medium 63 to the centrifuge vessel with the pellet, shake gently by the hand for 30 s, and then discard. The pellet should remain adhered to the centrifuge bottle wall (*see* **Note 2**).
9. Finally, the pellet is removed from the centrifuge vessel using a clean spatula and placed in the NMR sample tube as described above. The NMR tube top is wrapped with Teflon tape to avoid leakage.

3.3 Non-mutated E. coli JM109 Labeled with Deuterated Acid Chains

1. Prepare 0.25 mM oleic acid encapsulated in 1 mM DPC micelles:
2. In a 50 mL Falcon tube, mix 442.4 μL of the 63.85 mg/mL oleic acid stock, 2.5 mL of the 56 mg/mL DPC stock, and 7.058 mL of distilled water.
3. Using a 10 mL syringe, pass the solution through a 0.2 μm filter unit into a new 50 mL Falcon tube.
4. The tube with the mixture is placed in a water bath at 95 °C for 2 min and then submerged in liquid nitrogen for 2 min (*see* **Note 3**).
5. Finally, the tube is warmed in a water bath at room temperature (*see* **Note 4**).
6. The solution is used immediately.
7. To prepare 0.05 mM oleic acid encapsulated in 0.2 mM DPC micelles:
8. In a 50 mL Falcon tube, mix 88.5 μL of the 63.85 mg/mL oleic acid stock, 500 μL of the 56 mg/mL DPC stock, and 9.412 mL of distilled water.
9. Repeat **steps 2–5**.
10. To prepare 0.25 mM of deuterated palmitic acid encapsulated in 1 mM DPC micelles.
11. In a 50 mL Falcon tube, mix 3.97 mL of the 7.248 mg/mL PA-d31 stock, 140 mg of DPC, and 6.03 mL of distilled water.
12. Repeat **steps 2–5**.
13. To prepare 0.05 mM of deuterated palmitic acid encapsulated in 0.2 mM DPC micelles: In a 50 mL Falcon tube, mix 794 μL of the 7.25 mg/mL PA-d31 stock, 500 μL of the 56 mg/mL DPC stock, and 8.706 mL of distilled water.
14. Repeat **steps 2–5**.

3.4 Non-mutated E. coli JM109 Labeled with Deuterated Acid Chains

1. In a 50 mL Falcon tube, 10 mL of LB medium at room temperature is inoculated with a 1 mL aliquot of 50:50 glycerol: growth medium containing *E. coli* JM109. Grow overnight as detailed in Subheading 3.2, **step 3**.
2. A 4 L flask with 400 mL of LB medium enriched with the (0.25 mM/1 mM) oleic acid-DPC and (0.25 mM/1 mM) PA-d31-DPC micelle solutions preheated at 37 °C is inoculated with 4 mL of the overnight culture and placed in a 37 °C incubator with a shaking speed of 150 rpm (*see* **Note 5**).
3. Repeat Subheading 3.2, **steps 6–8**.
4. Add 60 mL of cold LB medium to the centrifuge vessel with the pellet, shake gently by the hand for 30 s, and then discard. The pellet should remain adhered to the centrifuge bottle wall (*see* **Note 2**).
5. Repeat Subheading 3.2, **step 9**.

3.5 Non-mutated B. subtilis Labeled with Deuterated Acid Chains

1. In a 50 mL Falcon tube, 10 mL of 2×YT medium at room temperature is inoculated with a 1 mL aliquot of 50:50 glycerol: growth medium containing *B. subtilis*. Grow overnight as detailed in Subheading 3.2, **step 3**.
2. A 4 L flask with 400 mL of 2×YT medium enriched with the (0.05 mM/0.2 mM) oleic acid-DPC and (0.05 mM/0.2 mM) PA-d31-DPC micelle solutions at 37 °C is inoculated with 4 mL of the overnight culture and placed in a 37 °C incubator with a shaking speed of 150 rpm. Note that this is a smaller concentration of oleic acid-DPC and PA/DPC than is used for *E. coli* (*see* **Notes 5** and **6**).
3. Repeat Subheading 3.2, **step 6**.
4. To harvest, the cells are centrifuged at 5670 × *g*, 4 °C for 10 min (*see* **Note 7**). The resulting pellet should have a whitish color and be adhered to the centrifuge vessel wall.
5. Add 60 mL of cold 2×YT medium to the centrifuge vessel with the pellet, shake gently by the hand for 30 s, and then discard. The pellet should remain adhered to the centrifuge bottle wall (*see* **Note 2**).
6. Repeat Subheading 3.2, **step 9**.

3.6 Treating 2H-Labeled Bacteria with Antimicrobial Peptide

Once the bacteria with deuterated lipid chains have been prepared, the protocol to treat them with antimicrobial peptide is the same regardless of the protocol followed for preparing the bacteria to this point:

1. The desired mass of AMP is calculated as a percentage of the dry weight of the cells, which could be found using the relation described below in **step 4**. The peptide is weighed and added to

60 mL of growth media (medium 63 for K1060 and LA8, LB for JM109, and 2×YT for *B. subtilis*) precooled to 4 °C.

2. The media + AMP is added to the centrifuge tube containing the deuterated bacterial cell pellet after harvest centrifugation (*see* Subheading 3.1, **step 5** for K1060, Subheading 3.2, **step 8** for LA8, and JM109 and Subheading 3.4, **step 4** for *B. subtilis*). Scrape the pellet with a spatula and put the centrifuge tube with the loosened pellet in an orbital shaker at 200 rpm for 20 min at room temperature. The pellet should be completely solubilized in the medium.
3. Centrifuge at 4100 × *g* (2590 × *g* in the K1060 case or 5670 × *g* in the *B. subtilis* case), 4 °C for 10 min. Remove excess medium and, using a spatula, transfer the resulting pellet into the NMR sample tube as described above. The NMR tube top is wrapped with Teflon tape to avoid leakage.
4. Determining the bacterial dry weight: The goal of this procedure is to establish a relationship between the bacterial culture's absorbance and the final pellet dry weight. This is useful to allow the appropriate amount of AMP to be added as a function of the bacteria dry weight.
5. Grow ~10 small cultures of bacteria exactly as for the NMR samples above (*see* Subheading 3.2). Harvest each at different time point of the exponential growth phase (*see* **Note 8**). The absorbance range selected should match the absorbance range normally used for harvesting cells for NMR studies. For example, in the protocols provided above, the target absorbances are $A_{600} \sim 2.0$ for *E. coli* K1060 and $A_{600} = 0.6$–1.0 for JM109, LA8, and *B. subtilis.*
6. Each resulting pellet is dissolved in 20 mL of the appropriate cold media for 20 min. Media + cells are centrifuged at the same conditions explained in Subheading 3.1, **step 5**.
7. Pellets are removed from the centrifuge tube using a clean spatula and placed in a pre-weighed vessel, which is then placed in a vacuum chamber for 48 h.
8. The vessels containing dry pellet are reweighed and the bacterial dry weight is obtained by subtracting the weight of the vessel from the weight of the vessel + dry bacteria.
9. At this point, an exponential relationship is expected between the absorbance at 600 nm and the cell dry weight. A plot of ln (dry weight) versus A_{600} should give a straight line as shown in Fig. 2. This relationship can be used to calculate the appropriate amount of AMP to be added to the bacteria for NMR experiments, using the absorbance measured just before the cells are harvested. For example, if the absorbance of a JM109 culture is 0.75, the dry weight of the pellet is (200 + 20) mg (*see* **Notes 9** and **10**).

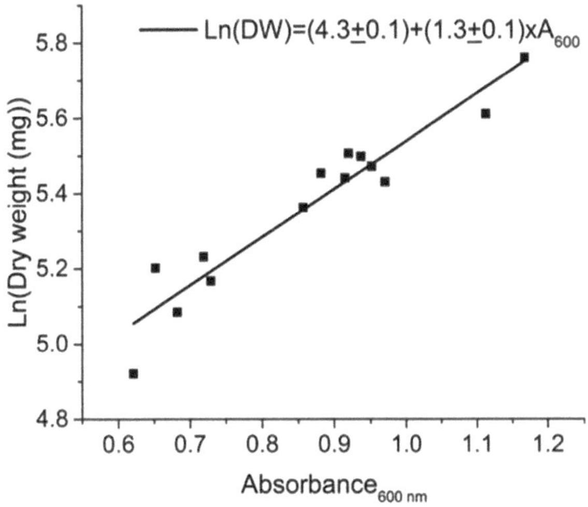

Fig. 2 ln (dry weight (mg)) cells vs. A_{600} for JM109. A linear regression was performed to obtain the relation shown in the graph

3.7 NMR Acquisition

1. The sample as prepared in Subheadings 3.1–3.6 is placed on ice and transported to the location of a wideline deuterium (^{2}H) nuclear magnetic resonance (NMR) spectrometer. The wideline ^{2}H NMR spectrometer used for this work was assembled in-house. The ^{2}H frequency for this system is 61.4 MHz. A 10 mm diameter transverse sample coil is contained within a copper oven on the NMR probe. The oven temperature is controlled by a Lakeshore Cryogenics Model 325 temperature controller.
2. The spectrometer frequency is set in advance of the experiment using a sample with similar dielectric properties but a stronger ^{2}H NMR signal.
3. The probe oven is preheated to 37 °C prior to insertion of the sample in the probe coil.
4. NMR data acquisition is started as soon as possible after insertion of the sample in the probe.
5. With the sample in place and equilibrated, probe tuning and matching are quickly performed, after which the NMR experiment is started immediately.
6. Time domain quadrupole echo signals are acquired using a quadrupole echo pulse sequence ($\pi/2$-τ -$\pi/2$-acq) [7] with a 30 μs pulse separation, a 5 μs $\pi/2$ pulse duration, and a 900 ms recycling delay.
7. For the acquisition of each transient, 8192 points in each of the real and imaginary channels are digitized with a dwell time of 1 μs. This results in oversampling [8] by a factor of 4 relative to the dwell time required to achieve the desired spectral width of 250 kHz.

8. Transients are averaged in four sequential blocks of 8000 scans (acquisition time ~2 h) and processed separately to monitor for time-dependent changes in the spectral shape.
9. Assuming no major changes in spectral shape occurs over the 8 h acquisition, the 32,000 transients are added together and processed.
10. If necessary, a baseline correction and a phase shift are applied to the acquired quadrupole echo to minimize signal in the imaginary channel.
11. The echo is contracted [8] by a factor of 4 to give an effective dwell time of 4 μs and the echo peak is symmetrized [8, 9], and points preceding the echo peak are discarded to leave a free induction decay (FID) which is zero-filled to 4096 points.
12. The FID is Fourier transformed to obtain the ^{2}H spectrum which is then baseline corrected and, if necessary, phase corrected.

3.8 NMR Analysis: Expected Results

Samples of processed spectra for the four bacterial strains examined in this work are shown in Fig. 3. In the absence of AMP, the spectra of all the saturated chain-labeled bacteria (Fig. 3b–d) are unresolved superpositions of Pake doublet powder patterns characteristic of axially symmetric reorientation of each chain segment about the bilayer normal. The prominent shoulders near ± 12 kHz correspond to deuterated segments near the headgroup end of the chain (the plateau region) where reorientation is most constrained. Segments closer to the bilayer center reorient more freely and contribute spectral components with smaller quadrupole splittings. The prominent doublet with the smallest splitting corresponds to deuterated methyl groups at the ends of the deuterated acyl chains. When the unsaturated chain is deuterated (panel A), the spectra display more intensity near the center of the spectrum, as expected given that unsaturated chains are more disordered than saturated chains. The strong peak in the center most likely arises from naturally occurring deuterated water although rapid isotropic reorientation of small particle containing deuterated components can also yield a narrow central peak.

The exact growth protocol used can have a large effect on the NMR spectra and thus care must be taken to keep the details of the preparation protocol the same when comparing the results with and without AMP. For example, K1060 cells grown with casamino acids produce a very different spectrum than K1060 cells grown without casamino acids (Fig. 3a). Spectra of *E. coli* prepared from a mutated strain grown with added palmitic and oleic acid (Fig. 3b) are quite different than those from unmutated *E. coli* grown with DPC-encapsulated palmitic and oleic acid (Fig. 3c). The spectra in Fig. 3c have a component with a large splitting, corresponding to deuterons in very rigid structures.

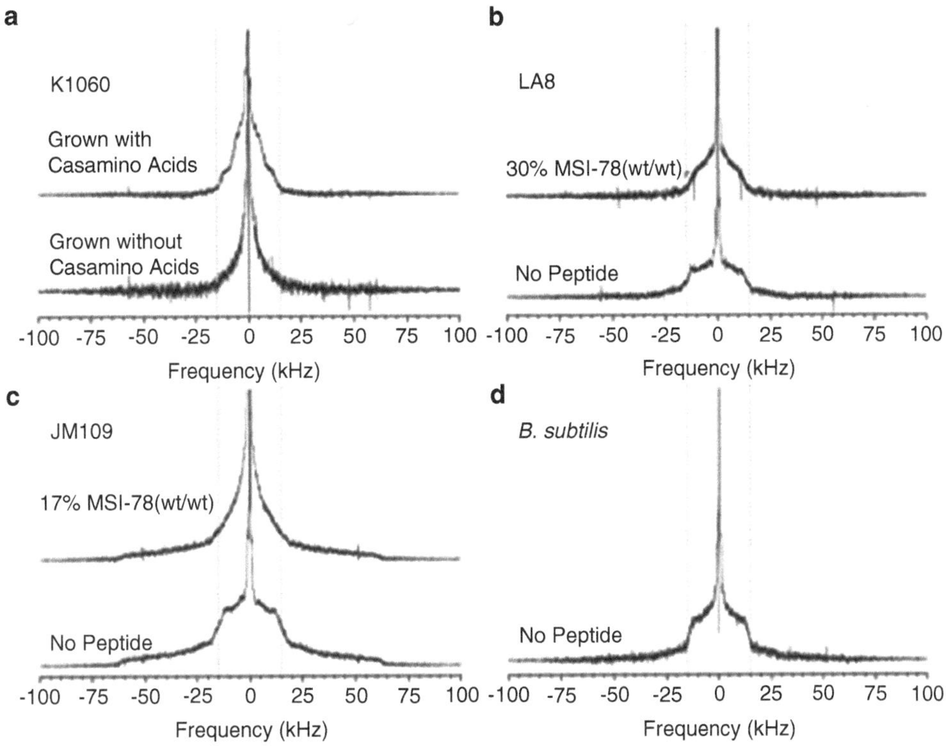

Fig. 3 Solid-state ^{2}H NMR spectra of ^{2}H-membrane-enriched bacteria. (**a**) ^{2}H-enriched K1060 *E. coli* grown with and without casamino acids in their growth media. Each spectrum represents 24,000 scans. (**b**) ^{2}H-enriched LA8 *E. coli* with and without the AMP MSI-78. Each spectrum represents 32,000 scans. (**c**) ^{2}H-enriched JM109 bacteria with and without the AMP MSI-78. Each spectrum represents 32,000 scans. (**d**) ^{2}H-enriched *Bacillus subtilis*. The spectrum represents 32,000 scans. All spectra were acquired at 37°C in a 9.4 T in-house-assembled wideline deuterium NMR spectrometer. *Gray vertical lines* at ±15 kHz are included to facilitate comparison of the spectra from panel to panel. Panels (**b**) and (**c**) include data replotted with permission from Plus, 2012 [5]

When treated with AMP (Fig. 3b, c), the intensity at larger splittings decreases and the intensity at smaller splittings increases. This reflects an AMP-induced decrease in acyl chain order corresponding to membrane disruption or perturbation resulting from interaction with the AMP.

There are a variety of parameters that can be calculated from the ^{2}H spectra in order to quantify and compare the experimental results [9]. Possible computed parameters include the first spectral moment, M_1; the second spectral moment, M_2; and the fractional mean squared width of the distribution of quadrupole splittings, Δ_2. The *n*th moment of the spectra is defined as:

$$M_n = \frac{\int_0^\infty \omega^n f(\omega)\, d\omega}{\int_0^\infty f(\omega)\, d\omega}$$

Table 2
First moments M_1 for different bacteria strains in the presence and absence of AMP

Bacterial strain	$M_1 \times 10^4$ (s^{-1})	Details
K1060	3.28	Grown without casamino acids (untreated)
	3.39	Grown with casamino acids (untreated)
LA8	4.84	(untreated)
	3.58	Treated with 30% MSI-78
JM109	10.03	(untreated)
	8.3	Treated with 17% MSI-78
B. subtilis	4.88	(Untreated)

where ω is the angular frequency with respect to the Larmor angular frequency ω_0, $f(\omega)$ is the spectrum intensity at ω frequency, and the integral is over positive ω in order to allow for the definition of odd moments. The first moment (M_1) is proportional to the average quadrupole spitting of deuterons contributing to the spectrum and thus to the average orientational order parameter of CD bonds on the deuterated acyl chain.

To calculate the first moment, it is necessary to select the area of the spectrum over which to do the integration. The integration region should cover, and be limited to, the range of positive frequencies (i.e., frequencies to the right of the spectral center) over which there is spectral intensity.

For chains undergoing axially symmetric reorientation, the upper limit of integration must be at least twice the frequency of the prominent spectral edges corresponding to the plateau deuterons. A segment of baseline (typically 100 points) beyond the integration limit is averaged to determine a baseline value for the integration. Any unsplit central peak corresponding to naturally deuterated water or small isotropically reorienting particles (typically within 0.7 kHz of the spectral center) is excluded from the integral by extending the intensity at the edge of that region into the spectral center before integration. Table 2 shows values of M_1 for all the spectra shown in Fig. 3.

4 Notes

1. Given the amount of NaOH needed to adjust the pH, we use a 5 M NaOH solution.
2. Before discarding the medium in the bottle, check that the pellet is still adhered to the vessel bottle. In case that the pellet

is loose, an extra centrifugation step with the same conditions can be used to separate the medium from the bacterial cells.

3. Submerging ~¾ of the 50 mL Falcon is enough to cool the mixture and allows the Falcon to be held in the hand.
4. At the end of this step, lipid crystal will be observed in the crystalline solution for a brief moment.
5. A change in the oleic acids concentration will affect the fluidity of the bacterial membrane and consequently the moment values. It is advisable not to change the concentration between experiments if a comparison is to be made.
6. The amount of oleic acid and PA-d_{31} encapsulated in DPC micelles used to label the *B. subtilis* cell envelope was reduced to 20% of the amount used with *E. coli* JM109. The reduction was necessary to optimize *B. subtilis* cell growth.
7. This temperature is important to obtain a good pellet to work with. The time and speed could be increased in order to allow a separation between the cells and the medium.
8. Time calculation is very important in this experiment. Because *E. coli* takes approximately 20 min to duplicate and *B. subtilis* 30 min, it is *not* advisable to start all cultures at the same time.
9. It is highly advisable to calculate the approximate amount of peptide needed for one experiment because the amount of peptide needed to perform one experiment is large in comparison with model lipid experiments.
10. It is highly recommended that the cells be harvested at an absorbance of approximately 0.6 because the culture should be in the exponential growth phase and the amount of peptide needed increases exponentially with the absorbance.

References

1. Pius J, Morrow MR, Booth V (2012) ^{2}H solid-state nuclear magnetic resonance investigation of whole *Escherichia coli* interacting with antimicrobial peptide MSI-78. Biochemistry 51: 118–125
2. Tardy-Laporte C, Arnold AA, Genard B et al (2013) A H-2 solid-state NMR study of the effect of antimicrobial agents on intact *Escherichia coli* without mutating. Biochim Biophys Acta 1828:614–622
3. Marcotte I, Booth V (2014) 2H solid-state NMR study of peptide-membrane interactions in intact bacteria. In: Advances in biological solid-state NMR: proteins and membrane-active peptides. The Royal Society of Chemistry, Cambridge, pp 459–475
4. Warnet XL, Laadhari M, Arnold AA et al (2015) A 2H magic-angle spinning solid-state NMR characterisation of lipid membranes in intact bacteria. Biochim Biophys Acta 1858:146–152
5. Pius J (2012) A novel approach to characterize the membrane - disrupting activity of antimicrobial peptides using 2 H solid - state NMR of whole *Escherichia coli*. Master Thesis Memorial University of Newfoundland, St. John's, NL, Canada
6. Konings AWT, Gipp JJ, Yatvin MB (1984) Radio- and thermosensitivity of E. coli K1060 after thiol depletion by diethylmaleate. Radiat Environ Biophys 23:245–253
7. Davis JH, Jeffrey KR, Bloom M et al (1976) Quadrupolar echo deuteron magne-

tic resonance spectroscopy. In: Ordered hydrocarbon chains. Chem Phys Lett 42: 390–394

8. Prosser RS, Davis JH, Dahlquist FW, Lindorfer MA (1991) Deuterium nuclear magnetic resonance of the gramicidin A backbone in a phospholipid bilayer. Biochemistry 30:4687–4696

9. Davis JH (1983) The description of membrane lipid conformation, order and dynamics by 2H-NMR. Biochim Biophys Acta 737:117–171

Chapter 16

Preparation of Membrane Models of Gram-Negative Bacteria and Their Interaction with Antimicrobial Peptides Studied by CD and NMR

Rickey Hicks

Abstract

The antibiotic activity of antimicrobial peptides is generally derived via some type of disruption of the cell membrane(s). The most common models used to mimic the properties of bacterial membranes consist of mixtures of various zwitterionic and anionic phospholipids. This approach works reasonably well for Gram-positive bacteria. However, since the membranes of Gram-negative bacteria contain lipopolysaccharides, as well as zwitterionic and anionic phospholipids, a more complex model is required to simulate the outer membrane of Gram-negative bacteria. Herein we present a protocol for the preparation of models of the outer membranes of the Gram-negative bacteria *Klebsiella pneumoniae* and *Pseudomonas aeruginosa*. This protocol can be used to prepare models of other Gram-negative bacteria provided the strain-specific lipopolysaccharides are available.

Key words Gram-negative bacteria membrane models, Phospholipids, Lipopolysaccharides, Circular dichroism, Nuclear magnetic resonance, *Klebsiella pneumoniae*, *Pseudomonas aeruginosa*

1 Introduction

Mechanistically the bactericidal activity of an antimicrobial peptide (AMP) is generally the result of the disruption of the cell membrane(s) resulting in lysis, thus causing cell death [1, 2]. In an effort to understand the mechanism(s) of membrane disruption, structure-activity relationship studies have been conducted using various combinations of zwitterionic and anionic phospholipids as simple models for bacterial membranes [3–6]. Bhunia and co-workers [7] have observed the protocol of employing various compositions of zwitterionic and anionic phospholipids as simple models for bacterial membranes [3–6] is relatively effective for the single phospholipid membrane found in Gram-positive bacteria. However, this approach fails to accurately model AMP interactions with Gram-negative bacteria since they contain two cell membranes of very different chemical compositions [7]. The outer leaflet of the outer membrane of

Paul R. Hansen (ed.), *Antimicrobial Peptides: Methods and Protocols*, Methods in Molecular Biology, vol. 1548,
DOI 10.1007/978-1-4939-6737-7_16, © Springer Science+Business Media LLC 2017

Gram-negative bacteria contains a very high concentration of lipopolysaccharides (LPSs) [8–10]. LPS molecules are highly negatively charged and function as a semipermeable membrane that modulates the transport of a variety of molecules across the membrane [11–14]. The differences in the chemical compositions of the membranes result in very different physicochemical surface properties for the membranes of Gram-negative bacteria compared to those of Gram-positive bacteria. It is the interactions that occur between the physicochemical properties of the membrane and the AMP that determine the mechanism and extent of membrane disruption.

LPSs consist of three major subcomponents [11, 15, 16]. A polysaccharide, referred to as the O-antigen, is the outermost component, the chemical composition of which varies as a function of the bacterial strain. The second component consists of the core oligosaccharide, the chemical composition of which also varies as a function of the bacterial strain. The inner component is the highly conserved lipid A, which consists of a polyacylated glucosamine-based bis-phospholipid. In order to design AMPs with increased antibacterial activity against Gram-negative bacteria, it is just as important to understand the physicochemical interactions that occur between an AMP and LPS as it is to understand the physicochemical interactions that occur between the AMP and the phospholipid components of the cell [7, 8, 15].

Figure 1 is a cartoon representation [3, 7, 8, 11, 15, 17–25] depicting many of the complex equilibria that exist during the

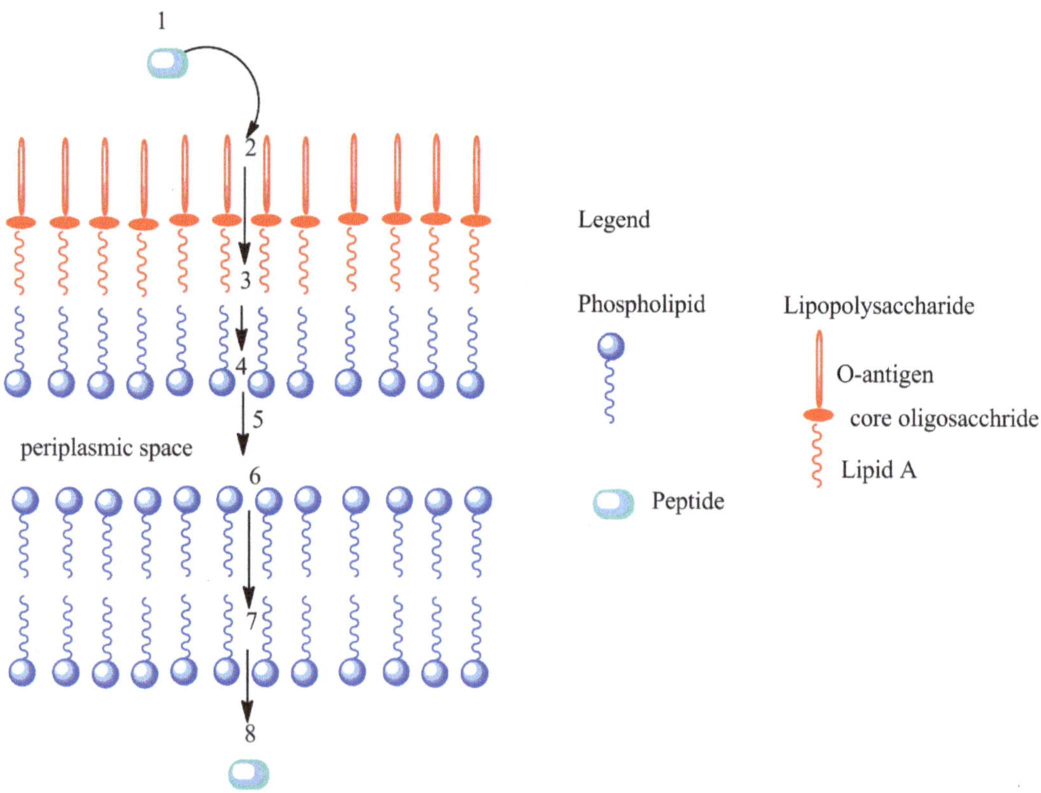

Fig. 1 A cartoon representation depicting many of the complex equilibria that exist during the multistep binding to, and the disruption of, the membranes of Gram-negative bacteria by AMPs

multistep binding to, and the disruption of, the membranes of Gram-negative bacteria by AMPs. It is very difficult to simulate each of these steps individually. Small unilamellar vesicles (SUVs) consisting of bacteria-specific LPS can be used to investigate the movement of the AMP thru the outer leaflet of the outer membrane (Fig. 1, steps 1–3). Here we present a protocol for studying membrane models of Gram-negative bacteria and their interaction with antimicrobial peptides studied by CD and NMR.

2 Materials

2.1 Chemicals and Apparatus

1. K_2HPO_4.
2. KH_2PO_4.
3. Sodium acetate-d_3.
4. D_2O.
5. 1-Palmitoyl-2-oleoyl-sn-glycero-3-phosphocholine (POPC).
6. 1-Hexadecanoyl-2-(9Z-octadecenoyl)-sn-glycero-3-phospho-(1′-rac-glycerol) (POPG).
7. 1-Palmitoyl-2-oleoyl-sn-glycero-3-phosphoethanolamine (POPE).
8. Lipopolysaccharide isolated from *Pseudomonas aeruginosa*.
9. Lipopolysaccharide isolated from *Klebsiella pneumoniae*.
10. Titanium-tip ultra-sonicator.
11. Sonicator.
12. CD spectrometer.
13. NMR spectrometer.

2.2 Preparation of Buffer

1. 40 mM potassium phosphate buffer: Dissolve 34.84 g of K_2HPO_4 (MW = 174.18 g/mol) in 500 mL of distilled water yielding a 0.4 M solution of K_2HPO_4; dissolve 27.22 g of KH_2PO_4 (MW = 136.09 g/mol) in 500 mL of distilled water yielding a 0.4 M solution of KH_2PO_4; take 49.7 mL of 0.4 M K_2HPO_4, add to 50.3 mL of 0.4 M KH_2PO_4, and dilute to a final volume of 1.0 L with distilled water to yield a 40 mM potassium phosphate buffer, pH = 6.8; and final pH is confirmed with a pH meter and adjusted as needed.
2. 150 mM perdeuterated sodium acetate buffer: Dissolve 12.75 mg of sodium acetate-d_3 in 10 mL of D_2O, and titrate with acetic acid-d_3 to a pH = 5.64 using a pH meter.
3. 2 mg/mL peptide stock solutions: The peptide solutions were prepared weighing approximately 2 mg of peptide dissolved in 1.0 mL of phosphate buffer. The peptides used in this study are shown in Table 1.

2.3 LPS-Mixed Phospholipid SUVs

1. 75%POPC/5%POPG/20% POPE SUV: Dry powder mixtures of a 5 mg sample of the lipopolysaccharide isolated from *Klebsiella pneumoniae* are mixed with 9.84 mg of POPC, 0.68 mg of POPG, and 2.53 mg of POPE yielding a lipid composition of 75%POPC/5%POPG/20% POPE; hydrate with 2 mL of buffer (40 mM sodium phosphate, pH = 6.8) and vortex to yield a milky suspension; SUVs are prepared by sonication of the suspension in an ice bath for approximately 40 min until the solution became transparent using a titanium-tip ultra-sonicator (*see* **Note 1**); centrifugation (8800 rev./min for 10 min using an Eppendorf tabletop centrifuge) is used to remove any titanium debris (*see* **Note 2**) resulting from sonication [26]. A similar procedure was used by Matsuzaki and co-workers to prepare LPS-POPC SUVs [23].
2. 59%POPC/21%POPG/20% POPE SUV: Dry powder mixtures of a 5.07 mg sample of the lipopolysaccharide isolated from *Pseudomonas aeruginosa* are mixed with 7.66 mg of POPC, 2.79 mg of POPG, and 2.52 mg of POPE yielding a lipid composition of 59%POPC/21%POPG/20% POPE; hydrate with 2 mL of buffer (40 mM sodium phosphate, pH = 6.8) and vortex to yield a milky suspension; SUVs are prepared by sonication of the suspension in an ice bath for approximately 40 min until the solution became transparent using a titanium-tip ultra-sonicator (*see* **Note 1**). Centrifugation (8800 rev./min for 10 min using an Eppendorf tabletop centrifuge) is used to remove any titanium debris (*see* **Note 2**) resulting from sonication [26]. A similar procedure was used by Matsuzaki and co-workers to prepare LPS-POPC SUVs [23].

Table 1
Amino acid sequences of the Tic-Oic containing AMPs referenced in this study

Comp #	Amino acid sequence
23	Ac-Gly-Phe-Tic-Oic-Gly-Lys-Tic-Oic- Gly-Phe -Tic-Oic- Gly-Lys -Tic-Lys-Lys-Lys-Lys-NH_2
43	Ac- Gly-Phe -Tic-Oic-Gly-Orn-Tic-Oic- Gly-Phe -Tic-Oic-Gly-Orn-Tic-Orn-Orn-Orn-Orn-NH_2
53	Ac- Gly-Phe -Tic-Oic-Gly-Dab-Tic-Oic- Gly-Phe -Tic-Oic-Gly-Dab-Tic-Dab-Dab-Dab-Dab-NH_2

Tic: tetrahydro-isoquinoline-3-carboxylic acid
Oic: cis-octahydroindole-2-carboxylic acid
Dab: l-2,4-diaminobutyric acid
Orn: l-ornithine

2.4 Preparation of Mixed POPC/POPE/POPG SUVs

35 mM POPC, POPG, and POPE SUV: The appropriate amount of dry POPC, POPG, and POPE is weighed out to yield a final lipid concentration of 35 mM with the desired percentage of each phospholipid; hydrate with 2 mL of buffer (40 mM sodium phosphate, pH = 6.8) and vortex to yield a milky suspension; SUVs are prepared by sonication of the suspension in an ice bath for approximately 40 min until the solution became transparent using a titanium-tip ultra-sonicator (*see* **Note 1**); centrifugation (8800 rev./min for 10 min using an Eppendorf tabletop centrifuge) is used to remove any titanium debris (*see* **Note 2**) resulting from sonication [26]. Final lipid concentration used for CD studies was 3.5 mM.

2.5 Preparation of LPS SUVs

1. LPS from *Klebsiella pneumoniae* SUV: A 4 mg sample of the lipopolysaccharide isolated from *Klebsiella pneumoniae* is hydrated with 4 mL of buffer (40 mM sodium phosphate, pH = 6.8) and vortexed to yield a milky lipid suspension; SUVs are prepared by sonication of the suspension for approximately 10 min at a temperature of 40 °C, or until the solution became transparent [3, 15, 18] using a titanium-tip ultra-sonicator; centrifugation (8800 rev./min for 10 min using an Eppendorf tabletop centrifuge) is used to remove any titanium debris (*see* **Note 2**) resulting from sonication.
2. LPS from *Pseudomonas aeruginosa* SUV: A 4 mg sample of the lipopolysaccharide isolated from *Pseudomonas aeruginosa* was hydrated with 4 mL of buffer (40 mM sodium phosphate, pH = 6.8) and vortexed to yield a milky lipid suspension; SUVs were prepared by sonication of the suspension for approximately 10 min at a temperature of 40 °C, or until the solution became transparent [3, 15, 18] using a titanium-tip ultra-sonicator; centrifugation (8800 rev./min for 10 min using an Eppendorf tabletop centrifuge) was used to remove any titanium debris (*see* **Note 2**) resulting from sonication.

3 Methods

3.1 Circular Dichroism Spectroscopy

CD spectroscopy is very sensitive and is commonly used to monitor conformational changes in peptides and proteins [27–29]. LPSs can exhibit strong CD spectra and thus requires careful subtraction of the LPS background spectrum before meaningful spectra of the AMPs bound to the LPS can be obtained [3, 15, 25]. All CD spectra were obtained by acquiring eight scans on a CD spectrometer using a 0.1 mm cylindrical quartz cell from 260 to 195 nm at 20 nm/min, 1 nm bandwidth, data pitch 0.2 nm, response time 2.0 s, and 5 mdeg sensitivity at room temperature (~25 °C *see* **Note 3**). Contributions due to liposomes or LPS were eliminated by subtracting the lipid or LPS spectra of the corresponding peptide-free solutions (*see* **Note 4**). All CD spectra should be smoothed (with a means-movement function) using the JASCO spectra analysis program before quantitative or qualitative analysis is performed (*see* **Note 5**) [17, 24, 30].

1. For LPS-lipid SUV studies, 100 μL of stock peptide solution is added to 20 μL of stock LPS-lipid solution and 80 μL of phosphate buffer.
2. For lipid SUV studies, 100 μL of stock peptide solution was added to 20 μL of stock lipid solution and 80 μL of phosphate buffer.
3. For LPS liposome studies, 350 μL of stock LPS solution was mixed with 50 μL of stock peptide solution.

3.2 Nuclear Magnetic Resonance Spectroscopy

1. All ^{1}H experiments were conducted on a Bruker Avance III 400 MHz NMR spectrometer equipped with a 5 mm direct observe broadband probe with a Z-gradient.
2. All spectra were collected at a temperature of 298 K. Data was collected with a spectral width of 4000 Hz, using 64 K data points per scan, 256 fids were collected per experiment.
3. Data was processed using exponential multiplication with a line-broadening function of 5 Hz.
4. The ^{1}H spectra were obtained using a 1.0 mg sample of the LPS *in* the presence of 0.1 mg of the AMP (*see* **Notes 6** and 7) in 600 μL of a 150 mM perdeuterated sodium acetate buffer at a pH of 5.64 in D_2O.

3.3 Data Analysis of CD Experiments

1. The CD spectra of three AMPs (**23**, **43**, **53**, amino acid sequences of these three AMPs are given in Table 1) in the presence of SUV consisting of LPS isolated from *K. pneumoniae* (solid lines) and *P. aeruginosa* (dashed lines) are shown in Fig. 2 and will be used to illustrate the type of data and information obtainable from these investigations.

   All six spectra as expected are different, reflecting the chemical differences between the two LPSs. The γ_{mim} values for all three AMPs are larger in the presence of LPS isolated from *K. pneumoniae* (solid lines) than those observed in the presence of *P. aeruginosa* (dashed lines). This suggests an increase in periodicity or structure (*see* **Note 8**) of the AMPs in the presence of LPS isolated from *K. pneumonia*. Due to amino acid composition of the AMPs (*see* **Note 9**) used in this investigation, spectral decomposition algorithms to determine relative percentages of secondary structures cannot be employed.

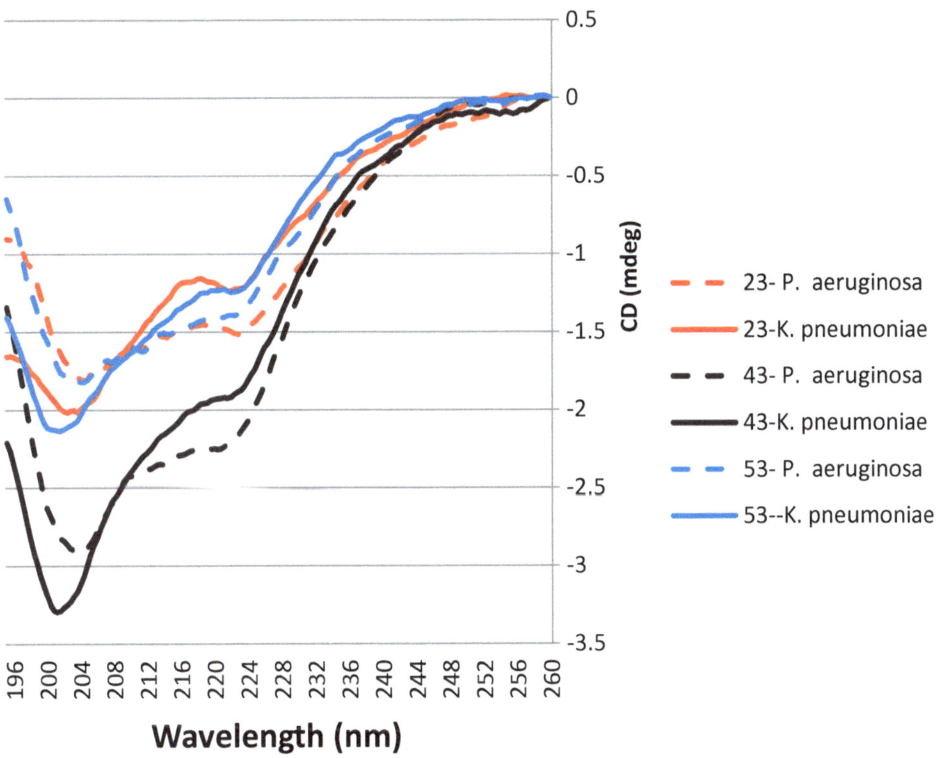

Fig. 2 Far-UV CD spectra of three AMPs (**23**, **43**, **53**) in the presence of SUV consisting of LPS isolated from *K. pneumoniae* (*solid lines*) and *P. aeruginosa* (*dashed lines*)

2. Mixed liposomes consisting of the LPS and phospholipids (associated with specific bacteria) were used as simple models to simulate the transport of an AMP across the outer membrane (Fig. 1, steps 1–5). The CD spectra of three AMPs (**23**, **43**, **53**) in the presence of mixed SUVs consisting of the phospholipids used to model the outer membrane and the LPS isolated from *K. pneumoniae* (solid lines) and *P. aeruginosa* (dashed lines) (*see* **Notes 10** and **11**) are shown in Fig. 3.

   The CD spectra of AMP **23** and **53** are different in the presence of the mixed SUVs representing the outer membranes of *K. pneumoniae* (solid lines) and *P. aeruginosa* (dashed lines). The γ_{mim} values for both AMPs are larger in the presence of the mixed SUV containing 21 % POPG (increasing the net positive charge of the membrane). This increase in γ_{mim} indicates an increase in the periodicity (structure) (*see* **Note 8**) of the AMPs in this environment. This observation suggests that the net positive charge of the SUV plays a major role in defining the interactions that occur between the AMPs and these two mixed SUVs. However, the CD spectra of AMP **43** are very similar in

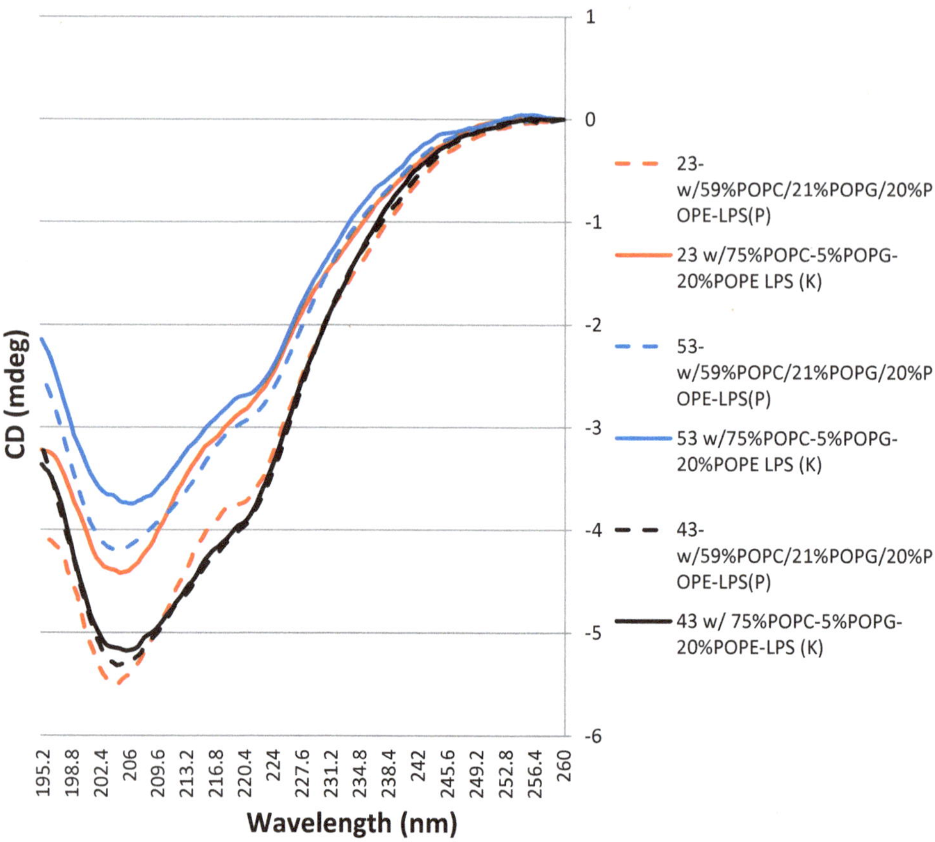

Fig. 3 Far-UV CD spectra of three AMPs (**23**, **43**, **53**) in the presence of mixed SUVs consisting of the phospholipids used to model the outer membrane and the LPS isolated from *K. pneumoniae* (*solid lines*) and *P. aeruginosa* (*dashed lines*)

the presence of the mixed SUVs representing the outer membranes of *K. pneumoniae* (solid lines) and *P. aeruginosa* (dashed lines). This observation suggests that the net positive charge of the SUV doesn't play a major role in defining the interactions that occur between AMP **43** and these two mixed SUVs.

3. Step 5 the conformational changes that may occur in the periplasmic space cannot be easily simulated. Steps 6–8 were collectively investigated using the following mixed phospholipid liposomes. The CD spectra of three AMPs (**23**, **43**, **53**) in the presence of SUV consisting of the phospholipids used to model the inner membrane of *K. pneumoniae* (solid lines) and *P. aeruginosa* (dashed lines) are shown in Fig. 4.

All six of the CD spectra are somewhat different. The γ_{mim} values for AMPs **43** and **53** are larger in the presence of the mixed SUV containing 5% POPG (POPG is a negatively charged phospholipid). This increase in γ_{mim} indicates an increase in the periodicity (structure) of the AMPs in this environment. This observation suggests that the net positive charge of the SUV plays a lesser role in defining the interactions that occur between these AMPs and the two mixed SUVs. However, the γ_{mim} values for AMP **23** are larger in the presence of the mixed SUV containing 21% POPG. This increase in γ_{mim} indicates an increase in the periodicity (structure) of the AMPs in this environment. This observation suggest that the net positive charge of this SUV plays a major role in defining the interactions that occur between AMP **23** and these two mixed SUVs.

3.4 Data Analysis of NMR Investigations

1. ^{1}H NMR spectra of the AMP **23** in the presence of LPS isolated from *K. pneumonia* and *P. aeruginosa* are used as two examples of the types of information that can be obtained from these studies. The region in the NMR spectrum between 1.5 and 0.5 ppm corresponds to the resonances associated with the side chain protons of lipid A of LPS.

As shown in Fig. 5, there is a dramatic reduction (67%) in the peak area of the resonances in this region of the ^{1}H spectrum of a 1.0 mg sample of LPS isolated from *P. aeruginosa* in the presence of 0.1 mg of AMP **23** (at this low a concentration of AMP, no NMR signals corresponding to the AMPs are observed) compared to the ^{1}H spectrum of the LPS alone. The other regions of the ^{1}H spectrum remain relatively unchanged on addition of the AMP. The reduction in peak area indicates a

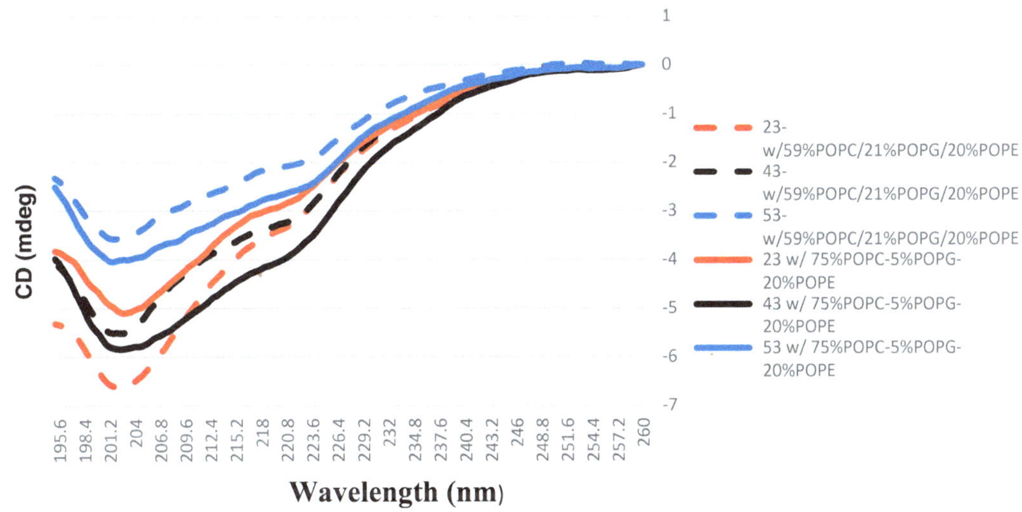

Fig. 4 Far-UV CD spectra of three AMPs (**23**, **43**, **53**) in the presence of SUV consisting of the phospholipids used to model the inner membrane of *K. pneumoniae* (*solid lines*) and *P. aeruginosa* (*dashed lines*)

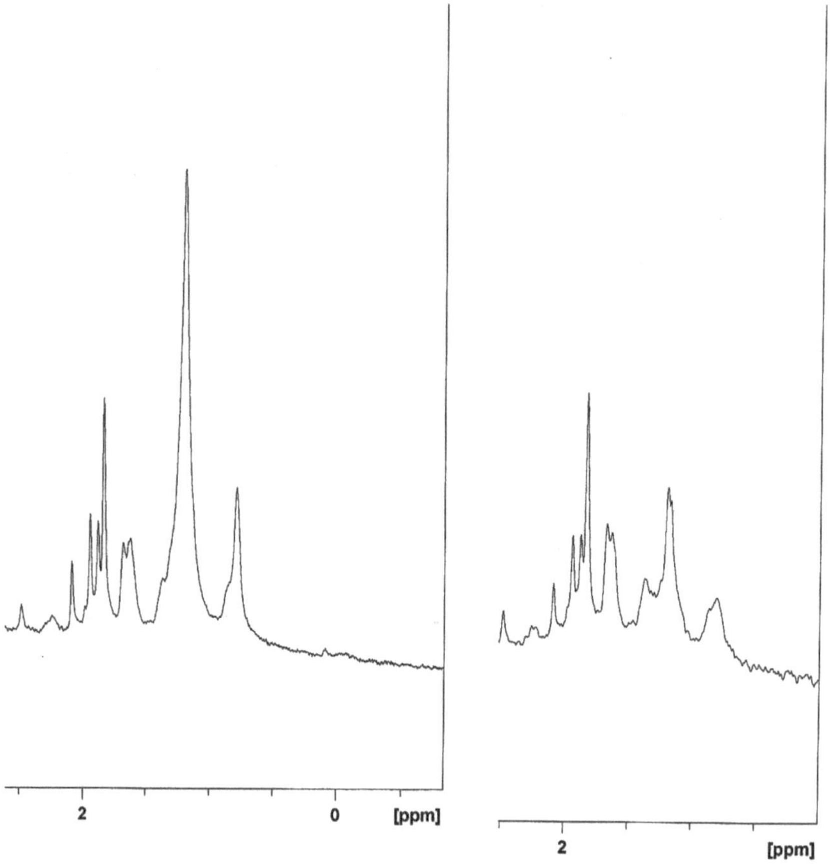

Fig. 5 There is a dramatic reduction in the peak heights of the resonances in this region of the ^1H spectrum (*right* spectrum) of a 1.0 mg sample of LPS isolated from *P. aeruginosa* in the presence of 0.1 mg of AMP **23** compared to the ^1H spectrum (*left* spectrum) of the LPS alone. The other regions of the ^1H spectrum remain relatively unchanged on addition of the AMPs. The reduction in peak area indicates a strong binding interaction of these AMPs with the lipid A region of the LPS isolated from *P. aeruginosa*

strong binding interaction of this AMP with the lipid A region of the LPS isolated from *P. aeruginosa*. The most likely reason for the reduction in ^1H signal intensity is an increase in the T_1 relaxation rates of these protons due to structural changes in the lipid side chains. Le Guemeve and Auger showed that the incorporation of cholesterol into DMPC has an effect on the high-frequency vibrations that contribute to T_1 relaxation inducing longer T_1 relaxation. They further explain that cholesterol restricts the high-frequency motions of the acyl chains, thus increasing the molecular order of these regions compared to the pure lipid bilayer [31].

2. The ^1H spectrum of a 1.0 mg sample of the LPS isolated from *K. pneumoniae* in the presence of 0.1 mg of AMP **23** (Fig. 6)

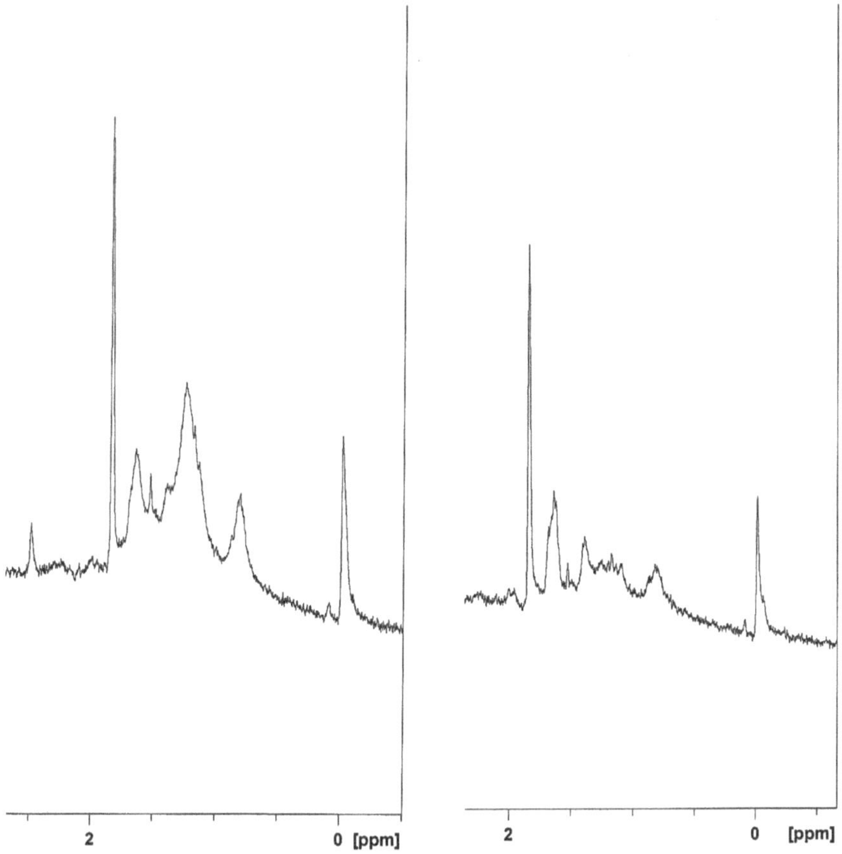

Fig. 6 The ^{1}H spectra (*right* spectrum) of a 1.0 mg sample of the LPS isolated from *K. pneumoniae* in the presence of 0.1 mg of AMP **23** exhibited a reduction in the signal intensity as well as changes in the observed chemical shifts in the region of 1.8–0.7 ppm compared to the ^{1}H spectrum (*left* spectrum) of the LPS alone. The other regions of the ^{1}H spectrum remained unchanged on addition of this AMP. The reduction in peak area was not as dramatic as was observed for LPS isolated from *P. aeruginosa*

exhibited a reduction (19 %) in peak area as well as changes in the observed chemical shifts in the region of 1.8–0.7 ppm compared to the ^{1}H spectrum of the LPS alone. The other regions of the ^{1}H spectrum remained unchanged on addition of this AMP. The reduction in peak area was not as dramatic as was observed for LPS isolated from *P. aeruginosa*. This data suggests that AMP **23** in the presence of both LPSs exhibits a higher partition coefficient for the lipid A region than for the polysaccharide or core oligosaccharide regions of the LPS. This observation is consistent with a proposed mechanism of AMP-LPS binding which involves hydrophobic interactions between the AMP and the hydrocarbon chain region of lipid A [32, 33].

4 Notes

1. During sonication the milky suspension will become a transparent solution with a very light blue color.
2. Centrifugation is required to remove any titanium debris generated during sonication because these particles will disperse light thus increasing the HT values of the observed CD spectra.
3. CD spectra that exhibited HT values of greater than 500 should not be used due to excessive light scattering and/or absorption which will distort the true CD spectrum.
4. LPS on its own can exhibit a strong CD spectrum which requires careful subtraction of the LPS background CD spectrum to yield meaningful spectra of the AMPs bound to LPS [3, 15, 25].
5. Care must be taken not to over-smooth the spectrum, which would result in the lost of spectral features.
6. Peptide concentrations higher than 0.1 mg in the presence of 1.0 mg/mL sample of the LPS in 600 μL of a 150 mM perdeuterated sodium acetate buffer at a pH of 5.64 in D_2O can cause precipitation of the peptide-LPS complex.
7. Peptide concentrations of 0.1 mg resulted in NMR signals which were difficult to observe in the presence of the much more intense NMR signals derived from the 1.0 mg/mL sample of the LPS.
8. An increase in the intensity of the CD spectrum on changing environments suggest the periodicity or structural ordering of the peptide or protein has increased. The net result is the greater the conformational homogeneity of the peptide (fewer different conformations contributing to the CD spectrum) will increase the intensity of the observed CD spectrum.
9. The X-ray and CD data used to develop algorithms to predict peptide and protein secondary structure based on deconvolution of CD spectra were obtained from proteins containing only 20 RNA-encoded amino acids. Since the AMPs used in this investigation contain unnatural amino acids, the use of such algorithms to predict secondary structure would be unreliable.
10. As shown below the CD spectrum of AMP **23** in the presence of the mixed SUV consisting of 75%POPC-5%POPG-20%POPE and the LPS isolated from *K. pneumonia* is not a linear combination of the CD spectra of AMP **23** in the presence of the SUV consisting of the LPS isolated from *K. pneumonia* and

AMP **23** in the presence of the mixed SUV consisting of 75%POPC-5%POPG-20%POPE.

These three CD spectra are given in Fig. 7, along with the 50% concentration linear combination spectra of AMP **23** in the presence of the SUV consisting of the LPS isolated from *K. pneumonia* and AMP **23** in the presence of the mixed SUV consisting of 75%POPC-5%POPG-20%POPE. Comparison of this spectrum with the CD spectrum of AMP **23** in the presence of the mixed SUV consisting of 75%POPC-5%POPG-20%POPE and the LSP isolated from *K. pneumonia* indicates it is not a linear combination of the two sub-spectra.

11. It is assumed, but not proven, that the outer leaflet of the SUVs used in this investigation will contain a higher percentage of LPS as compared to phospholipid and the inner leaflet will contain almost exclusively phospholipids. However, even if this is not the case and the LPS is randomly distributed over both the inner and outer leaflets meaning data concerning the combined effects of LPS and phospholipids on AMP, binding can be obtained.

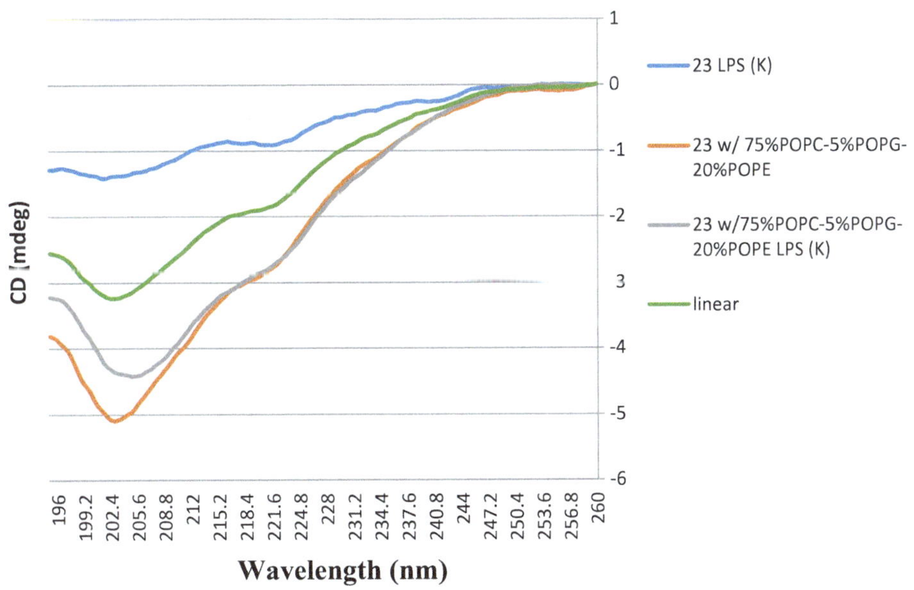

Fig. 7 Far-UV CD spectra of AMP **23** in the presence of the SUV consisting of the LPS isolated from *K. pneumonia* and in the presence of the mixed SUV consisting of 75%POPC-5%POPG-20%POPE and 75%POPC-5%POPG-20%POPE plus LPS. The CD spectrum resulting from the linear combination of the CD spectra (50% concentration level) AMP **23** in the presence of the SUV consisting of the LPS isolated from *K. pneumonia* and the phospholipids 75%POPC-5%POPG-20%POPE, with the CD spectra of AMP **23** in the presence of the mixed SUV consisting of 75%POPC-5%POPG-20%POPE and the LPS isolated from *K. pneumonia* are shown. These spectra clearly indicates the observed spectrum is not a linear combination of the two sub-spectra. Meaning the LPS and phospholipid are incorporated into one SUV

References

1. Bechinger B (2011) Insights into the mechanism of action of host defense peptides from biophysical and structural investigations. J Pept Sci 17:306–314
2. Brogden KA (2005) Antimicrobial peptides: pore formers or metabolic inhibitors in bacteria? Nat Rev Microbiol 3:238–250
3. Domadia PN, Bhunia A, Ramamoorthy A, Bhattacharjya S (2010) Structure, interactions, and antimicrobial activities of MSI-594 derived mutant peptide MSI 594F5A in lipopolysaccharide micelles, role of the helical hairpin conformation in outer-membrane permeabilization. J Am Chem Soc 132:18417–18428
4. Epand RM, Vogel HJ (1999) Diversity of antimicrobial peptides and their mechanisms of action. Biochim Biophys Acta 1462:11–28
5. Matsuzaki K (1998) Magainins as paradigm for the mode of action of pore forming polypeptides. Biochim Biophys Acta 1376:391–400
6. Shai Y (1999) Mechanism of the binding, insertion and destabilization of phospholipid bilayer membranes by α-helical antimicrobial and cell non-selective membrane-lytic peptides. Biochem Biophys Acta 1462:55–70
7. Bhunia A, Domadia PN, Torres J, Hallock KJ, Ramamoorthy A, Bhattacharjya S (2010) NMR structure of pardaxin, a pore-forming antimicrobial peptide, in lipopolysaccharide micelle mechanism of outer membrane permeabilization. J Biol Chem 285:3883–3895
8. Bhunia A, Mohanran H, Domadia PN, Torres J, Bhattacharjya S (2009) Designed b-boomerang antiendotoxic and antimicrobial peptides structures and activities in lipopolysaccharide. J Biol Chem 284:21991–22004
9. Raetz CR, Whitfield C (2002) Lipopolysaccharide endotoxins. Annu Rev Biochem 71:635–700
10. Rietschhel ET, Kirikae T, Schde FU, Mamat U, Schidt G, Loppnow H, Ulmer AJ, Zahringer U, Seydel U, DiPadova F, Schreuer M, Brade H (1994) Bacterial endotoxin: molecular relationships of structure to activity and function. FASEB J 8:217–225
11. Lad MD, Biirembaut F, Clifton LA, Frazier RA, Webster JRP (2007) Antimicrobial peptide-lipid binding interactions and binding selectivity. Biophys J 92:3575–3586
12. Delcour AH (2009) Outer membrane permeability and antibiotic resistance. Biochim Biophys Acta 1794:808–816
13. Hancock REW (1984) Alterations in outer membrane permeability. Annu Rev Microbiol 38:237–264
14. Hancock REW, Chapple DS (1999) Peptide antibiotics. Antimicrob Agents Chemother 43:1317–1323
15. Ding L, Yang L, Weiss TM, Warning AJ, Lehrer RI, Huang HW (2003) Interaction of antimicrobial peptides with lipopolysaccharides. Biochemistry 42:12251–12259
16. Bhattacharjya S (2010) De novo designed lipopolysaccharide binding peptides: structure based development of antiendotoxic and antimicrobial drugs. Curr Med Chem 17:3080–3093
17. Bringezu F, Wen S, Dante S, Hauss T, Majerowicz M, Waring A (2007) The insertion of the antimicrobial peptide dicynthaurin monomer in model membranes: thermodynamic and structural characterization. Biochemistry 46:5678–5686
18. Bruschi M, Pirri G, Giuliani A, Nicoletto SF, Baster I, Scorciapino MS, Casu M, Rinaldi AC (2010) Synthesis, characterization, antimicrobial activity and LPS-interaction properties of SB041, a novel dendrimeric peptide with antimicrobial properties. Peptides 31:1459–1467
19. Epand RM, Epand RF (2011) Bacterial membrane lipids in the action of antimicrobial agents. J Pept Sci 17:298–305
20. Hammer MU, Brauser A, Olak C, Brezesinski G, Goldmann T, Gutsmann T, Andra J (2010) Lipopolysaccharide interaction is decisive for the activity of the antimicrobial peptide NK-2 against *Escherichia coli* and *Proteus mirabilis*. Biochem J 427:477–488
21. Hartmann M, Berditsch M, Hawecker J, Ardakani MF, Gerthsen D, Ulrich AS (2010) Damage of the bacterial cell envelope by antimicrobial peptides Gramicidin S and PGLa as revealed by transmission and scanning electron microscopy. Antimicrob Agents Chemother 54:3132–3142
22. Junkes C, Harvey RD, Bruce KD, Dolling R, Bagheri M, Dathe M (2011) Cyclic antimicrobial R-, W-rich peptides: the role of peptide structure and *E. coli* outer and inner membranes in activity and mode of action. Eur Biophys J 40:515–528
23. Matsuzaki K, Sugishita K-I, Miyajima K (1999) Interactions of an antimicrobial peptide, magainin 2 with lipopolysaccharide-containing liposomes as a model for outer membranes of Gram-negative bacteria. FEBS Lett 449:221–224
24. Russell AL, Kennedy AM, Spuches A, Venugopal D, Bhonsle JB, Hicks RP (2010) Spectroscopic and thermodynamic evidence

for antimicrobial peptide membrane selectivity. Chem Phys Lipids 163:488–497

25. Singh S, Kasetty G, Schmidtchen A, Malmsten M (2012) Membrane and lipopolysaccharide interactions of C-terminal peptides from S1 peptidases. Biochim Biophys Acta 18:2244–2251
26. Wieprecht T, Apostolov O, Seelig J (2000) Binding of the antibacterial peptide magainin 2 amide to small and large unilamellar vesicles. Biophys Chem 85:187–198
27. Ladokhin AS, Vidal MF, White SH (2010) CD spectroscopy of peptides and proteins bound to large unilamellar vesicles. J Membr Biol 236:247–253
28. Glattli A, Daura X, Seebach D, van Gunsteren WF (2002) Can one derive the conformational preference of a β-peptide from its CD spectrum. J Am Chem Soc 124:12972–12978
29. Ladokhin AS, Selsted ME, White SH (1999) CD spectra of indolicidin antimicrobial peptides suggest turns, not polyproline helix. Biochemistry 38:12313–12319
30. Wei S-T (2006) Solution structure of a novel tryptophan-rich peptide with bidirectional antimicrobial activity. J Bacteriol 188:328–334
31. Le Guemeve C, Auger M (1995) New approach to study fast and slow motions in lipid bilayers: application to dimyristoylphosphatidylcholine-cholesterol interactions. Biophys J 68: 1952–1959
32. Andra J, Koch MH, Bartels R, Brandenburg K (2004) Antimicrob Agents Chemother 48: 1593–1599
33. Brandenburg K, Kusumoto S, Seyed U (1997) Conformational studies of synthetic lipid A analogues and partial structures by infrared spectroscopy. Biochim Biophys Acta 2:183–201

Chapter 17

Studying the Interaction of Magainin 2 and Cecropin A with *E. coli* Bacterial Cells Using Circular Dichroism

Concetta Avitabile, Luca Domenico D'Andrea, and Alessandra Romanelli

Abstract

The potential of antimicrobial peptides (AMPs) as an effective therapeutic alternative to classic and current antibiotics has encouraged studies to understand how they interact with the bacterial membrane. Here we describe how to detect, by circular dichroism (CD), the secondary structures of two antimicrobial peptides, magainin 2 and cecropin A, in the presence of *E. coli* bacterial cells.

Key words Antimicrobial peptides, Circular dichroism, *E. coli* cells

1 Introduction

The comprehension of the mechanism by which antimicrobial peptides (AMPs) kill bacteria requires a deep characterization of the interactions of such peptides with bacterial cells. A limited number of studies reported in the literature are focused on cell-peptide interactions, while the vast majority of investigations is aimed at the characterization of the interaction of AMPs with lipid mixtures or with the lipopolysaccharide (LPS), considered as a model systems for the bacterial outer membrane [1–5]. Urged by the need to develop methods to explore the structure of peptides when bound to bacterial cells, we have set up a protocol to determine the secondary structure of peptides in the presence of *E. coli* cells by circular dichroism (CD). The protocol has been successfully applied so far to the study of peptides of different length, including cecropin A, magainin 2, and peptides belonging to the temporin family, such as the temporin B and its analogue TBKK_G6A [2, 3]. Interestingly, the folding of peptides upon interaction with *E. coli* cells occurs only when the peptides are active against *E. coli*, while folding of peptides in the presence of *E. coli* lipopolysaccharide (LPS) seems to be independent on antimicrobial activity. In addition, we have recently reported the three-dimensional

Paul R. Hansen (ed.), *Antimicrobial Peptides: Methods and Protocols*, Methods in Molecular Biology, vol. 1548, DOI 10.1007/978-1-4939-6737-7_17, © Springer Science+Business Media LLC 2017

structure of the antimicrobial peptide TBKK_G6A either in the presence of *E. coli* lipopolysaccharide (LPS) or in the presence of *E. coli* cells [3, 6]. The two structures are different, further confirming that all the components of the bacterial leaflet, including lipid A, proteins, and LPS, are required for the specific peptide-cell recognition.

The case described refers to two natural peptides, magainin 2 and cecropin A, both cationic peptides, active against Gram-positive and Gram-negative bacteria. Cecropin A and magainin 2 are known to be random coils in buffer and fold into helices upon interaction with LPS, detergents, or vesicles [7, 8]. Interestingly, folding into helices is also observed in the presence of *E. coli* cells, and the appearance of the spectrum is slightly different from that obtained in the presence of LPS [2]. The protocol can be applied to analyze the secondary structure of other peptides with *E. coli* cells; it is worth highlighting that the concentration of the peptide needs to be optimized every time, depending on the length of the peptide.

2 Materials

Prepare all solutions using ultrapure water; filter buffer solutions on 0.22 μm filter and degas before use. Run CD experiments at 25 °C.

2.1 Cell Lines and Culture

1. BL21 (DE3) cells.
2. LB medium: Dissolve 10 g tryptone, 5 g yeast extract, and 10 g NaCl in 950 mL deionized water, adjust the pH of the medium to 7.0 using 1 N NaOH and bring volume up to 1 L, autoclave on liquid cycle for 20 min at 15 psi, and store at room temperature or +4 °C.
3. Sterile falcon tubes.
4. Sterile pipette or toothpick.
5. Refrigerated microcentrifuge.
6. Shaker.
7. Autoclave.
8. Circular dichroism spectrophotometer.

2.2 Peptide Solutions Components

1. 10 mM sodium phosphate buffer, pH 7: 154.7 mg $Na_2HPO_4 \cdot 7H_2O$ and 58.4 mg $NaH_2PO_4 \cdot H_2O$ in 100 mL.
2. Lyophilized pure magainin II peptide: Gly-Ile-Gly-Lys-Phe-Leu- His-Ser-Ala-Lys-Lys-Phe-Gly-Lys-Ala-Phe-Val-Gly-Glu-Ile-Met-Asn-Ser.

3. Lyophilized pure cecropin A peptide: Lys-Trp-Lys-Leu-Phe-Lys-Lys-Ile-Glu-Lys-Val-Gly-Gln-Asn-Ile-Arg-Asp-Gly-Ile-Ile-Lys-Ala-Gly-Pro-Ala-Val-Ala-Val-Val-Gly-Gln-Ala-Thr-Gln-Ile-Ala-Lys-NH_2.
4. 6 M guanidinium solution
5. Ultragradient water

3 Methods

3.1 Cell Culture of *E. coli* Cells

1. Day 1 (afternoon): Inoculate 5 mL of autoclaved LB medium in a sterile falcon with a small aliquot of bacteria, withdrawn scratching an *E. coli* glycerol stock with a sterile pipette or toothpick. Set the falcon into the shaker at 37 °C temperature and 180 rpm overnight.
2. Day 2 (morning): Refresh the bacterial culture, diluting 100 μL of the overnight culture into 10 mL of sterile LB medium, and set the suspension into the shaker at 37 °C, 180 rpm. Monitor the cell growth by UV, measuring the absorbance of the suspension at 600 nm (OD_{600}).
3. When the cell culture reaches the exponential phase (0.6–0.8 OD at 600 nm), centrifuge the bacterial suspension for 5 min at 5000 × *g* in a refrigerated microcentrifuge at 4 °C.
4. Discard the supernatant; resuspend the bacterial pellet in 10 mL of 10 mM phosphate buffer pH 7 and then centrifuge for 5 min at 5000 × *g* in a refrigerated microcentrifuge at 4 °C (*see* **Note 1**).
5. Repeat **step 4** twice.
6. Resuspend the bacterial pellet in 10 mM phosphate buffer pH 7, to have suspensions at the desired OD_{600}.

3.2 CD Experiments

3.2.1 Scanning the Conditions to Detect Secondary Structure

1. Dissolve lyophilized peptides in the minimum amount of ultra-gradient water (filtered to 0.2 μm) and estimate the concentration of the mother solutions by UV [9]. Measure the absorbance at 280 nm for cecropin A ($\varepsilon_{280} = 5500\ M^{-1}\ cm^{-1}$) and at 205 nm for magainin 2 ($\varepsilon_{205} = 94{,}390\ M^{-1}\ cm^{-1}$).
2. Record CD spectra for the peptides dissolved in buffer at a 5 μM concentration in the 260–201 nm measurement range, with 100 nm/min scanning speed, 1 nm bandwidth, 4 s response time, and 1.0 nm data pitch.
3. To determine the optimal peptide + cell ratio, start measuring CD spectra for mixtures at different peptide and cell concentrations. Prepare a cell suspension at 0.06 OD_{600}. This may be obtained diluting a 1 OD_{600} *E. coli* suspension in phosphate

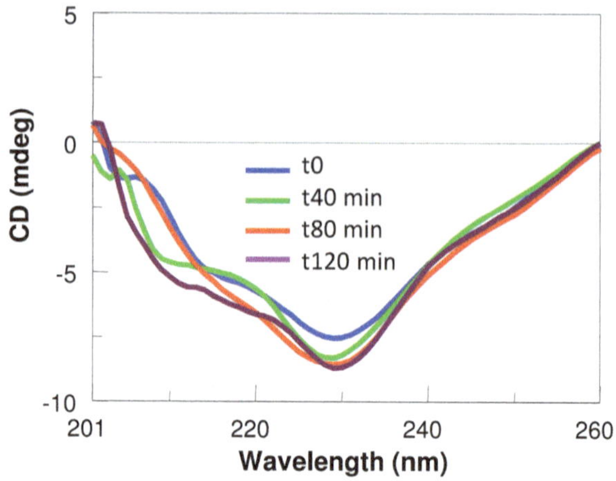

Fig. 1 CD spectra of *E. coli* cells suspended in phosphate buffer pH 7 at different times (Panels reported are a reproduction of those published in [2])

buffer 10 mM, pH 7. Run the blank using only cells in buffer. *E. coli* cells possess a characteristic signature (*see* **Note 2** and Fig. 1).

4. Soon after measure, in a different cuvette, the CD spectrum of the mixture cell + peptide. To start use a 5 μM peptide concentration and 0.06 OD_{600} cells (*see* **Notes 3** and **4**).
5. Record CD spectra for the peptide and the peptide + cell mixture at different times (0, 2, 4, 6 h) (*see* **Note 5**).
6. Repeat **steps 3–5** using suspensions prepared with different concentrations of cells (OD_{600}: 0.075, 0.1, 0.11, 0.13) and peptides (5 and 10 μM).
7. Analyze spectra obtained subtracting the cell blank to the cell + peptide mixture and choose the best conditions for recording more experiments.

3.2.2 Detect Folding of Peptides on Cells

1. Record CD spectra for the best peptide + cells mixture, chosen after the analysis of the spectra obtained at different cell and peptide concentrations. Run measurements every 20 min for 2 h, after 4 and 6 h. For magainin 2 and cecropin A, use *E. coli* cells at 0.1 OD_{600} and 5 μM peptide concentrations. Subtract the CD spectrum of the cell to the corresponding CD spectrum of the peptide + cell mixture (Fig. 2).
2. Repeat the experiment at least twice. Spectra obtained using different batches of cells are usually not perfectly superimposable, but the intensity and the position of the CD bands are reproducible.
3. Plot the intensity of the CD signal at 222 nm of the subtracted spectra (spectra of the cells + peptide mixture − spectra of the cells) vs time. Use prism or similar programs to calculate the

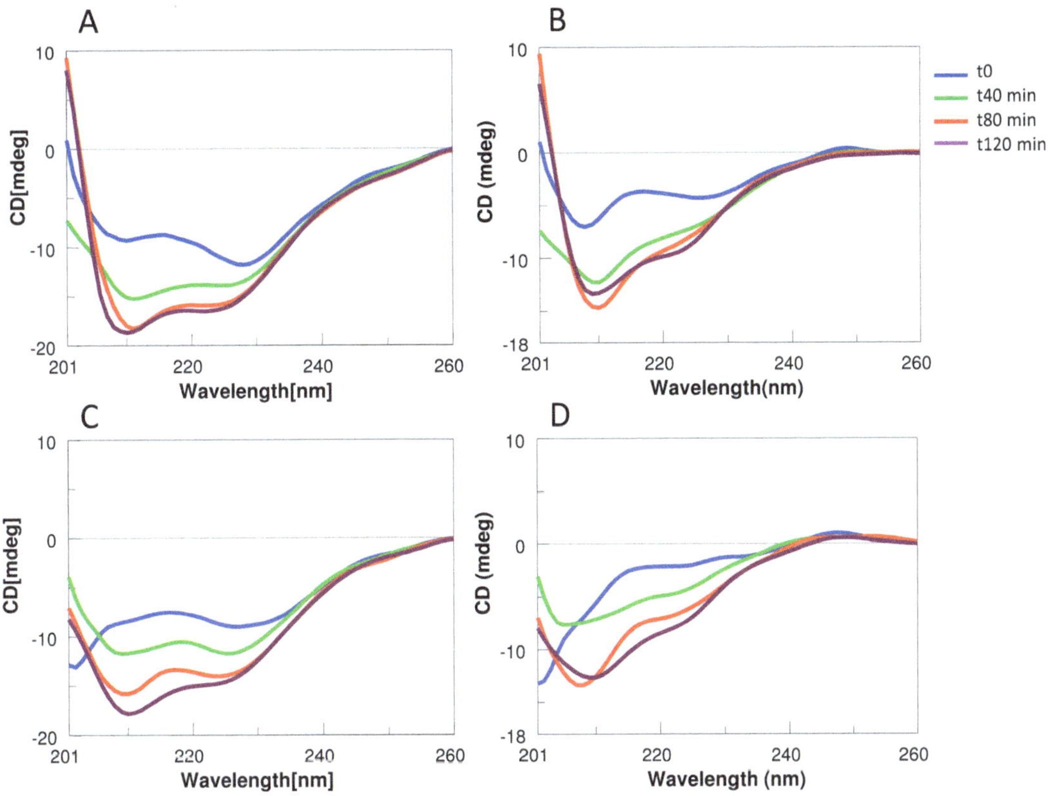

Fig. 2 (**a**, **c**) CD spectra of cell + peptide mixtures (**a**: magainin 2, **c**: cecropin A); (**b**, **d**) CD spectra obtained after subtraction of the cell spectra to the cell + peptide mixture spectra (**b**: magainin 2, **d**: cecropin A) (Panels reported are a reproduction of those published in [2])

error on each experimental point. This plot will help to understand how long it takes for your peptide to fold upon incubation with *E. coli* cells (Fig. 3).

The protocol described is optimized for peptides cecropin A and magainin 2 using a 10 mm quartz cuvette. In our experience, the optimal cell density is always between 0.08 and 0.1 OD_{600}, and the optimal peptide concentration depends on the length of the peptide. Shorter peptides are best analyzed at higher concentrations [3].

4 Notes

1. Collect the solutions containing cells, cell supernatants, or residual pellets in a falcon and add bleach to kill cells. Dispose the solutions following waste regulations.
2. Before running the experiments on cells + peptide mixtures, run CD spectra for the peptide and the cells separately and check the HT voltage value. The purity of the buffer is pivotal

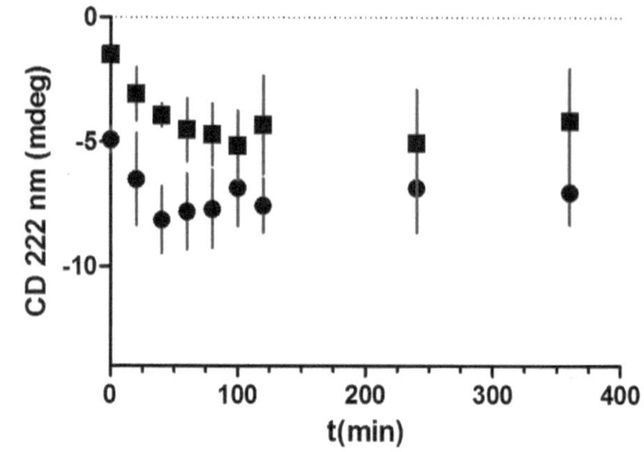

Fig. 3 Plot of the CD signal intensity at 222 nm (as obtained from subtracted spectra) vs time for magainin 2 (*circles*) and cecropin A (*squares*) (The figure reported is a reproduction from [2])

for the success of the experiment. The HT voltage value increases with cell concentration. Be careful when testing different cell concentrations.

3. Use the same batch of cells for the mixture peptide + cell and for the blank. The intensity of the signals, in fact, slightly changes with the batch of cells.
4. You will always work with two cuvettes, one for the blank and the other for the peptide + cell mixture. When you remove one cuvette (e.g., the cuvette with blank) from the spectrophotometer to run the other measurement (e.g., that on the cell + peptide mixture), keep always samples at 25 °C. If the temperature outside is not controlled, for example, if it is too hot, your cell will die very soon, and results will hardly be reproducible.
5. At the end of each experiment, the quartz cuvettes need to be carefully cleaned, as cells stick to the cuvette. Fill each cuvette with a 4 or 6 M guanidinium solution, remove the solution after 2/3 h. Rinse the cuvette ten times with ultrapure water, with ethanol and finally dry the cuvette by insufflating nitrogen.

References

1. Freire JM, Gaspar D, de la Torre BG, Veiga AS, Andreu D, Castanho MA (2015) Monitoring antibacterial permeabilization in real time using time-resolved flow cytometry. Biochim Biophys Acta 1848:554–560
2. Avitabile C, D'Andrea LD, Romanelli A (2014) Circular Dichroism studies on the interactions of antimicrobial peptides with bacterial cells. Sci Rep 4:4293
3. Malgieri G, Avitabile C, Palmieri M, D'Andrea LD, Isernia C, Romanelli A, Fattorusso R (2015) Structural basis of a temporin 1b analogue antimicrobial activity against Gram negative bacteria determined by CD and NMR techniques in cellular environment. ACS Chem Biol 10:965–969
4. Rosenfeld Y, Papo N, Shai Y (2006) Endotoxin (lipopolysaccharide) neutralization by innate immunity host-defense peptides. Peptide prop-

erties and plausible modes of action. J Biol Chem 281:1636–1643

5. Gee ML, Burton M, Grevis-James A, Hossain MA, McArthur S, Palombo EA, Wade JD, Clayton AH (2013) Imaging the action of antimicrobial peptides on living bacterial cells. Sci Rep 3:1557
6. Avitabile C, Netti F, Orefice G, Palmieri M, Nocerino N, Malgieri G, D'Andrea LD, Capparelli R, Fattorusso R, Romanelli A (2013) Design, structural and functional characterization of a Temporin-1b analog active against Gram-negative bacteria. Biochim Biophys Acta 1830:3767–3775
7. Silvestro L, Axelsen PH (2000) Membrane-induced folding of cecropin A. Biophys J 79: 1465–1477
8. Bechinger B, Zasloff M, Opella SJ (1993) Structure and orientation of the antibiotic peptide magainin in membranes by solid-state nuclear magnetic resonance spectroscopy. Protein Sci 2:2077–2084
9. Anthis NJ, Clore GM (2013) Sequence-specific determination of protein and peptide concentrations by absorbance at 205 nm. Protein Sci 22:851–858

Chapter 18

Studying the Mechanism of Membrane Permeabilization Induced by Antimicrobial Peptides Using Patch-Clamp Techniques

Giorgio Rispoli

Abstract

Many short peptides selectively permeabilize the bacteria plasma membrane, leading to their lyses and death: they are therefore a source of antibacterial molecules and inspiration for novel and more selective drugs, which may have wider application in many other fields, as selective anticancer drugs. In this chapter, it is presented a new method to investigate the permeabilization properties of antimicrobial peptides under strict physiological conditions, employing the patch-clamp technique coupled to a fast perfusion system.

Key words Pore-forming peptides, Pore-forming toxins, Ion channels, Photoreceptors, Rod outer segment, CHO-K1 cells, Whole-cell recording

1 Introduction

Most of the research on the pore-forming peptides has the long-term goal to understand at the molecular level how the peptide-lipid interactions produce structural changes in both, and optimal peptide orientation, to eventually produce a raise in membrane conductance [1–3]. A widely used technique to study this interaction consists in recording the ion current flowing through the pore formed by these peptides, inserted in an artificial lipid bilayer. However, these membranes have a symmetrical composition of one lipid kind, instead of an asymmetrical variety of many different lipids, typical of the natural membranes. Moreover, to have an acceptable signal-to-noise ratio in these experiments, it is necessary to apply nonphysiological voltages and ionic gradients that however could affect the peptides themselves or even the membrane in which they were inserted [4]. Finally, to our knowledge, no studies aimed to measure the time course of formation of peptidic pores in a membrane (either artificial or of a living cell) and the kinetics of their disaggregation.

Paul R. Hansen (ed.), *Antimicrobial Peptides: Methods and Protocols*, Methods in Molecular Biology, vol. 1548,
DOI 10.1007/978-1-4939-6737-7_18, © Springer Science+Business Media LLC 2017

Here, the biophysical characteristics of peptidic pores and their formation dynamics in the cell plasma membrane were studied by applying the peptide monomers to the extracellular side with a fast microperfusion system while recording the membrane current by using the whole-cell configuration of the voltage-clamp technique. The possible reversibility of the membrane permeabilization was assessed upon swiftly removing the peptide from the external solution and measuring the kinetic with which the current returned (or not) to the zero level.

To simplify the interpretation of the results, it is important to have most of the current flowing through the peptidic pores at any physiological voltage, rather than flowing through the endogenous ionic channels. Therefore, it was looked for a cell system having negligible endogenous conductances or having conductances that could be fully silenced by small concentrations of channel blockers (such as TTX, TEA, dihydropyridines, etc.), so to minimize the chance that these drugs may obstruct the peptidic pores and/or may interfere with pore formation. The first cell system we found satisfying these requirements is the outer segment of the photoreceptor rod mechanically isolated from low vertebrate retinae (Fig. 1a) [5]. The only channel type in the plasma membrane of this cell fragment can be fully closed by bright light, giving membrane resistances (recorded in whole cell) typically larger than 1 GΩ (Fig. 1b, c) at any physiological voltage (Fig. 1d). Moreover, the size of these cells is particularly large in low vertebrates, as in the *Rana esculenta* (Fig. 1a): this comestible species was the preferred choice because of its commercial availability and low cost. Recently, we found that the Chinese hamster ovary (CHO-K1) cells (Fig. 1e) are a suitable cell system as well, because they do not have any voltage- and/or calcium- and/or time-dependent conductances but only a very small background one (giving a whole-cell membrane resistance always larger than 3 GΩ; Fig. 1f, g), and they can be easily maintained and grown [6]. To further simplify the interpretation of the experiments, patch pipettes were usually filled with the same perfusion solution (that typically contained 130 mM of a monovalent cation) to ensure the current was only driven by holding potential.

To have a precise control of pore formation (and disaggregation), a custom-made, computer-controlled microperfusion system was employed to rapidly (~50 ms) apply and remove the peptides under study onto a cell [7]. This was accomplished by moving a multibarreled perfusion pipette in front of the patch-clamp recorded cell (Fig. 1h). Peptide formation and disaggregation dynamics were therefore gathered, respectively, from the time course of the development and fall of the exogenous current (at a given holding potential), following peptide application and removal. This system can apply up to six different solutions (but it can be easily scaled up to a much larger number of solutions); therefore, the peptide could be

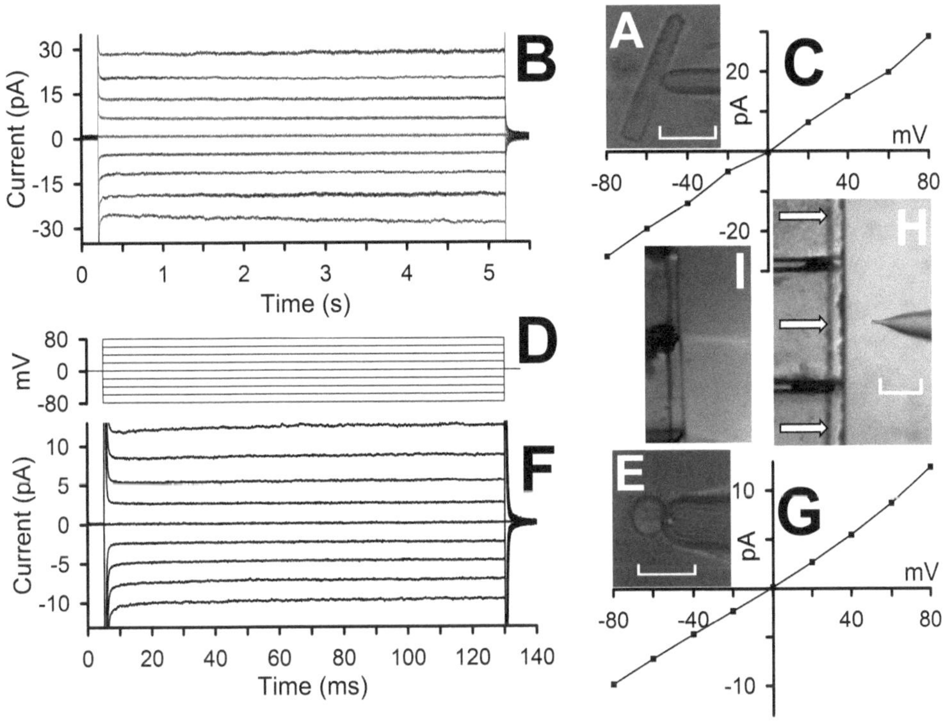

Fig. 1 Outline of the technique employed to investigate the permeabilization properties of the peptides inserted in a natural membrane. (**a**) A representative rod outer segment recorded in whole cell by means of a pressure-polished pipette (scale bar is 20 μm; pipette and external solution: 130 mM K^+ + 1 mM Ca^{2+}) was subjected to 5 s voltage steps from −80 mV to +80 mV in 10 mV increments starting from 0 mV (**d**) and repeated ten times; the average current amplitude (**b**) of each voltage step vs the voltage step amplitude (IV; **c**) was well fitted by a *straight line* (correlation coefficient ~0.99), whose slope gave a membrane resistance of ~3 GΩ. (**e**) A representative CHO-K1 cell recorded in whole cell by means of a pressure-polished pipette (scale bar is 20 μm; pipette solution, 130 mM K^+; external solution, 130 mM K^+ + 1 mM Ca^{2+}) was subjected to 125 ms voltage steps from −80 mV to +80 mV in 10 mV increments (**d**) starting from 0 mV (traces are the average of six cells); the average current amplitude (**f**) vs the voltage was well fitted by a straight line (correlation coefficient ~0.99; **g**), which gave a membrane resistance of ~6.1 GΩ. (**h**) The CHO-K1 cell of panel (**e**) was aligned in front of the multibarreled perfusion pipette (at low magnification; scale bar is 200 μm); horizontal *white arrows* denote perfusion flows. (**i**) Two solutions with different refraction index flowing through two adjacent barrels (each one of 500 μm of side) allowed to observe the interface between them (adapted from [6, 7, 10])

tested at five different concentrations (one tube is used to apply the control solution, with no peptide; *see* Fig. 2). Compounds that could accelerate [8] or slowdown (or preclude) the pore formation, or specific drugs affecting ion flow through the pore, could be tested by adding them at various concentrations to the solutions containing the peptide at a fixed concentration: this strategy would allow comparing the effects of the drugs at a constant membrane permeabilization. The pore selectivity could be instead readily evaluated by dissolving the peptide (at a fixed suitable concentration) in solutions containing equiosmotic concentrations of different monovalent or divalent ions.

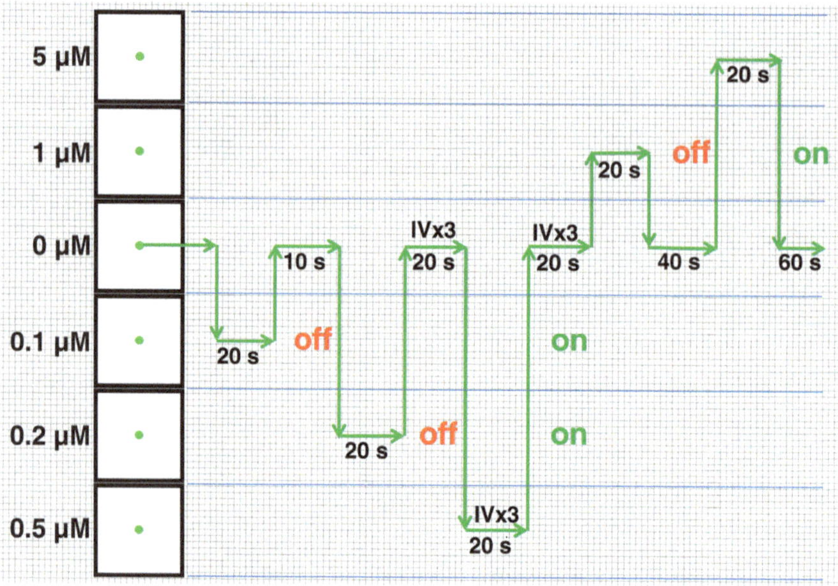

Fig. 2 Diagram of a sample protocol determining the stoichiometry and the voltage-dependence of the membrane permeabilization of a peptide. The *green dots* sketch a cell of 40 μm in diameter in scale with the 500 μm × 500 μm perfusion barrels, each one filled with the control solution having a peptide concentration reported on the *left*. The *green arrows* denote the temporal sequence of the solution changes; the *numbers* underneath each *horizontal arrow* denote the time that a cell should spend in a particular solution; these times are just an example. All perfusion lines are flowing at the beginning of the experiments and are turned "off" (in *red*) and back "on" (in *green*) when appropriate (see text). In this sample protocol three consecutive IVs are acquired in control solution ("IVx3"), three IVs during peptide application at 0.5 μM concentration, and three IVs once returned the cell in control solution

To faithfully record large currents (elicited by high concentrations of peptides and/or highly membrane permeabilizing ones), it is necessary to minimize access resistance, to reduce error in membrane potential control and time constant of charging the cell membrane capacitance. Moreover, the long tapered shank of the patch pipette may cause intracellular ion accumulation or depletion, and it slows down the rate of exogenous molecules incorporation via the patch pipette [9]. These problems can be circumvented, all at once, by widening the patch pipette shank (Fig. 1a, e) through the combination of heat (applied outside of the shank) and air pressure (applied to its lumen), as described previously [10]. Besides improving the electrical recordings, the enlarged tip geometry of the pressure polished pipettes could accommodate pulled quartz or plastic perfusion tubes close to the pipette tip, allowing the fast and controlled cytosolic incorporation of peptides and other exogenous molecules [11]. This is particularly important, because specific peptides may have different permeabilization properties depending upon the side of the membrane to which they are applied. For example, it is conceivable that viroporins [12, 13] are optimized to insert in the intracellular face of the plasma membrane, because they are synthesized in host cell cytosol.

2 Materials

2.1 Solutions

1. Always use ultrapure water (prepared by purifying or bidistilling deionized water to attain a conducibility no larger than ~0.06 μS/cm at 25 °C) and analytical grade reagents, but the reagents used for cell culture (*see* Subheading 2.2). Prepare all solutions at room temperature and store in refrigerator (4 °C) but the peptide stock solutions (stored at −80 °C) and unstable drugs (stored at −20 or −80 °C); the HEPES (*N*-(2-hydroxyethyl)piperazine-*N'*-(2-ethanesulfonic acid)), stock solution must be prepared fresh once a month.
2. The Ringer's solution: 115 mM NaCl, 3 mM KCl, 10 mM HEPES free acid 0.6 mM $MgCl_2$, 0.6 mM $MgSO_4$, 1.5 mM $CaCl_2$, 10 mM glucose (osmolality 260 mOsm/Kg, buffered to pH 7.6 with NaOH).
3. CHO-K1 solution: Ham's F12K medium supplemented with 10% fetal bovine serum and 1% penicillin/streptomycin
4. Phosphate-buffered saline buffer (PBS): 137 mM NaCl, 2.7 mM KCl, 10 mM Na_2HPO_4, 1.8 mM KH_2PO_4, pH = 7.4.
5. DMEM/F-12 medium (Dulbecco's Modified Eagle's Medium)
6. 0.25% trypsin-EDTA solution (Gibco™).
7. Control perfusion solution for both the isolated rod outer segments and the CHO-K1 cells:130 mM KCl (or an equiosmolar concentration of another monovalent or divalent cation), 1 mM $CaCl_2$, and 10 mM HEPES; osmolality 260 mOsm/Kg, buffered to pH 7.6 with KOH (or with a 1 N hydroxyl stock solution of the ion substituting K^+; *see* **Note 1**).
8. Peptide stock solution: 1 mM peptide stock solution in ultrapure water, or methanol, or ethanol, or DMSO (dimethyl sulfoxide) depending upon their solubility (*see* **Notes 2–4**).
9. Chloro-tri-*n*-butyl-silane.

2.2 Cells

1. All experiments on frogs (*Rana esculenta*) were performed in compliance with the European Communities Council Directive for animal use in science (86/609/EEC) and approved by a local ethical committee. Animals were purchased from a local supplier and were kept in filtered, running tap water in small tanks at room temperature (20–23 °C), fed two to three times a week with honey worms, and maintained on a 12 h light–12 h dark cycle.
2. A healthy specimen was anaesthetized by immersion in a tricaine methane sulfonate solution (1 g/l in water) during dark adaptation (≈4 h) and then decapitated and pithed in dim red light (generated by LEDs with wavelength of 660 nm). The head was swiftly transferred in a fully darkened box and immersed in a Petri dish containing Ringer solution.

3. The box was equipped with infrared LEDs (wavelength: 940 nm) and a high definition webcam that had its infrared filter removed and was connected to an external monitor.
4. The hands were inserted into the box through two light-tight holes to perform all the following procedures.
5. After focusing the head on the monitor, both eyes were removed from the head: one eye was kept intact in a vial filled with Ringer solution on ice for later use, and the other one was hemisected.
6. The back half of the hemisected eye was cut into pieces (up to four) with a razor blade fragment that were stored in oxygenated Ringer solution on ice for later use. The retina was "peeled" from an eyecup piece and was gently triturated in ~5 μl of Ringer, using a fire-polished Pasteur pipette to obtain the isolated rod outer segments.
7. CHO-K1 cells were purchased from the American Type Culture Collection (CCL61 ATCC) and were grown (following the manufacturer's directives) in Ham's F12K medium supplemented with 10 % fetal bovine serum and 1 % penicillin/streptomycin. Cell cultures were stored in humidified environment at 37 °C and in 5 % CO_2, according to manufacturer's instructions.
8. Cells were washed three times in PBS and detached from plate by trypsinization for 1 min in 0.25 % trypsin-EDTA solution. Trypsin enzyme was then inactivated by adding 1–2 ml of DMEM/F-12 medium and 2–3 ml of PBS to the cell suspension; the latter was centrifuged at ~5000 × *g* for 1 min, and finally the cells were dissolved in PBS for whole-cell experiments.

2.3 Whole-Cell Recording and Fast Perfusion

1. Patch pipettes with enlarged shank [10] were fabricated by using 50 μl borosilicate glass capillaries having a resistance in symmetrical K^+ solution that ranged from 1 to 4 MΩ (*see* **Note 5**).
2. The perfusion pipette was constituted of a three-barrel square glass capillary; a six-barreled pipette was obtained by carefully attach together two capillaries with a cyanoacrylate glue specific for glass (a three-barrel pipette is visible in Figs. 1h and 3a). A waterproof sheet of a 1000 grit sandpaper was necessary to even the pipette tip. A quartz needle was glued inside each barrel so that the solutions were fed to the perfusion pipette through the luer connector equipping each quartz needle.
3. The solutions flow was generated by means of 1 ml precision syringes, whose piston was moved by a DC motor controlled via a user-friendly computer interface. The pipette was moved on a horizontal plane in front of the cell with a precision step motor controlled by the same interface (Fig. 3b); the solution change typically occurs in ~50 ms, that is the time it takes the cell

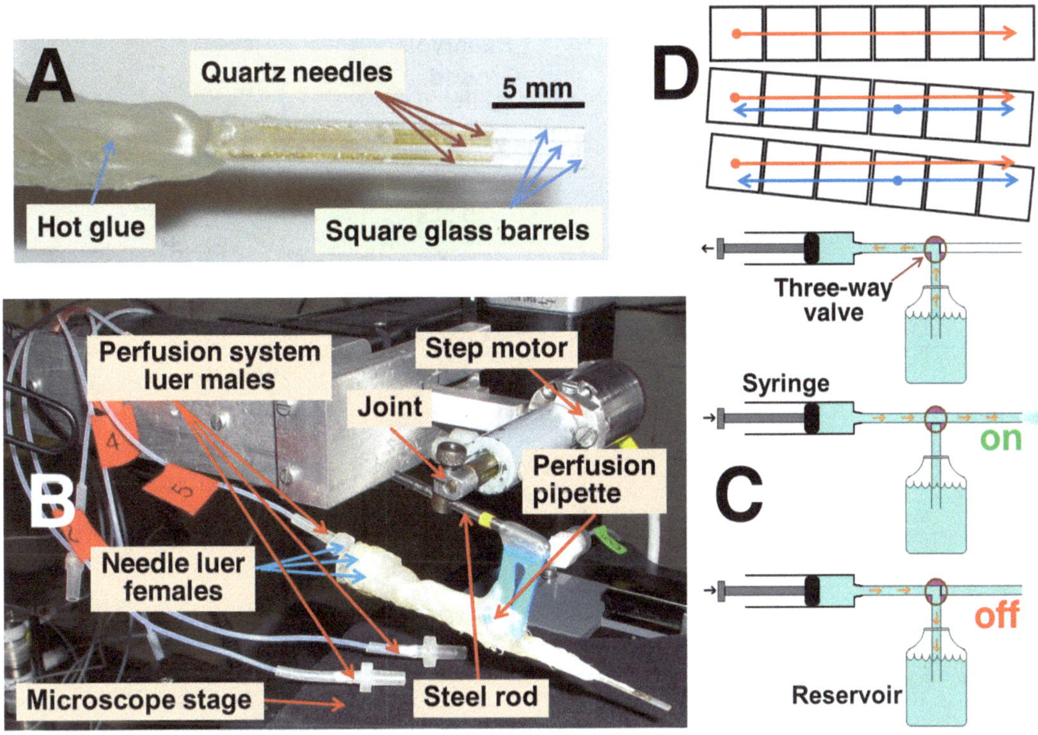

Fig. 3 Perfusion system. (**a**) The pipette tip and the quartz needles glued inside them. (**b**) The perfusion pipette coupled to the step motor and the perfusion system. (**c**) Scheme of one perfusion line (composed by a syringe, a three-way valve, a cylinder, and one perfusion pipette barrel) and perfusion flow (*orange arrows*) during syringe refilling (*upper panel*), during perfusion of a cell recorded in whole cell (*middle panel*; the perfusion is "on" as in Fig. 2), and when the cell perfusion is stopped and the solution is returned to the reservoir (*lower panel*; the perfusion is "off" as in Fig. 2); drawing not in scale (adapted from [7]). (**d**) *Upper panel*, the perfusion pipette perfectly aligned on a horizontal plane; *middle panel*, the perfusion pipette has a 3.4° angle with the horizontal plane; and *lower panel*, this angle is 5.1°. The horizontal *red arrow* denotes the trajectory followed by the cell (*red dot*) positioned in the far left barrel, while the perfusion pipette is moved on an horizontal plane; the horizontal *blue arrow* denote the trajectory followed by the cell (*blue dot*) positioned in a central barrel. The cell is sketched as a dot (*red* or *blue*) of a 80 μm in diameter in scale with the 500 μm × 500 μm perfusion barrel

to cross the sharp boundary between two adjacent streams. Every perfusion parameter was computer-controlled by this interface, as the rate of solution flow, the temporal sequence of solution changes and valve operation (a sample protocol is shown in Fig. 2), or the number of automatic, self-washing cycles. A key design of this system (described in details in [7]) was the possibility to direct the solution contained in each syringe, depending upon the position of a three-way solenoid equipping each perfusion line, to the perfusion pipette or to a reservoir (in which each solution was made; Fig. 3c; *see* **Note 6**).

4. The fluid drop containing the cells was held by surface tension between two parallel glass coverslips that were supported at one end by a teflon block and formed the floor and ceiling of

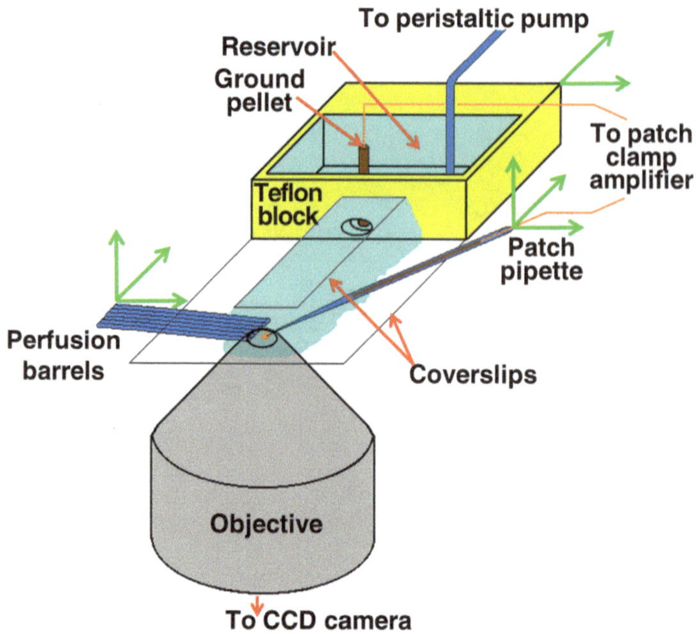

Fig. 4 Scheme of the recording chamber. Drawing not in scale; the cells are pipetted in between the two coverslips (the external solution is indicated in *blue*). *Green arrows* denote the *xyz* movements of the patch and the perfusion pipette and the *xy* movements of the recording chamber assembly

the recording chamber that was open on three sides (Fig. 4). This arrangement allowed the patch pipette and the perfusion pipette to enter the chamber from opposite sides and to keep the perfusion pipette almost horizontally (*see* **Note** 7). The coverslip that formed the floor was coated with chloro-tri-*n*-butyl-silane to prevent cell sticking. A reservoir was drilled in the teflon block that was filled with the same external solution of the chamber, with which was connected through an hole. The ground pellet and the tube for perfusion solution removal (Fig. 4), attained with a peristaltic pump, were housed in this reservoir.

5. The whole-cell recordings were performed employing Axopatch 200B under visual control, by focusing the cells on a fast digital camera coupled to the microscope and controlled by AquaCosmos software package. The microscope was equipped with 4×, 20×, and 60× objectives.
6. Electrophysiological data were low-pass filtered at 2 kHz by an eight-pole Butterworth filter sampled at 19.2 kHz/16 bits by a A/D board controlled by Clampex software and analyzed with Clampfit Sigmaplot and Mathcad software.

3 Methods

3.1 Perfusion

1. Build the perfusion pipette as follows. First, make sure that the three-barrel square glass capillary has a regular tip (Figs. 1h, i, and 3a): if one or more barrel has a dent, or one barrel is longer than an adjacent one, smooth the tip by sanding it very gently by holding the capillary by hand and keeping it orthogonally to the sandpaper sheet.
2. Use plenty of water or oil to damp the vibrations generated by the rubbing, which would chip the barrel tip. If more than three perfusion lines are required, glue together two or more three-barrel capillaries while pressing them against a flat surface to keep them aligned horizontally and while pressing their tip against another flat surface to have all the tips aligned on the same line.
3. Make sure that the glue does not accidentally run too close to the pipette tip; otherwise, the latter could not be clearly visible under the microscope, or in the worst case the glue could partially or totally clog one or more barrels.
4. Cut the quartz needles with sharp scissors so that when inserted completely inside each barrel, their tips are 3–5 mm back from the barrel opening (Fig. 3a).
5. Coat the needle shank with the cyanoacrylate glue as close as possible to its tip (but ensure not to clog it), and spin the needle while inserting it in the barrel to uniformly and completely fill the gap between the needle and the internal sides of the barrel. This ensures that the perfusion solution cannot flow back inside the barrel by capillarity, making deposits, once the perfusion solutions dry up, that are very difficult to clean up.
6. The solutions are fed to the perfusion pipette through the luer connector equipping each quartz needle: to avoid that it breaks off the needle, these connectors must be blocked together with hot glue (Fig. 3a, b; *see* **Note 7**).
7. A stainless steel rod is finally glued on the top of the assembled perfusion pipette with a spacer (the blue plastic rectangle of Fig. 3b) that is used to firmly couple the pipette to the step motor.
8. This rod is introduced into a mechanical joint, attached to the step motor axle, which allows rotating the pipette about the three axes and then tightening the rod in the right position with a single knob (Fig. 3b).
9. Make sure that the perfusion pipette moves on a horizontal plane, especially when using pipettes with many barrels: a tilt of a few degrees is enough to have a cell, perfectly centered in the first barrel, to find itself closer and closer to the edge of the

next barrels (where the perfusion solution could potentially mix with the bath solution), and even out of the perfusion flow in the most distal barrel (Fig. 3d).

10. Therefore, place the control solution in the barrel as close as possible to the center of the perfusion pipette: this circumvents to some extent the last problem (Fig. 3d), and it allows designing more reliable experimental protocols (*see* **Note 8**).
11. Check often the pipette horizontal alignment by focusing the bottom (or the top) of one barrel and ensure that the bottom (or the top) of all the other barrels have exactly the same focus.
12. All the perfusion lines should be built by using tubes and quartz needles of same length and diameter, in order to have the same hydraulic resistance in each perfusion line. This is particularly important in the case of gravity-fed solutions, because a solution flowing slower than the one flowing in an adjacent barrel could be contaminated by the latter and a cell could be blown away when switched from the former to the latter barrel. For this reason, it has been opted here for a perfusion system employing syringes driven by a single motor, which force all solutions to flow at the same rate (*see* **Note 9**).

3.2 Patch-Clamp Experiments

1. Patch-clamp experiments (included the ones on isolated rod outer segments) are carried out under room lights and at room temperature (20–24 °C). Transfer an aliquot (~2 ml) of the solution containing the cells (rod outer segments or CHO-K1) to the recording chamber placed on the microscope stage.
2. Choose a healthy-looking cell that does not adhere to the chamber bottom, seal on it, and gain the whole-cell configuration; the perfusion pipette should be kept outside of the bath during these operations.
3. Measure the cell capacitance and the seal, the access, and the membrane resistance from the current waveform elicited by repetitive −10 mV pulses. Set the holding potential in the range −20/−80 mV so to keep the size of the current (elicited, for instance, by the highest concentration of the peptide) enough restrained to avoid errors in membrane potential control. As a rule of thumb, a concentration of ~1 μM of a peptide that permeabilize the membrane as alamethicin F50 produces a current of ~0.7 nA at −20 mV in symmetric K^+, giving a small voltage drop (~2 mV) across a typical access resistance of 3 MΩ. The holding potential should be anyway kept large enough in order to detect the current flowing through the peptidic channels at the smallest concentrations. It is better to use negative holding potentials, because cell membranes do not usually tolerate long, continuous application of strong positive voltages and may develop a leakage.

4. Acquire a current to voltage relationship (IV; Fig. 1b, c, f, g) to ensure that the cell is electrically equivalent to a resistor. If the experiments require single channel recordings, the membrane resistance should be at least of 1 GΩ.
5. With the perfusion pipette outside of the bath, turn on the flow of all the perfusion lines to rinse the salts that could have accumulated on the perfusion pipette tip by evaporation. Stop all the perfusion flows, dry up any drop that could hang on the pipette bottom with paper tissue, and gently bring the pipette inside the bath by viewing it with the 4× objective (*see* **Note 10**).
6. With the perfusion pipette and the patch pipette in the field of view, move the barrel containing the control solution in the center of the viewfield, focus sharply the top of this barrel with the fine focus knob of the microscope, then focus its bottom, and finally turn the knob in between these two positions by helping with its graduated ring.
7. With the patch pipette micromanipulator, move vertically the cell until it appears sharp in this focal plane: the cell is so centered in the vertical plane with the pipette barrels. Next move the cell longitudinally to be in the middle of the "control" barrel and laterally to be no farther than 500 μm from the barrel opening.
8. Now the cell is in a region where the flow is laminar and there is a very sharp boundary between two adjacent solutions (Fig. 1i; *see* **Note 9**). Once properly aligned, the perfusion pipette and the patch pipette sealed on the cell should appear as in Fig. 1h: at this point, turn on all the perfusion flows (*see* **Note 11**).
9. Acquire another IV to ensure that the cell was not damaged during the alignment operations: if the membrane resistance is still very high at all voltages, runs the perfusion protocol. In Fig. 2, it is shown the diagram of a sample protocol designed to determine the stoichiometry and the voltage dependence of the membrane permeabilization of a (putative) peptide that gives a ~0.3 nA of current at −20 mV (the holding potential set in this experiment) at a concentration of 0.5 μM. The current at the largest voltages (±80 mV) for this peptide is therefore expected to be not larger than ~0.9 nA (in the case of a perfectly ohmic permeabilization) at 0.5 μM peptide concentration; therefore, it is better to measure the IV at this concentration to avoid errors in the voltage control (*see* Subheading 3.2, **step 3**). The IV can be determined by applying voltage steps or voltage ramps, but to minimize the duration of peptide perfusion a ramp (with a slope, for instance, of 0.25 mV/ms) is the best choice.

10. Acquire three consecutive IVs in control solution (indicate with "IVx3" in Fig. 2), then three IVs during peptide application once the current reached steady state, and finally a third set of three IVs after returning the cell in control solution (wait for the steady state).

    Subtract the average IV during peptide application with the average IV before peptide application, and check that the two average IVs in control solution before and after peptide application are similar.

11. When switching the cell between two nonadjacent barrels, the perfusion line(s) feeding the barrel(s) placed in between these two ones must be turned off (in red in Fig. 2), to avoid that the cell is transiently exposed to other undesired solution(s); this (these) line(s) is better to be turned on (in green in Fig. 2) as soon as possible (*see* **Note 12**).
12. The time with which the peptide must be kept in a particular solution depends by many factors as: (1) how long does it takes for the current to reach a stable amplitude (this time is usually longer for smaller concentrations of peptide and/or drugs interfering with peptidic pore); (2) how many IVs are acquired and if are performed with long voltage pulses or fast ramps; and (3) how long does it takes to recover the initial (high) membrane resistance once returning the cell in control solution (if the peptide permeabilization is reversible; this time is usually longer for larger concentrations of peptide and/or drugs).
13. Check every now and then that the pipette does not change its focus when moving from one end to the other one. If this is not the case, the pipette does not move horizontally; therefore, terminate the experiment and re-center it (*see* **Note 13**).

4 Notes

1. The presence of 1 mM Ca^{2+} in the control solution was necessary to avoid damaging the cell membrane, which would produce a current (leakage) that could be larger than the one flowing through the peptidic channels [7].
2. The stock solutions are aliquoted in vials of a volume large enough to complete one experiment. Peptides are dissolved in the perfusion solution to get a specific final concentration, varying from 10 nM (to get single channel events or when studying peptides at strong voltages having high capability to permeabilize the membrane [6, 12]; use the 10 μM stock) up to 10 μM (for these peptides with the smallest capability to permeabilize the membrane or to verify the inability of a peptide to permeabilize the membrane; use the 1 mM stock), and used within 1 h.

3. When the peptide is dissolved in solvents different than water, it is recommended to use the peptide stock concentration as high as possible, to minimize the solvent contamination in the external solution. Indeed, we found that isolated rod outer segments and CHO-K1 cells can tolerate contamination of ethanol, methanol, and DMSO in the external solution up to ~1 %; larger concentrations damage the cell membrane, producing a leakage current.
4. The patch pipette solution is better to be the same of the perfusion solution, so to have the current just driven by the holding voltage, but this can be performed on isolated rod outer segments only. Indeed, the CHO-K1 cells (as any other intact cell) do not tolerate the intracellular application of 1 mM Ca^{2+}, because this would activate the endogenous proteases and induce the cell apoptosis. In conclusion, the symmetric conditions can be used with CHO-K1 cells if the peptidic pore is not permeable to Ca^{2+}; therefore, the pipette Ca^{2+} can be removed and still the symmetric ionic conditions preserved; otherwise, the isolated rod outer segment is the alternative choice.
5. This glass type, if used appropriately, allows to consistently attaining seals not only on the cells described here but on an extremely wide variety of cell types, isolated from different amphibian, reptilian, fish, and mammalian tissues, on many different cultured cells, and on artificial membranes made with many different lipid mixtures [10].
6. This allowed to save the solutions that were not used, or to avoid that, upon switching between two nonadjacent barrels, the cell was transiently exposed to another undesired solution in between. Once emptied, the syringes could be refilled from the reservoirs (Fig. 3c). All tubing, valves, and connectors in contact with the perfusion solutions were made in teflon.
7. When gluing together the female luer connectors of the quartz needles and the glass barrels with the hot glue, keep a barrel region of 2.5/3.5 cm from the tip free of glue, and keep the glue lump underneath the pipette as small as possible. Otherwise, when moving down the pipette, this lump could touch the microscope stage first (Fig. 3b), making it impossible to bring the pipette tip further down in focus. This problem could be circumvented by entering the pipette into the recording chamber with a strong angle, but this makes more problematic the view of the pipette tip. Even worst, the perfusion flows may lose their laminarity when they hit the chamber floor, especially if the perfusion pipette tip is very close to the latter.
8. When performing experiments aimed at determining the dose–response effect of drugs and/or peptides, keep the smallest concentrations of these compounds close to the barrel containing the control solution, as shown in Fig. 2. Otherwise,

when the cell is made to pass in front of a barrel with a very high concentration of a test compound while moving to a barrel with a smaller concentration of this compound, a possible small leak from the former barrel could modify the kinetics at early time of the effect of the solution of the latter barrel.

9. The perfusion speed must be slow enough to have a laminar flow, to avoid blowing the cell away, and to save precious compounds but must be fast enough to have a sharp boundary between two adjacent solutions (Fig 1i). The optimal flow speed is 15 μl/min, ensuring stable recordings and complete solution changes in ~50 ms and allowing to use minimal amounts of peptide/drug solutions (<500 μl) to perform perfusion experiments lasting more than half an hour.
10. Be very careful when entering the perfusion pipette in the bath, because it displaces a large fluid volume and may blow the cell away. A safe maneuver is to position the cell at the top right edge of the viewfield of the 4× objective and enter the perfusion pipette in its lower left edge. Then, perfusion and patch pipettes could be safely moved to obtain the alignment of Fig. 1h.
11. Make sure to purge all the bubbles that may be trapped in the perfusion lines. It is better to design protocols having all the perfusion flows on when the cell is in front of the barrel with the control solution: if a bubble happens to be in one barrel (but the control one), it can flow away while the cell is (safely) standing in control solution.
12. Avoid turning on or off the valve controlling the flow in the barrel in front of which is placed the cell, because its operation causes a flow pulse that may blow the cell away; this flow pulse is always harmless if it occurs in an adjacent barrel.
13. Many compounds tend to stick on plastic surfaces; therefore, it is better to use a perfusion system made with teflon tubes, valves, and connectors. At the end of the experiment, it is necessary to clean thoroughly the perfusion system with tap water, and finally rinse few times with ultrapure water. Periodically, it is useful to wash all the perfusion lines with a sequence of solutions: 50% of the solvent used to make the stock solution (i.e., ethanol, methanol, DMSO, etc.) diluted in 50% of water; water; 0.1 M HCl; water; 0.1 M KOH; water; finally rinse several times with ultrapure water.

Acknowledgment

Many thanks to Marco Aquila, Mascia Benedusi, Anna Fasoli, and Natascia Vedovato for helping in applying and refining this technique; Andrea Margutti manufactured many custom-made mechanical pieces. This work was supported by Ministero dell'Università e della Ricerca (MIUR), Roma, Italy, Project PRIN 2006 and 2008 (Progetti di Rilevante Interesse Nazionale), and by University of Ferrara, Project FAR 2006–2014 (Fondi di Ateneo per la Ricerca).

References

1. Gilbert RJ, Dalla Serra M, Froelich CJ, Wallace MI, Anderluh G (2014) Membrane pore formation at protein-lipid interfaces. Trends Biochem Sci 39:510–516
2. Rawat A, Nagaraj R (2014) Peptide self-assembly: from toxins to amyloid fibrils and nanotubes. Curr Top Med Chem 14:740–746
3. Fuertes G, Giménez D, Esteban-Martín S, Sánchez Muñoz OL, Salgado J (2011) A lipocentric view of peptide-induced pores. Eur Biophys J 40:399–415
4. Bockmann RA, Hac A, Heimburg T, Grubmuller H (2003) Effect of sodium chloride on a lipid bilayer. Biophys J 85: 1647–1655
5. Rispoli G, Sather WA, Detwiler PB (1993) Visual transduction in dialyzed detached rod outer segments from lizard retina. J Physiol 465:513–537
6. Fasoli A, Salomone F, Benedusi M, Boccardi C, Rispoli G, Beltram F, Cardarelli F (2014) Mechanistic insight into CM18-Tat11 peptide membrane-perturbing action by whole-cell patch-clamp recording. Molecules 19: 9228–9239
7. Vedovato N, Rispoli G (2007) A novel technique to study pore-forming peptides in a natural membrane. Eur Biophys J 36:771–778
8. Aquila M, Benedusi M, Koch KW, Dell'Orco D, Rispoli G (2013) Divalent cations modulate membrane binding and pore formation of a potent antibiotic peptide analog of alamethicin. Cell Calcium 53:180–186
9. Pusch M, Neher E (1988) Rates of diffusional exchange between small cells and a measuring patch pipette. Pflugers Arch 411:204–211
10. Benedusi M, Aquila M, Milani A, Rispoli G (2011) A pressure-polishing set-up to fabricate patch pipettes that seal on virtually any membrane, yielding low access resistance and efficient intracellular perfusion. Eur Biophys J 40:1215–1223
11. Aquila M, Benedusi M, Fasoli A, Rispoli G (2015) Characterization of zebrafish green cone photoresponse recorded with pressure-polished patch pipettes, yielding efficient intracellular dialysis. PLoS One 10:e0141727
12. Madan V, Sánchez-Martínez S, Vedovato N, Rispoli G, Carrasco L, Nieva JL (2007) Plasma membrane-porating domain in poliovirus 2B protein. A short peptide mimics viroporin activity. J Mol Biol 374:951–964
13. Royle J, Dobson SJ, Müller M, Macdonald A (2015) Emerging roles of viroporins encoded by DNA viruses: novel targets for antivirals? Viruses 7:5375–5387

Chapter 19

Assays for Identifying Inducers of the Antimicrobial Peptide LL-37

Frank Nylén, Peter Bergman, Gudmundur H. Gudmundsson, and Birgitta Agerberth

Abstract

One promising approach to meet the growing problem of antibiotic resistance is to modulate host defense mechanisms, i.e., host-directed therapy (HDT), in the fight against infections. Induction of endogenous antimicrobial peptides (AMPs) via small molecular compounds, such as 1,25-dihydroxyvitamin D_3 or phenylbutyrate, could provide one such HDT-based approach.

We have developed a cell-based screening assay for the identification of novel compounds with the capacity to induce AMP expression and here follows the detailed protocol.

Key words Chemical library screening, *CAMP* gene, *Z'*-factor, Cell-based reporter system

1 Introduction

Innate immunity, the front line of our defense against pathogens, relies to a great extent on antimicrobial peptides (AMPs). Many AMPs are synthesized at epithelial surfaces, where the initial contact with microbes takes place [1], but are also a part of the microbicidal arsenal of phagocytes. In humans, AMPs include the defensins (α-and ß-families) and a single cathelicidin, LL-37 [2, 3]. These peptides exhibit a broad antimicrobial activity [4] and act as chemoattractants for cells of both the adaptive and innate immune system [5–7].

The development of bacterial resistance against classical antibiotics is a growing problem, and novel antimicrobial strategies are urgently needed [8]. Our concept is to strengthen the immune defense against invading pathogens by inducing endogenous AMP expression, utilizing small molecules that can be taken orally or applied topically, such as 1,25-dihydroxyvitamin D_3 and/or phenylbutyrate [9–12].

Paul R. Hansen (ed.), *Antimicrobial Peptides: Methods and Protocols*, Methods in Molecular Biology, vol. 1548, DOI 10.1007/978-1-4939-6737-7_19,

In order to find additional inducers of AMPs in a non-biased fashion, we decided to take advantage of the high-throughput screening (HTS) concept, which is the preferred method to identify new lead molecules or potential repurposed drugs from the pharmaceutical industry. We therefore set out to establish a cell-based assay for screening of compounds with potential AMP-inducing properties [13].

At this point, we would like to stress a few key aspects of our assay:

First, given our focus on LL-37 (LLGDFFRKSKEKIGKEFKRIVQRIKDFLRNLVPRTES), we selected this peptide to start with. There are several features of this AMP that makes it a good target: (1) there is only one cathelicidin gene (*CAMP*) in humans and also in mice, rats, and rabbits, making these animals suitable animal models. Thus, findings related to the human gene can be confirmed in an animal model—and vice versa. In contrast, there are many defensin genes, which make similar approaches more complicated for this gene family.

Second, antimicrobial peptides have been shown to be regulated and released in a coordinated fashion, leading to a massive attack from multiple peptides with different mechanisms of action. This is probably the reason why bacterial resistance has not developed against AMPs [14, 15].

Third, we took the decision to design a cell line with a stable transfection. This way we avoid the direct effect of inducers on the transient transfection. Butyrate, for example, is a potent inducer of both transfection and LL-37, and any potential effect on LL-37 induction would be difficult to distinguish from effects on transfection efficacy [16, 17]. Therefore, the construction of a novel cell line with stable integration of the plasmid construct would eliminate problems with transfection efficiency, despite the more laborious approach.

Furthermore, a stable cell-based reporter assay will reduce the time and increase the reproducibility in future experiments and even more for HTS purposes. HT-29 (ATCC® HTB-38™), a colonic epithelial cell line known to express LL-37 and respond to several known inducers, was chosen to host the plasmid vector construct.

2 Materials

1. BAC clone RP11_502L5.
2. Seamless cloning kit.
3. pGL-4.26 Luciferase Reporter Vector.
4. DNA polymerase.

5. Competent cells for transformation.
6. Culture medium: RPMI 1640, supplemented with 10% fetal calf serum (FCS), 100 μg/ml streptomycin, and 100 U/ml penicillin.
7. Phosphate-buffered saline (PBS): 137 mM NaCl, 2.7 mM KCl, 4.3 mM Na_2HPO_4, 1.4 mM KH_2PO_4, pH 7.4.
8. *SpeI* restriction enzyme.
9. PCR clean up kit.
10. Cell culture flasks and 24 well plates.
11. Transfection reagent.
12. 800 μg/ml hygromycin.
13. Trypsin-EDTA (0.05%).
14. 96 well plates preferably with white wells and "see-through" bottom (*see* **Note 1**).
15. Stock 100 mM sodium phenyl butyrate (PBA) in PBS.
16. 2 mM sodium phenyl butyrate (PBA) in PBS.
17. Stock 100 μM 1,25-dihydroxyvitamin D_3 solution in EtOH.
18. 100 nM 1,25-dihydroxyvitamin D_3 in EtOH.
19. Passive lysis buffer (PLB).
20. DNA extraction kit.
21. 0.8% agarose gel: Mix 0.8 g of agarose with 100 ml H_2O.
22. Tecan M200 plate reader with an injector module.
23. Luciferase assay reagent: luciferase assay substrate and luciferase assay.
24. The Prestwick Chemical Library of 1200 FDA-approved compounds was kindly provided by Chemical Biology Consortium, Sweden, and is commercially available.
25. Dimethyl sulfoxide (DMSO).

3 Methods

3.1 Vector Construct

1. Amplify the CAMP gene, including 3 kilo base pairs (kb) of DNA upstream of the ATG start codon [18] but without the endogenous stop codon in exon 4, with the following primers: forward 5′-**ATTTCTCT**GAATTCTCATGATATCTTCCGTCG (named A1; *see* Fig. 1a) and reverse 5′-**GCATCTTC**GGACTCTGTCCTGGGTACAAGATT (A2). The template for amplification was the BAC clone RP11_502L5.
2. Clone the resulting product using the seamless cloning kit into a pGL-4.26 Luciferase Reporter Vector. For

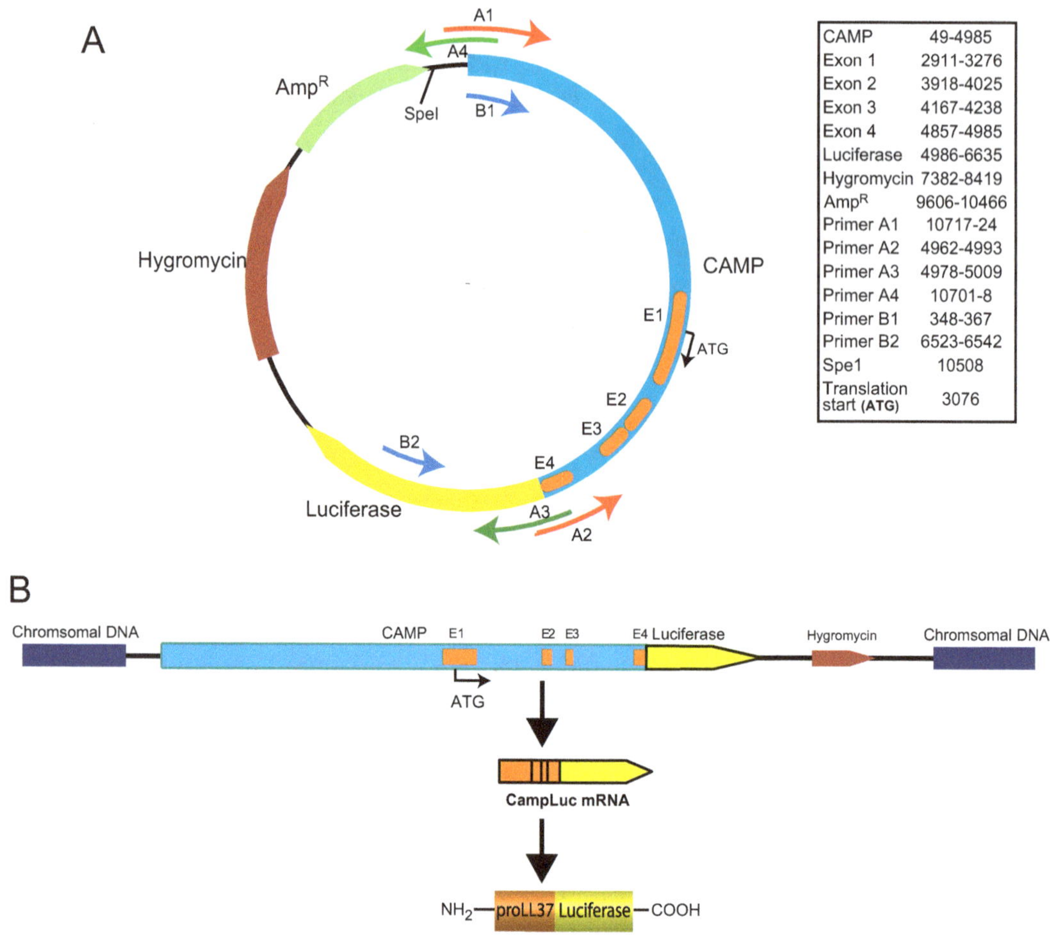

Fig. 1 Schematic representation of the vector construct and of the stable transfection of HT-29 cells. (**a**) The *CAMP* gene, shown in *blue* (the *CAMP* gene exons are indicated in *orange*, E1, E2, E3, and E4), was inserted into the pGL-4.26 vector in frame with the *firefly luciferase* gene (shown in *yellow*). The vector also contains a hygromycin-resistance gene, shown in *red*, and ampicillin-resistance gene in *green*. The primers utilized for cloning of the *CAMP* gene into the pGL4-26 vector are indicated in the figure: A1 and A2 for the *CAMP* gene (*red*) and A3 and A4 for the pGL4-26 plasmid (*green*). Primers used to assess the construct integration of the entire *CAMP* gene were B1 and B2 (*blue*). (**b**) The genomic insert is transcribed into *CAMP*–luciferase fusion mRNA that is translated into the fusion protein proLL37-luciferase

amplification of the vector with excision of the luciferase starting codon, we used the following primers: forward 5′-**CAGAGTCC**GAAGATGCCAAAAACATTAAGAAG (A3) and reverse 5′-**GAGAATTC**AGAGAAATGTT CTGGCACCTGCAC (A4). Bolded letters denote primer overhang that was utilized for recombinant construction (*see* **Note 2** and Fig. 1a).

3. Propagate the plasmids in *E. coli*.

4. Sequence the entire plasmid to confirm a correct insert and to investigate if any single nucleotide mutations were introduced by PCR (*see* **Note 3**).

   We termed the final vector construct pGL-CampLuc.

3.2 Cell Cultures and Stable Transfection

1. Grow HT-29 cells in culture medium in a 5% carbon dioxide atmosphere at 37 °C. Use as low passage as possible.
2. For transfection, HT-29 cells will be grown to 70% confluence in a 10 cm dish.
3. Linearize 15 μg of pGL-CampLuc with *SpeI* restriction enzyme and perform a "PCR clean up" using a kit (*see* **Note 4**).
4. Transfect the purified and linearized pGL-CampLuc plasmid in to the HT-29 cells.
5. Replenish the culture medium 24 h post transfection.
6. Add 800 μg/ml hygromycin supplemented culture medium to the cells 72 h post transfection (*see* **Note 5**).
7. After 3–4 weeks, clones resistant to the selection medium can be isolated (*see* **Note 6**).
8. Clones from now on will be maintained in culture medium containing 400 μg/ml hygromycin (*see* **Note** 7).
9. Allow the isolated clones to adhere and propagate in a 24-well plate for 3–4 days or until 70–90% confluency is reached (*see* **Note 8**).
10. Trypsinize the cells using 1× Trypsin-EDTA (0.05%) for approximately 5 min or until the majority of the cells are de-attached after tapping the container. Reseed in three wells, one for maintenance and the other two for luciferase expression assessment.
11. Stimulate one of the two wells with the combination of (final concentration 2 mM) PBA and (final concentration 100 nM) 1,25-dihydroxyvitamin D_3 and treat the other well with vehicle (negative control).
12. Assay both wells after 24 h for luciferase activity. First, lyse the cells using 100 μl of PLB and allow plate to shake for 20 min.
13. Transfer 20 μl of cell lysate to a white-welled 96-well plate and measure luminescence according to methods in Subheading 3.3, **steps** 7 and **8**.
14. Discard all clones not expressing luciferase or not able to express more luciferase upon co-treatment with PBA and 1,25-dihydroxyvitamin D_3.
15. For all other clones, the integrity of the *CAMP* gene insertion will be assessed by PCR.

16. Extract genomic DNA from each clone using a "DNA extraction kit" and perform PCR with the following primers. Forward primer 5′-AGGTAGTGGACACCGACCTG (B1 *see* Fig. 1a), located at 2728 bp upstream of the start codon of the *CAMP* gene, and the reverse primer 5′-GTCCACGAACAC AACACCAC (B2 *see* Fig. 1a) located at 1557 bp into the luciferase gene.
17. Visualize the 6195 bp PCR product on a 0.8% agarose gel (*see* **Note 9**).

3.3 Cell-Based Reporter Assay

1. Pick one clone, which fulfills a complete genomic integration of the *CAMP* gene and its promotor, along with the expected response to known stimuli coupled to a moderate nonstimulated expression of ProLL37-luciferase, e.g., background luminescence (*see* Fig. 1b for a schematic representation of a successful genomic integration and translation to ProLL37-luciferase).
2. We have termed our clone MN8CampLuc and have optimized seeding in a 96-well plate to 6×10^4 cells/well.
3. Culture the cells for 48 h in normal culture medium, e.g., without hygromycin.
4. Treat with test compounds or the vehicle as the negative control.
5. After 24 h, remove medium and wash the cells once with PBS.
6. Add 20 μl of passive lysis buffer (*see* **Note 10**).
7. Use a luminometer with an injector for measuring of luminescence.
8. Treat each well according to the following sequence; program the injector to add 50 μl (*see* **Note 11**) of luciferase assay reagent to each well and followed by a 2 s halt period before measurement of luminescence for 10 s (*see* **Note 1**).

3.4 Z′-Factor Measurement

Before attempting any large experiment, such as a library screen, make sure to validate your assay to be sensitive enough for screening purposes. We validated our assay by determining the *Z*′-factor of our assay as described in [19] (*see* **Note 12**).

1. Seed and treat MN8CampLuc cells according to Subheading 3.3, **step 2** and onward. Incubate the cells with vehicle (negative control) or with PBA in combination with 1,25-dihydroxyvitamin D_3 (positive control) in technical replicates of 12 wells per condition.
2. Calculate *Z*′-factor according to the following formula:

$$Z' = 1 - \frac{(3\sigma_N + 3\sigma_P)}{|\mu_N - \mu_P|}$$

where σ denotes standard deviation and μ mean value. N and P indicate negative and positive controls, respectively.

3.5 Prestwick Chemical Library Screening

The library compounds were provided at a stock concentration of 10 mM dissolved in DMSO in 96 well plates with row 1 and 12 empty for internal controls.

1. Dilute the 10 mM stocks, 1:100, in normal culture medium and apply vehicle and positive controls to the empty rows in four technical replicates for each plate.
2. Add the library compounds from these plates to the MN8CampLuc cells, prepared according to Subheading 3.3, **steps 3–8** at a 1:10 dilution, giving a final concentration of 10 μM for the compounds and 0.1 % of the solvent.
3. Measure luciferase activity in the cell lysates after 24 h, as described under Subheading 3.3.
4. Calculate the fold induction per 96-well plate, i.e., as the ratio between the signal for each library compound and the average signal of the vehicle control as measured per plate.
5. Then calculate the average fold induction for all test compounds ± three standard deviations ($\mu \pm 3\sigma$).
6. Compounds with a fold induction higher than $\mu + 3\sigma$ are considered to be inducers while compounds with a fold induction lower than $\mu - 3\sigma$ are considered to be inhibitors within a 99.7% confidence interval [19] (*see* Fig. 2).
7. Assess the statistical robustness of the screen, plate by plate, by calculating the *Z'*-factor, emanating from the positive and vehicle controls.
8. Also consider machine errors, e.g., increased or decreased signal specific to wells, rows, or columns. Calculate the average signal for each well, row, and column for all plates used in the screen. The average signal should not deviate from that of the total average control.
9. Confirm positive hits and at the same time measure the concentration curve by treating MN8CampLuc cells with each of the positive compounds obtained from the initial screen at 11 concentrations, ranging from 100 μM to 1.7 nM (serial dilutions of 1–3) (*see* **Note 13**).
10. To allow for this concentration range, the final concentration of DMSO needs to be adjusted to 1 % in all wells.

4 Notes

1. Wells with "see-through" bottom allows for visual inspection of seeding and to assess toxicity of the test compounds. By

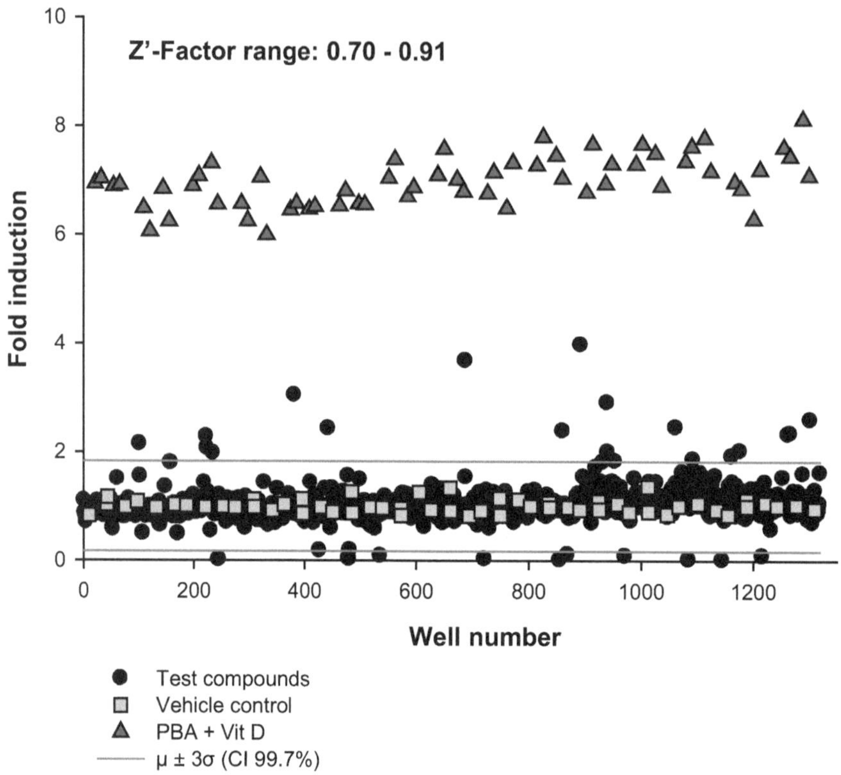

Fig. 2 Screen of the Prestwick Chemical Library. The 1200 compounds of the Prestwick Chemical Library were screened as described in Subheading 3.5. The screen spanned 15 plates and in each plate four vehicle controls (0.1 % DMSO) and four positive controls [vitamin D_3 (Vit D; 100 nM)] in combination with sodium PBA (2 mM) were included. The signal recorded in each well was normalized to the average vehicle control of the corresponding plate. Using the formula $\mu \pm 3\sigma$, the hit thresholds were calculated (*gray lines*) in our experiment, giving 18 positive and 14 negative hits. The *Z'*-factor was calculated for each plate, giving the range of 0.70–0.91

applying white tape on the bottom of the plates before measurement, you allow for the signal to "bounce" toward the detector rather than dispersing underneath the plate. This increases the luminescent signal significantly.

2. Utilizing the seamless cloning kit allowed us to remove the stop codon from the CAMP gene and also the start codon from the luciferase gene. Moreover, we could paste these genes together without codons for any extra amino acids and thus keep the construct as "natural" as possible. By removing the start codon from the luciferase, we positively removed all background luminescence.

   Note that the minimal promotor from the pGL-4.26 plasmid was removed together with the start codon of the luciferase gene.

3. Either design sequencing primers with a spacing of 600–700 bp over the plasmid or assign your sequencing provider with the task to sequence the entire plasmid.
4. Because of the circular nature of a plasmid, it needs to be linearized prior to genomic integration. Actively determining where the plasmid is linearized increases the chances of a successful integration and also reducing the number of false-positive clones. The plasmid construct utilized here only allowed for linearization within 3.1 kb out of 10.7 kb. This means that if linearization would occur spontaneous, 71 % of all integration events would be useless and that 68 % of hygromycin resistant clones would be of false-positive character.
5. A concentration curve would be performed prior to Subheading 3.2, **step 8,** to determine the minimum concentration that kills all cells within 5–7 days.
6. For isolation of the colonies, we found that HT-29 colonies can be picked with a pipette under a microscope due to their firm attachment to each other. This quickened the procedure but one should emphasize that this increase the risk for contamination of the cells. We also promote the use of a colony cylinder and trypsin.
7. The 800 μg/ml of hygromycin supplemented culture medium is used for the establishment of resistant clones and at this concentration the cells propagate at a very slow rate. Therefore, we maintain the cells at 400 μg/ml of hygromycin. If hygromycin is completely removed, the luminescence drops significantly within a couple of passages of the cell cultures.
8. If colonies are picked according to the first part of **Note 6**, trypsinization of the colony with re-seeding will greatly increase the proliferation time.
9. The hygromycin and luciferase cassette must be intact since the clones are hygromycin resistant and the luminescence was detected; therefore, only the CAMP gene with its promotor needs to be confirmed for full integrity.
10. By placing the plates in a −80 °C freezer, you assure complete lysis and also allow analysis at a later time with no significant decrease in signal for at least 72 h. Plates must be at room temperature at the time of measurement. Optimal temperature for luciferase is around 25 °C.
11. The manufacturer promotes the use of 100 μl of luciferase assay reagent per well. We tested both 50 and 100 μl and we found the signal to be comparative.
12. The *Z′*-factor gives you an indication of the span where you can find hits and as such gives you an indication of your assay's suitability for screening purposes. *Z′*-factor calculation gives a number between 0 and 1 where 0.5–1 indicates an excellent assay.

13. To avoid crystallization of the library compounds the dilution series would be prepared in organic solvent, e.g., DMSO, before addition of RPMI medium.

Acknowledgment

We are grateful to Hanna Axelsson, at CBCS and Karolinska Institutet, for discussions and advice regarding assay formatting and data analysis.

"The authors are supported by the Swedish Research Council (56X-1217-01-6 (BA), 20854 (BA) and 2013-2709 (PB), a Linneus grant CERIC, BA), The Swedish Strategic Foundation: RBd08-0014 (BA), Karolinska Institutet (BA, PB), the Icelandic Centre for Research (RANNIS, GHG), and the University of Iceland Research Fund (GHG). Groschinsky's Foundation (PB), Scandinavian Society for Antimicrobial Therapy Foundation (PB), Swedish Society of Physicians (PB), and Heart and Lung Foundation (BA, PB).

Gudmundur H. Gudmundsson is a visiting scientist at Karolinska Institutet supported by the Wenner-Gren foundation. The Prestwick Chemical Library and plating thereof was provided by the Chemical Biology Consortium Sweden (CBCS).

References

1. Cederlund A, Gudmundsson GH, Agerberth B (2011) Antimicrobial peptides important in innate immunity. FEBS J 278:3942–3951
2. Ganz T (2003) Defensins: antimicrobial peptides of innate immunity. Nat Rev Immunol 3:710–720
3. Kai-Larsen Y, Agerberth B (2008) The role of the multifunctional peptide LL-37 in host defense. Front Biosci 13:3760–3767
4. Brogden KA (2005) Antimicrobial peptides: pore formers or metabolic inhibitors in bacteria? Nat Rev Microbiol 3:238–250
5. Hancock RE, Sahl HG (2006) Antimicrobial and host-defense peptides as new anti-infective therapeutic strategies. Nat Biotechnol 24:1551–1557
6. Agerberth B, Charo J, Werr J, Olsson B, Idali F, Lindbom L, Kiessling R, Jornvall H, Wigzell H, Gudmundsson GH (2000) The human antimicrobial and chemotactic peptides LL-37 and alpha-defensins are expressed by specific lymphocyte and monocyte populations. Blood 96:3086–3093
7. Yang D, Chertov O, Oppenheim JJ (2001) The role of mammalian antimicrobial peptides and proteins in awakening of innate host defenses and adaptive immunity. Cell Mol Life Sci 58:978–989
8. Czaplewski L, Bax R, Clokie M, Dawson M, Fairhead H, Fischetti VA, Foster S, Gilmore BF, Hancock RE, Harper D, Henderson IR, Hilpert K, Jones BV, Kadioglu A, Knowles D, Olafsdottir S, Payne D, Projan S, Shaunak S, Silverman J, Thomas CM, Trust TJ, Warn P, Rex JH (2016) Alternatives to antibiotics-a pipeline portfolio review. Lancet Infect Dis 16:239–251
9. Steinmann J, Halldorsson S, Agerberth B, Gudmundsson GH (2009) Phenylbutyrate induces antimicrobial peptide expression. Antimicrob Agents Chemother 53:5127–5133
10. Wang TT, Nestel FP, Bourdeau V, Nagai Y, Wang Q, Liao J, Tavera-Mendoza L, Lin R, Hanrahan JW, Mader S, White JH (2004) Cutting edge: 1,25-dihydroxyvitamin D3 is a direct inducer of antimicrobial peptide gene expression. J Immunol 173:2909–2912
11. Gombart AF, Borregaard N, Koeffler HP (2005) Human cathelicidin antimicrobial peptide (CAMP) gene is a direct target of the

vitamin D receptor and is strongly up-regulated in myeloid cells by 1,25-dihydroxyvitamin D3. FASEB J 19:1067–1077

12. Weber G, Heilborn JD, Chamorro Jimenez CI, Hammarsjo A, Torma H, Stahle M (2005) Vitamin D induces the antimicrobial protein hCAP18 in human skin. J Invest Dermatol 124(5):1080–1082
13. Nylen F, Miraglia E, Cederlund A, Ottosson H, Stromberg R, Gudmundsson GH, Agerberth B (2014) Boosting innate immunity: development and validation of a cell-based screening assay to identify LL-37 inducers. Innate Immun 20:364–376
14. Marr AK, Gooderham WJ, Hancock RE (2006) Antibacterial peptides for therapeutic use: obstacles and realistic outlook. Curr Opin Pharmacol 6:468–472
15. Hancock RE, Nijnik A, Philpott DJ (2012) Modulating immunity as a therapy for bacterial infections. Nat Rev Microbiol 10:243–254
16. Gorman CM, Howard BH, Reeves R (1983) Expression of recombinant plasmids in mammalian cells is enhanced by sodium butyrate. Nucleic Acids Res 11:7631–7648
17. Schauber J, Svanholm C, Termen S, Iffland K, Menzel T, Scheppach W, Melcher R, Agerberth B, Luhrs H, Gudmundsson GH (2003) Expression of the cathelicidin LL-37 is modulated by short chain fatty acids in colonocytes: relevance of signalling pathways. Gut 52:735–741
18. Gudmundsson GH, Agerberth B, Odeberg J, Bergman T, Olsson B, Salcedo R (1996) The human gene FALL39 and processing of the cathelin precursor to the antibacterial peptide LL-37 in granulocytes. Eur J Biochem 238:325–332
19. Zhang JH, Chung TD, Oldenburg KR (1999) A simple statistical parameter for use in evaluation and validation of high throughput screening assays. J Biomol Screen 4:67–73

Chapter 20

Methods for Elucidating the Mechanism of Action of Proline-Rich and Other Non-lytic Antimicrobial Peptides

Monica Benincasa, Giulia Runti, Mario Mardirossian, Renato Gennaro, and Marco Scocchi

Abstract

A distinct group of antimicrobial peptides kills bacteria by interfering with internal cellular functions and without concurrent lytic effects on cell membranes. Here we describe some methods to investigate the mechanisms of action of these antimicrobial peptides. They include assays to detect the possible temporal separation between membrane permeabilization and bacterial killing events, to assess the capacity of antimicrobial peptides to cross the bacterial membranes and reside in the cytoplasm, and later to inhibit vital cell functions such as DNA transcription and protein translation.

Key words Antimicrobial peptide, Proline-rich peptide, Mechanism of action, Membrane permeabilization, Cell uptake, Flow cytometry

1 Introduction

Antimicrobial peptides (AMPs) are a large class of innate immunity effectors with a remarkable capacity to inactivate microorganisms. Most of them kill microorganisms by a selective membrane-interacting activity leading to a lethal permeabilization of the microbial envelope [1]. However, some AMPs affect microbial viability by mechanisms different from membrane permeabilization [2–5]. These AMPs may recognize and inactivate cellular targets in vitro, such as nucleic acids, proteins, and enzymes, and their modes of action are mediated by translocation across the plasma membrane in a nonlethal manner [4, 5].

The group of proline-rich AMPs (PR-AMPs) is the best characterized example of AMPs showing non-membranolytic effects [6, 7]. They are cationic peptides with a high content of proline residues that are present in mammalians [7] as well as in several species of insects and crustaceans [6]. PR-AMPs are mainly active against Gram-negative bacteria sharing a similar mode of bacterial cell internalization involving the inner membrane protein SbmA [8, 9]. The

Paul R. Hansen (ed.), *Antimicrobial Peptides: Methods and Protocols*, Methods in Molecular Biology, vol. 1548,
DOI 10.1007/978-1-4939-6737-7_20, © Springer Science+Business Media LLC 2017

molecular chaperone DnaK has been proposed as a first cytoplasmic target for a PR-AMP [10]. Recently, different peptides including both mammalian [Bac7(1–35)] and insect (oncocin, apidaecins) PR-AMPs have been demonstrated to specifically and completely inhibit in vitro translation in the micromolar concentration range [11, 12]. Then it has been shown that these peptides specifically interact with ribosomes, indicating that the killing mechanism of PR-AMPs is based on a specific block of protein synthesis [13–15].

In this chapter, we report methods to investigate the mechanism of action of PR-AMPs and other AMPs predicted to have a mode of action different from membrane lysis. Firstly, we describe an assay to evaluate the possible temporal dissociation between bacterial cell death and changes in membrane permeability, a hallmark of this kind of AMPs; secondly, we describe a method to evaluate the capacity of these peptides to cross the bacterial membranes and to measure their concentration within the cytoplasm; and, lastly, we suggest a strategy to establish whether these peptides kill bacteria by inhibiting vital cellular functions such as DNA transcription and protein translation.

2 Materials

2.1 Membrane Permeabilization and Bacterial Killing

1. Mueller-Hinton Broth (MHB) or other culture media (*see* **Note 1**).
2. Mueller-Hinton agar (MHA) plates containing 1.5 % agar.
3. Buffered saline (BS): 10 mM Na-phosphate buffer, 150 mM NaCl, pH 7.4.
4. Stock solution in sterile water of each AMP to be tested: 1–10 mg/ml. In this study, Bac7(1–35) RRIRPRPPRLPRPRPRPLPFPRPGPRPIPRPLPFP is used.
5. Overnight culture of the bacterial strains in the medium of choice (*see* **Note 2**).
6. Propidium iodide (PI) stock solution: 1 mg/ml in 10 mM Na-phosphate buffer, 150 mM NaCl, pH 7.4 (BS), filtered using a 0.2 μm membrane filter.
7. Thermostatic bath set at 37 °C (or a different temperature depending on the bacterial species).
8. Flow cytometer (*see* **Note 3**).
9. Incubator set at 37 °C (or a different temperature depending on the bacterial species).
10. Inoculation spreader.

2.2 Peptide Internalization into Bacterial Cells

1. Mueller-Hinton Broth (MHB) or other culture media (*see* **Note 1**).
2. Buffered high salt solution (BHSS): 10 mM Na-phosphate, 400 mM NaCl, 10 mM $MgCl_2$, and pH 7.4.

3. Stock solution of fluorescently labeled peptides in sterile water (peptide concentrations ranging from 1 to 10 mg/ml) (*see* **Note 4**).
4. Overnight culture of the bacterial strains in the medium of choice (*see* **Note 2**).
5. Trypan blue stock solution: 10 mg/ml in BS solution, filtered using a 0.2 μm membrane filter.
6. Thermostatic bath set at 37 °C (or a different temperature depending on the bacterial species).
7. Flow cytometer (*see* **Note 3**).
8. Centrifuge for Eppendorf tubes.
9. Cover glasses.
10. Confocal microscope with an oil immersion objective lens (100×).

2.3 Evaluation of Cytosolic Peptide Concentration

1. Bacterial culture of the desired strain in Mueller-Hinton Broth (MHB) or in other growth medium of choice.
2. Buffered high salt solution (BHSS): 10 mM Na-phosphate, 400 mM NaCl, 10 mM $MgCl_2$, and pH 7.4.
3. A peptide labeled with a radioisotope, e.g., alkylated with radioactive (^{14}C) iodoacetamide [12].
4. Flexible-24, clear PET 24-well flexible microplate.
5. Scintillation fluid OptiPhase Supermix (for liquid samples and ^{14}C).
6. Scintillation β-counter.
7. Spectrophotometer.
8. Disposable polystyrene 1.6 ml cuvettes.

2.4 In Vitro Evaluation of Transcription/Translation Inhibition

1. Commercial kit S30 T7 High-Yield Protein Expression System for in vitro cell-free coupled transcription/translation (T7/S30 Extract for Circular DNA, S30 Premix Plus, S30 T7 Control DNA, nuclease-free water).
2. RNase inhibitor (e.g., RNase inhibitor murine).
3. Solution of PR-AMPs to be tested in sterile water (possibly RNase free).
4. RNase-free tips and tubes.
5. Thermomixer comfort.

2.5 Evaluation of Results by Luciferase Activity Assay and Evaluation of Results by SDS-PAGE

1. Commercial kit Renilla Luciferase Assay System (Renilla Luciferase Assay Substrate, Renilla Luciferase Assay Buffer).
2. Black 96 multiwell microtiter plates.
3. Luminometer.
4. Resolving gel buffer 4×: 1.5 M Tris–HCl pH 8.8.

5. Stacking gel buffer 4×: 0.5 M Tris–HCl pH 6.8.
6. Running buffer 5×: 0.125 M Tris-HCl, 0.96 M glycine pH 8.3, 0.5 % SDS.
7. Sample Buffer 4×: 8 % SDS (w/v), 0.4 M DTT, 40 % glycerol (w/v), 0.04 % bromophenol blue in 0.25 M Tris–HCl pH 6.8.
8. Tris-buffered saline (TBS): 50 mM Tris–HCl, 150 mM NaCl, pH 7.5.
9. Coomassie staining solution: 0.1 % (w/v) Coomassie Brilliant Blue R250, 10 % (v/v) acetic acid, 40 % (v/v) methanol in MQ water.
10. Destaining solution: 10 % (v/v) acetic acid in milli-Q water.
11. Tris-glycine polyacrylamide mini-gel (12.5 % running gel, 4 % stacking gel, 0.75 mm thickness).

3 Methods

3.1 Membrane Permeabilization and Bacterial Killing

Temporal dissociation between cell death and changes in membrane permeability is a common feature of AMPs having a non-lytic mode of action. Since the lethal step is different from membrane damage, cell permeabilization is usually observable only after a lag period and as a secondary effect due to the dead cells that over time lose membrane integrity. On the contrary, in the case of lytic AMPs, membrane damage represents per se the lethal event, so that cell killing and membrane permeabilization are usually observed as concomitant events.

Membrane integrity could be assessed by measuring the time course of propidium iodide (PI) uptake in bacterial cells treated with the AMP of interest. PI enters the cells and emits fluorescence only if the cell membranes are damaged. Therefore, the percentage of PI-positive cells after treatment with the peptide correlates to the number of damaged cells.

1. Dilute a mid-log phase bacterial culture at 1×10^6 CFU/ml in the appropriate medium (*see* **Note 1**).
2. Aliquot 1 ml of bacterial suspension directly into the tubes used for the cytometric sample acquisition, and add 10 μl of PI (final concentration 10 μg/ml) to each aliquot. Prepare one tube for each concentration of peptide (*see* **Note 5**).
3. Add the peptide to the desired concentration and incubate in a thermostatic bath at 37 °C (*see* **Note 6**). Add an equal volume of water to a sample as negative control (untreated cells). Add an equal volume of a solution of a molecule with well-known lytic activity as positive control for permeabilization (*see* **Note 7**).
4. Acquire each sample with the flow cytometer every 15 min for 2 h (*see* **Note 8**).

5. Plot the number of events (cells) counted as a function of the fluorescence intensity of the detected signals (PI fluorescence) (Fig. 1).
6. Calculate the percentage of permeabilized cells by subtracting the fluorescence of the untreated control (cell autofluorescence) after each incubation time (*see* Fig. 1 and **Notes 9** and **10**).
7. Dilute a mid-log phase bacterial culture at 1×10^6 CFU/ml in the appropriate medium (*see* **Note 1**).
8. Prepare tenfold serial dilutions in BS of the initial suspension (*see* **Note 11**).
9. Aliquot 1 ml of bacterial suspension in Eppendorf tubes, and add the peptide at different concentrations based on the permeabilization assay (*see* **Note 6**). Use one tube for each peptide concentration and time point, as a negative control water can be used.
10. Incubate in a thermostatic bath at 37 °C.
11. For each peptide concentration, remove the corresponding Eppendorf tube from the thermostatic bath after the desired incubation time, and prepare tenfold serial dilutions in BS (*see* **Note 12**).
12. Plate 50 μl of each dilution in duplicate by spreading it onto a Mueller-Hinton agar plate (*see* **Note 13**).
13. Incubate the plates for approximately 24 h at 37 °C, and count the number of CFUs at each time point and for each peptide

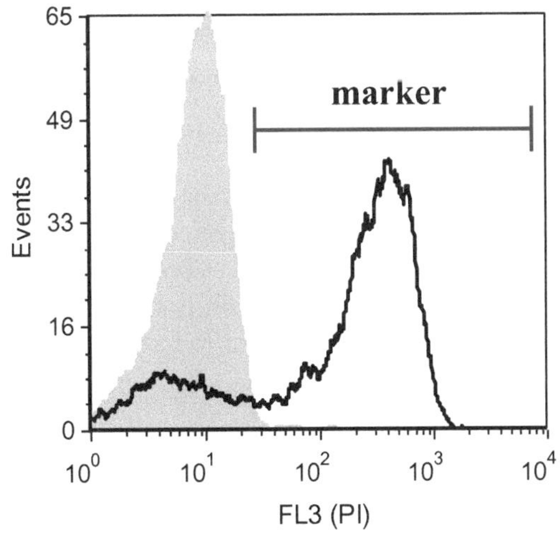

Fig. 1 Example of superimposed histograms of untreated (*full gray*) and peptide-treated (*black line*) *E. coli* cells. The figure shows the shift at a higher fluorescence intensity after treatment with a cell membrane damaging agent. The *marker window* indicates the interval of fluorescence intensity at which the cells are considered PI+

concentration. Plot the number of CFUs for each sample on a log scale.

14. Compare, for each time and each peptide concentration assayed, the number of CFUs and the percentage of permeabilized cells as reported above (*see* **step 6**).
15. A non-lytic AMP is expected to show a significant reduction of CFUs (>90%) after a precise time of incubation without concurrent membrane permeabilizing effects (<5–10% of PI-positive cells) or to see signs of permeabilization only at concentrations higher than the bactericidal concentrations (*see* **Note 14**). On the contrary, a lytic peptide is expected to cause significant membrane permeabilization at shorter incubation times (e.g., 30 min) and produce at the same concentration a significant decrease in the number of CFUs.

3.2 Peptide Internalization into Bacterial Cells

Uptake of an AMP in bacteria could be investigated by a cytofluorimetric method that couples the use of the fluorescently labeled AMP to a quencher, which is excluded from the interior of intact cells, and bleaches only the fluorescence of the peptide that remains extracellular or bound on the bacterial surface.

1. Dilute a mid-log phase bacterial culture at 1×10^6 CFU/ml in the appropriate medium (*see* **Note 1**).
2. Aliquot 1 ml of bacterial suspension in Eppendorf tubes.
3. Add the fluorescently labeled peptide to the desired concentration, and incubate at 37 °C in a thermostatic bath (*see* **Note 6**). It is necessary to use only non-permeabilizing concentrations of the AMP, previously evaluated by the PI-uptake assay (see above). Use one tube for each peptide concentration and time point, as a negative control water can be used.
4. At the end of incubation, wash bacterial samples three times in BHSS in order to remove the fraction of peptide slightly bound to the surface (*see* **Note 15**). Initially, the internalization should be determined after a short incubation time (e.g., 10 min); subsequently, the incubation could be prolonged up to a maximum of 2 h if the internalization is not detectable after a short treatment.
5. Analyze all samples by the flow cytometer.
6. Plot the number of events (cells) counted as a function of the fluorescence intensity due to the labeled peptide.
7. If the cell population appears fluorescent, add 10 μl of trypan blue (final concentration 1 mg/ml) in each bacterial suspension, incubate 10 min at room temperature, and analyze again all samples.
8. Compare the mean fluorescence intensity (MFI) values obtained in the absence and in the presence of TB of each sample.

9. Little difference between these two values (low quenching) indicates that the peptide is mainly internalized into bacterial cells (e.g., in *E. coli*, the percentage of quenching for Bac7(1–35) is less than 10%). A large difference between MFI values of cells treated with or without TB may indicate that the peptide is mainly located on the surface of the cells or in both the cytoplasm and membranes. In this case (high quenching) further investigations are required.
10. Confocal laser scanning microscopy (CLSM) may be used to complete the analysis on peptide internalization and/or eventually confirm the data collected by the cytofluorimetric analyses. Bacterial cells treated with the different peptides are prepared without any fixation by following the same protocol used for the flow cytometric assay.
11. Prepare the samples following the protocol used for the internalization analysis (*see* **steps 1–4**).
12. Place 10 μl of each bacterial suspension, after washing it in BHSS, between two cover glasses to obtain an unmovable monolayer of cells.
13. Observe at the microscope, and analyze the image stacks collected by the CLSM using an appropriate software (*see* **Note 16**).
14. Evaluate the distribution of fluorescence into bacterial cells.

3.3 Evaluation of Cytosolic Peptide Concentration

Determination of the concentration reached by an AMP inside the bacteria may help to understand the mechanism of action of the peptide. This concentration can be estimated by using a radioactive derivative of the AMP of interest and measuring the radioactivity associated to the peptide-treated bacteria. This procedure is not affected by the strategy chosen to make the peptide radioactive; however, the smaller the modification, the lower the expected perturbation of the AMP activity. Times of incubation and peptide concentrations could require optimization depending on the bacterial species used. Data here reported concern the proline-rich peptide Bac7(1–35) and the *E. coli* BW25113 cells.

1. Grow bacteria overnight in the appropriate liquid medium at 37 °C under agitation. For a detailed description, *see* Chap. 29, Subheading 3.1.2.
2. The day after, add 3 ml of this culture to 120 ml of fresh medium, and incubate at 37 °C under agitation until they reach the mid-log phase [optical density at 600 nm (OD_{600}) of approximately 0.3].
3. Centrifuge the bacterial suspension at 2000 ×*g* for 20 min, discard the supernatant, and resuspend the pellet in 5 ml of medium.

4. Dilute bacteria in the same medium to the final concentration of 4×10^8 CFU/ml. Then incubate at least 4 ml of the diluted bacteria at 37 °C under agitation for 10 min with a non-lytic concentration (*see* Subheading 3.1) of the radioactive peptide.
5. After the incubation, collect 1 ml of treated culture (sample A). Collect another 1 ml of treated culture, centrifuge it at $6200 \times g$ for 10 min and resuspend the pellet in 1 ml of fresh Mueller-Hinton medium (sample B). Collect a third 1 ml of treated culture, centrifuge it at $6200 \times g$ for 10 min, wash the pellet three times with 1 ml of BHSS to remove the peptide from the bacterial external surface, and then resuspend the pellet in 1 ml of fresh MH Mueller-Hinton medium (sample C).
6. Add 100 μl of each sample (samples A, B, and C) to 900 μl of fresh MH medium, mix them, and check the $OD_{600.}$ Use these data to normalize the measurement of radioactivity with the number of bacteria for each sample.
7. Mix in triplicate 200 μl of each sample with 500 μl of scintillation fluid OptiPhase Supermix in a Flexible-24, Clear PET 24-well Flexible Microplate, and pipette avoiding bubbles (*see* **Note 17**).
8. Measure the radioactivity of each sample for 1 min by using a β-counter (*see* **Note 18**).
9. Use the OD_{600} values previously collected to normalize the amount of radioactivity with the number of bacteria present in each sample.

Estimate the volume occupied by the bacterial pellet based on the dimension of a single cell (*see* **Note 19**). Calculate the intracellular amount of peptide by dividing the radioactivity associated with the bacterial pellet by its volume.

3.4 *In Vitro* Evaluation of Transcription/Translation Inhibition (*See* Note 20)

1. Gently thaw T7/S30 Extract for Circular DNA, S30 Premix Plus, S30 T7 Control DNA, nuclease-free water, and PR-AMP solution. Mix vigorously S30 Premix Plus and mix gently T7/S30 extract.
2. Set reactions as follows in 1.5 ml RNase-free tubes (Table 1) (*see* **Note 21**).
3. Incubate the samples at 37 °C for 60 min in a thermomixer under vigorous agitation. To have reproducible results, strictly control the temperature and the mixing; anyhow, the temperature is the most important parameter for a good reaction yield.
4. Stop the reaction by cooling down the tubes on ice for 10 min and keep the samples on ice during the further steps (*see* **Note 22**).

Table 1
Reaction mix for in vitro transcription/translation assay

Reagents	Volumes (μl)		
	Ctrl –	Ctrl+	PR-AMP
S30 Premix Plus	27	27	27
T7/S30 Extract for Circular DNA	18	18	18
RNase inhibitor	1	1	1
S30 T7 Control DNA	/	2	2
PR-AMP solution	/	/	x
Nuclease-free H_2O	4	2	$y-x$
Total (μl)	50	50	50

3.5 Evaluation of Results by Luciferase Activity Assay and SDS-PAGE

1. Prepare an appropriate amount of Renilla reaction mix by adding one volume of 100× Renilla Luciferase Assay Substrate to 100 volumes of Renilla Luciferase Assay Buffer according to the supplier instructions. Equilibrate reagents at room temperature before preparing the mix.
2. Quickly vortex your transcription/translation reactions, then dilute 2.5 μl of each sample in 97.5 μl of Renilla Luciferase Assay Buffer and mix by vortexing.
3. Transfer 50 μl of previously prepared Renilla reaction mix into a well of a black multiwell microplate, add 50 μl of the sample previously diluted in Renilla Assay Buffer. Mix quickly by pipetting and avoiding bubbles formation, and then immediately read light intensity for 5 s using a luminometer for multiwell plates (*see* **Note 23**).
4. Prepare a Tris-glycine polyacrylamide mini-gel.
5. Mix 5 μl of the undiluted transcription/translation reaction with 15 μl TBS buffer and 5 μl of 4×Laemli Sample Buffer. Mix by pipetting avoiding bubbles and denature samples by heating at 90 °C for 5 min.
6. Load 10–12 μl of each denatured sample, and then separate them by using constant voltage (stacking 50 V; separation 200 V) until the bromophenol blue reaches the bottom of the gel.
7. Stain the gel with Coomassie Brilliant Blue for at least 1 h under agitation, and then destain with 10% acetic acid.
8. Visually compare the bands corresponding to the luciferase, or, if needed, perform a densitometric analysis of the gel by using a scanner and a devoted software.

4 Notes

1. The choice of the medium depends on the peptide used. It is desirable to use the same medium to perform the killing kinetics and membrane permeabilization kinetics assays.
2. These permeabilization and internalization assays have successfully been tested on *E. coli*, *S. typhimurium*, *K. pneumoniae*, *A. baumannii*, *P. aeruginosa*, *S. aureus*, and *S. epidermidis*. In principle these assays could be used also with other bacterial species.
3. The flow cytometer should be equipped with an argon laser (488 nm, 5 mW) and photomultiplier tube fluorescence detectors for filtered light set at 610 nm for PI detection and 525 nm for BODIPY (BY) detection. Each detector should be set to logarithmic amplification; 10,000 events are usually acquired for each sample. Data analysis may be performed with the FCS Express 3 software or similar software.
4. The choice of the fluorophore depends on the characteristics of the cytometer and confocal microscope (laser source, detectors, and filters). Fluorophores commonly used to label peptides are BODIPY_FL [16], carboxyfluorescein, etc. Before using labeled AMPs for internalization studies, it is advisable to verify that the antimicrobial activity of the fluorescent peptide derivative is not changed after labeling. Both fluorescent labeling and MIC determination are described in Chap. 22, Subheading 3.2.
5. One milliliter of the bacterial suspension is generally sufficient for four acquisitions one after another of the same sample tube at different times of peptide treatment. During the acquisition, it is important that the volume in the tube does not drop below a certain value [e.g., ~250 μl with Cytomics FC 500].
6. The peptide's working solutions must be prepared at such a concentration that can be used in a volume range of 2–20 μl for each tube. In general, the concentrations used for permeabilization and killing assays are based on the MIC value (e.g., 1×MIC, 2×MIC, 4×MIC, and 8×MIC).
7. Peptide antibiotics such as polymyxin B or other lytic agents such as ethanol, SDS, and Triton X-100 may be used as a positive control for permeabilization assay.
8. It is better to analyze samples for short times, and anyway not exceeding 2 h, so that only potential direct lytic effects caused by the AMP are detected. Lytic AMPs generally affect membrane integrity within minutes. Prolonging the treatment for longer times could allow to detect membrane lysis as indirect effects of cellular death.

9. Based on the histogram obtained for the untreated cell population, trace a marker to establish the range of fluorescence intensity in which cells are considered PI positive. The percentage of untreated cells within the range established by the marker should not exceed 2%. Some analysis software electronically subtracts the histogram of the negative control to those of the treated samples.
10. It is recommended to verify in advance that the bacterial cells incubated in the medium chosen for the assay do not change their morphology and do not alter their status of membrane integrity even in the absence of the AMP. This is important to correctly evaluate the extent of membrane damage truly caused by the AMP of interest. For example, it has been observed that certain bacterial species show changes in morphological parameters (FS and SS) and/or tend to be partially permeabilized when incubated in poor or diluted culture medium or in buffer (e.g., BS or BHSS buffer) already in the absence of the tested peptide (unpublished data).
11. This step allows to determine the real cellular concentration of the initial bacterial suspension before the addition of the peptide. Decide the dilution factor taking into account that in order to have 500, 50, and 5 CFUs on each plate, it should be necessary to plate a 10^{-2}, 10^{-3}, and 10^{-4} dilution (starting from 1×10^6 CFU/ml).
12. For longer incubation times (more than 2 h), it could be necessary to increase the dilution factor for the untreated sample with respect to the samples treated with the peptide, in order to obtain a countable number of CFUs on the plate.
13. In order to reduce the number of agar plates, it is possible to divide one plate in four sectors and to spot 25 μl of a sample in each sector. This method is preferable for a raw initial screening, although the count of CFUs may be more difficult due to the reduced space.
14. To unravel the mechanism of action of an AMP, it is important to carry out both assays and, whenever possible, to use the same assay conditions. However, as a first step, it is convenient to perform the membrane permeabilization assay since its setup is less time- and cost-consuming than the killing kinetics test. This allows to optimize and to restrict the conditions to be used in the killing kinetics experiments. Insights on the succession of events from membrane permeabilization to bacterial cells death can be obtained by killing kinetics assay that should be performed using the same bacteria, medium, peptide concentrations, and incubation times chosen for the membrane permeabilization assay.

15. Check the morphologic parameters and the membrane integrity of the untreated cells subjected to the described washing protocol, to verify that the procedure does not cause any damage or changes to the cell population. This is important to properly assess the internalization of the peptide.
16. The EZ-C1 FreeViewer and the ImageJ 1.40 g software are suitable for this kind of analysis. Peptides that are internalized are visible for their fluorescence that is usually homogeneously distributed into the cytoplasm. On the contrary, peptides interacting with the membranes are detectable as they accumulate on the cell surfaces.
17. The choice of the scintillation liquid depends on the radioisotope used to label the peptide. Select the best reagent for your needs according to the instructions of the supplier. For example, the OptiPhase Supermix is compatible for liquid samples and for measurements of ^{14}C.
18. The protocol used to analyze the radioactivity data depends on the radioisotope used to label the peptide, e.g., for experiments with a ^{14}C-labeled Bac7(1–35), we measured the radioactivity of each well for 1 min, and the reported result was the average automatically calculated by the instrument.
19. For this calculation relatively to *E. coli*, *see* Volkmer et al. [17].
20. The cell-free in vitro transcription/translation systems are a powerful tool to simulate the intracellular conditions and the peptide concentration (if known) that the AMP of interest, e.g., a PR-AMPs, reaches within bacteria. Another advantage of in vitro cell-free systems is that they allow to avoid possible toxic or lethal lytic side effects that the peptide may exert on whole living bacteria if used at high concentrations. Moreover, since PR-AMPs need to be internalized by bacteria to exert their antimicrobial activity, it is often difficult to distinguish if these peptides are inactive on bacteria because of poor internalization or because of an intrinsic poor activity. The use of these in vitro cell-free systems allows to investigate the effects of PR-AMPs directly on the molecular machinery dedicated to the transcription/translation processes. During the whole setting up of the in vitro transcription/translation reaction, always work on ice and under RNase-free conditions.
21. The final volume for the reactions can be scaled up to test multiple samples. Prepare a premix for better reproducibility of results.
22. It is also recommended to check the amount of produced luciferase by SDS-PAGE rather than measure only its activity. Poor luminescence is not sufficient to discriminate between no luciferase production and produced inactive luciferase (e.g., because of luciferase unfolding) or errors in the light quantification assay.

Results of the transcription/translation reactions can be evaluated by quantifying the activity of the produced reporter gene (*see* Subheading 3.4, **step 1**) and/or the amount of protein produced by the reporter gene (*see* Subheading 3.4, **step 2**).

23. The light production drops very quickly. Read each sample singularly and immediately after its mixing with the Reaction mix. Standardize the procedure for sample handling in order to read each sample exactly after the same time.

References

1. Epand R, Vogel HJ (1999) Diversity of antimicrobial peptides and their mechanisms of action. Biochim Biophys Acta 1462:11–28
2. Hale JD, Hancock RE (2007) Alternative mechanisms of action of cationic antimicrobial peptides on bacteria. Expert Rev Anti Infect Ther 5:951–959
3. Li W, Tailhades J, O'Brien-Simpson NM, Separovic F, Otvos L Jr, Hossain MA, Wade JD (2014) Proline-rich antimicrobial peptides: potential therapeutics against antibiotic-resistant bacteria. Amino Acids 46:2287–2294
4. Nicolas P (2009) Multifunctional host defense peptides: intracellular-targeting antimicrobial peptides. FEBS J 276:6483–6496
5. Scocchi M, Mardirossian M, Runti G, Benincasa M (2016) Non-membrane permeabilizing modes of action of antimicrobial peptides on bacteria. Curr Top Med Chem 16:76–88
6. Cudic M, Otvos L Jr (2002) Intracellular targets of antibacterial peptides. Curr Drug Targets 3:101–106
7. Scocchi M, Tossi A, Gennaro R (2011) Proline-rich antimicrobial peptides: converging to a non-lytic mechanism of action. Cell Mol Life Sci 68:2317–2330
8. Berthold N, Hoffmann R (2014) Cellular uptake of apidaecin 1b and related analogs in gram-negative bacteria reveals novel antibacterial mechanism for proline-rich antimicrobial peptides. Protein Pept Lett 21:391–398
9. Mattiuzzo M, Bandiera A, Gennaro R, Benincasa M, Pacor S, Antcheva N, Scocchi M (2007) Role of the Escherichia coli SbmA in the antimicrobial activity of proline-rich peptides. Mol Microbiol 66:151–163
10. Kragol G, Lovas S, Varadi G, Condie BA, Hoffmann R, Otvos L (2001) The antibacterial peptide pyrrhocoricin inhibits the ATPase actions of DnaK and prevents chaperone-assisted protein folding. Biochemistry 40:3016–3026
11. Krizsan A, Volke D, Weinert S, Sträter N, Knappe D, Hoffmann R (2014) Insect-derived proline-rich antimicrobial peptides kill bacteria by inhibiting bacterial protein translation at the 70 S ribosome. Angew Chem Int Ed 53:12236–12239
12. Mardirossian M, Grzela R, Giglione C, Meinnel T, Gennaro R, Mergaert P, Scocchi M (2014) The host antimicrobial peptide Bac71-35 binds to bacterial ribosomal proteins and inhibits protein synthesis. Chem Biol 21:1639–1647
13. Krizsan A, Prahl C, Goldbach T, Knappe D, Hoffmann R (2015) Short proline-rich antimicrobial peptides inhibit either the bacterial 70S ribosome or the assembly of its large 50S subunit. ChemBioChem 16:2304–2308
14. Roy RN, Lomakin IB, Gagnon MG, Steitz TA (2015) The mechanism of inhibition of protein synthesis by the proline-rich peptide oncocin. Nat Struct Mol Biol 22:466–469
15. Seefeldt AC, Nguyen F, Antunes S, Pérébaskine N, Graf M, Arenz S, Inampudi KK, Douat C, Guichard G, Wilson DN et al (2015) The proline-rich antimicrobial peptide Onc112 inhibits translation by blocking and destabilizing the initiation complex. Nat Struct Mol Biol 22:470–475
16. Guida F, Benincasa M, Zahariev S, Scocchi M, Berti F, Gennaro R, Tossi A (2015) Effect of size and N-terminal residue characteristics on bacterial cell penetration and antibacterial activity of the proline-rich peptide Bac7. J Med Chem 58:1195–1204
17. Volkmer B, Heinemann M (2011) Condition-dependent cell volume and concentration of *Escherichia coli* to facilitate data conversion for systems biology modeling. PLoS One 6:e23126

Chapter 21

The Interaction of Antimicrobial Peptides with the Membrane and Intracellular Targets of *Staphylococcus aureus* Investigated by ATP Leakage, DNA-Binding Analysis, and the Expression of a LexA-Controlled Gene, *recA*

Sanne Gottschalk and Line Elnif Thomsen

Abstract

The analysis of how antimicrobial peptides (AMPs) interact with bacterial membranes and intracellular targets is important for our understanding of how these molecules affect bacteria. Increased knowledge may aid the design of AMPs that work on their target bacterium without inducing bacterial resistance. Here, we describe different methods to investigate the mode of action of peptides against the Gram-positive bacterium *Staphylococcus aureus*. ATP leakage analysis can be used to evaluate the ability of AMPs to perturb bacteria. DNA-binding and SOS response induction can be analyzed to investigate intracellular targets.

Key words Antimicrobial peptides, Mode of action, DNA binding, SOS response, ATP leakage, Membrane perturbation

1 Introduction

Identifying the mode of action of antimicrobial peptides (AMPs) is important to understand how natural peptides attack microorganisms. However, it also provides knowledge that can be used to optimize the structure of the AMPs in order to improve activity and prevent development of bacterial resistance. It is believed that most AMPs, due to their cationic and amphipathic nature, selectively kill bacteria by penetrating the anionic cell membrane by membrane disintegration or pore formation [1]. The assumption that AMPs kill the bacteria by membrane disruption can be tested by measuring the amount of ATP leaking from the bacteria as a measure for disrupted bacteria. This will furthermore give the possibility of investigating the correlation between AMP concentration and membrane disruption. In addition to causing membrane

Paul R. Hansen (ed.), *Antimicrobial Peptides: Methods and Protocols*, Methods in Molecular Biology, vol. 1548,
DOI 10.1007/978-1-4939-6737-7_21, © Springer Science+Business Media LLC 2017

disruption, several AMPs are also known to be able to traverse the cytoplasmic membrane and target intracellular molecules, including DNA [2–7]. The binding of AMPs to DNA can affect the DNA replication machinery due to competition between replication proteins and AMPs for the same DNA sequence. A simple in vitro binding assay can indicate whether the AMP is able to bind DNA. AMP interaction with DNA has previously been suggested to have an effect on replication and DNA repair [2, 7, 8]. DNA damage elicits the SOS response, a conserved pathway essential for DNA repair and restart of stalled or collapsed replication forks. This process is regulated by the repressor LexA and the activator recA. Thus, methods that monitor changes in *recA* expression can support a hypothesis stating that the AMPs interact with DNA and hence induce the SOS response. Methods to investigate whether *recA* expression is induced include quantitative real-time PCR and northern blotting, but using a bacterial strain harboring a *recA::lacZ* fusion makes it possible to visualize changes in *recA* expression by the use of a simple color plate assay [9]. This method detects accumulated expression and is therefore less sensitive to transient expression than qRT-PCR and northern blotting.

Thus using these methods, we can obtain valuable information about the mode of action of selected peptides. Furthermore, the experiments can be performed with peptide variants to evaluate the effect of smaller structural changes and thereby shed light on how changes in peptide structure can affect the function. By obtaining increased knowledge on mode of action, it will be possible to construct peptides with improved function and reduced risk of resistance development.

2 Materials

2.1 ATP Leakage Components

1. Bacterial strain, e.g., *Staphylococcus aureus* 8325-4 [10].
2. Brain Heart Infusion (BHI) Agar (BHI + 1.5 % agar).
3. Tryptone Soy Broth (TSB).
4. Incubator 37 °C.
5. Spectrophotometer with absorbance of OD_{546}.
6. Centrifuge.
7. Phosphate buffer: 50 mM potassium phosphate, pH 7.0.
8. HEPES buffer: 50 mM HEPES (4-(2-hydroxyethyl)-1-piperazineethanesulfonic acid), pH 7.0.
9. Antimicrobial peptides (AMPs) 10× the desired concentrations.
10. 20 % (w/V) glucose.
11. Vortexer.
12. Sterile water.

13. Dimethyl sulfoxide (DMSO).
14. ATP assay: Adenosine 5′-triphosphate (ATP) Bioluminescent Assay Kit.
15. BioOrbit 1253 luminometer.
16. Reaction vial for BioOrbit 1253 luminometer.

2.2 DNA Binding Components

1. Binding buffer: 5% glucose, 10 mM Tris–HCl pH 8.0, 1 mM EDTA (ethylenediaminetetraacetic acid), 1 mM DTT (dithiothreitol), 20 mM KCl, 50 μg/ml BSA (bovine serum albumin, *see* **Note 1**).
2. Agarose gel (1%): 1 g agarose in 100 ml Tris-acetate-EDTA (TAE) buffer (40 mM Tris–HCl, 20 mM acetic acid, and 1 mM EDTA) including ethidium bromide (*see* **Note 2**).
3. Loading buffer: 10 mM Tris–HCl (pH 7.6), 0.03% bromophenol blue, 0.03% xylene cyanol FF, 60% glycerol, and 60 mM EDTA.

2.3 SOS Regulation Components

1. Tryptone Soy Agar (TSA) plates (Tryptone Soy Broth 1.5% agar) containing 5 μg/ml tetracycline.
2. HI2682, *Staphylococcus aureus* 8325-4, which contains a *recA*::*lacZ* plasmid fusion and a tetracycline resistance gene [2].
3. Tryptone Soy Broth (TSB).
4. X-gal (5-bromo-4-chloro-3-indolyl-β-d-galactopyranoside, *see* **Note 3**).
5. Sodium chloride solution: 0.9% NaCl.
6. Tetracycline.

3 Methods

3.1 ATP Leakage

3.1.1 Plating and Single-Colony Incubation

Day 1:

1. The bacterial strain *Staphylococcus aureus* 8325-4 (from −80 °C freezer) is plated on a BHI agar plate. The plate is incubated overnight at 37 °C.

Day 2:

2. A single colony from the BHI agar plate is inoculated into 10 ml TSB and is incubated overnight at 37 °C while shaking (≈200 rounds per minute (rpm)).

3.1.2 Bacterial Cells

1. Bacterial cells from overnight culture (from Day 2) are diluted 1:100 in 25 ml of TSB and are grown at 37 °C while shaking to mid-exponential phase (e.g., OD_{546}≈2.5 for *S. aureus* 8325-4).
2. Bacterial cells are harvested by centrifugation at 2000×*g* for 10 min (*see* **Note 4**).

3. Bacterial cells are washed in 1/5 volume of 50 mM potassium phosphate buffer (pH 7.0).
4. Bacterial cells are washed in 50 mM HEPES buffer (pH 7.0).
5. Bacterial cells are resuspended in 50 mM HEPES buffer (pH 7.0) to $OD_{546} \approx 10$.
6. Bacterial cells are stored on ice and used within 5 h.
7. Bacterial cells in 50 mM HEPES (pH 7.0) are energized with 0.2 % (w/V) glucose at 37 °C for 10 min (cell suspension) (*see* **Note 5**).
8. Cell suspension is treated with antimicrobial peptide (AMP) (*see* **Note 6**) immediately before ATP measurement (see below).

3.1.3 ATP: Measurement of Blank

1. 100 μl ATP assay mix solution is added to a glass vial (reaction vial) (*see* **Note 7**) and incubated for 3 min at room temperature.
2. Add 100 μl sterile water.
3. Mix the solution vigorously using a vortex mixer.
4. Measure the light produced immediately (*see* **Note 8**). Subtract the mean value of the blank from all later ATP measurements.

3.1.4 Measurement of ATP Standards

1. Dilution series of ATP standard stock solution; from a 1 mM stock solution, make the following concentrations: 0.005, 0.01, 0.02, 0.05, 0.1, 0.2, 0.5, and 1.0 μM.
2. 100 μl ATP assay mix solution is added to each glass vial (reaction vial) of the dilution series and stands and is incubated for 3 min at room temperature.
3. Add 100 μl ATP standard.
4. Mix the solution vigorously.
5. Measure the light produced immediately.

3.1.5 Measurements of Total ATP

1. 100 μl ATP assay mix solution is added to a glass vial (reaction vial) and incubated for 3 min at room temperature.
2. 20 μl cell suspension is permeabilized by adding 80 μl dimethyl sulfoxide (DMSO).
3. The mixture is diluted in 4.9 ml sterile water.
4. 100 μl diluted cell suspension is added to the ATP assay mix solution.
5. Mix the solution vigorously.
6. Measure the light produced immediately.

3.1.6 Measurements of Extra Cellular ATP

1. 100 μl ATP assay mix solution is added to a glass vial (reaction vial) and incubated for 3 min at room temperature.
2. 20 μl cell suspension is added to 80 μl sterile water.

3. The mixture is diluted in 4.9 ml sterile water.
4. 100 μl diluted cell suspension is added to the ATP assay mix solution.
5. Mix the solution vigorously.
6. Measure the light produced immediately.

In order to calculate the intracellular ATP concentration, the extracellular ATP concentration is subtracted from the total ATP concentration. ATP concentration in μM of intracellular ATP and extracellular ATP is plotted for each concentration of AMP investigated (Fig. 1).

3.2 DNA Binding

1. Purify plasmid DNA from *S. aureus* (*see* **Note 9**).
2. Make twofold dilutions of AMP in binding buffer (stock), and follow the scheme below as an example to get the desired concentration in 20 μl (Table 1).
3. Add 1 μl of purified 100 ng/μl plasmid DNA to all of the tubes with 20 μl diluted AMP.
4. Add 1 μl of purified 100 ng/μl plasmid DNA to 20 μl binding buffer (control without AMP).
5. Incubate 1 h at room temperature.
6. Add 4 μl 6 × loading buffer to each tube.
7. Load 12 μl from each tube to a 1 % agarose gel.
8. Run the gel, and after the electrophoresis is complete, visualize the molecules in the gel using UV light (Fig. 2).

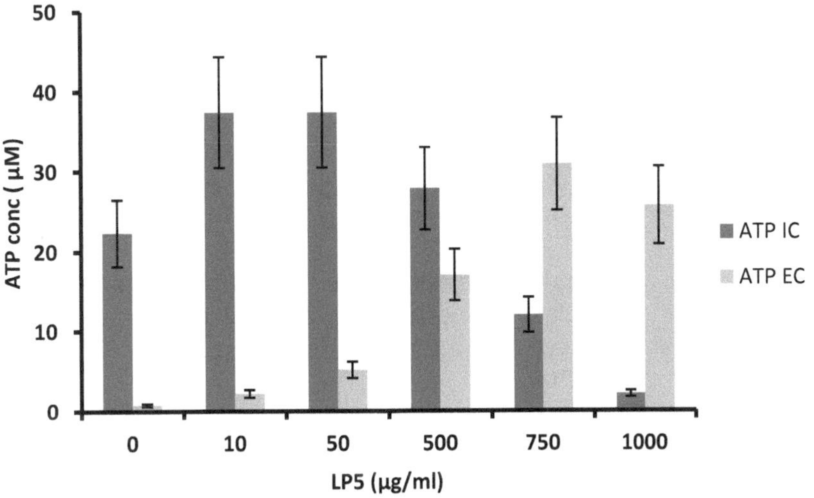

Fig. 1 Measurement of intracellular (IC) and extracellular (EC) ATP after treatment of bacterial cells with increasing concentration of the antimicrobial peptide, LP5 (0–1000 μg/ml) [2]

Table 1
Dilution scheme for AMPs used for DNA binding assay

AMP stock	μl AMP stock	μl binding buffer	AMP concentration in 20 μl
1000 μg/ml	1.6 μl	18.4 μl	80 μg/ml
500 μg/ml			40 μg/ml
250 μg/ml			20 μg/ml
125 μg/ml			10 μg/ml
62.5 μg/ml			5 μg/ml
31.25 μg/ml			2.5 μg/ml

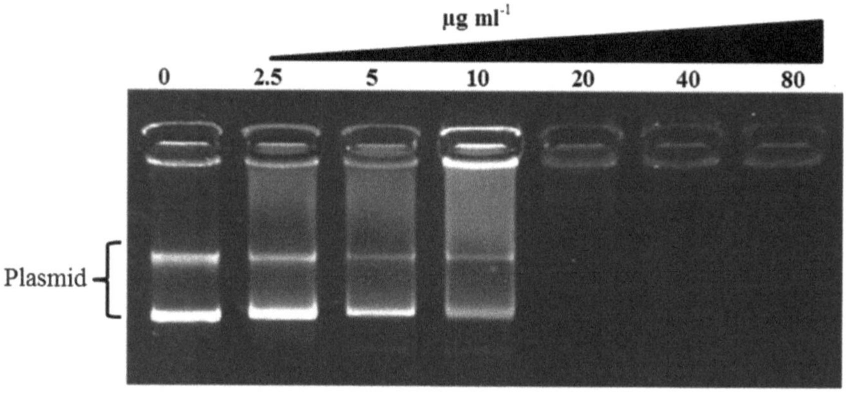

Fig. 2 Gel retardation with *S. aureus* DNA shows how increasing concentrations of an AMP bind to the plasmid and prevent it from running down in the gel. Increasing amounts of the AMP are incubated with 100 ng pRMC2 plasmid DNA and run on an agarose gel. Lane 1: negative control containing binding buffer. Lane 2–7: containing increasing amounts of the antimicrobial peptide, LP5 (2.5, 5, 10, 20, 40, and 80 μg/ml). Modified from [2]

3.3 Well Plate Assay with HI2682, S. aureus 8325-4 recA::lacZ

Day 1:

1. Strain HI2682 from −80 °C freezer is streaked out on TSA plates containing 5 μg/ml tetracycline in order to select for the *lacZ* fusion. The plates are incubated overnight at 37 °C.

Day 2:

2. Single colonies from the TSA plates containing 5 μg/ml tetracycline are inoculated into 10 ml TSB containing 5 μg/ml tetracycline and are incubated overnight at 37 °C while shaking (≈200 rpm).

Day 3:

3. Bacterial cells from overnight culture are diluted 1000-fold in 0.9% NaCl.

4. 1 ml diluted bacterial cells are transferred to a 9 cm petri dish.
5. Liquid TSA is cooled to approximately 40 °C.
6. 150 μg/ml X-gal and 5 μg/ml tetracycline (final concentration) are added to the TSA and are mixed well (avoid bobbles in the agar).
7. Approximately 25 ml agar is immediately poured into a petri dish containing the diluted bacterial cells.
8. Agar and bacterial cells are mixed by gently making circular movements in both directions until the agar is almost solidified.
9. Agar plates are dried for 45 min in a laminar air flow (LAF) unit.
10. Wells are made manually with a sterile sharp iron tube/drill to make a ring-shaped cut through the agar. The little piece of agar in the middle of the ring is then removed using a sterile scalpel.
11. 30 μl of AMP in the desired concentration and sterile water (used as a negative control) is added to the wells. *Use above the MIC since the compound will be diluted out in the agar.* MIC determination is described in Chap. 22, Subheading 3.2.
12. 30 μl of positive control like, e.g., ciprofloxacin known to induce the SOS response in *S. aureus*. *Use above the MIC since the compound will be diluted out in the agar.*
13. The plates are incubated at 37 °C until blue color appears on the plate (Fig. 3, *see* **Note 10**).

4 Notes

1. Prepare, e.g., 2 ml as the solution is stable at room temperature.
2. A 50× TAE stock solution can be prepared by dissolving 242 g Tris base in water, adding 57.1 ml glacial acetic acid and 100 ml of 500 mM EDTA (pH 8.0) solution and bringing the final volume up to 1 l. This stock solution can be diluted 50:1 with water to make a 1× working solution. After boiling of agarose in the TAE buffer, the gel is casted. DNA can be visualized using ethidium bromide which, when intercalated into DNA, fluoresce under ultraviolet light.
3. X-gal is hydrolyzed by the β-galactosidase. X-gal, when cleaved by β-galactosidase, yields galactose and 5-bromo-4-chloro-3-hydroxyindole. The latter then spontaneously dimerizes and is

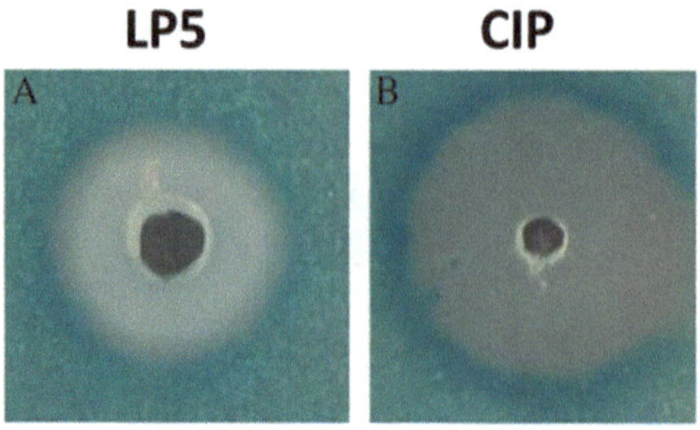

Fig. 3 The antimicrobial peptide, LP5, induces *recA* expression in *S. aureus*. (**a**) LP5 or (**b**) ciprofloxacin (positive control) was added to wells in TSB agar plates containing the *S. aureus* 8325–4 derived *lacZ* reporter strain HI2682 (*recA*::*lacZ*). Incubation time was 18 h [2]

oxidized into 5,5′-dibromo-4,4′-dichloro-indigo, an intensely blue product which is insoluble. X-gal itself is colorless; the presence of blue-colored product therefore may be used as a test for the presence of an active β-galactosidase. This easy identification of an active enzyme allows the gene for β-galactosidase (the *lacZ* gene) to be used as a reporter gene in various applications.

4. Harvest 15 ml of bacterial cells from mid-exponential phase. Between each washing step, the cells are harvested by centrifugation at 2000 × *g*.
5. Use 50 μl cells per reaction and add 0.5 μl glucose (20 %) in an Eppendorf tube
6. Add 5 μl AMP to 50 μl cells, thus make AMP stocks that are 10× more concentrated than the desired concentration.
7. Belonging to, e.g., a BioOrbit 1253 luminometer.
8. Quickly make measurements three times and calculate the mean.
9. Any plasmid can be used. Purification of plasmid from bacterial strain can be done using any plasmid purification kit.
10. Incubation time depends on cell density and copy number of the reporter fusion, plus strength and regulation of the promoter fused to the *lacZ* gene.

Acknowledgment

This work was supported by the Lundbeck Foundation (grant R13-A1193) and the University of Copenhagen.

References

1. Zasloff M (2002) Antimicrobial peptides of multicellular organisms. Nature 415:389–395
2. Gottschalk S, Ifrah D, Lerche S, Gottlieb CT, Cohn MT, Hiasa H, Hansen PR, Gram L, Ingmer H, Thomsen LE (2013) The antimicrobial lysine-peptoid hybrid LP5 inhibits DNA replication an induces the SOS response in *Staphylococcus aureus*. BMC Microbiol 13:192
3. Jenssen H, Hamill P, Hancock REW (2006) Peptide antimicrobial agents. Clin Microbiol Rev 19:491–511
4. Brogden KA (2005) Antimicrobial peptides: pore formers or metabolic inhibitors in bacteria? Nat Rev Microbiol 3:238–250
5. Makobongo MO, Gancz H, Carpenter BM, McDaniel DP, Merrell DS (2012) The oligo-acyl lysyl antimicrobial peptide C12K-2beta12 exhibits a dual mechanism of action and demonstrates strong in vivo efficacy against *Helicobacter pylori*. Antimicrob Agents Chemother 56:378–390
6. Peschel A, Sahl HG (2006) The co-evolution of host cationic antimicrobial peptides and microbial resistance. Nat Rev Microbiol 4:529–536
7. Gottschalk S, Gottlieb CT, Vestergaard M, Hansen PR, Gram L, Ingmer H, Thomsen LE (2015) The amphibian antimicrobial peptide fallaxin analogue, FL9, affects virulence gene expression and DNA replication in *Staphylococcus aureus*. J Med Micro 64:1504–1513
8. Gunderson CW, Segall AM (2006) DNA repair, a novel antibacterial target: Holliday junction-trapping peptides induce DNA damage and chromosome segregation defects. Mol Microbiol 59:1129–1148
9. Nielsen A, Nielsen KF, Frees D, Larsen TO, Ingmer H (2010) Method for screening compounds that influence virulence gene expression in *Staphylococcus aureus*. Antimicrob Agents Chemother 54:509–512
10. Novick R (1967) Properties of a cryptic high-frequency transducing phage in *Staphylococcus aureus*. Virology 33:155–166

Part III

Assaying Selected Biological Activities of AMPs

Chapter 22

Methods for Investigating Biofilm Inhibition and Degradation by Antimicrobial Peptides

Li-av Segev-Zarko and Yechiel Shai

Abstract

Multidrug-resistant bacteria are a growing problem worldwide. One extensively studied resistance mechanism is biofilm colonization—microbial colonies formed by many Gram-positive and Gram-negative bacteria species. Cationic antimicrobial peptides (AMPs) are innate immune system molecules serving as a first line of defense in fighting invading pathogens. The AMPs' underlying mechanism and biophysical properties required for anti-biofilm activity are not fully known. Here we present protocols for investigating AMPs' biological activity against major stages of biofilm life cycle, namely, planktonic stage (MIC assay), initial adhesion to surfaces (bacterial attachment assay), and formation or degradation of sessile microcolonies (biofilm formation and degradation assays). Furthermore, we demonstrate experiments that allow determination and comparison between peptide biophysical properties (secondary structure, hydrophobicity, and oligomerization) and how they affect their mechanism (peptide-binding assays) of anti-biofilm activity.

Key words Biofilm, Antimicrobial peptides, Anti-biofilm activity, Biophysics

1 Introduction

Biofilms are sessile communities formed by many bacterial species to cope with unfavorable surroundings [1]. Biofilm-embedded bacteria are much less susceptible to antibiotics since the dense colonization enables bacterial communication (quorum sensing) [2], horizontal gene transfer, and formation of synergistic micro-consortia [3]. Furthermore, the complex architecture of the extracellular polymeric substance (EPS) reduces the penetration of harmful agents [4]. Embedded bacteria often enter a "stationary-like" phase making them less susceptible to antibiotics that target growing cells [5]. All of the above can explain why clinical therapy fails to treat biofilm infections [6]. Therefore, new therapeutic agents and their mechanism should be investigated [7]. Antimicrobial peptides (AMPs) are attractive alternatives to conventional antibiotics. They are part of the innate immune system,

Paul R. Hansen (ed.), *Antimicrobial Peptides: Methods and Protocols*, Methods in Molecular Biology, vol. 1548,
DOI 10.1007/978-1-4939-6737-7_22, © Springer Science+Business Media LLC 2017

serving as a first line of defense in fighting invading pathogens in all life forms [7, 8]. Cationic AMPs are short, hydrophobic, and amphipathic and have a net positive charge [9]. For most of these peptides, the mode of action involves bacterial membrane disruption, causing bacteria to burst [10]. Their biophysical properties along with their mode of action make them a potential therapy against biofilms.

In the search for new anti-biofilm therapy strategies, we referred to the different stages of biofilm formation. Biofilms start from planktonic bacteria, followed by adhesion to organic or abiotic surfaces. Following the initial adhesion begins the formation of sessile microcolonies and secretion of EPS. As the biofilm matures, a natural degradation process occurs, and bacteria disperse from the biofilm in order to re-localize in other locations [5, 11]. We tested the AMPs' activity against several stages of biofilm maturation. Since planktonic bacteria have the potential to create a biofilm, we tested the minimal peptide concentration needed to prevent planktonic bacterial growth using the minimal inhibitory concentration (MIC) assay. The effect of nonlethal peptide concentration on bacterial adhesion to polystyrene and biofilm formation was used to learn about its mechanism of activity [12]. Checking biofilm degradation after peptide addition allows us to see whether the peptides can deal with mature biofilm, penetrate the biofilm, and kill its embedded bacteria [13].

Here we describe a series of experiments to determine what are the biophysical properties that contribute to the anti-biofilm activity of peptides. To determine the peptides' hydrophobicity, we used reverse-phase high-performance liquid chromatography (RP-HPLC), which relies on the concept that as the peptide is more hydrophobic, the retention time for its release is longer. Secondary structure in lipopolysaccharide (LPS) environment was determined using circular dichroism and Fourier transform infrared spectroscopy (FTIR). Peptides' binding to polystyrene and bacterial cells [14], along with peptide oligomerization state [13], which can dramatically affect its biological activity, was all evaluated using rhodamine-labeled peptides.

2 Materials

Use ultrapure water (18 MΩ cm at 25 °C) for all solutions and preparation of peptide stocks. Peptides and resin-bound peptides may be synthesized as described in Chap. 3 or 4 or obtained commercially.

2.1 Peptide Labeling and Purification

1. 5-(and-6)-carboxytetramethylrhodamine succinimidyl ester (5(6)-TAMRA SE), mixed isomers.
2. AMP and resin-bound AMP (*see* Chap. 3 or 4 for synthesis).

3. *N*,*N*-Diisopropylethylamine (DIEA).
4. Dichloromethane (DCM).
5. DCM–DIEA solution; DCM:DIEA (95:5 v/v).
6. Dimethylformamide (DMF).
7. Trifluoroacetic acid (TFA).
8. Triethylsilane (Tis).
9. Acetonitrile (AcN).

2.2 Bacterial Growth and Biofilm Assays

1. Growth medium, Luria–Bertani (LB) broth or BM2 minimal medium: 62 mM potassium phosphate buffer, pH 7.0, 7 mM ammonium sulfate, 1 mM magnesium sulfate, 10 μM ferrous sulfate, 0.2% (w/v) glucose, 0.5%(w/v) casamino acids.
2. Polystyrene 96-well u-bottom plates (not tissue culture treated).
3. Biofilm dye: 0.5% crystal violet (*see* **Note 1**).
4. Acetic acid (30%).
5. Viability dye: 0.5% 3-(4,5-dimethylthiazol-2-yl)-2,5-diphenyltetrazolium bromide (MTT) (*see* **Note 2**).
6. PBS buffer: 137 mM NaCl, 2.7 mM KCl, 10 mM Na_2HPO_4, 1.8 mM KH_2PO_4.
7. Isopropanol containing 0.4% HCl.
8. Multichannel pipette.
9. LB agar plates.
10. Cell culture slides (eight chambers).
11. Bacterial strain, e.g., *Pseudomonas aeruginosa* PAO1.

2.3 Hydrophobicity Comparison and Secondary Structure Determination

1. Acetonitrile (AcN) supplemented with 0.1% trifluoroacetic acid (TFA).
2. Dissolving buffer: 5 mM 4-(2-hydroxyethyl)-1-piperazineethanesulfonic acid (HEPES).
3. Lipopolysaccharide (LPS) of the relevant bacteria.
4. Circular dichroism thermostatic quartz cuvette with a path length of 1 mm.
5. Fourier transform infrared spectroscopy calcium fluoride cuvette, 5 μm spacer (Miller plastic) and a liquid transmission cell.

2.4 Peptide Binding Assay and Oligomerization State

1. Polystyrene 96-well u-bottom plates (not tissue culture treated).
2. Polystyrene 96-well flat-bottom **black** plates for fluorescence reading.
3. GdnHCl buffer: guanidine hydrochloride 6 M, 50 mM Tris(hydroxymethyl)aminomethane, 20 mM EDTA, pH 6.5.

3 Methods

Carry out all procedures at room temperature unless indicated otherwise.

3.1 Peptide Labeling with 5(6)-TAMRA SE and Purification

For 20 mg of peptide-bound resin, use 1 mg of 5(6)-TAMRA SE powder.

1. Wash the resin-bound peptide with 200 μL DCM for three times followed by one wash with DCM supplemented with 5 % DIEA and three times with DMF.
2. Dissolve the 5(6)-TAMRA SE powder in 98 μL DMF and add it to the washed resin.
3. Add 2 μL DIEA and incubate with agitation protected from light for 16–72 h.
4. After labeling, wash the resin with 500 μL DMF for three times followed by one wash in DCM. Dry the resin using N_2 or in air.
5. For cleavage of 50 mg resin, synthesized by Fmoc SPPS, add 1 mL trifluoroacetic acid (TFA) and 25 μL triethylsilane (Tis). Protect the tube from light and incubate at room temperature for 2 h (agitate all time).
6. Filter the resin and separate the solvent to five 2 mL tubes.
7. Precipitate the crude peptide by adding 1.5 mL cold Ether.
8. Repeat the ether precipitation three times and then dry the powder using N_2 or in air.
9. Dissolve the crude peptide in 50% acetonitrile (0.1% TFA (v/v)).
10. Purify by reverse-phase HPLC (RP-HPLC). RP-HPLC conditions are as follows: C4 column and a linear gradient (10–90%) of acetonitrile in water [both containing 0.1% TFA (v/v)] for 40 min.

3.2 Minimum Inhibitory Concentration (MIC) Assay

Unless indicated otherwise, grow bacteria overnight at 37 °C in Luria–Bertani (LB) broth followed by three washes and resuspension in BM2 minimal medium.

1. Prepare a 96-well plate containing serial dilutions (Fig. 1) of the desired peptide in the final volume of 50 μL (*see* **Note 3**). This is the final goal. **Steps 2–10** are how to reach this goal.
2. Dissolve your peptide in 160 μL BM2 medium to reach the final concentration of 200 μM.
3. For each peptide you check, fill three columns of the 96-well plate with 50 μL BM2 medium, per well.

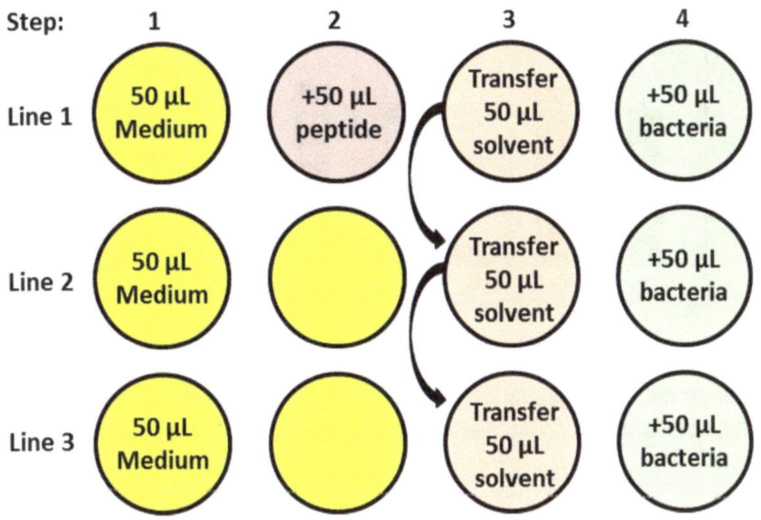

Fig. 1 MIC dilution scheme—peptide serial dilution steps for MIC, biofilm formation, and biofilm degradation assays (with small adjustments)

4. Add 50 µL of your dissolved peptide to three wells in the first row to reach a final concentration of 100 µM peptide in 100 µL.
5. Using a multichannel pipette, pipette the first row, and transfer 50 µL of the liquid to the second row.
6. Repeat the pipetting and dilution until the last row.
7. Discard the remaining 50 µL of excess liquid.
8. Leave wells without peptide as a control.
9. Dilute the overnight culture to ~1×10^6 CFU/mL (*see* **Note 4**).
10. Add 50 µL of diluted bacteria to each well.
11. Incubate the plate for 18 h at 37 °C with agitation.
12. After incubation, gently agitate the plate, and read the OD at 600 nm to evaluate the percentage of growth.

3.3 Biofilm Formation and Degradation Assay

Biofilm formation assay tests the ability of the bacteria to create biofilm in the presence of an antimicrobial agent. It is used to determine if the peptide can prevent biofilm formation or reduce the final biofilm biomass created (Fig. 2).

1. Follow instructions in Subheading 3.2 **steps 2–10** of the MIC assay.
2. Incubate the plate for 24 h at 37 °C **without** agitation.
3. After incubation, remove all liquid and unattached bacteria by inverting the plate and blotting it against clean paper towels.
4. Wash the plate by soaking it in a tap water bath and removing liquid as in **step 3**.
5. Repeat the washes three times.

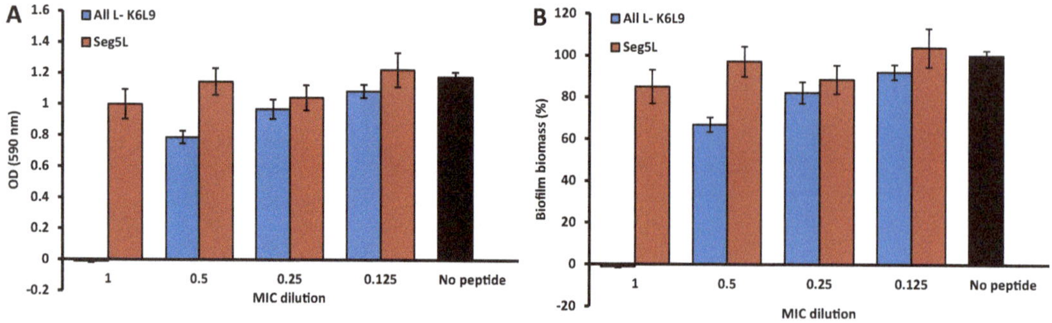

Fig. 2 Biofilm formation assay—*P. aeruginosa* bacteria were incubated for 12 h in the presence of K_6L_9 peptides (at MIC dilutions). (**a**) Posttreatment surface-associated biofilm, examined using 0.1 % crystal violet staining followed by OD measurements at 590 nm. (**b**) Percentage of biofilm formation after fitting to untreated bacteria

6. Add 120 μL crystal violet dye per well, and incubate the plate for 20 min at room temperature.
7. Remove the dye by inverting the plate, followed by three washes (*see* **step 4**), and leave plate to dry completely at room temperature.
8. Dissolve the crystal violet dye by adding 125 μL of 30 % acetic acid, and incubate for 20 min at room temperature.
9. Gently agitate the plate and read OD at 590 nm to evaluate the biofilm biomass.

3.4 Bacterial Attachment to Polystyrene

The attachment of bacteria to a substratum is the first step in biofilm formation. The bacterial attachment to polystyrene plates after a 1 h incubation is used to evaluate how peptides affect this step.

1. Dissolve your peptide in 160 μL BM2 medium (*see* **Note 5**). Fill three wells with dissolved peptide, 50 μL per well.
2. Dilute the overnight culture to ~1×10^9 CFU/mL (*see* **Note 4**).
3. Add 50 μL of diluted bacteria to each well.
4. Incubate the plate for 1 h at 37 °C **without** agitation.
5. To evaluate the bacterial attachment, follow the biofilm formation assay Subheading 3.3, **steps 3–9**.

3.5 Biofilm Degradation

Biofilm degradation assay tests the ability of the antimicrobial agent to degrade preformed biofilm. It allows evaluating the biofilm biomass and cell viability after treatment (Fig. 3).

1. Dilute the overnight culture to ~5×10^5 CFU/mL.
2. Add 100 μL of diluted bacteria to each well of the 96-well plate.
3. Incubate the plate for 24 h at 37 °C **without** agitation.

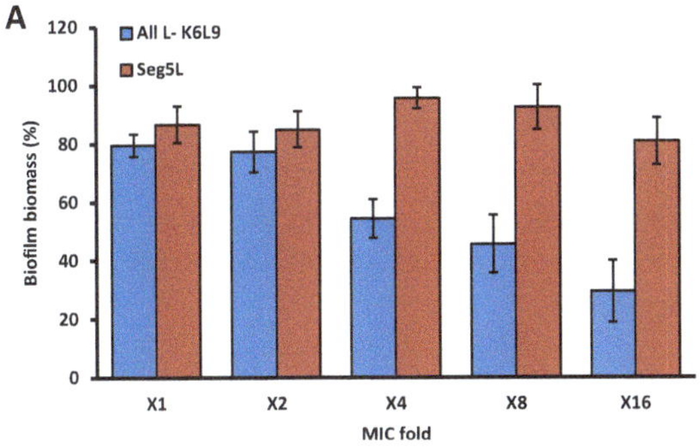

Fig. 3 Biofilm degradation assay—*P. aeruginosa* 12 h biofilms were treated with different peptide concentrations for 1 h. (**a**) Biofilm biomass, examined using 0.1 % crystal violet staining followed by OD measurement at 590 nm. (**b**) Biofilm-embedded bacteria viability, examined using 3-(4,5-dimethylthiazol-2-yl)-2,5-diphenyltetrazolium bromide (MTT) followed by OD measurements (570 nm, reference at 630 nm)

4. After 24 h, aspirate each well and wash three times with fresh BM2 medium.
5. Aspirate each well (*see* **Note 6**).
6. Dissolve your peptide in 175 μL BM2 medium to reach the final concentration of 200 μM.
7. For each peptide you check, fill three columns of the 96-well plate with 55 μL BM2 medium, per well.
8. Add 55 μL of your dissolved peptide to three wells in the first row to reach a final concentration of 100 μM peptide in 110 μL.
9. Use a multichannel pipette, to pipette the first row and transfer 55 μL of the liquid to the second row.
10. Repeat the pipetting and dilution until the last row.
11. Discard the remaining 55 μL of excess liquid.
12. Add 50 μL of BM2 medium to all wells.
13. Leave wells without peptide as a control.
14. Incubate the plate for 1 h at 37 °C **without** agitation.
15. To evaluate the viability of detached bacteria, proceed as follows; otherwise, continue to **step 19** or **20**.
16. Collect 50 μL medium from each well to sterile 1.7 mL tubes containing 450 μL fresh BM2 medium (diluting the bacteria 1:10).
17. Repeat diluting the bacteria as necessary.
18. Seed 100 μL of diluted bacteria on LB agar plates to calculate CFU/mL.
19. To evaluate the biofilm biomass, follow Subheading 3.3 **steps 3–9** of the biofilm formation assay.

20. To evaluate bacterial viability, aspirate each well, and wash three times with fresh BM2 medium then proceed as follows:
21. Add 110 μL 0.5% MTT per well, and incubate the plate protected from light for 20 min at 37 °C **without** agitation.
22. Aspirate each well and add 120 μL isopropanol containing 0.4% HCl. Incubate the plate protected from light for 10 min at 37 °C **without** agitation.
23. Gently agitate the plate, and read OD at 570 nm (reference at 630 nm) to evaluate the biofilm viability.

3.6 Confocal Imaging

To monitor the distribution and relative number of rhodamine-labeled peptides in the biofilm milieu, measure the fluorescence intensity through the biofilm layers (Fig. 4).

1. Dilute the overnight culture to ~5×10^5 CFU/mL.
2. Add 200 μL of diluted bacteria to each well of the eight-chamber cell culture slides.
3. Incubate the culture slides for 24 h at 37 °C **without** agitation, leaning at ~45 °C.
4. After 24 h, aspirate each chamber and wash three times with fresh BM2 medium.
5. Aspirate each chamber, and add 200 μL of your tested rhodamine-labeled peptide at the minimal inhibitory concentration.
6. Incubate the plate for 1 h at 37 °C **without** agitation, leaning at ~45 °C.
7. Aspirate each chamber and wash three times with fresh BM2 medium.
8. Use the confocal microscope for z-section measurements at 590 nm fluorescence intensity (Fig. 4c).

3.7 Peptide Hydrophobicity Comparison

Hydrophobicity is a known requirement for perforation of bacterial membranes. Using RP-HPLC retention time allows the comparison of peptides' hydrophobicity.

1. Dissolve 10 μg of pure peptide in 20 μL 50% acetonitrile in DDW (both containing 0.1% TFA (v/v)) (*see* **Note 7**).
2. Inject the dissolved peptide into the RP-HPLC system. Use a C4 column and a linear gradient (10–90%) of acetonitrile in water [both containing 0.1% TFA (v/v)] for 40 min (*see* **Note 8**).

3.8 Peptide Secondary Structure Determination Using Circular Dichroism Spectroscopy

Secondary structure is a known requirement for interaction with and perforation of bacterial membranes. Using CD and FTIR (especially for d amino acid enantiomers) allows determination of the secondary structure of peptides in relevant environments.

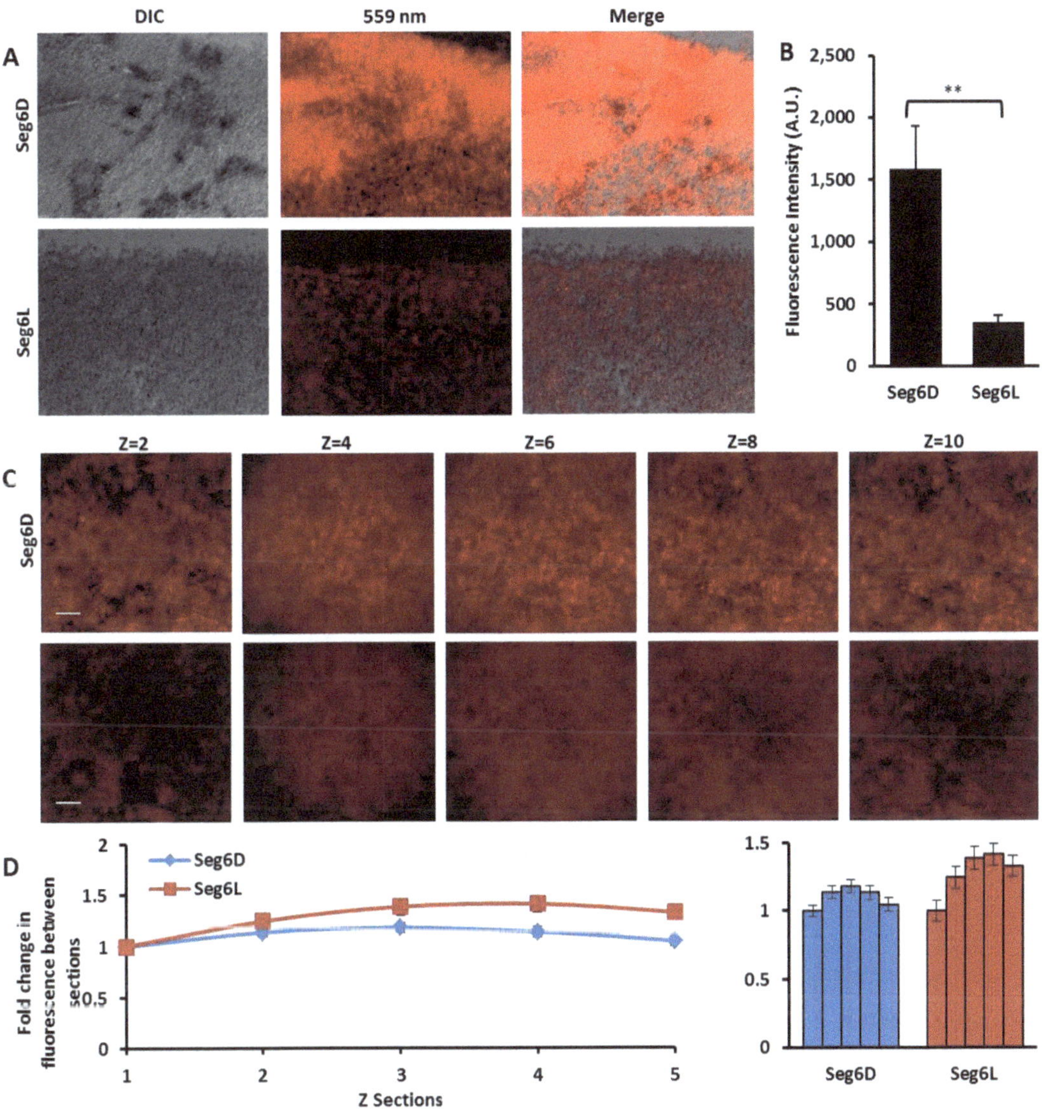

Fig. 4 Peptide dispersion over the biofilm milieu—representative images of a 24 h preestablished *P. aeruginosa* biofilm (**a**) treated for 1 h with 3 μM Seg6D (*top*) and Seg6L (*bottom*) rhodamine-labeled peptides (559 nm). White bars = 40 μm. (**b**) Fluorescence intensity quantification of Seg6D and Seg6L. (**c**) Z-section images of a 24 h preestablished *P. aeruginosa* biofilm treated for 1 h with 3 μM Seg6D (*top*) and Seg6L (*bottom*) rhodamine-labeled peptides (559 nm). Fluorescence intensity was manually increased in this figure for better visualization of the labeled peptides. (**d**) Fold change in fluorescence between the sections (two options of presentation)

1. Dissolve your peptide in 220 μL 5 mM HEPES for the final concentration of 100 μM.
2. Dissolve LPS powder in 110 μL 5 mM HEPES for the final concentration of 100 μM.

3. To determine the secondary structure of the peptide, mix 105 μL peptide with 105 μL 5 mM HEPES. To determine the secondary structure of the peptide in LPS presence, mix 105 μL peptide with 105 μL dissolved LPS.
4. To ensure homogeneity, vortex, sonicate, and spin down the mixture.
5. Transfer 200 μL of the sample to a 1 mm pathlength quartz cuvette, and read the wavelength in the range 190–260 nm in 1 nm steps.
6. To subtract the background, measure the spectrum of 200 μL 5 mM HEPES or 200 μL LPS (50 μM in 5 mM HEPES) (Fig. 5).

3.9 Fourier Transform Infrared Spectroscopy

1. Prepare aliquots of 150–250 μg peptide in 1.7 mL tubes.
2. To determine the secondary structure of the peptide, dissolve the peptide in 50 μL 5 mM HEPES. To determine the secondary structure of the peptide in the presence of LPS, add 50 μL of dissolved LPS for 1:1 concentration.
3. To ensure homogeneity, vortex, sonicate, and spin down the mixture.
4. Place a spacer on the calcium fluoride cuvette, then put 30 μL of the sample (make sure no bubbles are trapped), and read the amide I region (1700–1550/cm).
5. To subtract the background, measure the spectrum of 30 μL 5 mM HEPES or 30 μL LPS in 5 mM HEPES.

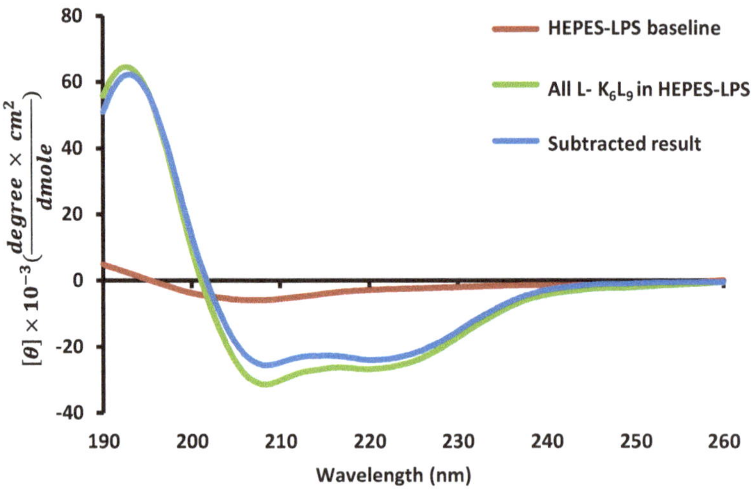

Fig. 5 Circular dichroism baseline correction—CD spectra taken from 50 μM. All L-K_6L_9 peptides dissolved in 5 mM HEPES buffer with 50 μM LPS. *Red line*, baseline of HEPES with LPS; *green line*, peptide dissolved in HEPES with LPS (1:1 ratio); *blue line*, subtracted result

3.10 Peptide Adhesion to Polystyrene

When using uncharged polystyrene plates, adhesion of AMPs and bacteria depends mainly on hydrophobic interactions. Peptides can act as coating agents that cover biomaterial surfaces to reduce bacterial attachment. Using rhodamine labeling allows testing the peptide adhesion to polystyrene in the presence or absence of bacteria (Fig. 6a).

1. Dissolve rhodamine-labeled peptide in 310 μL BM2 (*see* **Note 5**).
2. Add 50 μL peptide per well to six wells of a 96-well plate.
3. To evaluate the binding of the peptide to the polystyrene, add 50 μL of BM2 to three wells. To evaluate the binding of the peptide in the presence of bacteria, add 50 μL of 1×10^6 CFU/mL bacteria.
4. Incubate the plate protected from light for 1 h at 37 °C **without** agitation.
5. Aspirate the liquid and transfer it to a black 96-well plate.

Add 100 μL fresh BM2 medium per well, gently agitate the plate, and transfer the liquid to clean wells in the black plate.

6. Repeat this step three times (*see* **Note 9**).
7. Add 200 μL of 6 M guanidine hydrochloride, gently agitate the plate, and read fluorescence (excitation 528 nm and emission 590 nm).

3.11 Peptide Binding to Bacterial Cells

Peptides can act as coating agents that cover the bacterial surface to reduce its attachment to biomaterials. Using rhodamine labeling allows testing the peptide adhesion to bacterial cells (Fig. 6b).

1. Dissolve rhodamine-labeled peptide in 160 μL BM2 (*see* **Note 10**).
2. Dilute the overnight culture to five different concentrations of ~1–5×10^8 CFU/mL (*see* **Note 4**).

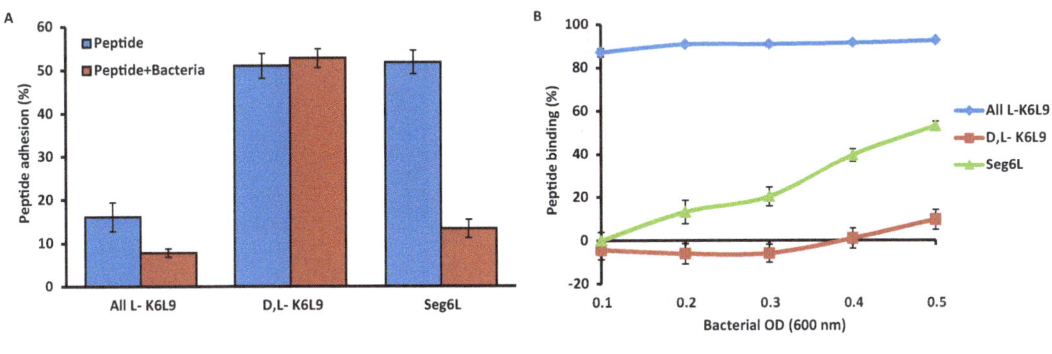

Fig. 6 Peptide adhesion assay—(**a**) peptide adhesion to polystyrene with or without 1×10^6 CFU/mL *P. aeruginosa*. Results are reported relative to initial peptide concentration (¼ MIC). (**b**) Binding of K_6L_9 peptides to different loads of *P. aeruginosa*. Results are reported relative to initial peptide concentration (1 μM)

3. Transfer 190 μL of each bacterial concentration to a 500 μL tube, in triplicates.
4. Add 10 μL of dissolved peptide, per tube.
5. Incubate the plate protected from light for 10 min at 37 °C with agitation.
6. Centrifuge the tubes for 1 min.
7. Aspirate 100 μL from each sample and transfer it to a black 96-well plate.
8. Add 200 μL of 6 M guanidine hydrochloride, gently agitate the plate, and read fluorescence (excitation 528 nm and emission 590 nm).

3.12 Peptide Oligomerization in Growth Medium

Self-assembly and oligomerization can contribute or disrupt antimicrobial peptide activity. Rhodamine probes are highly sensitive to self-quenching when in close proximity; therefore, using labeled peptides helps indicate oligomerization (Fig. 7).

1. Dissolve rhodamine-labeled peptide in 610 μL BM2 medium (*see* **Note 11**).
2. Fill six wells of a black 96-well plate with 50 μL of the peptide, per well.
3. To measure the maximal fluorescence of the folded peptide, add 100 μL of DDW, per well, to three wells. Gently agitate the plate and read fluorescence (excitation 528 nm and emission 590 nm).

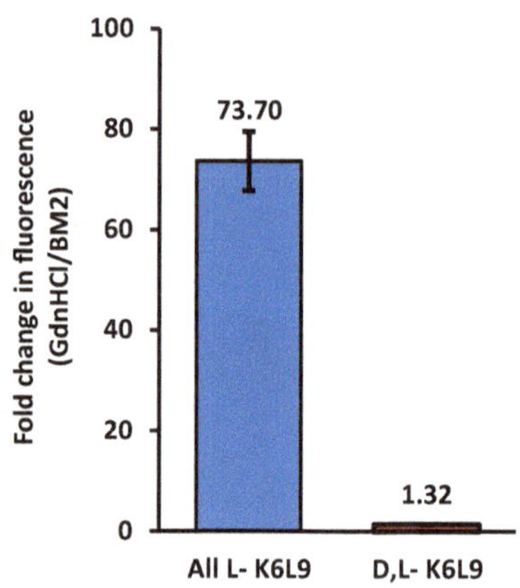

Fig. 7 Peptide oligomerization in growth medium—peptide oligomerization state as evaluated by fluorescence measurements. Rhodamine fluorescent probes are highly sensitive to self-quenching when in close proximity (e.g., labeled peptide oligomerization). Results represent the fold change in fluorescence between the folded peptide (in BM2 medium) and its denatured form (in GdnHCl, 6 M)

4. To measure the maximal fluorescence of the denatured peptide, add 100 μL of guanidine hydrochloride, per well, to three wells. Gently agitate the plate and read fluorescence (excitation 528 nm and emission 590 nm).

4 Notes

1. Take precautions by wearing a particulate matter-absorbing mask, goggles, and a lab coat, and work in a chemical flow hood with a disposable diaper underneath. Dissolve 10 g of crystal violet in 1.4 L of 96% ethanol (technical grade). Dissolve the crystal violet by swirling, and then add 0.6 L DDW. Leave overnight to ensure complete dissolution. Filter the solution using a disposable 0.22 filter. Transfer the solution to a bottle and close tightly to prevent evaporation.
2. Dissolve 5 mg MTT in 1 mL PBS. Filter the solution using a disposable 0.22 μm filter.
3. Minimal inhibitory concentration assay evaluates the lowest concentration of an antimicrobial agent that inhibits growth of bacteria. It is used to confirm resistance of bacteria to an antimicrobial peptide and to determine the concentrations that will be used for further assays. It is recommended to start the MIC assay at 50 μM peptide concentration in triplicates. *See* Fig. 1 for schematic presentation of the serial dilution method.
4. After washing the overnight culture, read the OD of the bacteria at 600 nm. For *Pseudomonas aeruginosa* PAO1, calculate the amount of bacteria according to the following equation: OD_{600} of $1.0 = 1 \times 10^9$ cells/mL.
5. The final concentration of the peptide should be two times higher than the non-bactericidal concentrations.
6. Prepare the plate before starting **step 4**. It is recommended to start the biofilm degradation assay at 50 μM peptide concentration in triplicates.
7. The Agilent 1100 Series HPLC Value System requires a minimum of 10 μg peptide for detection. Use the minimum amount your system requires for detection.
8. Use the appropriate column to fit your peptide hydrophobicity. The comparison between peptides should be performed on the same day with the same column.
9. This step ensures that you measure the whole amount of unbound peptide. When calculating the unbound peptide fluorescence, sum the reads.
10. The final concentration of the peptide should be 20 times higher than the non-bactericidal concentrations.

11. The final concentration of the peptide should be three times higher than the non-bactericidal concentrations. The volume mentioned is not obligatory, but the final volume you read should be at least 100 μL.

Acknowledgment

This study was supported by The Pasteur-Weizmann Foundation, Israel Science Foundation (ISF) and the German-Israel Foundation (GIF).

References

1. Parsek MR, Singh PK (2003) Bacterial biofilms: an emerging link to disease pathogenesis. Ann Rev Microbiol 57:677–701
2. de Kievit TR (2009) Quorum sensing in Pseudomonas aeruginosa biofilms. Environ Microbiol 11:279–288
3. Keller L, Surette MG (2006) Communication in bacteria: an ecological and evolutionary perspective. Nature Reviews Microbiol 4:249–258
4. Mah TF, O'Toole GA (2001) Mechanisms of biofilm resistance to antimicrobial agents. Trends in microbiol 9:34–39
5. Costerton JW, Stewart PS, Greenberg EP (1999) Bacterial biofilms: a common cause of persistent infections. Science 284:1318–1322
6. Hall-Stoodley L, Costerton JW, Stoodley P (2004) Bacterial biofilms: from the natural environment to infectious diseases. Nature Reviews Microbiol 2:95–108
7. Jorge P, Lourenco A, Pereira MO (2012) New trends in peptide-based anti-biofilm strategies: a review of recent achievements and bioinformatic approaches. Biofouling 28:1033–1061
8. Rossi LM, Rangasamy P, Zhang J, Qiu XQ, Wu GY (2008) Research advances in the development of peptide antibiotics. J Pharmaceuti Sci 97:1060–1070
9. Lehrer RI, Ganz T (1999) Antimicrobial peptides in mammalian and insect host defence. Curr Opin Immunol 11:23–27
10. Shai Y (2002) Mode of action of membrane active antimicrobial peptides. Biopolymers 66:236–248
11. Hall-Stoodley L, Stoodley P (2005) Biofilm formation and dispersal and the transmission of human pathogens. Trends in microbiology 13:7–10
12. Overhage J, Campisano A, Bains M, Torfs EC, Rehm BH, Hancock RE (2008) Human host defense peptide LL-37 prevents bacterial biofilm formation. Infection Imm 76:4176–4182
13. Segev-Zarko L, Saar-Dover R, Brumfeld V, Mangoni ML, Shai Y (2015) Mechanisms of biofilm inhibition and degradation by antimicrobial peptides. Biochem J 468:259–270
14. Saar-Dover R, Bitler A, Nezer R, Shmuel-Galia L, Firon A, Shimoni E, Trieu-Cuot P, Shai Y (2012) D-alanylation of lipoteichoic acids confers resistance to cationic peptides in group B streptococcus by increasing the cell wall density. PLoS pathogens 8:e1002891

Chapter 23

Protocols for Studying Inhibition and Eradication of Bacterial Biofilms by Antimicrobial Peptides

Vijayalekshmi Sarojini

Abstract

Many pathogenic microorganisms have the ability to form biofilms that are impervious to conventional antibiotics making these pathogens resistant to multiple antibiotics. This necessitates the development of novel antimicrobial compounds with less chance of resistance development and the ability to penetrate the extracellular polymer matrix of bacterial biofilms. In this report, simple assays to test the antibiofilm potential of antimicrobial peptides are described.

Key words Antimicrobial peptides, Bacterial biofilms, Antibiotic resistance, Confocal microscopy, Live/Dead stain, Crystal violet stain

1 Introduction

Biofilm formation is a key factor that contributes to the multidrug resistance (MDR) of bacterial pathogens [1]. Biofilms are implicated in clinical, industrial, and natural settings (such as stream water). Eradicating bacterial biofilms using conventional antibiotics is extremely difficult [2]. In the context of the ever increasing problem of global antibiotic resistance, there is an urgent need to develop novel antimicrobial compounds. Antimicrobial peptides (AMPs) are attractive alternatives to conventional antibiotics for several reasons [3–5]. The chances of resistance development to antimicrobial peptides, even though not nil, are much rare. AMPs are part of the innate immune systems in all forms of life. The most common mechanism of action of AMPs, which involves the lysis of microbial membranes, requires that the microbes have to go through multiple rounds of mutations to gain resistance [5, 6]. The main drawback of AMPs as therapeutics comes from their sensitivity to proteases. Advances in chemical and biological syntheses and computational modeling have helped to conquer this problem to a great extent through the use of structural modifications that make the peptides evade attack by proteases [7–9]. There has been

Paul R. Hansen (ed.), *Antimicrobial Peptides: Methods and Protocols*, Methods in Molecular Biology, vol. 1548, DOI 10.1007/978-1-4939-6737-7_23, © Springer Science+Business Media LLC 2017

significant progress in the development of AMPs as therapeutics, particularly against MDR pathogens [10–21]. These include both laboratory scale research and those that have progressed into different stages of clinical trials. Research on the use of antimicrobial peptides against bacterial biofilms is quite recent [22–30].

Quantification of bacterial biofilms in microtiter plates has been achieved following different colorimetric assays such as the crystal violet (CV) assay [31–33], the Live/Dead staining assay, the resazurin assay, the fluorescein diacetate (FA) assay, the XTT assay, and the dimethyl methylene blue (DMMB) assay. The method described in this protocol is based on biofilm staining and visualization using specific stains and confocal laser scanning microscopy. Theses assays are typically done in 96-well microtiter plates. Even though these assays are robust and give an accurate indication of the antibiofilm potential of the test compounds, for the purpose of gaining commercial registration for an antibiofilm agent, the assays will have to be conducted on biofilms grown using biofilm reactors as approved by the US standard setting organization, American Society for Testing and Materials (ASTM) International [34]. The assays discussed in this chapter can be used for assessing the antibiofilm potential of antimicrobial peptides as well as conventional antibiotics.

2 Materials

1. 96-well microtiter plates with air-permeable plate seal (polypropylene, flat bottomed; *see* **Note 1**).
2. Pipettes (preferably multichannel pipette).
3. Autoclave.
4. Incubator shaker.
5. Class II biological safety cabinet inside a physical containment level 2 laboratory setting.
6. Inverted confocal laser scanning microscope.
7. Cell staining reagents, such as crystal violet.
8. BacLight Kit (*see* **Note 2**).
9. Spectrophotometer (to read absorbance in the 600 nm range).
10. Vortex mixer.
11. Reagents for growing bacteria as liquid cultures and also on solid medium: (e.g., Mueller Hinton Broth media or Luria broth, agar).
12. EnSpire Multimode Plate reader.
13. Standard antibiotic to serve as positive controls (streptomycin, gentamicin, etc.).

3 Methods

Bacterial cultures should be stored in glycerol stock at −80 °C for longer-term use. For routine use, bacteria can be stored on agar plates in a refrigerator.

The protocol involves preparing bacterial pre-cultures, growing the biofilms in 96-well plates using the pre-cultures in the presence of various concentrations of the antimicrobial peptides, and staining using appropriate stains, workup, and visualization of the biofilms.

All operations should be performed in a biological safety cabinet under sterile conditions.

3.1 Pre-culture Preparation

1. Preparation of Mueller Hinton Broth (MHB) media agar plates (*see* **Note 3**). Dissolve 21 g of Mueller Hinton Broth in 1 L of milli-Q water, and add 17 g of agar into it. Autoclave the suspension at 121 °C for 20 min.
2. Remove the flask from the autoclave and allow it to cool down to ~60 °C. Pour approximately 25 mL into a 90 mm circular plate, gently swirl the plate to ensure uniform thickness across, and allow the plate to stand still until the agar solidifies.
3. Streak the bacterial culture from the −80 °C stock into the agar plates (*see* **Note 4**), and incubate the plates overnight to obtain single colonies. This might require slightly longer incubation time depending on the bacterium. Temperature suitable for the particular bacterium should be set on the incubator using the temperature set function (*see* **Note 5**).
4. Prepare MH Broth following manufacturer's instructions, i.e., dissolve 21 g of Mueller Hinton Broth in 1 L of water, and autoclave the solution at 121 °C for 20 min. Allow the medium to cool to room temperature in a laminar flow hood. Inoculate a single colony from the agar plate (in **step 3** above) into 5 mL of the autoclaved MHB medium, and incubate the flask overnight at the appropriate temperature with shaking (180 rpm at ~16 h).
5. Prepare stock solutions of the antimicrobial peptide in freshly autoclaved MHB medium, and then prepare 10–12 two-fold serial dilutions of the stock. An example is shown in Table 1 (*see* **Note 6**).
6. On the following day, prepare a subculture of the 5 mL culture from **step 4** by diluting in the autoclaved (fresh) MHB medium to adjust the OD_{600} to 0.3.

3.2 Growing the Biofilms for the Assay

1. Place the required number of 96-well microtiter plates (typically these have 12 columns and 8 rows) inside the biosafety cabinet. Remove the plates from the packing.
2. Label each row (on the side of the plate) with the codes for the different peptides to be tested.

Table 1
Scheme for preparing antimicrobial peptide stock solutions and dilutions (*see* Note 8)

	Stock solution (source)	To prepare sub-concentrations in Eppendorf tubes			Into each well of the microtiter plate	
S. No.	Antimicrobial concentration (μg/mL) of the source	Volume (μL) of source	Volume (μL) of MHB medium	Concentration obtained (μg/mL)	Volume (μL) from the previous column	Final concentration in microtiter plate (μg/mL)
1	2000	500	500	1000	50	500
2	1000	200	800	200	50	100
3	200	500	500	100	50	50
4	100	500	500	50	50	25
5	50	600	400	30	50	15
6	200	100	900	20	50	10
7	20	500	500	10	50	5
8	10	500	500	5	50	2.5
9	5	400	600	2	50	1
10	2	500	500	1	50	0.5

3. Label the first well of each row as growth control and the last well as sterility control. Place circular coverslips (5 mm diameter, 1 mm thick; *see* **Note 7**) into each well. (These will sit horizontally at the bottom of the wells.)
4. Add 50 μL of the diluted bacterial subculture (*see* Subheading 3.1, **step 5**) into each well of the 96-well microtiter plate using a multipipette, except into the sterility control well. Then add 50 μL of the sterile MHB medium into the growth control well and 100 μL of sterile MHB medium into the sterility control well.
5. Add 50 μL of the antimicrobial peptide solution at the different concentrations (as in the last but one column of Table 1) into each well of the 96-well microtiter plate using a multipipette, except the growth and sterility control wells. (*Note*: all wells now contain a total of 100 μL liquid.) The use of three replicates of each peptide concentration per plate and three such plates per assay is recommended. Cover the plates with the top seal A and carefully transport the plates into the incubator and allow the biofilms to grow for 48 h at 50 rpm.
6. Transfer the plates from the incubator carefully into the biosafety cabinet. Remove the coverslips from the wells, discard the cultures, rinse the cover slips with sterile water to remove any planktonic cells, and dry these at 50 °C for 20 min.

3.3 Biofilm Staining (to Be Performed in the Dark, if Using the BacLight Stain) and Preformed Biofilm Assay

1. Prepare BacLight stock solution for staining by (3:1000 dilution) mixing 1.5 μL each of SYTO 9 (green fluorescent nucleic acid stain) and propidium iodide (red fluorescent nucleic acid stain) solution in 997 μL of sterile water.
2. Dispense 60 μL of the BacLight solution onto each of the coverslips, and incubate them in the dark for 25 min at room temperature. Using sterile water, wash off the excess stain twice and air-dry the coverslips.
3. Invert the coverslips onto a drop of mounting oil (supplied with BacLight kit) on a fresh glass slide.
4. View the biofilms separately using FITC 2 and Tx Red filters at 1000× magnification on an inverted confocal laser scanning microscope, and capture the images using the inbuilt digital sight DS-U1 camera. Merge the images using the inbuilt ACT-2U software.
5. For the preformed biofilm assay, perform all steps as above, except for adding the antimicrobial peptide solutions at the time of incubation. The biofilms should be allowed to be formed for 48 h, planktonic cells washed off, and then these preformed biofilms incubated for another 48 h with the peptides at the various concentrations. The rest of the procedures is to be followed as above.

3.4 Semiquantitative Estimation of Biofilm Biomass

1. Prepare biofilms in 96-well plates following exact same protocols as above (*see* Subheading 3.2, but do not use cover slips).
2. Transfer the plates into the biosafety cabinet, and discard the cultures (either pipette out or invert the plates).
3. Wash the plates with distilled water to remove all planktonic cells. Stain the wells with 100 μL of 1% crystal violet for 5 min.
4. Remove the stain carefully, and wash the wells twice with 100 μL milli-Q water.
5. Add 100 μL of 96% ethanol and allow to stand for 15 min (to solubilize the stain) and read the absorbance of this ethanol solution at 560 nm using the plate reader.
6. Compare the absorbance values from the treated wells with the growth control well, and deduce the percentage of biofilm inhibition.

3.5 Anticipated Results from a Live/Dead Staining Experiment

The Live/Dead BacLight viability kit utilizes a mixture of SYTO 9 (the green fluorescent nucleic acid stain) and propidium iodide (the red fluorescent nucleic acid stain). If the antimicrobial peptide has antibiofilm potential, the proportion of dead cells (red) should increase and green cells (live) decrease with increasing concentrations of the peptide. A typical set of images of *P. aeruginosa* preformed biofilms before and after treatment with an antibiofilm peptide performed in our laboratory is shown in Fig. 1 [30].

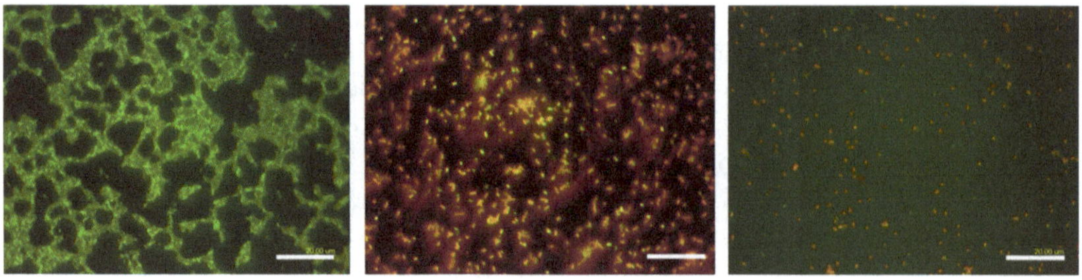

Fig. 1 Effect of treating *Pseudomonas aeruginosa* preformed biofilms with an antibiofilm peptide for 48 h. From *left* to *right*: Untreated *P. aeruginosa* biofilm architecture observed after Live/Dead staining. Biofilm architecture observed after treating these biofilms with 50 and 100 μM of an antibiofilm peptide for 48 h

4 Notes

1. Microtiter plates made of different materials such as polystyrene (PS), polyvinyl chloride (PVC), and polycarbonate plastic or glass (borosilicate) have also been used successfully for forming biofilms of various bacteria [35]. However, it is to be noted that majority of AMPs are cationic in nature and selectively interact with the negatively charged bacterial membranes. Therefore, the use of polystyrene or other negatively charged materials for the assay plates should be avoided to minimize the interaction of the cationic peptides with the plate material.
2. Other colorimetric or metabolic stains could be used as well.
3. Luria broth (LB) can be used instead of MHB
4. Use at least 2–3 plates every time bacterial cultures are getting revived from −80 °C stock.
5. 37 °C is suitable for growing *Pseudomonas aeruginosa*, *Staphylococcus aureus*, and *Escherichia coli*.
6. The minimal inhibitory concentrations (MICs) of the AMPs should be determined previously. This is described in Chap. 22, Subheading 3.2. Recommended starting concentration for the antibiofilm assay would be ~100 times the MICs against planktonic cells of any bacterium.
7. These assays can be conducted in 24- or 12-well plates as well depending on the number of samples to be tested. Coverslips of 12 or 15 mm diameters, suitable for the 12- and 24-well plates, respectively, are available from vendors. To provide a surface for biofilm formation amenable to microscopy, the use of these coverslips is recommended.
8. This is just an example. Stock concentrations and volumes can be any multiples of these numbers depending on the MIC of the peptides under investigation and the number of tests to be performed.

References

1. Xu KD, McFeters GA, Stewart PS (2009) Biofilm resistance to antimicrobial agents. Microbiology 146:547–549
2. Parsek MR, Singh PK (2003) Bacterial biofilms: an emerging link to disease pathogenesis. Annu Rev Microbiol 57:677–701
3. Hancock RE (2001) Cationic peptides: effectors in innate immunity and novel antimicrobials. Lancet Infect Dis 1:156–164
4. Hancock RE (1997) Peptide antibiotics. Lancet 349:418–422
5. Zasloff M (2002) Antimicrobial peptides of multicellular organisms. Nature 415:389–395
6. Brogden KA (2005) Antimicrobial peptides: pore formers or metabolic inhibitors in bacteria? Nat Rev Microbiol 3:238–250
7. Grauer A, Konig B (2009) Peptidomimetics - a versatile route to biologically active compounds. European J Org Chem 30:5099–5111
8. Adessi C, Soto C (2002) Converting a peptide into a drug: strategies to improve stability and bioavailability. Curr Med Chem 9:963–978
9. Gentilucci L, De Marco R, Cerisoli L (2010) Chemical modifications designed to improve peptide stability: incorporation of non-natural amino acids, pseudo-peptide bonds, and cyclization. Curr Pharm Des 16:3185–3203
10. Jenssen H, Hamill P, Hancock RE (2006) Peptide antimicrobial agents. Clin Microbiol Rev 3:491–511
11. Yeung ATY, Gellatly SL, Hancock REW (2011) Multifunctional cationic host defence peptides and their clinical applications. Cell Mol Life Sci 68:2161–2176
12. Afacan NJ, Yeung ATY, Pena OM, Hancock REW (2012) Therapeutic potential of host defense peptides in antibiotic-resistant infections. Curr Pharm Des 186:807–819
13. Choi KY, Chow LNY, Mookherjee N (2012) Cationic host defence peptides: multifaceted role in immune modulation and inflammation. J Innate Immun 4:361–370
14. Mok WWK, Li YF (2014) Therapeutic peptides: new arsenal against drug resistant pathogens. Curr Pharm Des 20:771–792
15. Mansour SC, Pena OM, Hancock REW (2014) Host defense peptides: front-line immunomodulators. Trends Immunol 35:443–450
16. Dutta P, Das S (2015) Mammalian antimicrobial peptides: promising therapeutic targets against infection and chronic inflammation. Curr Top Med Chem 16:99–129
17. Mohammad H, Thangamani S, Seleem MN (2015) Antimicrobial peptides and peptidomimetics - potent therapeutic allies for staphylococcal infections. Curr Pharm Des 21:2073–2088
18. Barrett JF (2001) Oritavancin Eli Lilly & Co. Curr Opin Investig Drugs 2:1039–1044
19. Toney JH (2002) Iseganan (intrabiotics pharmaceuticals). Curr Opin Investig Drugs 3:225–228
20. Sajjan US, Tran LT, Sole N, Rovaldi C, Akiyama A, Friden PM et al (2001) P-113D, an antimicrobial peptide active against *Pseudomonas aeruginosa*, retains activity in the presence of sputum from cystic fibrosis patients. Antimicrob Agents Chemother 45:3437–3444
21. Rothstein DM, Spacciapoli P, Tran LT, Xu T, Roberts FD, Dalla Serra M et al (2001) Anticandida activity is retained in P-113, a 12-amino-acid fragment of histatin 5. Antimicrob Agents Chemother 45:1367–1373
22. de la Fuente-Nunez C, Reffuveille F, Mansour SC, Reckseidler-Zenteno SL, Hernandez D, Brackman G et al (2015) D-enantiomeric peptides that eradicate wild-type and multidrug-resistant biofilms and protect against lethal *Pseudomonas aeruginosa* infections. Chem Biol 22:196–205
23. de la Fuente-Nunez C, Reffuveille F, Haney EF, Straus SK, Hancock RE (2014) Broad-spectrum anti-biofilm peptide that targets a cellular stress response. PLoS Pathog 10, e1004152
24. Minardi D, Ghiselli R, Cirioni O, Giacometti A, Kamysz W, Orlando F et al (2007) The antimicrobial peptide tachyplesin III coated alone and in combination with intraperitoneal piperacillin-tazobactam prevents ureteral stent Pseudomonas infection in a rat subcutaneous pouch model. Peptides 28:2293–2298
25. Mataraci E, Dosler S (2012) *In vitro* activities of antibiotics and antimicrobial cationic peptides alone and in combination against methicillin-resistant *Staphylococcus aureus* biofilms. Antimicrob Agents Chemother 56:6366–6371
26. Overhage J, Campisano A, Bains M, Torfs EC, Rehm BH, Hancock RE (2009) Human host defense peptide LL-37 prevents bacterial biofilm formation. Infect Immun 76: 4176–4182
27. Hell E, Giske CG, Nelson A, Romling U, Marchini G (2010) Human cathelicidin peptide LL37 inhibits both attachment capability and biofilm formation of Staphylococcus epidermidis. Lett Appl Microbiol 50:211–215
28. Costerton JW, Stewart PS, Greenberg EP (1999) Bacterial biofilms: a common cause of persistent infections. Science 284:1318–1322

29. Costerton JW, Lewandowski Z, Caldwell DE, Korber DR, Lappin-Scott HM (1995) Microbial biofilms. Annu Rev Microbiol 49:711–745
30. De Zoysa GH, Cameron AJ, Hegde VV, Raghothama S, Sarojini V (2015) Antimicrobial peptides with potential for biofilm eradication: synthesis and structure activity relationship studies of battacin peptides. J Med Chem 58:625–639
31. Kjaergaard K, Schembri MA, Hasman H, Klemm P (2000) Antigen 43 from Escherichia coli induces inter- and intraspecies cell aggregation and changes in colony morphology of Pseudomonas fluorescens. J Bacteriol 182:4789–4796
32. Christensen GD, Simpson WA, Younger JJ, Baddour LM, Barrett FF, Melton DM et al (1985) Adherence of coagulase-negative staphylococci to plastic tissue culture plates: a quantitative model for the adherence of staphylococci to medical devices. J Clin Microbiol 22:996–1006
33. Stepanovic S, Vukovic D, Dakic I, Savic B, Svabic-Vlahovic M (2002) A modified microtiter-plate test for quantification of staphylococcal biofilm formation. J Microbiol Methods 40:175–179
34. ASTM International E2196 - 12 (2002) Standard test method for quantification of a Pseudomonas aeruginosa biofilm grown with shear and continuous flow using a rotating disk reactor. Annual Book of ASTM Standards, ASTM International, West Conshohocken, PA
35. O'Toole GA, Pratt LA, Watnick PI, Newman DK, Weaver VB, Kolter R (1999) Genetic approaches to study of biofilms. Methods Enzymol 310:91–109

Chapter 24

Protocols for Studying Antimicrobial Peptides (AMPs) as Anticancer Agents

Laurence Madera and David W. Hoskin

Abstract

Antimicrobial peptides (AMPs) are a class of small cationic peptides that are important for host defense. In a manner that is similar to AMP-mediated destruction of microbial pathogens, certain AMPs can physically associate with the anionic lipid membrane components of cancer cells, resulting in destabilization of the lipid membrane and subsequent peptide binding to intracellular targets, which ultimately leads to the death of the cancer cell. In comparison, normal healthy cells possess a neutral membrane charge and are therefore less affected by AMPs. Based on the selective cytotoxicity of certain AMPs for cancer cells, these peptides represent a potential reservoir of novel anticancer therapeutic agents. The development and improvement of AMPs as anticancer agents requires appropriate methods for determining the effects of these peptides on the viability and function of cancer cells. In this chapter, we describe methods to assess the ability of AMPs to cause cell membrane damage (measured by propidium iodide uptake), apoptosis and/or necrosis (measured by annexin V-FLUOS/propidium iodide staining), and mitochondrial membrane destabilization (measured by 3,3′-dihexyloxacarbocyanine iodide staining), as well as reduced motility (measured by a migration and invasion assay) of cancer cells growing in suspension or as monolayers. We also describe a tubule-forming assay that can be used to assess the effect of AMPs on angiogenesis.

Key words Angiogenesis, Antimicrobial peptide, Apoptosis, Cancer, Cytolysis, Metastasis, Mitochondria

1 Introduction

Considering the diverse mechanisms of actions employed by cationic antimicrobial peptides (AMPs) in their antimicrobial and immunoregulatory functions, it is not surprising to find that AMPs can reduce tumor cell growth through multiple routes [1, 2]. The best characterized anticancer effect of AMPs is their destabilization and disruption of the cancer cell's plasma membrane, which can be measured by the uptake of the DNA-intercalating red-fluorescent dye propidium iodide (PI) [3]. PI is a membrane-impermeable dye that is not able to penetrate cells with intact lipid membranes; thus, PI can be used as an effective indicator of plasma membrane damage [4]. Once bound to DNA, the fluorescence of PI is significantly

Paul R. Hansen (ed.), *Antimicrobial Peptides: Methods and Protocols*, Methods in Molecular Biology, vol. 1548, DOI 10.1007/978-1-4939-6737-7_24, © Springer Science+Business Media LLC 2017

increased with an excitation/emission maximum of 535/617 nm that is detectable by flow cytometry and fluorescent microscopy applications.

Certain AMPs induce cancer cell death through activation of apoptotic machinery, rather than through a direct lytic mechanism [5]. In these cases, combinatorial staining with propidium iodide and fluorescently labeled annexin V can be done to distinguish between the two. Annexins are a class of phospholipid-binding proteins that associate strongly with negatively charged phospholipids, such as phosphatidylserine. In healthy cells, phosphatidylserine is predominantly localized to the cytosolic side of the plasma membrane where it is unable to bind extracellular annexin V; however, at the onset of the apoptotic program, phosphatidylserine is translocated to the extracellular side of the intact plasma membrane, thereby allowing phosphatidylserine to interact with annexin V [6]. Cells in the early stages of apoptosis will therefore bind an annexin V probe while excluding cell-impermeable dyes such as PI.

AMPs that enter cancer cells as a result of plasma membrane damage or via delivery vehicles can interact with anionic intracellular targets, leading to cell death. For example, upon internalization, certain AMPs interact with mitochondria, resulting in the destabilization of mitochondrial membranes, loss of transmembrane potential, and release of apoptosis-promoting molecules [5]. 3,3′-Dihexyloxacarbocyanine iodide ($DiOC_6$) is a cationic lipophilic green-fluorescent dye that is cell-permeable and accumulates in the mitochondria, allowing it to be used as an indicator of mitochondrial membrane integrity and transmembrane potential [7].

Cancer cell motility, which is reflected by the increased ability of cancer cells to migrate and transverse through extracellular matrix components (ECM), is often correlated to tumor cells undergoing epithelial-to-mesenchymal transition and metastasis [8]. These properties can be investigated with migration and invasion assays. The use of a micro-chemotaxis chamber allows one to assess the effects of AMPs on cancer cell migration toward various chemoattractants and the ability of cancer cells to transverse ECM or basement membrane components, such as fibronectin or laminin.

Finally, an angiogenesis assay based on tubule formation can be used to assess the effects of AMPs on blood vessel formation, which is essential for the continued growth of a tumor and eventual metastasis. Human umbilical vein endothelial cells (HUVECs) that are grown on a growth factor-enriched matrix will form tubules [9]. Changes in tubule complexity and formation in the presence of an AMP are an indicator of the AMP's impact on the reorganization stage of angiogenesis.

2 Materials

2.1 Cell Culture

1. Dulbecco's Modified Eagle's Medium (DMEM).
2. RPMI-1640 medium.
3. Clonetics EGM-2 endothelial cell growth medium.
4. Heat-inactivated (at 56 °C for 30 min) fetal bovine serum.
5. Trypsin (0.25 %) in 1 mM ethylenediaminetetraacetic acid (EDTA).
6. Penicillin (10,000 U/mL) and streptomycin (10 mg/mL) solution.
7. 1 M HEPES (4-(2-hydroxyethyl)-1-piperazineethanesulfonic acid) solution.
8. 200 mM l-glutamine solution.
9. T75 cell culture flasks.
10. Polystyrene flat-bottom 24-well cell culture plates.

2.2 Peptides

1. AMPs are obtained commercially or synthesized by Fmoc chemistry (Chap. 3 or 4) and purified (>95 %) by high-performance liquid chromatography, with trifluoroacetic acid desalting (*see* Chap. 10, **Note 1**).
2. Lyophilized peptides are reconstituted with sterile deionized distilled water or phosphate-buffered saline (PBS, pH 7.4; *see* Chap. 26, Subheading 2.1) to the desired concentration. Aliquots of peptide solutions are frozen at −20 °C and thawed directly prior to use.

2.3 PI Staining to Measure Membrane Permeability by Flow Cytometry

1. 10 mM EDTA in PBS: 10 mM EDTA in 137 mM NaCl, 2.7 mM KCl, 10 mM Na_2HPO_4, and 1.8 mM KH_2PO_4.
2. Staining buffer: 10 mM HEPES, 10 mM NaCl, 5 mM $CaCl_2$ (pH 7.4).
3. Propidium iodide (PI).
4. Polystyrene 5 mL round-bottom tubes.

2.4 Annexin V-FLUOS/PI Staining to Assess Apoptosis/Necrosis by Flow Cytometry

1. 10 mM EDTA in PBS.
2. Staining buffer: 10 mM HEPES, 10 mM NaCl, 5 mM $CaCl_2$ (pH 7.4).
3. Propidium iodide.
4. Annexin V-FLUOS.
5. Polystyrene 5 mL round-bottom tubes.

2.5 $DiOC_6$ Staining to Measure Mitochondrial Membrane Potential by Flow Cytometry

1. 10 mM EDTA in PBS.
2. DMEM supplemented with 100 U/mL penicillin, 100 μg/mL streptomycin, 5 mM HEPES, and 2 mM l-glutamine.
3. $DiOC_6$.
4. Polystyrene 5 mL round-bottom tubes.

2.6 Migration and Invasion Assay to Measure Cell Motility

1. 48-well micro-chemotaxis chamber (NeuroProbe).
2. Polycarbonate membranes with 8 μm pores.
3. 10 mM EDTA 10 mM in PBS.
4. DMEM or RPMI-1640 medium supplemented with 100 U/mL penicillin, 100 μg/mL streptomycin, 5 mM HEPES, and 2 mM l-glutamine.
5. DMEM or RPMI-1640 medium supplemented with 10% (v/v) FBS, 100 U/mL penicillin, 100 μg/mL streptomycin, 5 mM HEPES, and 2 mM l-glutamine.
6. Diff-Quik™ stain set.
7. Rubber scraper.
8. Permount® histological mounting medium.
9. Microscope slides and glass covers.
10. Polystyrene 5 mL round-bottom tubes.

2.7 Angiogenesis Assay

1. In vitro angiogenesis kit (Chemicon International).
2. Polystyrene flat-bottom 96-well tissue culture plates.
3. Microfuge tubes.
4. Polystyrene 5 mL round-bottom tubes.

3 Methods

3.1 Cell Culture

1. All cells are obtained from American Type Culture Collection (ATCC) and cultured according to supplied recommendations. In brief, non-adherent and adherent cancer cells are cultured in RPMI-1640 medium and DMEM, respectively, supplemented with 10% (v/v) FBS, 100 U/mL penicillin, 100 μg/mL streptomycin, 5 mM HEPES, and 2 mM l-glutamine.
2. HUVECs are grown in Clonetics EGM-2 endothelial cell growth medium. Non-adherent and adherent cells are cultured in T75 cell culture flasks in a humidified 37 °C incubator under 5% or 10% CO_2, respectively. Adherent cells are allowed to grow as a monolayer until 80–85% confluency is reached.
3. The growth status of suspension cultures is indicated by the pH-dependent change in color of the phenol red-containing culture medium, with an orange color indicating exponential

cell growth and a yellow color indicating medium depletion and excessive acidity. Cell cultures are routinely tested for mycoplasma contamination using the MycoAlert® mycoplasma detection kit (Lonza).

4. Once the monolayer of adherent cells (e.g., MDA-MB-231 breast cancer cells) reaches 80–85 % confluency, the spent culture medium is aspirated and discarded. A 2.5 mL volume of trypsin-EDTA is added to the flask for 5 min, ensuring that the entire surface is covered by solution. Sharply rapping the side of the flask may encourage cell detachment.
5. Trypsin-EDTA is then neutralized by the addition of 7.5 mL DMEM with 10 % FBS. A small volume of cell suspension (the amount depends on the specific cell line) is then added to a new T75 cell culture flask containing fresh DMEM. Cells are then allowed to grow until 80–85 % confluence is reached.
6. Suspension cells (e.g., RPMI-8266 multiple myeloma cells) are passaged by removing a small volume of the cell suspension and transferring it to a new T75 flask containing fresh RPMI-1640 medium with 10 % FBS.
7. To count cells, cell suspensions are centrifuged at $500 \times g$ for 5 min. Pellets are resuspended in 5 mL of DMEM or RPMI-1640 medium containing 2.5 % FBS. The use of a lower serum concentration minimizes AMP adsorption to anionic serum components.
8. A 25 μL volume of cell suspension is added to 75 μL of 0.1 % trypan blue dye, and a 10 μL aliquot of the cells is then counted on a hemocytometer. Dead cells will be stained blue. Only cultures with >90 % viable cells are used. The original cell suspension is then diluted to the desired concentration, usually 1×10^5 cells per mL, unless otherwise indicated.
9. Prior to treatment with AMPs, 1 mL aliquots of non-adherent cells in suspension are distributed into 24-well cell culture plates (1×10^5 cells per mL per well), unless otherwise indicated. For adherent cell lines, cells are incubated overnight in the plate to allow for adherence prior to their use in assays.

3.2 Peptides

1. Peptides are diluted from stock solutions into serum-free medium prior to the addition of AMPs to cell cultures. The treatment dose is highly dependent on the AMP of choice. The toxic range of most anticancer AMPs is between 50 and 200 μM, with cell lysis usually evident within 10–30 min of treatment. Sub-cytotoxic effects of these AMPs typically occur within a 1–20 μM dose range.

3.3 PI Staining to Measure Membrane Permeability by Flow Cytometry

Cancer cells are exposed to various concentrations of AMPs for the desired length of time. Cells are then washed with PBS, detached if necessary, pelleted by centrifugation, and then stained with PI. Uptake of PI is determined by measuring cellular fluorescence by flow cytometry. Cells that were not exposed to AMPs are used to determine baseline fluorescence.

3.3.1 PI Staining of Adherent Cells (e.g., MDA-MB-231)

1. Remove medium from the overnight cultures and discard. Replace with fresh DMEM containing 2.5 % FBS.
2. Incubate cells on the plate with the desired concentrations of AMP for 1–4 h at 37 °C with 5 % CO_2. Mix well by pipetting or shaking the plate.
3. After peptide treatment, carefully remove the medium from each well, and wash the cell monolayer with PBS.
4. Add 500 μL of 10 mM EDTA to each well and incubate at room temperature for 5 min.
5. Pipette vigorously to detach cells from the plate.
6. Transfer the contents of each well to a round-bottom tube, and centrifuge at 500 × *g* for 5 min. Discard the supernatant (*see* **Note 1**).
7. For each treatment condition, prepare 55 μL of a 1 μg/mL PI solution in staining buffer.
8. Resuspend the cell pellet in 50 μL of PI solution, and mix well by pipetting (*see* **Note 2**).
9. Incubate cells in the PI solution for 15 min at 37 °C with 5 % CO_2 in a humidified incubator.
10. After incubation, add 400 μL staining buffer to each tube.
11. Read the samples on a flow cytometer, measuring the fluorescence on the FL-2 channel. Increased fluorescence is indicative of increased PI uptake. The percentage of PI-stained cells can be quantified using untreated cells as a negative control.

3.3.2 PI Staining of Non-adherent Cells (e.g., RPMI-8266)

1. Incubate cells on the plate with the desired concentrations of AMP for 1–4 h at 37 °C with 5 % CO_2 in a humidified incubator. Mix well by pipetting or shaking the plate.
2. After treatment, transfer the contents of each well into a round-bottom tube, and centrifuge at 500 × *g* for 5 min. Discard the supernatant.
3. Resuspend the pellet with 500 μL of PBS and centrifuge again, discarding the supernatant.
4. Follow Subheading 3.3.1, **steps 7–11** as listed above.

3.4 Annexin V-FLUOS/PI Staining to Assess Apoptosis/Necrosis by Flow Cytometry

Cancer cells are exposed to various concentrations of AMPs for the desired length of time. Cells are then washed with PBS, detached if necessary, pelleted by centrifugation, and then stained with annexin V-FLUOS/PI. Flow cytometry is used to measure annexin V-FLUOS binding and/or PI uptake on a cell per cell basis. Cells that were not exposed to AMPs are used to determine baseline fluorescence.

3.4.1 Annexin V-FLUOS/PI Staining of Adherent Cells (e.g., MDA-MB-231)

1. Remove medium from the overnight cultures and discard. Replace with fresh DMEM containing 2.5 % FBS.
2. Incubate cells on the plate with the desired concentration of AMPs for 1–4 h at 37 °C with 5 % CO_2 in a humidified incubator. Mix well by pipetting or shaking the plate.
3. After treatment, carefully remove the medium from each well, and wash the cell monolayer with PBS.
4. Add 500 μL of 10 mM EDTA to each well and incubate at room temperature for 5 min.
5. Pipette vigorously to detach cells from the plate.
6. Transfer the contents of each well to round-bottom tubes and centrifuge at 500 × *g* for 5 min. Discard the supernatant (*see* **Note 1**).
7. For each treatment condition, prepare 55 μL of a 1 μg/mL PI solution in staining buffer. Dilute annexin V-FLUOS into this solution following manufacturer's instructions.
8. Resuspend the cell pellet by adding 50 μL of annexin V-FLUOS/PI solution, and mix well by pipetting (*see* **Note 2**).
9. Incubate cells with the annexin V-FLUOS/PI solution for 15 min at 37 °C with 5 % CO_2 in a humidified incubator.
10. After incubation, add 400 μL staining buffer to each tube.
11. Read the samples on a flow cytometer, measuring fluorescence using the FL-1 and FL-2 channel as an indicator of annexin V-FLUOS staining and PI uptake, respectively. The population of annexin V-FLUOS-positive PI-negative cells consists of cells undergoing early apoptosis, whereas double positive cells are in late-stage apoptosis or are necrotic.

3.4.2 Annexin V-FLUOS/PI Staining of Non-adherent Cells (e.g., RPMI-8266)

1. Incubate cells on the plate with the desired concentration of AMPs for 1–4 h at 37 °C with 5 % CO_2 in a humidified incubator. Mix well by pipetting or shaking the plate.
2. After treatment, transfer the contents of each well to a round-bottom tube, and centrifuge at 500 × *g* for 5 min. Discard the supernatant.

3. Resuspend the cell pellet in 500 μL of PBS and centrifuge again, and then discard the supernatant (*see* **Note 1**).
4. Follow **steps 7–11** as listed above.

3.5 DiOC$_6$ Staining to Measure Mitochondrial Membrane Potential by Flow Cytometry

Cancer cells are exposed to various concentrations of AMPs for the desired length of time. Cells are then washed with PBS, detached if necessary, pelleted by centrifugation, and stained with fluorescent DiOC$_6$ to measure changes in the mitochondrial membrane potential (Δψm). Cells that were not exposed to AMPs are used to determine baseline levels of Δψm.

3.5.1 DiOC$_6$ Staining of Adherent Cells (e.g., MDA-MB-231)

1. Remove medium from the overnight cultures and discard. Replace with fresh DMEM containing 2.5 % FBS.
2. Incubate cells on the plate with the desired concentration of AMPs for 1–4 h at 37 °C with 5 % CO_2. Mix well by pipetting or shaking the plate.
3. After treatment, carefully remove the medium from each well, and wash the cell monolayer with PBS.
4. Add 500 μL of 10 mM EDTA to each well and incubate at room temperature for 5 min.
5. Pipette vigorously to detach cells from the plate.
6. Transfer the contents of each well to round-bottom tubes, and centrifuge at 500 × *g* for 5 min. Discard the supernatant (*see* **Note 1**).
7. For each treatment condition, prepare a 110 μL working solution of 1 μM DiOC$_6$ in serum-free DMEM.
8. Resuspend the cell pellet in 100 μL of the DiOC$_6$ solution. Mix vigorously by pipetting (*see* **Note 2**).
9. Incubate cells with the DiOC$_6$ solution for 30 min at 37 °C with 5 % CO_2 in a humidified incubator.
10. After incubation, add 900 μL of serum-free DMEM to each tube. Centrifuge the cell suspension for 5 min at 500 × *g*. Discard the supernatant.
11. Resuspend the cell pellet in PBS and centrifuge again, and then discard the supernatant.
12. Resuspend the cell pellet in 500 μL of PBS.
13. Read the samples on a flow cytometer, measuring fluorescence on the FL-1 channel as an indicator of cellular Δψm. Loss of fluorescence in comparison to untreated cells indicates a decrease in Δψm and mitochondrial membrane integrity.

3.5.2 DiOC$_6$ Staining of Non-adherent Cells (e.g., RPMI-8266)

1. Incubate cells on the plate with various concentrations of AMPs for 1–4 h at 37 °C with 5 % CO_2 in a humidified incubator. Mix well by pipetting or shaking the plate.

2. After treatment, transfer the contents of each well to a round-bottom tube, and centrifuge at 500 × *g* for 5 min. Discard the supernatant.
3. Resuspend the cell pellet in 500 μL of PBS and centrifuge again, and then discard the supernatant (*see* **Note 1**).
4. Follow **steps 7–13** as listed above, using serum-free RPMI-1640 medium instead of DMEM as the assay diluent.

3.6 Migration and Invasion Assay to Measure Cell Motility

Cancer cells are exposed to sub-cytotoxic concentrations of AMPs for 24 h in serum-free medium to enhance subsequent serum-dependent cell migration. Cells are then added to the upper compartment of a 48-well chemotaxis chamber, the lower compartment of which contains 10% FBS as a chemoattractant. Cells are allowed to transverse the porous polycarbonate membrane that separates the upper compartment from the lower compartment. In invasion assays, polycarbonate membranes are pre-coated with basement membrane components or an extracellular matrix component such as collagen IV. Cells that do not migrate are scraped off the membrane, and migrated cells on the underside of the membrane are stained before quantification by light microscopy.

3.6.1 Preparation and Treatment of Cells Prior to Migration/Invasion

1. Harvest cells from tissue culture flasks, and resuspend adherent cells (e.g., MDA-MB-231) or non-adherent cells (e.g., RPMI-8226) in serum-free DMEM or RPMI-1640 medium, respectively (*see* **Note 3**).
2. Count the cells and aliquot into wells of a 24-well plate at 5×10^5 cells per mL per well.
3. Incubate cells on the plate with various concentrations of AMPs for 24 h at 37 °C and 5% CO_2 in a humidified incubator.
4. After 24 h, for adherent cells, carefully remove medium, and wash the cell monolayer with PBS. Then detach cells by incubation with 10 mM EDTA in PBS for 5 min at room temperature followed by vigorous pipetting. For non-adherent cells, wash once with PBS.
5. Transfer cells from each treatment condition into a 5 mL round-bottom tube. Centrifuge the cell suspension for 5 min at 500 × *g*. Discard the supernatant.
6. Resuspend the cell pellets in 500 μL of PBS. Centrifuge for 5 min at 500 × *g* and discard the supernatant.
7. Resuspend the cell pellets in 300 μL of serum-free medium and count the cells.
8. Adjust volume so that the cell concentration in each tube is 1×10^6 cells per mL. Keep the cells at 37 °C until they are used.

3.6.2 Preparation of the Chemotaxis Chamber

Quadruplicate samples are used for each treatment condition. Thus, eight wells are required for each peptide treatment condition: four wells containing DMEM or RPMI-1640 medium with 10% FBS as a chemoattractant and four wells containing DMEM or RPMI-1640 medium without a chemoattractant as a negative control to determine baseline migration. In a 48-well chamber, this allows for five treatment conditions, including the untreated controls. Refer to Fig. 1 for a sample layout. The end columns (shaded in Fig. 1) are not used to allow for membrane processing. For invasion assays, coat the polycarbonate membrane with a collagen IV solution (1 mg/mL in PBS) overnight at 4 °C. Prior to use, wash with PBS and allow to air dry completely.

1. Prewarm 1 mL of serum-containing medium and 1 mL of serum-free medium to 37 °C.
2. Add 26 μL of serum-containing medium or serum-free medium to the bottom wells of the chemotaxis chamber (*see* **Note 4**). A positive meniscus should be present. Add serum-free medium to the wells in the unused columns.
3. Evenly layer the polycarbonate membrane over the bottom wells of the chemotaxis chamber. Take care not to shift or readjust the membrane to avoid cross contamination between wells.
4. Place the chemotaxis chamber's rubber gasket over the polycarbonate membrane. Next place the upper chamber over the gasket and evenly tighten the screws.
5. Add 50 μL of the cell suspension to the top well of the chemotaxis chamber, taking care not to introduce any bubbles. A

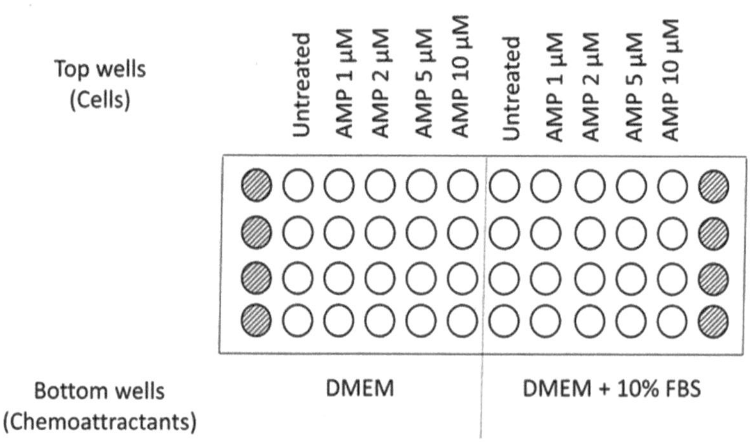

Fig. 1 Sample layout of a 48-well chemotaxis chamber. The migration of control and AMP-treated cells toward the chemoattractant or medium alone are tested in quadruplicate. Columns on either edge (*shaded*) are not used, as this area of the polycarbonate membrane is used during processing

negative meniscus should be present. Note that a positive meniscus indicates the presence of a bubble.

6. Place the chemotaxis chamber in a humidified incubator at 37 °C and 5 % CO_2 for 4–24 h to allow cell migration to occur. The incubation period used will depend on the inherent motility of the cell type being examined. For longer incubation times, loosely wrap the chemotaxis chamber with tin foil to prevent any excessive loss of moisture from the wells.

3.6.3 Staining of Migrated Cells

1. Remove the medium from the top wells by inverting the chemotaxis chamber.
2. Loosen the screws of the chemotaxis chamber, and remove the polycarbonate membrane, making careful note of which side has the cells that did not migrate. Using clips, clamp either side of the polycarbonate membrane using the area that covered the unused wells.
3. Place the polycarbonate membrane into a container of PBS, making sure that the side containing the cells that did not migrate is uppermost (*see* **Note 5**).
4. Use a rubber scraper to remove the cells that did not migrate.
5. Repeat **steps 3** and **4** two more times.
6. Fill a 50 mL tube with Diff-Quick™ staining solution 1 and a second 50 mL tube with Diff-Quick™ staining solution 2. Fill a third 50 mL tube with Diff-Quick™ fixative solution.
7. Dip the polycarbonate membrane into the fixative solution for 10 s. Dip the membrane into staining solution 1 for 10 s and then into staining solution 2 for 10 s.
8. Rinse the polycarbonate membrane by dipping it into a container of distilled water.
9. Allow the polycarbonate membrane to air dry completely.

3.6.4 Quantification of Migrated Cells

1. Place the dry polycarbonate membrane onto a microscope slide.
2. For each well, count the number of cells in five high-powered fields (HPF; 400× magnification). Determine the average number of cells per HPF for each replicate (*see* **Note 6**).
3. For each treatment condition, determine the average cells per HPF over the four replicates.
4. Fold change in migration can be calculated by comparing the cells per HPF values of each treatment to that of untreated cells that migrated toward the serum-free medium (negative control).

3.7 Angiogenesis Assay

The tubule formation assay is widely used as an in vitro model of the reorganization stage of angiogenesis. This assay can be used to determine the effect of AMPs on the ability of endothelial cells such as HUVECs to form capillary-like structures. Endothelial cells that are plated onto the appropriate extracellular matrix support create tracks that guide cellular migration. Inhibition of tubule formation is indicative of antiangiogenic activity, which interferes with tumor growth by denying tumor cells blood-borne oxygen and nutrients.

1. Thaw ECMatrix™ Solution and Diluent Buffer on ice (*see* **Note 7**).
2. For each treatment condition, dilute 198 μL of ECMatrix™ Solution into 22 μL of 10× Diluent Buffer, and mix well. Keep the solution on ice to prevent polymerization of ECMatrix™.
3. Transfer 50 μL of the ECMatrix™ solution into each well of a 96-well tissue culture plate. Each treatment condition should be performed in quadruplicate.
4. Incubate the tissue culture plate for 1 h at 37 °C to allow the ECMatrix™ solution to polymerize.
5. Harvest HUVECs, centrifuge, and resuspend in serum-free DMEM. Count the HUVECs.
6. For each treatment condition, add 3×10^4 HUVECs in 300 μL of serum-free DMEM in a 5 mL round-bottom tube.
7. Add the desired concentrations of AMPs to the tubes, and incubate for 30 min at 37 °C and 5% CO_2 in a humidified incubator.
8. Overlay 50 μL of cell solution onto each ECMatrix™-containing well in the 96-well tissue culture plate. Incubate the plate for 16 h at 37 °C and 5% CO_2 in a humidified incubator.
9. Examine tubule formation by HUVECs under light microscopy at 200× magnification.
10. The complexity of tubule formation can be scored using the numerical grading scale provided by the manufacturer.

4 Notes

1. For a large number of samples, staining and washing can also be done in a 96-well round-bottom plate instead of tubes. Samples can be transferred to tubes by a multichannel pipettor directly prior to the use of the flow cytometer.
2. To facilitate the resuspension of the cell pellet, run the bottom of the tube along a tube rack to agitate.

3. If significant migration is not detected, a longer duration of serum starvation (24–48 h) may improve cell migration toward the chemoattractant.
4. Recombinant chemokines (e.g., CXCL12) can also be used as a chemoattractant, instead of serum, to determine chemokine-specific migration/invasion.
5. Ensure that PBS does not make contact with the upper side of the polycarbonate membrane containing the migrated cells.
6. The number of cells added to the upper wells can be adjusted to ensure a sufficient number of cells are detected per field (20–200). This will vary depending on the cancer cell line.
7. Keep ECMatrix™ solution on ice to avoid premature polymerization.

References

1. Hoskin DW, Ramamoorthy A (2008) Studies on anticancer activities of antimicrobial peptides. Biochim Biophys Acta 1778:357–375
2. Gaspar D, Veiga AS, Castanho MA (2013) From antimicrobial to anticancer peptides. A review. Front Microbiol 4:1–16
3. Krishan A (1975) Rapid flow cytofluorometric analysis of mammalian cell cycle by propidium iodide staining. J Cell Biol 66:188–193
4. Van Zoggel H, Carpentier G, Dos Santos C, Hamma-Kourbali Y, Courty J, Amiche M, Delbé J (2012) Antitumor and angiostatic activities of the antimicrobial peptide dermaseptin B2. PLoS One 7:e44351
5. Mader JS, Salsman J, Conrad DM, Hoskin DW (2005) Bovine lactoferricin selectively induces apoptosis in human leukemia and carcinoma cell lines. Mol Cancer Ther 4:612–624
6. Koopman G, Reutelingsperger CP, Kuijten GA, Keehnen RM, Pals ST, van Oers MH (1994) Annexin V for flow cytometric detection of phosphatidylserine expression on B cells undergoing apoptosis. Blood 84: 1415–1420
7. Johnson LV, Walsh ML, Bockus BJ, Chen LB (1981) Monitoring of relative mitochondrial membrane potential in living cells by fluorescence microscopy. J Cell Biol 88:526–535
8. Clark AG, Vignjevic DM (2015) Modes of cancer cell invasion and the role of the microenvironment. Curr Opin Cell Biol 36: 13–22
9. Donovan D, Brown NJ, Bishop ET, Lewis CE (2001) Comparison of three in vitro human 'angiogenesis' assays with capillaries formed in vivo. Angiogenesis 4:113–121

Chapter 25

Using Oral and Colon Cancer Cells for Studying the Anticancer Properties of Antimicrobial Peptides

Teerakul Arpornsuwan, Wimolpak Sriwai, Hathaitip Sritanaudomchai, and Sittiruk Roytrakul

Abstract

Antimicrobial peptides (AMPs) are of importance in defense mechanism of many organisms and are potential candidate for treatment of infections in animals and humans. AMPs exhibit a wide range of immunomodulatory activities related to innate immunity, wound healing, and inflammation. AMPs also serve as drug delivery vectors, antitumor agents, and mitogenic agents. Here, we describe the investigation of anticancer and cytotoxic activities of antimicrobial peptides by colorimetric MTT assay using smooth muscle, dental pulp stem cell, human colon cancer cell line (SW620), and human oral squamous carcinoma cell line (HSC4).

Key words Antimicrobial peptide, Human oral squamous carcinoma cell line (HSC4), Human colon cancer cell line (SW620), Smooth muscle cell, Dental pulp stem cell

1 Introduction

The preliminary screening of anticancer activity of antimicrobial peptides has been developed using numerous bioassays [1]. The cell viability using colorimetric MTT assay is the most used for preliminary screening as it was first developed by Mosmann [2]. This method is convenient and simple enumeration of live and dead cells present. The percent of cell survival or dead cell can be calculated as a result of cytotoxicity or anticancer activity. The first homogeneous cell viability assay developed for a 96-well format using MTT (3-(4,5-dimethylthiazol-2-yl)-2,5-diphenyltetrazolium bromide) tetrazolium reduction assay was suitable for high-throughput screening (HTS) [3]. This assay has been widely used for in vitro cytotoxicity and anticancer activity of antibiotic peptides [4, 5].

Cell-based assays are often used to screen for in vitro cytotoxicity and anticancer activity; many cell types can be carried out on these purposes. For example, 12 human cancer cell lines

Paul R. Hansen (ed.), *Antimicrobial Peptides: Methods and Protocols*, Methods in Molecular Biology, vol. 1548, DOI 10.1007/978-1-4939-6737-7_25, © Springer Science+Business Media LLC 2017

(SMMC-7721 human hepatocellular carcinoma, BEL-7402 human hepatocellular carcinoma, Bcap-37 human breast carcinoma, MDA-MB-231 human breast Caucasian adenocarcinoma, MCF-7 human breast carcinoma, LK-2 human squamous cell carcinoma, A-549 human lung adenocarcinoma, NCI-H446 human small cell lung carcinoma, BGC-823 human gastric carcinoma, HeLa human cervical carcinoma, HO-8910 human ovarian carcinoma, and HT-29 human colon carcinoma) and normal human umbilical vein smooth muscle cells (HUVSMCs) were demonstrated for the cytotoxicity and cell selectivity of the antimicrobial peptide temporin-1CEa precursor cloned from the Chinese brown frog *Rana chensinensis* [6]. The development and implementation of a pilot scale, in vitro, anticancer drug screen utilizing a panel of 60 human tumor cell lines organized into subpanels representing leukemia, melanoma, and cancers of the lung, colon, kidney, ovary, and central nervous system were demonstrated by Monks et al. in 1991 [7]. The anticancer activities of differently charged chitooligosaccharide (COS) derivatives were investigated by neutral red and MTT cell viability studies using three cancer cell lines, HeLa, Hep3B, and SW480 [8].

Here we demonstrate the anticancer activity of antimicrobial peptide on smooth muscle, dental pulp stem cell, human colon cancer cell line (SW620), and human oral squamous carcinoma cell line (HSC4) by means of a 96-well format colorimetric MTT assay.

2 Materials

2.1 Setting Up the Lab for Cell Culturing

In general, the maintenance of cells should ideally be conducted in a laboratory established solely for sterile cell culture. Standard sterile tissue culture procedures should be employed throughout. The majority of tissue culture is performed within the vertical flow cabinet. With care, microbial contamination can be kept to a minimum. Testing for mycoplasma-infected cells should be periodically undertaken.

1. Vertical flow tissue culture laminar flow hood with a *HEPA* filter.
2. Tissue culture incubator: All parameters (CO_2 concentration (5% in air)), humidity, and temperature (37 °C) need to be checked daily.
3. Bench-top centrifuge.
4. Water bath.
5. Inverted, phase-contrast microscope with phase objectives of 4, 10, 20, and 40×.
6. Disposables pipettes (pipette sizes: 5, 10, 25, and 50 ml).

7. Handheld Pipet-Aid.
8. Vacuum pump, with two waste traps, to aspirate spent medium through a tube connected to disposable, heat-sterilized, long-form glass Pasteur pipettes.
9. Bone forceps.
10. Blade No. #15.
11. Hemocytometer.
12. Flat-bottom 96-well microtiter plate.
13. 0.22 μm SFCA membrane syringe filter.
14. Tissue culture incubator.
15. Orbital shaker.
16. 96-well plate reader capable of measuring the absorbance.
17. Kimwipe.

2.2 Tissue Culture Reagents

1. Dulbecco's Modified Eagle Medium (DMEM) high glucose.
2. Roswell Park Memorial Institute (RPMI) 1640.
3. Fetal bovine serum (FBS): each lot needs to be tested and freeze in aliquots.
4. Penicillin-streptomycin 100× solution (kept in frozen single-use aliquots).
5. Trypsin-EDTA (kept in frozen single-use aliquots).
6. Phosphate-buffered saline (DPBS): 137 mM NaCl, 10 mM phosphate, 2.7 mM KCl, pH 7.4, Ca^{2+}, and Mg^{2+}-free.
7. Collagenase (300 U/ml).
8. Soybean trypsin inhibitor.
9. Smooth muscle buffer: 120 mM NaCl, 4 mM KCl, 2.6 mM KH_2PO_4, 2.0 mM $CaCl_2$, 0.6 mM $MgCl_2$, 25 mM HEPES (4-(2-Hydroxyethyl)piperazine-1-ethanesulfonic acid), *N*-(2-Hydroxyethyl)piperazine-*N′*-(2-ethanesulfonic acid), 14 mM glucose, and 2.1 % essential amino acid mixture, pH 7.4.
10. Stadie-Riggs tissue slicer.

2.3 Media Recipes and Cell Lines

Bottles of media that are opened frequently become alkali rapidly; we suggest making smaller quantities that would last approximately 2 weeks.

1. Dental pulp stem cell and human oral squamous carcinoma cell line (HSC4) are cultivated in high-glucose DMEM supplemented with 1 % (v/v) penicillin-streptomycin and 10 % (v/v) FBS.
2. Human colon cancer cell line (SW620) is cultivated in RPMI 1640, supplemented with 10 % inactivated fetal bovine serum (FBS), penicillin (100 U/ml), and streptomycin (100 μg/ml).

3. Smooth muscle cells are cultivated in DMEM-10 culture medium, sterile containing 200 U/ml penicillin, 200 μg/ml streptomycin, 100 μg/ml gentamycin, 2.5 μg/ml amphotericin B, and 10% fetal bovine serum.
4. Completed media is sterilized by filtering through 0.22 μm microbiological filters and kept at 4 °C before use. Media are warmed to 37 °C immediately before use.

2.4 Cytotoxicity Assay In Vitro

1. Flat-bottom 96-well microtiter plate.
2. 15 ml centrifuge tubes.
3. 50 ml centrifuge tubes.
4. 0.22 μm SFCA membrane syringe filter.
5. Disposables pipettes (pipette sizes: 5, 10, 25, and 50 ml).
6. Handheld Pipet-Aid.
7. Autopipettes and tips.
8. Tissue culture incubator.
9. Orbital shaker.
10. 96-well plate reader capable of measuring the absorbance.
11. Aluminum foil.

2.5 Reagents

1. Growth medium: high-glucose DMEM supplemented with 1% (v/v) penicillin-streptomycin and 10% (v/v) FBS.
2. Synthetic peptides.
3. 3-(4,5-dimethylthiazol-2yl)-2,5-diphenyltetrazolium bromide (MTT) reagent.
4. Dimethyl sulfoxide (DMSO).
5. Dulbecco's phosphate-buffered saline (DPBS): 138 mM NaCl, 1.47 mM KH_2PO_4, 8.1 mM Na_2HPO_4, 2.7 mM KCl, pH 7.4, Ca^{2+}, and Mg^{2+}-free.
6. Trypsin-EDTA: 0.05% trypsin-ethylenediaminetetraacetic acid.

2.6 Peptides and MTT Solution

1. Stock 10 mg/ml of synthetic peptide: Weigh 1 mg of each synthetic peptide powder into a 1.5 ml low-binding microtube and dissolve in 100 μl of dimethyl sulfoxide (DMSO). Prepare freshly before use.
2. Working 10 mg/ml of synthetic peptide: Add 90 μl of RPMI cell culture media without 10% inactivated fetal bovine serum (FBS) into a new 1.5 ml low-binding microtube containing 10 μl of 10 mg/ml of synthetic peptide and then vigorously shake (*see* **Note 1**). The final concentration of synthetic peptide is 1 mg/ml. Convert mg/ml to molarity (*see* **Note 2**).
3. Prepare the serial dilution of synthetic peptide to make the final concentrations of 100 μg/ml, 50 μg/ml, 10 μg/ml, and 5 μg/ml, respectively.

4. MTT 3-(4,5-dimethylthiazol-2-yl)-2,5-diphenyltetrazolium bromide: Weigh 50 mg of the tetrazolium dye MTT into a falcon tube and dissolve in 10 ml of PBS (*see* Chapter 26, Subheading 2.1); mix vigorously and filter-sterilize the MTT solution through a 0.2 μM filter into a sterile, light-protected container; MTT solution should appear bright yellow in color; store at 4 °C for frequent use or at −20 °C for long-term storage.

3 Methods

3.1 Cell Counting

1. Carefully clean the chamber and the cover glass using a Kimwipe. Place the cover glass over the cell counting chamber. Shake and pipette up and down the cell suspension to get an accurate count. Using a P20, place 10 μl of the cell suspension onto one of the V-shaped sample points on either side of the chamber. The liquid will spread evenly throughout the chamber through capillary action. Repeat on the opposite side of the camber (Fig. 1a).
2. Place the slide on a microscope, and bring one side of chamber into focus under low power (4× objective). Each chamber is divided into nine large (1 mm^2) squares. The center square is divided into 25 squares (5 across by 5 down), each of which is further divided into 16 smaller squares (4×4) (Fig. 1b).
3. Using a higher power objective (10×), focus on the center square. Count the cells in 5 of the 25 squares. To obtain a random sample, the four corner squares and the one in the middle are used. For cells that overlap a ruling, count two sides of the square as in and don't count two sides. In order to be more accurate, at least 100 cells should be counted. Repeat the counting process for the opposite side of the chamber. Take the average of the two counts (n). Divide the average count (n) by 20:

   Number of cells ($\times 10^6$)/ml = $n/20$ × volume of cell × dilution factor

   For example, if five of the 25 squares had 18, 22, 17, 22, and 21 cells on one side, and 21, 15, 18, 25, and 21 on the other, then $n = [(18+22+17+22+21)+(21+15+18+25+21)]/2 = 100$.

   Number of cells ($\times 10^6$)/ml = 100/20 = 5 million cells/ml (5×10^6). If original volume of cell suspension, for example, was 3 ml, then final count is 15 million cells.
4. Pipette up and down and add the appropriate amount of medium/cell mixture to each culture flask.
5. Repeat pipetting and plating until all of the flasks have been plated.

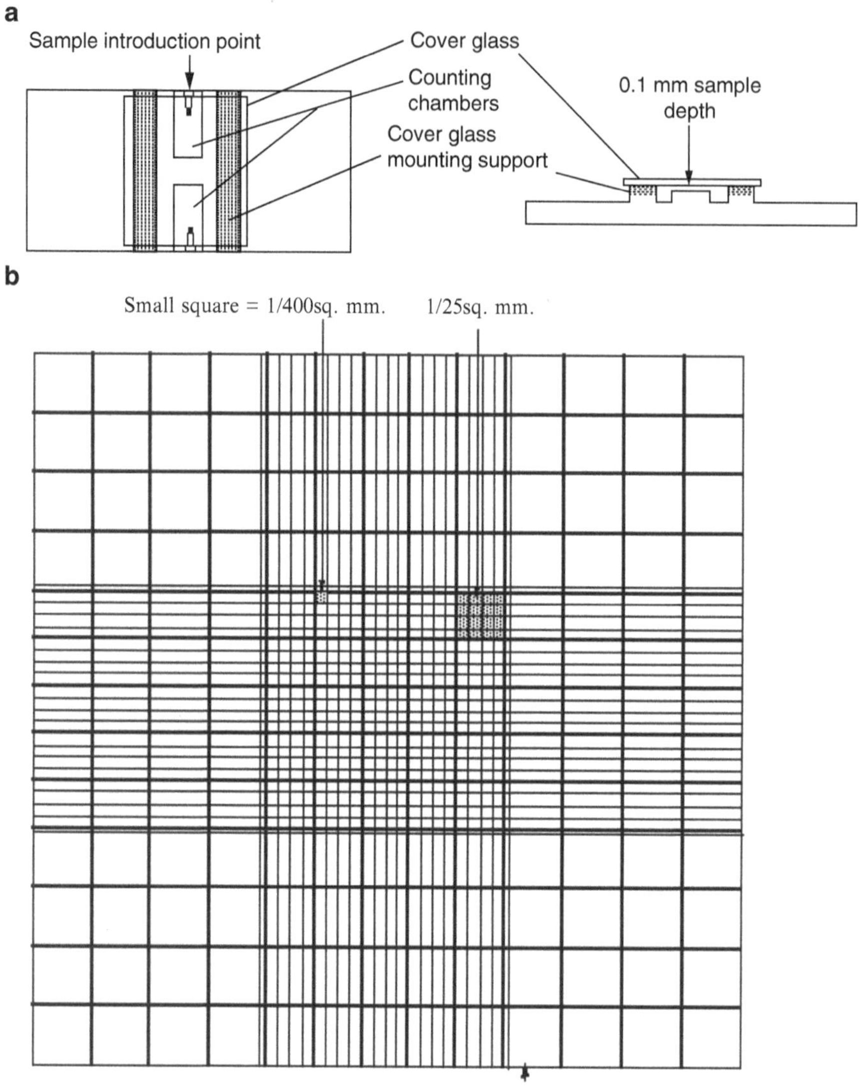

Fig. 1 The parts of the hemocytometer (as viewed from the side) are identified: (**a**) hemocytometer gridlines and (**b**) hemocytometer diagram indicating one of the sets of 16 squares that should be used for counting

6. Place all culture containers in an incubator at 37 °C, 5 % CO_2. Gently shake flasks to evenly distribute cells. Practice shaking the dish under the microscope to determine what movements will get the most evenly distributed cells.

3.2 Culture of Dental Pulp Stem Cell

The dental pulp stem cell is isolated from human teeth by modified outgrowth method [9].

1. Keep everything as sterile as possible while collecting extracted caries-free permanent teeth. Put teeth immediately into growth medium and transport to laboratory within 24 h.

2. Take the teeth into the laminar flow hood and wash them with 10 ml DPBS buffer for ten times. Cut the teeth with sterile bone forceps, carefully remove and place the pulpal tissue with forceps in 60 mm dish without containing any solution.
3. Cut the pulpal tissue with a sharp blade No. 15 into tiny pieces and transfer them into T25 tissue culture flask. Separate the pulpal tissue far from another piece approximately 5 mm, and place the flask upright to allow tissue pieces to attach to the flask wall for about 1 min.
4. Add 1.5–1.8 ml growth medium into the flask in order to cover all pulpal tissue pieces and all flask walls. Label the flasks with cell type, passage number (P0), and date.
5. Place the flasks in 37 °C, 5 % CO_2, >95 % humidified air incubator. Change the medium every other day. Check the outgrowth of cells from tissue pieces daily. The cells generally reach confluency in 21–30 days. Once the cells have reached confluency, freeze or split the cells. Use the expression level of mesenchymal stem cell surface markers CD44, CD73, CD90, CD105, and CD146 and multipotent differentiation capacity to verify the isolated dental pulp stem cells phenotype and biological properties.
6. Count the cells using the cell counting chamber (hemocytometer). *See* Subheading 3.1. Notice the confluence of cells in the flask, amount of pellet they made after trypsin treatment, and centrifugation, and correlate these measurements with actual cell count in order to be able in the future to predict the cell count.

3.3 Culture of Smooth Muscle Cells

Gastric smooth muscle cells have been isolated from antrum of rabbit stomach. The techniques for isolating smooth muscle cells from antrum involves mechanical dissection to separate muscle layers from the mucosa, sequential enzymatic digestion of muscle strips, filtration, and centrifugation as described previously [10–12].

1. Obtain the specimen, wash the tissue several times with smooth muscle buffer, and place tissue in 100 ml of ice-cold smooth muscle buffer in a plastic disposable container.
2. Remove the specimen from buffer. Place tissue on a flat surface covered with dry paper towels, and remove fat, connective tissue, and gross debris.
3. Identify the structure. The pyloric antrum is the portion between the sulcus intermedius and the duodenopyloric constriction. The sulcus intermedius and incisura angularis are the landmarks used to separate the pyloric antrum from the rest of the stomach.
4. Carefully cut lower region of the stomach to free the antrum from the body with surgical scissors or other scalpel.

5. Using scissors cut lengthwise along the antrum. Place slightly stretched tissue flat on paper towels, mucosal side up. Using curved fine forceps, gently rub or scrap mucosal layers off.
6. Place stretched tissue flat on a Stadie-Riggs tissue slicer, mucosal side up. Further remove the mucosal layer by sharp dissection (*see* **Note 3**).
7. Following the removal of the mucosa, cut the antrum into thin slices using a Stadie-Riggs tissue slicer, and place muscle strips in a 50 ml Erlenmeyer flask. Add 30 ml of a smooth muscle buffer containing 0.1 % collagenase (300 U/ml) and 0.01 % (w/v) soybean trypsin inhibitor. Digest the muscle strips by incubation at 31 °C for 30 min. Continuously gas the tissue with 100 % O_2 throughout the isolation procedure.
8. Remove enzymatic buffer by careful aspirate with disposable transfer pipette. Wash the partly digested tissues twice with 50 ml of collagenase-free smooth muscle buffer. Leave the tissue in the flask.
9. Add 30 ml of collagenase-free smooth muscle buffer to each flask.
10. Using wide bore transfer pipettes, pass the tissue and buffer through the pipette repeatedly (about 10–20 times). Continue agitation and aeration of the buffer in the flask to allow smooth muscle cells to disperse spontaneously for 30 min.
11. Harvest freshly dispersed smooth muscle cells by filtration through 500 μm Nitex filter in 50 ml tube. Collect the cells by centrifugation twice at 350 × *g* for 10 min at room temperature to eliminate broken cells and organelles.
12. After removing supernatant, resuspend the cell pellet in 30 ml of DMEM, switch the cells to a new 50 ml conical tube, and wash the cells twice by centrifugation at 350 × *g* for 10 min at room temperature.
13. Aspirate off the supernatant, resuspend the recovered cells in 10 ml of DMEM-10.

3.4 Determination of Cell Number

1. Dilute the cells in a volume appropriate for counting.
2. Add trypan blue solution to make a cell dilution of 1:20.
3. Fill the hemocytometer with the diluted cell suspension.
4. Count the number of live smooth muscle cells at least four squares. Viable or intact cells exclude the dye and appear unstained, while dead cells take up the dye and appear blue stained.
5. Transfer the culture to a T25 flask. Place the smooth muscle cell suspension at a concentration of 5×10^5 cells/ml in a culture disk and incubate the cultures in a humidified 5 % carbon dioxide 95 % air incubator at 37 °C.
6. Change DMEM-10 medium every 3 days. The cells should reach confluency after 2–3 weeks (*see* **Note 4**).

7. The in vitro cytotoxicity assay should be done on smooth muscle cells in the first passage.
8. Trypsinize (0.5 mg trypsin/ml) 70–80% confluent smooth muscle cells in primary cultures and replate at 1×10^5 cells/ml in six-well plate.
9. Place the plate in a humidified 5% carbon dioxide 95% air incubator at 37 °C for 1 day.
10. Change the culture medium and continue experiment.

3.5 Culture of Human Oral Squamous Carcinoma Cell Line (HSC4)

1. Remove desired vial (s) (1 or 2 vials per flask depending on concentration of cells) from liquid nitrogen storage tank. A vial containing 1 °C 10^6 cells thawed into a T75 tissue culture flask will generally be confluent in 3–4 days.
2. Vent the vial making sure to close it tightly before placing it into the 37 °C H_2O water bath. Do not submerge the vial rather hold it such that the water level in water bath is even with the frozen liquid level in the vial. The vial can be twirled during thawing until all ice crystals have disappeared.
3. Spray the vial with 70 % ethanol to minimize contamination. Under the laminar flow hood, transfer the cell suspension from 1 vial into a 15 ml centrifuge tube. Very slowly, over a total of 2 min, add 9 ml growth medium dropwise while gently mixing the tube, bringing the total volume to 10 ml. This decreases the DMSO concentration gradually.
4. Centrifuge the cells at $157 \times g$ for 5 min, remove supernatant and resuspend the pellet with 1–2 ml of growth medium, and transfer the cell suspension into T75 tissue culture flasks containing 10 ml growth medium and place into a 5% CO_2, 37 °C incubator. Medium is changed every other day.
5. Once cells reach confluency, discard medium and rinse the cell layer 2–3 times with DPBS in order to remove any traces of FBS which would inhibit trypsin reaction, tip the flask side to side once, and remove. Vigorous swirling may cause the cells to lift.
6. Add 4 ml of prewarmed trypsin-EDTA solution to the cell layer in the flask. Close the flask, and place it onto a 37 °C warming dish for 2–5 min. The trypsin solution will work quickly to detach the cells from the bottom of the flask. Then gently shake the flask, and check under the microscope that most of the cells are detached and floating.
7. Back under the hood, using a 10 ml serological pipette, quickly disaggregate cells into a single-cell suspension, and add 10 ml of growth medium to each flask. The FBS in the medium will deactivate the trypsin. Transfer the cell suspension (cells/trypsin/medium) from the flask to a 15 ml centrifuge tube and centrifuge at $157 \times g$ for 5 min.

8. Work under the hood and without disturbing the cell pellet, remove and discard the supernatant. Resuspend the cell pellet with 1–2 ml of growth medium.
9. Determine the total number of cells in the cell suspension by hemocytometer. *See* above on Subheading 3.1. Once the total cell number in the tube is known, calculate the exact volume needed to dilute cells to a desired concentration for in vitro cytotoxicity assay.

3.6 Culture of Human Colon Cancer Cell Line (SW620)

1. Remove desired vial (s) (1 or 2 vials per flask depending on concentration of cells) from liquid nitrogen storage tank.
2. Culture cells in T-75 flasks with completed media, and maintain in a humidified atmosphere of 5 % CO_2—95 % air at 37 °C until 80–90 % confluent.
3. Aspirate media from flask and wash cells twice with 10 ml PBS (*see* **Note 5**).
4. Trypsinize cells with 1 ml of 0.05 % trypsin-EDTA and leave the flask at 37 °C for 5 min.
5. Add 5 ml of fresh medium (RPMI plus 10 % FBS). Mix by pipette up and down gentle to disperse cells.
6. Centrifuge the cell suspension at 157 × *g* for 5 min and discard the supernatant (*see* **Note 6**).
7. Resuspend the cell pellet with 1 ml of serum-free media for cell counting by trypan blue (*see* **Note** 7). Add 50 μl of the cell suspension in 400 μl of PBS containing 50 μl of 0.04 % trypan blue (*see* **Note 8**). Load 10 μl of staining cells into a hemocytometer (*see* **Note 9**). Count the cells in four separate fields of hemocytometer using light microscope. Count viable, clear cells, and calculate the number of viable cells/ml as described in Subheading 3.1, **steps 1–6**, before in vitro cytotoxicity assay.

3.7 Cytotoxicity Assay In Vitro

Each condition should be done in triplicate or more.

1. Day 1: Harvest the cells as explained in Subheading 3.1. Seed 100 μl of cell suspension (2×10^4 total cells) for dental pulp stem cell and human oral squamous carcinoma cell line (HSC4) and 2×10^4 total cells/200 μl for smooth muscle cell and SW620 cell line into each 96-well plate and incubate overnight.
2. Day 2: Prepare the serial dilution of synthetic peptide in media to make the final concentrations of 100 μM, 50 μM, 25 μM, 12.5 μM and 6.25 μM, respectively. Aspirate the supernatant and add 200 μl of fresh medium containing synthetic peptide (ranging from 100 to 6.25 μM) into each well (*see* **Note 10**). Except column B1-H1 for control columns containing cells without peptides. Incubate at 37 °C for overnight. Cells treated with 0.1% DMSO were used as a solvent control, whereas untreated cells were used as an OD control.

3. Day 3: Add 1.2 ml of 5 mg/ml of the tetrazolium dye MTT 3-(4,5-dimethylthiazol-2-yl)-2,5-diphenyltetrazolium bromide in 20 ml of PBS for working solution of MTT. Discard the supernatant from microtiter plate of SW620 cell line. Add 200 μl of working solution of MTT into each well (*see* **Note 11**).
4. Add 100 μl of 0.5 mg/ml MTT to each well of microtiter plate for smooth muscle cell, dental pulp stem cell and human oral squamous carcinoma cell line (HSC4) (*see* **Note 12**). All should be done aseptically and wrap the plate with aluminum foil to avoid exposure to light.
5. Incubate for 2–3 h at 37 °C in tissue culture incubator.
6. Carefully remove media. Do not disturb cells and do not rinse with PBS. Discard the MTT and dissolve the purple formazan product with 100 μl and 200 μl of DMSO for smooth muscle cell, dental pulp stem cell, HSC4, and SW620 cell line, respectively (*see* **Note 12**). Cover plate with aluminum foil and agitate cells on orbital shaker for 30 min at room temperature (*see* **Note 13**).
7. Read absorbance at 570 nm with a reference filter of 650 nm.
8. Calculate cytotoxicity assay in vitro as follows:

   % Cytotoxicity = 1 − [(OD peptide treated − OD blank)/(OD control − OD blank)] × 100.

4 Notes

1. The serial dilution of synthetic peptide was curious step to maintain the activity and stability of peptide by adding the media into synthetic peptide.
2. Convert mg/ml to μM, divide the concentration in mg/ml by the molecular weight of the peptide, and multiply by 1000 × 1000.
3. Never let tissues dry out during dissection. Always keep them immersed in a smooth muscle buffer.
4. Instead of flasks, Petri dishes of various sizes can be used for culture.
5. It is critical to remove all traces of fetal bovine serum in complete media in washing step; otherwise it will be difficult for trypsin solution to remove cells from the flask.
6. Centrifuge the cell suspension at 157 × *g* for 5 min to remove trypsin-containing media and replace with fresh media.
7. Use PBS or a serum-free medium for the cell suspension. Serum proteins may stain with trypan blue, resulting in falsely depressed viable counts.

8. 0.4% (w/v) trypan blue in PBS could be stored up to 1 year at room temperature in the dark. Filter if a precipitate forms.
9. Do not overfill the surface for loading the hemocytometer. Load two samples on one hemocytometer, one into each of the two grids.
10. Carefully remove media. The most variation in data may occur in this step.
11. The concentration of MTT and the amount of formazan signal generated may need to be adjusted for optimal condition. The cell type, number of cell/well, and the cellular metabolic activity can cause the reduction of MTT.
12. Dissolve the formazan product in 96-well plate by dispenses gently to ensure complete solubilization. Occassionally, pipetting up and down may be required to completely dissolve the MTT formazan crystals especially in dense cultures.
13. Gentle stirring in a shaker will enhance dissolution.

Acknowledgment

This work was supported by TU Research Grant, Thammasat University. Thanks are also extended to the Faculty of Science, Ramkhamhaeng University, and the Faculty of Dentistry, Mahidol University, for their kind support in providing SW620 cell line, oral cancer cell line, and dental pulp stem cells.

References

1. Cetin Y, Bullerman LB (2005) Cytotoxicity of Fusarium mycotoxins to mammalian cell cultures as determined by the MTT bioassay. Food Chem Toxicol 43:755–764
2. Mosmann T (1983) Rapid colorimetric assay for cellular growth and survival: application to proliferation and cytotoxicity assays. J Immunol Methods 65:55–63
3. Riss TL, Moravec RA, Niles AL, Benink HA, Worzella TJ, Minor L (2004) Cell viability assays. In: Sittampalam GS, Coussens NP, Nelson H et al (eds) Assay guidance manual [Internet]. Eli Lilly & Company and the National Center for Advancing Translational Sciences, Bethesda, MD
4. Lin WJ, Chien YL, Pan CY, Lin TL, Chen JY, Chiu SJ, Hui CF (2009) Epinecidin-1, an antimicrobial peptide from fish (*Epinephelus coioides*) which has an antitumor effect like lytic peptides in human fibrosarcoma cells. Peptides 30:283–290
5. Chu HL, Yip BS, Chen KH, Yu HY, Chih YH, Cheng HT, Chou YT, Cheng JW (2015) Novel antimicrobial peptides with high anticancer activity and selectivity. PLoS One 10:e0126390
6. Wang C, Li HB, Li S, Tian LL, Shang DJ (2012) Antitumor effects and cell selectivity of temporin-1CEa, an antimicrobial peptide from the skin secretions of the Chinese brown frog (*Rana chensinensis*). Biochimie 94:434–441
7. Monks A, Scudiero D, Skehan P, Shoemaker R, Paull K, Vistica D, Hose C, Langley J, Cronise P, Vaigro-Wolff A et al (1991) Feasibility of a high-flux anticancer drug screen using a diverse panel of cultured human tumor cell lines. J Natl Cancer Inst 5(83):757–766
8. Huang R, Mendis E, Rajapakse N, Kim SK (2006) Strong electronic charge as an important factor for anticancer activity of chitooligosaccharides (COS). Life Sci 11(78):2399–2408
9. Gronthos S, Arthur A, Mark Bartold P, Shi S (2011) A method to isolate and culture expand

human dental pulp stem cells. Methods Mol Biol 698:107–121

10. Murthy KS, Makhlouf GM (1991) Phosphoinositide metabolism in intestinal smooth muscle: preferential production of Ins (1,4,5) P3 in circular muscle cells. Am J Physiol 261:G945–G951
11. Murthy KS, Makhlouf GM (1996) Opioid mu, delta, and kappa receptor-induced activation of phospholipase C-beta 3 and inhibition of adenylyl cyclase is mediated by Gi2 and G (o) in smooth muscle. Mol Pharmacol 50:870–877
12. Sriwai W, Mahavadi S, Al-Shboul O, Grider JR, Murthy KS (2013) Distinctive G protein-dependent signaling by protease-activated receptor 2 (PAR2) in smooth muscle: feedback inhibition of RhoA by cAMP-independent PKA. PLoS One 8:e66743

Chapter 26

Using Disease-Associated Enzymes to Activate Antimicrobial Peptide Prodrugs

Éanna B. Forde, Graeme Kelly, Hisham Makki, Zahraa Al-Sharshahi, Deirdre Fitzgerald-Hughes, and Marc Devocelle

Abstract

Prodrugs of antimicrobial peptides can be generated by modifying their sequences at their N-termini with a linker and a negatively charged promoiety. These modifications can be selectively reversed by a disease-associated enzyme, thereby confining the activity of the peptide to pathologically affected body parts. A general method for the generation of prodrug candidates, based on a linker constituting the substrate of a disease-associated protease and an oligo-glutamic acid promoiety, as well as a protocol to validate the activation of the prodrug, are described herein.

Key words Active parent AMP, Disease-associated protease, Oligo-glutamic acid promoiety, Prodrug activation

1 Introduction

Prodrugs are inactive precursors of pharmaceutical agents that are activated in vivo [1]. If this activation is mediated by a chemical and/or biochemical reaction confined to a specific body site, targeted delivery of the active parent drug can be achieved [2]. A prodrug approach can therefore overcome a toxicity barrier in drug development and could be applied to antimicrobial peptides (AMPs) to increase their selectivity toward bacterial or cancer cells. Most AMPs have a net positive charge which contributes to their biological activities [3]. Consequently, their antimicrobial and anticancer activities, but also off-target effects, can be modulated by covalently, but reversibly, linking their sequences to a negatively charged promoiety (Fig. 1). A promoiety is a functional group or molecular entity which modifies the physico-chemical and pharmacokinetic/pharmacodynamic properties of a parent drug. The modification aims to overcome a barrier to the clinical use of the drug, such as toxicity as discussed here for AMPs. A promoiety for their prodrugs can be the target of an activating enzyme [4] or

Paul R. Hansen (ed.), *Antimicrobial Peptides: Methods and Protocols*, Methods in Molecular Biology, vol. 1548, DOI 10.1007/978-1-4939-6737-7_26,

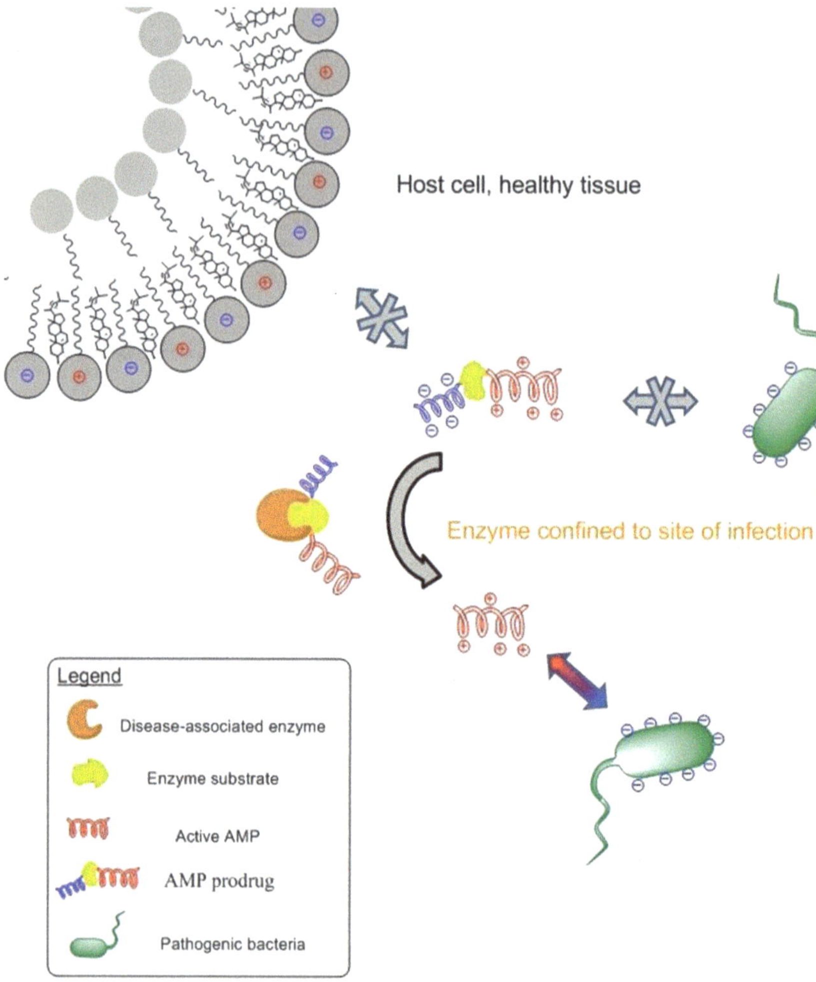

Fig. 1 Overview of AMP prodrug approach. Masking the net positive charge of the active AMP by linking it to a negatively charged promoiety, which can be selectively removed at the site(s) of infection, can generate an AMP candidate targeted at specific pathogens and devoid of activity (toxicity) at non-infected body parts

attached to the parent peptide through a linker which constitutes a substrate for an activating enzyme [5]. In the former case, the promoiety is a substrate of the activating enzyme and must be linked to the AMP sequence through a linker which spontaneously decomposes upon catalytic activation (Table 1, entry 1). In the latter case, the promoiety is either a biologically inert constituent (Table 1, entry 2) or another therapeutic agent also rendered inactive by its linkage to the AMP (Table 1, entry 3). In every case, the promoiety should be non-toxic, have a net negative charge and not affect the activity of the AMP when non-covalently bonded to it.

The activating enzyme selected determines the confinement of the active peptide and the associated targeted antibiotic strategy. Among suitable enzymes, disease-associated proteases provide a

Table. 1
Characteristics and representative structures of different types of AMP prodrugs

Entry	Prodrug type	Promoiety	Linker	Activating enzyme (target bacteria)	Representative structure[a]	Ref.
1	Peptide conjugate	Cephalosporin[b]	Carbamate	β-Lactamase (Gram-negative bacteria expressing extended spectrum β-lactamases (ESBLs))	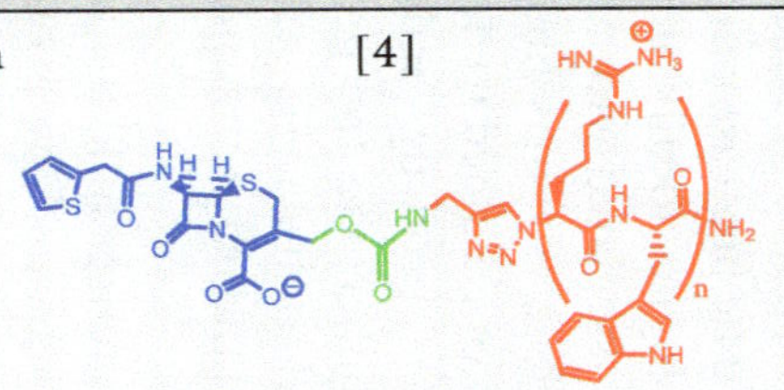	[4]
2	Fusion peptide	Oligo-glutamic acid sequence	Peptidic substrate[c]	Protease (narrow to broad spectrum of activity, depending on the protease selected)	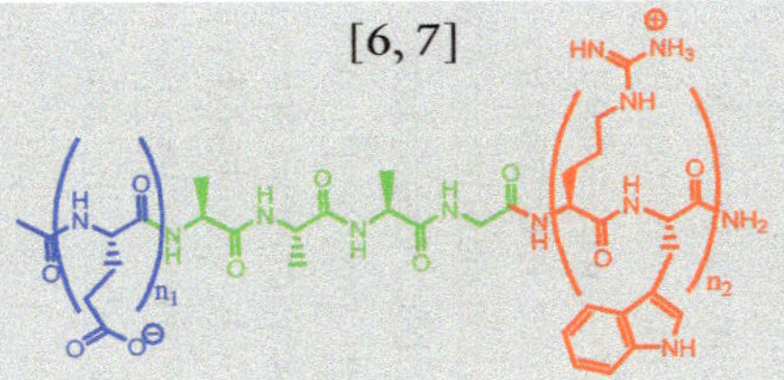	[6, 7]
3	Co-drug	Anti-inflammatory drug	Azo bond	Azo reductase (anaerobic pathogens, including *Clostridium difficile*)	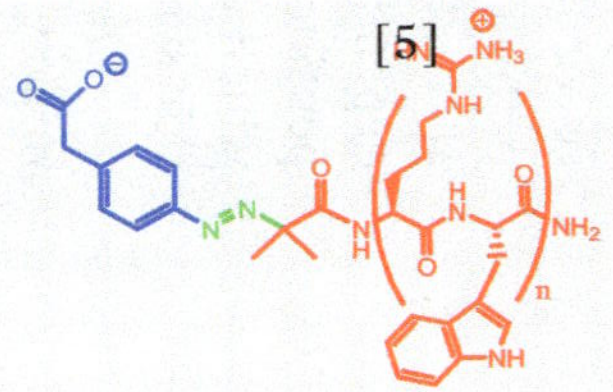	[5]

[a]RW repeat as an illustrative AMP sequence
[b]Other β-lactam antibiotics which allow the elimination of a substituent upon hydrolysis of the β-lactam bond, such as carbapenems and penems, are also eligible promoieties [8]
[c]AAAG (neutrophil elastase substrate) as an illustrative linker sequence

convenient and general opportunity to achieve this targeted delivery of a parent peptide [9]. The protease can be a bacterial enzyme with no mammalian homologue, ideally an enzyme specific to a bacterial pathogen and/or associated with an antibiotic resistance or virulence mechanism. Host enzymes highly up-regulated in a pathological condition and/or confined to diseased cells/tissue are also eligible candidates for prodrug activation. In protease-dependent prodrugs, the linker and the promoiety can both be peptide sequences appended to the AMP, yielding a fusion peptide which can be prepared by standard techniques of solid-phase peptide synthesis (SPPS) [10].

Oligo-glutamic acid sequences fulfil the requirements of a promoiety in AMP prodrugs activated by proteases. Each glutamic acid reduces the positive net charge of the peptide by 1 unit. A sequential addition can be implemented to investigate the effect of the progressive reduction of the charge on the biological activity of the parent peptide and to determine the minimal number of glutamate residues to generate a prodrug. N-terminal acetylation of the oligo-glutamic acid promoiety can further reduce the net charge of the parent peptide by 1 unit. This promoiety can also be placed at the C-terminus of the AMP, but as peptides are synthesised from the C- to N-termini and as AMPs have a high hydrophobic amino acid content (generally 40–50 %), assembling first the active parent AMP sequence limits the interchain association effects which may otherwise hinder the elongation of the complete sequence [11]. This initial assembly of the active parent AMP sequence can also reduce the quantities of amino acids required for its elongation, if done from more expensive d-amino acids. Finally, it allows the synthesis of prodrug candidates varying by the number of glutamate residues in the promoiety, from the same batch of resin, by sampling the latter after each glutamic acid addition. Introduction of a short spacer (e.g. glycine of β-alanine) after four consecutive glutamic acid residues can increase the reactivity of the N-terminal amino groups and favour further elongation of the promoiety [6]. Other acidic amino acids can be used for the promoiety, but among common amino acids, glutamic acid should be preferred to aspartic acid for peptides synthesised by the Fmoc/*t*-Bu strategy, because of the aspartimide-mediated formation of by-products [12]. The promoiety can be assembled from l- or d- glutamic acids, but l-oligo-glutamate promoieties can avoid the isolation of the linker between heterochiral residues, when the AMP sequence itself is synthesised from d-amino acids. In the latter case, the inclusion of one glycine residue between the linker and the active AMP sequence is recommended. The promoiety and the residual amino acids from the linker (at its N-terminal side) can be synthesised separately from a resin such as a preloaded Wang resin (*see* Chapter 3, Fig. 2). This will yield the peptide acid corresponding to the promoiety released from the prodrug upon activation. This peptide can be tested for toxicity against relevant host cells [7]. It can also be evaluated, individually and in combination with the active AMP, to ensure that it doesn't affect the antimicrobial activity of the latter peptide [6].

The linker in a protease-dependent AMP prodrug should be an exclusive peptide motif directed at a particular extracellular protease. The latter should preferably have narrow substrate specificity. Secreted proteases which constitute virulence factors in pathogenic bacteria are ideal targets. Examples of bacterial proteases which, in principle, could be targeted are V8 from *Staphylococcus aureus* [13] and CD2830 from *Clostridium difficile* [14]. The optimal linker sequences are reported for these two enzymes. For newly identified proteases, the amino acids at substrate positions (e.g. P_3, P_2, P_1 | P'_1, P'_2, P'_3), associated with the highest cleavage kinetics and specificity, can be identified using synthetic FRET peptides [15].

This protease-dependent prodrug approach is broadly applicable to any AMP which can be assembled by SPPS, does not require post-cleavage modification using active ester-based conjugation chemistry and is, or can be made, resistant to cleavage by the targeted protease. Among suitable candidates, linear (α-helical or extended) AMPs, with primary membranolytic activities which are independent of the stereochemistry of their constitutive amino acids, are attractive sequences. As the proteolytic activation of the AMP prodrug might leave residual amino acids from the linker at the N-terminus of the active sequence, the processed peptide rather than the parent peptide should be used as a positive control. It should also be ensured that these residual amino acids do not adversely affect the antimicrobial activity of the parent AMP and that the processed peptide is not an inhibitor of the activating protease. A negative control can generally be designed by omitting the sequence of the linker and directly elongating the oligo-glutamic acid promoiety at the N-terminus of the AMP sequence.

The design strategy presented above allows the preparation of protease-dependent AMP prodrugs as targeted antibiotics with narrow or broad spectrum of activity, depending on the protease (from the host or bacteria) selected. They can be conveniently prepared by manual or automated SPPS [16, 17] as described in Chapter 3 and 4, respectively. This strategy also facilitates the parallel production of various candidates based on different active AMP sequences for combination studies. Moreover, it can be employed to target the delivery of AMPs in any condition (e.g. infection, inflammation and cancer) which may be addressed by these peptides. The protocols to test these prodrug candidates, with either exogenous or endogenous protease, in the absence or presence of bacteria, to validate their activation, are presented below. The target outcomes of these assays are the quantitative release of the active AMP in the former case (no bacteria) and the acquisition of similar minimum inhibitory/bactericidal concentrations between the active AMP and its prodrug, in the latter case. This validation is one of the most critical steps in the successful development of AMP prodrugs.

2 Materials

2.1 Activation Assay in the Absence of Bacteria

1. Incubator maintained at 37 °C in a 5% CO_2 humidified atmosphere.
2. Phosphate buffered saline (PBS): dissolve 8 g of NaCl, 0.2 g KCl, 1.42 g of Na_2HPO_4, and 0.24 g of KH_2PO_4 in 1 l of deionised water (dH_2O), adjust to pH 7.4 and sterilise by autoclaving (121 °C, 15 PSI, 15 min), which gives a concentration of 137 mM, 2.7 mM, 10 mM, and 1.8 mM, respectively.
3. Purified enzyme, e.g. human neutrophil elastase (NE), at 1 mg/ml prepared in PBS
4. Sterile 96-well plate.
5. Acetonitrile (HPLC grade).
6. Water (HPLC grade).
7. *p*-Hydroxycinnamic acid.
8. C18 analytical HPLC column.
9. Peptide stock 1: 1 mg/ml in PBS.
10. Matrix-assisted laser desorption ionisation-time of flight mass spectrometry (MALDI-TOF-MS) solution: 10 mg/ml *p*-hydroxycinnamic acid in acetonitrile/H_2O (1:1).
11. RP-HPLC analysis of the purified peptide with C18 analytical column at a flow rate of 1 ml/min with a linear gradient over 30 min from 5 to 65% solvent B. UV detection at 214 nm (amide bond) or a dual wave wavelength at 214 and 280 nm for peptides containing amino acids with aromatic side chains.
12. Solvent A: water containing 0.1% TFA
13. Solvent B: acetonitrile containing 0.1% TFA.

2.2 Activation Assay in the Presence of Bacteria

1. Shaking incubator.
2. DensiCHEK instrument with McFarland standards and glass vials.
3. *Pseudomonas aeruginosa* reference strain PAO1 (or other test organism).
4. Mueller-Hinton agar plates.
5. Solution 1: 0.9% w/v solution of NaCl in dH_2O. Sterilise by autoclaving.
6. Solution 2: 10 mM solution of potassium phosphate in dH_2O pH 7.4. Sterilise by autoclaving (121 °C, 15 PSI, 15 min).

7. Solution 3: Dissolve 2 g of BSA in dH_2O by rigorous agitation for 10 min, filter sterilise the solution as above, aseptically add 100 μl of the filtered solution to 9.9 ml of Solution 2 and mix.
8. Peptide stock 2: 1 mg/ml peptide stock solution in Solution 2. Filter sterilise the solution in a laminar airflow cabinet using a disposable syringe and a 0.2 μm syringe filter (this can be further diluted to the required concentration for assay in Solution 3).
9. Purified human neutrophil elastase (NE) [6, 7] in Solution 2.

   or
10. Human bronchoalveolar lavage (BAL) fluid confirmed to contain μg/ml quantities of NE (such as from cystic fibrosis patients) (*see* **Note 1**).
11. Sterile microcentrifuge tubes.
12. Sterile inoculating loops.
13. Sterile L-shaped cell spreaders.
14. Pipette with sterile pipette tips.
15. Vortex mixer.
16. Bunsen burner.

3 Methods

3.1 Activation Assay in the Absence of Bacteria

1. Record the retention time (RT) of the peptide prodrug purified by RP-HPLC. The RT of the active peptide, processed by the activating enzyme from the prodrug, will generally be lower.
2. From the Peptide stock 1, prepare a serial doubling dilution from 1 mg/ml (stock solution) to 0.007 mg/ml in PBS.
3. To quantify the reactivation of the peptide, prepare a standard curve of the peptide prodrug concentration versus peak area using RP-HPLC.
4. Pipette 100 μl of the NE stock solution into a well of a 96-well pate and place in the incubator.
5. Pipette 100 μl of the peptide prodrug stock solution to the NE solution, and incubate the combined solutions.
6. At hourly intervals, mix the solution with a pipette and sample 25 μl. Analyse the removed sample by RP-HPLC, monitoring the percentage of processed peptide(s) until quantitative, or near quantitative, reactivation is reached, or until no significant increase is reached between multiple time points. The peptide prodrug peak will decrease in magnitude, while one or more product peaks will form. The extent of cleavage of the peptide

prodrug can be calculated by comparing the peak area to the standard curve. The reactivation can be confirmed by complete shift in RT of the peptide prodrug. For enzymes with high kinetics of activation, the assay might need to be repeated with 10 min (or lower) intervals between samplings (*see* **Note 2**).

3.2 Activation Assay in the Presence of Bacteria

1. Aseptically inoculate PAO1 on Mueller-Hinton agar using the streak-plating method for isolation of single colonies, and incubate overnight.
2. Aseptically take three or four colonies and suspend in 5 ml of Solution 1.
3. Calibrate the DensiCHEK meter using the 1.0 McFarland standard and blank with 1 ml of Solution 1 in a glass vial. Aseptically remove 1 ml of the bacterial suspension and transfer to another glass vial and measure using the DensiCHEK meter.
4. If the reading is below 1.0 McFarland, aseptically add more colonies to the suspension. If the reading is above 1.0 McFarland, aseptically dilute the suspension with Solution 1. Repeat until a suspension of 0.95–1.05 McFarland is achieved (measured three times).
5. Aseptically transfer 100 μl of the bacterial suspension to 9.9 ml of Solution 3. This will be the bacterial inoculum.
6. To a microcentrifuge tube, aseptically add each component of the susceptibility assay (the required quantity of Peptide stock 2, NE solution or human BAL fluid, 10% v/v of the bacterial inoculum and Solution 3 to the final volume) and vortex for 30 s. Controls (*see* **Note 3**) should include AMP prodrug with bacteria, NE solution/BAL fluid with bacteria, BAL fluid alone and bacteria alone (positive growth control). Multiple peptide and enzyme concentrations may be investigated, and it is advisable to repeat experiments on three separate occasions at least in duplicate.
7. Incubate the microcentrifuge tubes in the shaker incubator at 37 °C on medium speed (approximately 200 rpm) for 60 min. Remove and aseptically add 9 volumes of Solution 1 to each tube and vortex mix for 30 s (*see* **Note 4**).
8. Aseptically pipette 100 μl of each bacterial suspension onto a Mueller-Hinton agar plate, and spread evenly on the plate with a disposable cell spreader and incubate plates inverted overnight at 37 °C. Refrigerate the bacterial suspensions for later use if required.
9. Count the colony forming units (CFU) on the plates. If the colony count is too high, dilute the bacterial suspensions further, spread on a new plate and incubate for counting the next day.
10. Express the colony counts as % reduction in CFU on test plates (from assays containing AMP prodrug, or AMP prodrug and

NE/human BAL) compared to the bacteria alone plates. This will allow comparison between the bactericidal effects of AMP prodrug alone to the AMP prodrug in the presence of NE/human BAL fluid.

4 Notes

1. While neutrophil elastase can be replaced as targeting enzyme, consideration must be given to the levels of activating enzyme available in clinical samples. For example, human BAL fluid contains many different pulmonary enzymes but none are likely to be present in the same order of magnitude as neutrophil elastase.
2. The reactivation product(s) can be identified by MALDI-TOF MS. The molecular ions of the incubation products are likely to correspond to the active AMP peptide sequence with one or more residual amino acids from the linker. In the case of AMPs modified with a acetylated tetraglutamic acid promoiety and a tri-alanyl-glycine linker as a neutrophil elastase substrate [6, 7], two main cleavage products are obtained, corresponding to the active AMP sequence with a remaining glycine and one or two residual alanine(s). Synthesis of these processed peptides will be required to validate their identification by comparison of their RP-HPLC RTs and for use as positive controls (*see* Subheading 1).
3. The use of controls is important as high concentrations of AMP prodrug may display bactericidal activity that will need to be taken into account. Similarly, NE at a high concentration and some human BAL fluids will have some bactericidal activity. Experiments should be designed to minimise these effects so that activity due to the activation of AMP prodrug alone is investigated.
4. The amount of Solution 1 can be increased or decreased proportional to the assay volume. Typically, assay volumes of 100 μl are used, which require 900 μl of Solution.

Acknowledgements

This material is based upon works supported by the Science Foundation Ireland under Grants 05/RFP/CHE0063, 06/RFP/CHO024, 06/RFP/CHO024/EC07 and SFI 08/RFP/CHE1563, Enterprise Ireland under Grant PC/2005/164, RCSI under the RCSI Student Research Programme and the Higher Education Authority, Ireland under the BioAT programme in Cycle 5 of the Programme for Research in Third-Level Institutions.

References

1. Rautio J, Kumpulainen H, Heimbach T et al (2008) Prodrugs: design and clinical applications. Nat Rev Drug Discov 7:255–270
2. Stella VJ (2004) Prodrugs as therapeutics. Exp Opin Ther Patents 14:277–280
3. Zasloff M (2002) Antimicrobial peptides of multicellular organisms. Nature 415:389–395
4. Desgranges S, Ruddle CC, Burke LP et al (2012) β-Lactam-host defence peptide conjugates as antibiotic prodrug candidates targeting resistant bacteria. RSC Adv 2:2480–2492
5. Kennedy DA, Vembu N, Fronczek FR et al (2011) Synthesis of mutual azo prodrugs of anti-inflammatory agents and peptides facilitated by α-aminoisobutyric acid. J Org Chem 76:9641–9647
6. Desgranges S, Le Prieult F, Daly A et al (2011) *In vitro* activities of synthetic host defense propeptides processed by neutrophil elastase against cystic fibrosis pathogens. Antimicrob Agents Chemother 55:2487–2489
7. Forde É, Schütte A, Reeves E et al (2016) Differential in vitro and in vivo toxicities of antimicrobial peptide prodrugs for potential use in cystic fibrosis. Antimicrob Agents Chemother 60:2813–2821
8. Smyth TP, O'Donnell ME, O'Connor MJ et al (2000) β-lactamase-dependent prodrugs - recent developments. Tetrahedron 56:5699–5707
9. Hu Q, Katti PS, Gu Z (2014) Enzyme-responsive nanomaterials for controlled drug delivery. Nanoscale 21:12273–12286
10. Merrifield RB (1986) Solid phase synthesis. Science 232:341–347
11. Hyde C, Johnson T, Owen D et al (1994) Some 'difficult sequences' made easy. A study of interchain association in solid phase peptide synthesis. Int J Pept Protein Res 43:431–440
12. Mergler M, Dick F, Sax B et al (2003) The aspartimide problem in Fmoc-based SPPS. Part II. J Pept Sci 9:518–526
13. Dubin G (2002) Extracellular proteases of *Staphylococcus* spp. Biol Chem 383:1075–1086
14. Hensbergen PJ, Klychnikov O, Bakker D et al (2014) A novel secreted metalloprotease (CD2830) from *Clostridium difficile* cleaves specific proline sequences in LPXTG cell surface proteins. Mol Cell Proteomics 5:1231–1244
15. Fields GB (2010) Using fluorogenic peptide substrates to assay matrix metalloproteinases. Methods Mol Biol 622:393–433
16. Hansen PR, Oddo A (2015) Fmoc solid-phase peptide synthesis. Methods Mol Biol 1348: 33–50
17. Pedersen SL, Jensen KJ (2013) Instruments for automated peptide synthesis. Methods Mol Biol 1047:215–224

Chapter 27

Anti-inflammatory Properties of Antimicrobial Peptides and Peptidomimetics: LPS and LTA Neutralization

Sarah Line Skovbakke and Henrik Franzyk

Abstract

Lipopolysaccharide (LPS) and lipoteichoic acid (LTA) neutralization constitute potential non-antibiotic treatment strategies for sepsis - a systemic infection-induced inflammatory response. Studies on LPS- and LTA-neutralizing compounds are abundant in literature, and a number of peptides and peptidomimetics appear to display promising activity. However, in this ongoing search for potential antisepsis drug leads, it will be preferable that the assays used by different research groups lead to readily comparable data for the most efficient compounds. Here, we propose and describe standardized methods to be used for testing of novel compounds for their LPS- and LTA-neutralizing capacity with a focus on functional suppression of pro-inflammatory responses in cell-based systems. To best mimic the human in vivo conditions, we suggest the use of freshly isolated human leukocytes combined with an appropriate method for the chosen cytokine (e.g., IL-6 or TNF-α). The described protocols comprise isolation, stimulation, and viability test of the human leukocytes.

Key words LPS neutralization, LTA neutralization, Peptides, Peptidomimetics, Cytokines

1 Introduction

Continuous emergence and spreading of multidrug-resistant bacterial strains of human pathogens combined with the fact that only few novel classes of antibiotics or treatment strategies are in the pipeline for entering the clinic constitute major healthcare problems worldwide [1–3]. Especially in the case of sepsis, a critically exacerbated systemic host response to infection, the available treatment options remain very limited [4]. Unless the rise in antibiotic resistance can be reversed, the number of incurable infections and fatalities will keep increasing.

Alternatives to traditional antibiotics comprise host-directed approaches by which natural defense mechanisms in the host are augmented by intervention with non-antibiotic drugs. Endogenous human host defense (antimicrobial) peptides (HDPs) comprise the cathelicidin LL-37 and several defensins produced by

Paul R. Hansen (ed.), *Antimicrobial Peptides: Methods and Protocols*, Methods in Molecular Biology, vol. 1548,
DOI 10.1007/978-1-4939-6737-7_27, © Springer Science+Business Media LLC 2017

epithelial cells and innate immune cells in the mucosa [5, 6]. Besides direct antimicrobial effects, such HDPs exert strong triggering and/or modulation of host responses giving them an intriguing role in the regulation of innate immunity involved in combating infections [6–8]. Thus, the concept of modulation of the innate immune system by peptides is a well-known natural regulatory process that may be exploited as a non-antibiotic strategy for treatment of bacterial infections [9]. Synthetic peptides termed innate defense regulatory (IDR) peptides have already shown protective effects in animal models of infections [10]. Importantly, such therapies aimed at controlling fundamental innate host responses to bacterial infections constitute novel non-antibiotic treatment regimens that minimize the risk of resistance development due to the lack of a direct selection pressure. Therefore, immune modulation that enhances anti-infective responses (e.g., by promoting recruitment of immune-competent cells) while reducing harmful pro-inflammatory responses appears to be a promising potential treatment strategy that has been reviewed comprehensively elsewhere [11–13].

By nature, infection-triggered immune responses are the result of an immensely complex network of pattern recognition with ensuing signaling pathways that need to be tightly balanced in order to ensure effective neutralization of the invading microbes without collateral damage to host tissues. In fact, it is the exacerbated inflammatory responses that cause severe life-threatening infection-related effects with sepsis being a prime example. The option of partial/selective inhibition of the initial signaling pathways in the infection-related host response thus appears more beneficial than specific inhibition of the effector molecules in sepsis [14–16]. Hence, focus should be on means of preventing excessive induction of pro-inflammatory mediators due to bacteremia and high levels of bacterial components released during antibiotic treatment [17]. In this context the discussion will be limited to selected representatives of peptides and peptidomimetics that show potential for intervention.

The bacterial cell wall components lipopolysaccharide (LPS) and lipoteichoic acid (LTA) constitute important immunostimulating virulence factors in Gram-negative and Gram-positive infections, respectively [18, 19]. Also, other bacterial membrane components such as peptidoglycan and lipoproteins/peptides possess an ability to elicit innate immune responses [20, 21]. They are recognized by Toll-like receptors (TLRs) and other pattern-recognition receptors (PRRs) that are highly expressed on monocytes and other host defense cells. Upon activation PRRs initiate pro-inflammatory signaling that lead to release of cytokines, chemokines, complement factors, reactive oxygen species, and lipid mediators that are important for the anti-infective immune response but, if uncontrolled, may promote development of sepsis [22–24].

Despite some recent failures in using specific TLR4 antagonists as antisepsis agents, exploration of this treatment strategy is ongoing with other compounds under clinical investigation [12].

Several peptides/peptidomimetics capable of reducing undesired excessive pro-inflammatory effects have been identified. Such LPS- and/or LTA-neutralizing activity typically involves one of two general modes of actions: (1) a direct binding to the pro-inflammatory membrane constituents, whereby their ability to interact with the respective PRRs is eliminated, or (2) inhibitory interactions either directly on the PRRs or via interference with components in their signaling cascades. Here, we focus on the ability of peptides/peptidomimetics to inhibit the LPS/LTA-induced pro-inflammatory response, and thus the methods described do not identify their specific molecular mechanisms.

Studies concerning neutralization of LPS-induced excessive inflammation during infection with Gram-negative bacteria are abundant in literature as previously reviewed [25–27]. Thus, in the search for potential antisepsis drug leads for Gram-negative infections LPS-neutralizing peptides and peptidomimetics appear to possess promising properties [28–31], and in Table 1 several of the most potent peptides and peptidomimetics are listed [25, 32–44]. Bacterial killing during antibiotic treatment induces inflammation via initial binding of released LPS to LPS-binding protein (LBP) forming an LPS-LBP complex that interacts with CD14 to give an LPS-LBP-CD14 complex, which in turn activates TLR4-MD2 (MD2 being the myeloid differentiation factor 2) present on macrophages [27]. Thus, the murine macrophage cell line RAW 264.7 together with human mononuclear immune cells (hMNCs) have been extensively used as preliminary cellular test models in the identification of compounds with potential applications in therapies combining antibacterial activity with neutralization of LPS-mediated excessive inflammation also required for successful resolution of sepsis.

As evident from Table 1, so far quite different test conditions have been employed when estimating the efficacy of LPS-neutralizing compounds. To achieve more reliable comparison of novel and previously reported compounds, we recommend a higher degree of consensus regarding these assays as discussed below.

A major limitation in development of peptides for therapeutic use is their inherent susceptibility to degradation by proteases with ensuing reduced bioavailability. Thus, exploration of peptidomimetics, partly composed of unnatural residues that confer stability toward enzymatic processing, appears more promising. Designed peptidomimetics containing unnatural residues, e.g., β-amino acids, α-peptoid, or *N*-acylated residues, have proved capable of imitating the antibacterial activity of natural AMPs [45–47]. Notably, while cationic α-peptide/β-peptoid hybrid peptidomimetics display broad-spectrum antibacterial effects including anti-biofilm activity [48–50],

Table 1
Peptides and peptidomimetics with potent LPS-neutralizing properties

Peptides	Cell type	Cytokine	LPS conc. (ng/mL)	IC_{50} (μM)	Ref.
LKWLKKLLKKL-NH_2 (WALK11.3)	RAW 264.7	IL-6	100	<1	[32]
		TNF-α		<1	
RKCNFLCKLKEKLRTVITSHIDKVLRPQG	RAW 264.7	IL-6	20	~1–2.5	[33]
		TNF-α		~2.5–5	
igkkfkrivlrikkwlrkl[a] (a4-W2-E)	RAW 264.7	TNF-α	20	1–2.5	[34]
LLGDFFRKSKEKIGKEFKRIVQRIKDFLRNLVPRTES (LL-37)	RAW 264.7	TNF-α	20	~0.25	[35]
KWFRVYRGIYRRR-NH_2 (CDT)	RAW 264.7	TNF-α	100	0.1–1	[35]
GCKKYRRFRWKFKGKFWFWG (Pep19-2.5)	hMNCs	TNF-α	100	~0.01	[36]
Pep-Lys(Pep)-Lys(Pep)-Lys(Pep)-βAla	RAW 264.7	TNF-α	20	0.04–0.3	[37]
Pep = KKIRVRLSA					
Pam-IKISGKWKAQKRFLKM-NH_2 (C_{16}-BPI)	RAW 264.7	TNF-α	0.5	0.06	[38]
Lauryl-FQWQRNIRKVR-NH2 (Lau-LF11)	hMNCs	TNF-α	0.5	~0.01–0.03	[39]
LKLLKKLLKKLLKLL-NH_2 (amphipathic-L)	RAW 264.7	TNF-α	10	<0.5	[25]
Oct-KKllKLLlKlLK-NH_2 (C8-K5L7)	RAW 264.7	TNF-α	10	<0.5	[40]
Oct-KKllKKLkKlLK-NH_2 (C8-K7L5)	RAW 264.7	TNF-α	10	<0.5	[40]
IGKEFKRIVQRIKDFLRNLVPRTES-NH_2	Human leukocytes	IL-6	1 EU/mL	0.09	[41]
Peptidomimetics					
RH([$(CH_2)_7$Ph]$_2$)R-NH_2 (HDAMP-1)	RAW 264.7	TNF-α	20	4–8	[42]
α-AApeptide[b] 3	RAW 264.7	TNF-α	20	~3	[43]
γ-AApeptide YL-36	RAW 264.7	TNF-α	20	2.1	[44]
Pam-(Lys-βNSpe)$_6$-NH_2[c]	Human leukocytes	IL-6	1 EU/mL	0.06	[41]
Polymyxin B (PMB)	Human leukocytes	IL-6	1 EU/mL	0.04	[41]

[a]Non-capitalized letters indicate D-amino acids
[b]AA = *N*-acylated *N*-aminoethyl
[c]βNSpe denotes the β-peptoid residue derived from (*S*)-2-phenylethylamine

related lipidated subclasses possess potent immunomodulatory properties. Thus, we have identified members that exert LPS neutralization comparable to that found for polymyxin B [41]. Similarly, hydrophobic AA-peptides constitute an interesting novel class of peptidomimetics exhibiting dual antibacterial and immunomodulatory properties [42, 43]. Inspection of Table 1 reveals that lipidation of cationic peptides and peptidomimetics in several cases was crucial for obtaining high LPS-neutralizing activity [38, 40–44, 51]. Besides possessing increased stability, it is advantageous for immunomodulating peptidomimetics that their chemical synthesis is simple and that their direct antibacterial activity is negligible in order to avoid a selection pressure. This is indeed the case for Pam-(Lys-βNspe)$_6$-NH$_2$ for which immunomodulating activity is seen at concentrations 100-fold lower than the minimal inhibitory concentration [41].

By contrast, LTA neutralization has been investigated more sporadically, albeit a handful of examples involving proteins and peptides have been reported. Thus, a fragment of MD-2, apolipoprotein A-I, and the mannopeptimycin glycopeptide antibiotics all exhibited LTA binding [52–54], while gelsolin gave rise to partial inhibition of LTA-induced release of IL-8 from human neutrophils [55]. A structure–activity study has inferred that truncated analogs of LL-37 may be interesting leads for peptides displaying both LTA- and LPS-neutralizing properties [56]. Some AMPs and peptidomimetics have been found to bind directly to LTA [57–59], while other compounds neutralize the pro-inflammatory effects of

Table 2
Peptides and peptidomimetics displaying LTA-neutralizing properties

Peptides	Cell type	Cytokine or chemokine	LTA conc. (μg/mL)	IC$_{50}$ (μM)	Ref.
LL-37	Whole blood	IL-8	50	0.36	[56]
IGKEFKTIVERIKRFLRELVRPLP (P60.4)	Whole blood	IL-8	50	0.59	[56]
LL-37	RAW 264.7	TNF-α	20	~3[a]	[60]
Human β-defensin 3 (HBD-3)	THP-1	IL-8	5	<1.0	[61]
		TNF-α		<1.0	
Peptidomimetics					
SMAMPs 01 to 03	RAW 264.7	IL-6	10	~1.0	[62]
		TNF-α		<1.0	
Pam-(Lys-βNSpe)$_6$-NH$_2$	Human leukocytes	IL-6	1	0.85	[41]

[a]At higher conc. (ca. 15 μg/mL) potentiation of LTA was observed instead of inhibition

LTA as shown in Table 2. Interestingly, the anti-inflammatory activity of the cationic SMAMPs, based on aromatic scaffolds, proved specifically to neutralize the effects of LTA, while no effects on the responses to the TLR4 agonist LPS or even to other TLR2 agonists were observed [62].

Comparison of the potencies of LPS- and LTA-neutralizing compounds reported in different studies (e.g., as compiled in Tables 1 and 2) is complicated by the different type(s) and amount(s) of pyrogens used for stimulation. In particular, LPS from different bacterial species and even strains differ in their biological activity and potency [63]. Depending on the mode of action, the amount and type of LPS used may thus influence the apparent potency of an LPS-neutralizing compound. Using various types of LPS may serve other purposes (as discussed in Chapter 29), but with respect to determination of potency in LPS neutralization for a given compound, we recommend the use of ultrapure high-potency endotoxin standards from a certified supplier as they are well-defined in both chemical structure and biological activity. Furthermore, we propose that the amount(s) of LPS used should be measured as endotoxin units (EU) instead of by weight. EU was initially established by the FDA relative to the pyrogenic potential contained in 0.2 ng of the US Reference Standard Endotoxin Lot EC-2, evaluated by the rabbit pyrogen test, and it is now accepted as the standard unit for endotoxin [64]. Thus, EU represents the biological activity of LPS independently of the source and purity better than any weight unit, and general use of this unit will facilitate comparison of the data obtained in experimental settings used across studies. If, for some scientific reason, less well-characterized LPS is used, we suggest its pyrogenic potential to be compared to a reference standard for which the potency in endotoxin units is already known.

Likewise, the quality and source of LTA influence the reliability of the assay. The extraction method has been shown to be of major importance for the biological activity of the resulting LTA product. Previously, LTA from some commercial suppliers were found to be fractionated and/or contaminated with LPS and other pyrogens (reviewed in [65]). Furthermore, as for LPS, the bacterial strain and species as well as their culture conditions may affect the structure and thus the biological activity of the isolated LTA [65, 66]. Therefore, it is important to consider the LTA purification method as well as the possible degree of LPS contamination in the LTA used.

2 Materials and Preparations

All glassware, utensils, and liquids must be sterile and free of pyrogens (*see* **Note 1**).

2.1 Isolation of Human Primary Leukocytes from Venous Blood or Buffy Coats

1. Human venous blood or buffy coat (*see* **Note 2**).
2. Sterile endotoxin-free red blood cell (RBC) lysis buffer: 8.26 mg/mL NH_4Cl, 1 mg/mL, $KHCO_3$, 37 μg/mL EDTA.
3. Sterile endotoxin-free Hank's balanced salt solution (HBSS) without Ca^{2+} and Mg^{2+}.

2.2 Assay for LPS and LTA Neutralization

1. Complete culture medium: low endotoxin RPMI-1640 supplemented with 5 % heat-inactivated fetal bovine serum (FBS), 1% GlutaMax, 100 U/mL penicillin, and 100 μg/mL streptomycin.
2. 24- and 96-well sterile flat-bottomed plates for cell culture.
3. Sealing tape for 96-well plates.
4. Endotoxin-free water for intermediate dilutions of LPS and LTA.
5. Ultrapure high-potency endotoxin standard.
6. Endotoxin-free LTA.

2.3 Measurement of Pro-inflammatory Cytokine Content in Cell Culture Supernatants by DELFIA

1. White FluoroNunc™ MaxiSorp™ surface microtiter plates or other plates suitable for fluorometry.
2. Monoclonal capture antibodies recommended for immunoassays such as ELISA in phosphate buffer (0.2 M Na_2HPO_4 pH 8.0). Here we used anti-human interleukin (IL)-6 antibody and anti-human IL-8 antibody.
3. Plate shaker.
4. Single-use plate sealers.
5. Automated multi-well pipette or DELFIA® Columbus M8/R2 platewash, Wallac
6. Wash buffer: 0.05 % Tween 20 in phosphate-buffered saline pH 7.4.
7. Blocking buffer: 1 % bovine serum albumin (BSA), 5 % sucrose in phosphate buffer (0.2 M Na_2HPO_4 pH 8.0).
8. Cytokine standards: recombinant human IL-6 and recombinant human IL-8.
9. Biotinylated polyclonal detection antibody in Tris-buffered sodium chloride (20 mM Trizma base, 150 mM NaCl, pH 7.4) with 0.1 % BSA. Here we used goat anti-human IL-6 antibody and goat anti-human IL-8 antibody.
10. DELFIA® Eu-labeled streptavidin (100 ng/mL; 100 μL/well).
11. Assay buffer: 0.1 % BSA, 7.866 μg/mL diethylenetriaminepentaacetic acid, 0.2 % azide in phosphate-buffered saline pH 7.75
12. DELFIA® enhancement solution (100 μL/well).

13. SpectraMax M5 fluorescence plate reader.
14. Analysis software such as SoftMax Pro.

2.4 Measurement of Leukocyte Cytotoxicity by MTT Assay

1. Phosphate-buffered saline (PBS): 137 mM NaCl, 2.7 mM KCl, 10 mM Na_2HPO_4, 1.8 mM KH_2PO_4, pH = 7.4.
2. 5 mg/mL MTT (3-[4,5-dimethylthiazol-2-yl]-2,5 diphenyl tetrazolium bromide) solution in PBS.
3. Lysis buffer: 10% sodium dodecyl sulfate, 50% *N*,*N*-dimethylformamide, pH 4.5.
4. Single-use plate-sealing tape.
5. Multi-well plate reader capable of detecting absorbance at 590 and 635 nm.

3 Methods

To investigate biologically relevant levels of neutralization of the pro-inflammatory actions of LPS/LTA, the freshly isolated leukocytes are stimulated with LPS/LTA, and the effect of the test compound(s) on secretion of pro-inflammatory cytokines/chemokines is evaluated. Since a successful neutralization of LPS/LTA results in lowered cytokine/chemokine responses, which also might result from cytotoxicity, it is important also to address the effect of the test compound(s) on the viability of the leukocytes used.

It is very important to avoid contamination with pyrogens, and thus all work should be performed aseptically.

3.1 Isolation of Human Primary Leukocytes from Venous Blood or Buffy Coats

As an alternative to human primary leukocytes, a specific leukocyte fraction such as isolated human monocytes, an immunologically competent cell line such as the human monocyte cell line Mono Mac 6 [67], or the murine macrophage cell line RAW264.7 [68] may be used (as described in Chapter 29).

1. Prewarm HBSS and RBC lysis buffer to near 37 °C.
2. Mix one part buffy coat with four parts RBC lysis buffer in a conical tube, and gently turn the tube until lysis was visible.
3. Gently spin down the cells (7 min; 100 × *g*) and discard the supernatant.
4. Repeat the lysis by gently resuspending the pellet in RBC lysis buffer, spin down the cells (7 min; 100 × *g*), and discard the supernatant.
5. Wash the cells by resuspending the pellet in HBSS, spin down the cells (7 min; 100 × *g*), and discard the supernatant.
6. If the supernatant appears clear, proceed to **step** 7. If not, repeat the washing in **step 5**.
7. Resuspend the pellet in complete culture medium and then count the cells.

3.2 Initial Determination of Appropriate Pyrogen Doses

To be able to identify LPS/LTA-neutralizing compounds with both high and low potency, the final concentration of LPS and/or LTA used in neutralization studies should optimally induce a significant, but submaximal, cytokine response. Since the response to LPS and LTA is dependent on the source and purity of these substances as well as the source of cells, the choice of pyrogen dose should be based on preceding dose–response experiments using identical experimental conditions. When using primary cells isolated from human donors, the sensitivity toward pyrogens may also differ, and we therefore suggest that the dose–response investigations are made using cells from several different donors.

Figure 1 shows an example of LPS and LTA dose–response curves for leukocytes isolated from three different healthy donors. The leukocytes from the three different donors respond with different magnitudes in the release of IL-6; however, the dose–response curves display similar profiles.

3.2.1 Stimulation of Leukocytes

1. Resuspend the freshly isolated leukocytes in complete culture medium to a density of 2×10^6 cells/mL, and seed 300 μL/well in 24-well culture plates.
2. Prepare solutions of different LPS and/or LTA concentrations in complete culture medium (*see* **Note 3**); 300 μL is required for each sample. The concentration of LPS, LTA, and the compounds of interest in the solutions should be two times the intended final concentration.
3. Add 300 μL LPS/LTA solution or complete culture medium (unstimulated control) to the leukocytes to obtain a final leukocyte density of 1×10^6 cells/mL.
4. Incubate the leukocytes in a humidified atmosphere (5 % CO_2, 95 % air) at 37 °C. After the relevant incubation time (*see* Subheading 3.3), cell-free supernatants can be harvested by

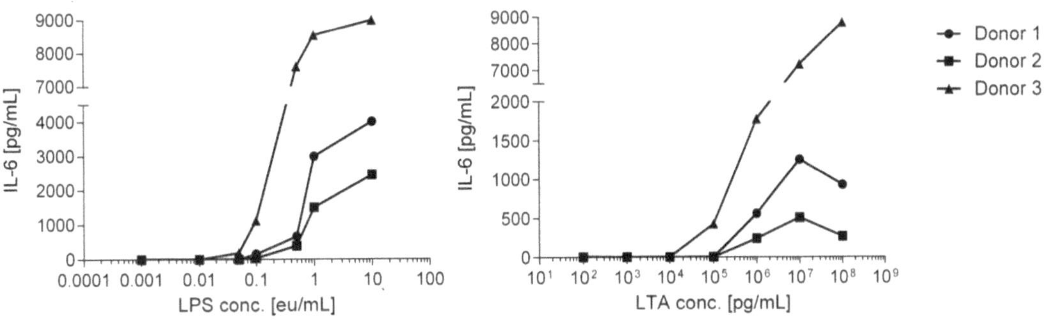

Fig. 1 Dose–response curves of freshly isolated leukocytes from three different donors stimulated with *E. coli* 055:B5 LPS (Lonza) and *S. aureus* ultrapure LTA (InVivogen). The IL-6 concentration in the culture medium was measured after 20 h incubation by time-resolved dissociation-enhanced lanthanide fluoroimmunoassay (DELFIA; a noncompetitive sandwich immunoassay) as previously described [43]

centrifugation (2000×g for 3 min). The supernatants can either be stored by freezing at −80 °C or analyzed for cytokine/chemokine content immediately (see below). In Fig. 1, the incubation time was 20 h.

3.2.2 Measurement of Pro-inflammatory Cytokine Content in Cell Culture Supernatants by DELFIA

Here we describe how to use time-resolved dissociation-enhanced lanthanide fluorescence immunoassay (DELFIA®) to measure the cytokine content in cell culture supernatants as previously described for IL-6 in [69]. DELFIA® is a noncompetitive sandwich solid-phase immunoassay like the TNF-α ELISA described in Chapter 29; however, in DELFIA® europium fluorescence is used for detection by time-resolved fluorometry instead of the colorimetric enzyme assay used in ELISAs, resulting in a wider determination range [69] (*see* **Note 4**).

1. Add 100 μL, 2 μg/mL capture antibody solution to each well in a white FluoroNunc™ MaxiSorp™ surface microtiter plate. Seal the plate using single-use sealing tape, and then incubate at room temperature in a plate shaker (950 horizontal shakes/min) for 30 min. Leave the plate at room temperature overnight without shaking.
2. Wash the wells with 3× 300 μL wash buffer using an automated multi-well pipette or a Columbus plate washer.
3. Block residual binding sites in the wells by adding 300 μL blocking buffer to each well. Seal the plate with single-use sealing tape, and incubate at room temperature in a plate shaker (950 horizontal shakes/min) for 60 min.
4. Repeat the washing procedure in **step 2**.
5. Prepare cytokine standard solutions of 0, 5, 10, 25, 50, 100, 250, 500, 1000, 2000, and 4000 pg/mL by dilution in complete culture medium.
6. Prepare dilutions of cell culture supernatants in complete culture medium. Here, samples for IL-6 determinations were used undiluted, whereas samples for IL-8 determinations were diluted 1:25.
7. Add 100 μL cytokine standard or sample to each well. All samples are run in triplicate. Seal the plate with single-use sealing tape, and then incubate 2 h at room temperature in a plate shaker (950 horizontal shakes/min).
8. Repeat the washing procedure in **step 2**.
9. Add 100 μL biotinylated detection antibody solution (50 ng/mL) to each well. Seal the plate with single-use sealing tape, and then incubate for 2 h at room temperature in a plate shaker (950 horizontal shakes/min).
10. Repeat the washing in **step 2**.

11. Add 100 μL europium-labeled streptavidin solution (100 ng/mL) to each well. Seal the plate with single-use sealing tape and incubate 30 min at room temperature in a plate shaker (950 horizontal shakes/min).
12. Wash the wells with 6 × 300 μL wash buffer using an automated multi-well pipette or a Columbus plate washer.
13. Add 100 μL DELFIA® enhancement solution to each well. Seal the plate with single-use sealing tape, and incubate 5–10 min at room temperature in a plate shaker (1150 horizontal shakes/min).
14. Measure fluorescence using the SpectraMax M5 fluorescence plate reader.

3.3 Initial Determination of Incubation Times

Based on the data depicted in Fig. 2, the sampling time in determination of IL-6 and IL-8 was chosen to be 20 h, whereas for TNF-α it was chosen to be 5 h (*see* **Note 5**).

1. Resuspend the freshly isolated leukocytes in complete culture medium to a density of 2×10^6 cells/mL, and seed 300 μL/well in 24-well culture plates. Make sure to prepare sufficient plates in order to withdraw an appropriate number of samples during the entire incubation time (*see* **steps 4** and **5**).
2. Prepare solutions of LPS and/or LTA (the concentrations should be based on dose–response experiments; *see* Subheading 3.2) in complete culture medium (*see* **Note 3**); 300 μL is required for each sample. The concentration of LPS, LTA, and the test compounds in the solutions should be two times the intended final concentration.
3. Add 300 μL LPS/LTA solution, or complete culture medium as unstimulated control to separate samples of leukocytes to obtain a final leukocyte density of 1×10^6 cells/mL.
4. Immediately withdraw three samples (one unstimulated, one LPS-stimulated, and one LTA-stimulated control) at time $(t) = 0$. Harvest the cell-free supernatant by centrifugation (2000 × *g* for 3 min), and store by freezing at −80 °C.
5. Incubate the leukocytes in a humidified atmosphere (5 % CO_2, 95 % air) at 37 °C, and withdraw samples at 30 min–1 h intervals for the first 30 h or as convenient using the same procedure as in **step 4**.
6. Analyze the frozen samples for cytokine content by DELFIA® (*see* Subheading 3.2.2).

3.4 LPS and LTA Neutralization: Cytokine and Viability Assays

As described previously, the assay for LPS and LTA neutralization consist of two parts: (1) leukocyte cytokine secretion assay and (2) leukocyte viability assay.

In Fig. 3 a screening of 20 compounds for LPS- and LTA-neutralizing activity is depicted (*see* **Note 6**).

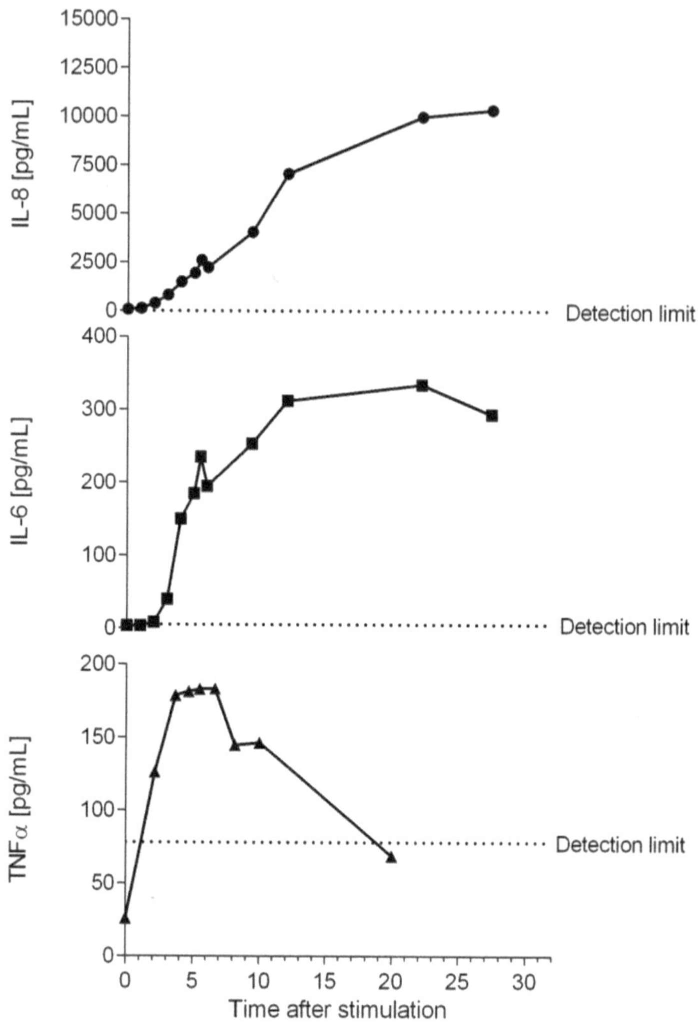

Fig. 2 Kinetics of cytokine/chemokine release from freshly isolated leukocytes from one representative donor after stimulation with 1 EU/mL *E. coli* 055:B5 LPS (Lonza). The concentrations of IL-6 and IL-8 were measured by DELFIA, and the concentration of TNF-α was determined by using DIAPLEX multiplex for flow cytometry (Diaclone) as previously described [41]

1. Resuspend the freshly isolated leukocytes in complete culture medium to a density of 2×10^6 cells/mL, and seed 300 μL/well in 24-well culture plates (for the neutralization assay) and 50 μL/well in 96-well culture plates (for the MTT cytotoxicity assay).
2. Incubate the leukocytes in a humidified atmosphere (5 % CO_2, 95 % air) at 37 °C until stimulation and addition of test compounds.
3. Prepare solutions of LPS, LTA, and of the test compounds in complete culture medium (*see* **Note 2**); 150 μL is required for each neutralization sample and 75 μL for cytotoxicity assays.

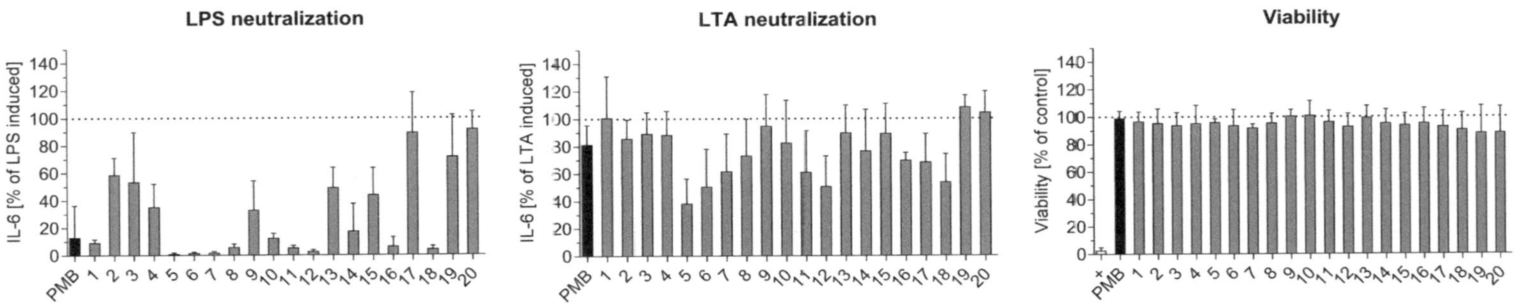

Fig. 3 LPS and LTA neutralization as well as viability screening of 20 peptides/peptidomimetics. The compounds were tested in a concentration of 1 μM in all assays. The bar graphs show mean + confidence intervals of six independent experiments using freshly isolated leukocytes from six different donors. PMB: polymyxin B, +: positive control for cytotoxicity by addition of lysis buffer

The concentration of LPS, LTA, and the test compounds in the prepared solutions should be four times the desired final concentration.

4. Add 150 μL LPS/LTA solution and 150 μL test compound solution to the leukocytes to obtain a final leukocyte density of 1×10^6 cells/mL. Remember to include control samples for which only complete culture medium, compound solution(s), or LPS/LTA alone are added. Also include a positive control compound. For positive LPS neutralization, polymyxin B is often used. For positive LTA neutralization, we suggest to choose one or more of the compounds listed in Table 2.
5. Incubate the leukocytes in a humidified atmosphere (5% CO_2, 95% air) at 37 °C until harvest.
6. After the relevant incubation time (determined as described in Subheading 3.3), cell-free supernatants are harvested by centrifugation (2000 × *g* for 3 min). The supernatants can either be stored by freezing at −80 °C or analyzed for cytokine/chemokine content immediately using DELFIA® (*see* Subheading 3.2.2).
7. Add 25 μL/well complete culture medium to all samples in the 96-well plate.
8. Add 25 μL/well of the test compound solutions. All samples should be assayed in triplicate. Remember to include positive and negative controls for which a cytotoxin (e.g., lysis buffer) or complete culture medium, respectively, is added instead of test compound solution. Also include a blank only containing culture medium. In case of autofluorescent test compounds, interference controls should also be included.
9. Add 20 μL/well of the MTT solution, and incubate the cells in humidified atmosphere (5% CO_2, 95% air) at 37 °C. Use the same incubation time in the cytotoxicity assay as in the cytokine secretion assay (*see* **steps 4–6**).
10. Add 100 μL/well lysis buffer, seal the plate to avoid vaporization of the sample, and incubate at 37 °C for 6 h or until the formazan crystals have been solubilized.
11. Read the plate at 570 and 635 nm. Calculate the percentage of viable cells using the formula in Chapter 29, Subheading 3.2.2.

4 Notes

1. To remove contaminating pyrogens including endotoxin, filter all "homemade" solutions through a 20 kDa cutoff filter immediately after preparation. Glassware and other laboratory utensils are rendered free of pyrogens by heating to 250 °C for a minimum of 30 min in accordance with the European Pharmacopeia.

2. The buffy coat is the fraction of an anticoagulated blood sample containing the majority of the leukocytes and platelets following a density gradient centrifugation, and it is a waste product in the purification of medical blood components. Therefore buffy coats can often be acquired for research purposes. Alternatively, a fresh anticoagulated blood sample may be used.
3. When diluting LPS and LTA, be aware that they are both amphiphiles and will adsorb to most types of glass and plastic. Therefore, vigorous vortexing of all stock solutions is of major importance for the final measurement. Also, solutions containing LPS and/or LTA should always be made immediately before use.
4. Several different quantitative methods for measuring cytokines and chemokines in cell-free supernatants exist. If only a few cytokines/chemokines are to be measured with high sensitivity, a solid-phase immunoassay is preferred as it often gives the widest determination range. However, several different assays for multiplex cytokine/chemokine measurements in a single small sample also exist.
5. Stimulation of immune-competent cells with LPS and LTA induces release of various pro-inflammatory cytokines and chemokines such as IL-6, TNF-α, and IL-8. The relevant incubation time of the cells with pyrogens depends on which cytokine(s) and/or chemokine(s) that should be measured due to variation in their release kinetics. Therefore, initial studies of the kinetics of cytokine/chemokine release upon stimulation with LPS and LTA should be performed. Figure 2 shows an example of an experiment investigating the kinetics of the release of IL-8, IL-6, and TNF-α from freshly isolated leukocytes stimulated with LPS. Optimally, the incubation time used should allow for sampling when the cytokine/chemokine content is high and relatively stable in order to minimize variation caused by inaccuracy in timing.
6. Several of the compounds depicted in Fig. 3 are likely hits for LPS neutralization as they inhibit LPS- and/or LTA-induced IL-6 secretion, but have no apparent effect on the viability of leukocytes in the tested concentration. To investigate these compounds further, their potency for LPS neutralization (IC_{50}) and LC_{50} should be determined by obtaining dose–response curves.

References

1. Wellington EMH, Boxell ABA, Cross P, Feil EJ, Gaze WH, Hawkey PM, Johnson-Rollings AS, Jones DL, Lee NM, Otten W (2013) The role of the natural environment in the emergence of antibiotic resistance in Gram-negative bacteria. Lancet Infect Dis 13:155–165
2. Theuretzbacher U (2013) Global antibacterial resistance: the never-ending story. J Glob Antimicrob Resist 1:63–69
3. Jabes D (2011) The antibiotic R&D pipeline: an update. Curr Opin Microbiol 14: 564–569

4. Uppu DS, Ghosh C, Haldar J (2015) Surviving sepsis in the era of antibiotic resistance: are there any alternative approaches to antibiotic therapy? Microb Pathog 80:7–13
5. Wang G (2014) Human antimicrobial peptides and proteins. Pharmaceuticals 7:545–594
6. Jarczak J, Kosciuczuk EM, Lisowski P, Strzalkowska N, Jozwik A, Horbanczuk J, Krzyzewski J, Bagnicka E (2013) Defensins: natural component of human innate immunity. Human Immunol 74:1069–1079
7. Scott MG, Davidson DJ, Gold MR, Bowdish D, Hancock REW (2002) The human antimicrobial peptide LL37 is a multifunctional modulator of innate immune responses. J Immunol 169:3883–3891
8. Hancock REW, Stahl H-G (2006) Antimicrobial and host-defense peptides as new anti-infective therapeutic strategies. Nat Biotechnol 24:1551–1557
9. Scott MG, Dullaghan E, Mookherjee N, Glavas N, Waldbrook M, Thompson A, Wang A, Lee K, Doria S, Hamill P, Yu JJ, Li Y, Domini O, Guarna MM, Finlay BB, North JR, Hancock REW (2007) An anti-infective peptide that selectively modulates innate immune responses. Nat Biotechnol 25:465–472
10. Nijnik A, Madera L, Ma S, Waldbrook M, Elliot MR, Easton D, Mayer ML, Mullaly SC, Kindrachuk J, Jenssen H, Hancock REW (2010) Synthetic cationic peptide IDR-1002 provides protection against bacterial infections through chemokine induction and enhanced leukocyte recruitment. J Immunol 184:2539–2550
11. Yeung ATY, Gellatly SL, Hancock REW (2011) Multifunctional cationic host defence peptides and their clinical applications. Cell Mol Life Sci 68:2161–2176
12. Hancock REW, Nijnik A, Philpott DJ (2012) Modulating immunity as a therapy for bacterial infections. Nat Rev Microbiol 10:243–254
13. Hilchie AL, Wuerth K, Hancock REW (2013) Immune modulation by multifaceted cationic host defense (antimicrobial) peptides. Nat Chem Biol 9:761–768
14. Cohen J (2009) Non-antibiotic strategies for sepsis. Clin Microbiol Infect 15:302–307
15. Alejandria MM, Lansang MA, Dans LF, Mantaring JB (2013) Intravenous immunoglobulin for treating sepsis, severe sepsis and septic shock. Cochrane Database Syst Rev (9):CD001090
16. Leitman IM (2011) Modulating the inflammatory response in sepsis. J Surg Res 171:e183–e185
17. van Langevelde P, Kwappenberg KM, Groeneveld PH, Mattie H, van Dissel JT (1998) Antibiotic-induced lipopolysaccharide (LPS) release from Salmonella typhi: delay between killing by ceftazidime and imipenem and release of LPS. Antimicrob Agents Chemother 42:739–743
18. Heumann D, Roger T (2002) Initial responses to endotoxins and Gram-negative bacteria. Clin Chim Acta 323:59–72
19. Ginsburg I (2002) Role of lipoteichoic acid in infection and inflammation. Lancet Infect Dis 2:171–179
20. Rockel C, Hartung T (2012) Systematic review of membrane components of Gram-positive bacteria responsible as pyrogens for inducing human monocyte/macrophage cytokine release. Front Pharmacol 3:56
21. de Tejada GM, Heinbockel L, Ferrer-Espada R, Heine H, Alexander C, Bárcena-Valera S, Goldmann T, Correa W, Wiesmüller K-H, Gisch N, Sánchez-Gómez S, Fukuoka S, Schürholz T, Gutsmann T, Brandenburg K (2015) Lipoproteins/peptides are sepsis-inducing toxins from bacteria that can be neutralized by synthetic anti-endotoxin peptides. Sci Rep 5:14292
22. Remick DG (2007) Pathophysiology of sepsis. Am J Pathol 170:1435–1444
23. Castellheim A, Brekke OL, Espevik T, Harboe M, Mollnes TE (2009) Innate immune responses to danger signals in systemic inflammatory response syndrome and sepsis. Scand J Immunol 69:479–491
24. Kimbrell MR, Warshakoon H, Cromer JR, Malladi S, Hood JD, Balakrishna R, Scholdberg TA, David SA (2008) Comparison of the immunostimulatory and proinflammatory activities of candidate Gram-positive endotoxins, lipoteichoic acid, peptidoglycan, and lipopeptides, in murine and human cells. Immunol Lett 118:132–141
25. Rosenfeld J, Papo N, Shai Y (2006) Endotoxin (lipopolysaccharide) neutralization by innate immunity host-defence peptides. Peptide properties and plausible modes of action. J Biol Chem 281:1636–1643
26. Pulido D, Nogués MV, Boix E, Torrent M (2012) Lipopolysaccharide neutralization by antimicrobial peptides: a gambit in the innate host defense strategy. J Innate Immun 4:327–336
27. Sun Y, Shang D (2015) Inhibitory effects of antimicrobial peptides on lipopolysaccharide-induced inflammation. Mediators Inflamm 2015:167572

28. Wong K-F, Luk JM (2009) Endotoxin-neutralizing peptides as Gram-negative sepsis therapeutics. Protein Pept Lett 16:539–542
29. Suzuki MM, Matsumoto M, Yamamoto A, Ochiai M, Horiuchi Y, Niwa M, Omi H, Kobayashi T, Takagi T (2010) Molecular design of LPS-binding peptides. J Microbiol Methods 83:153–155
30. Schuerholz T, Brandenburg K, Marx G (2012) Antimicrobial peptides and their potential application in inflammation and sepsis. Crit Care 16:207
31. Shukla P, Rao GM, Pandey G, Sharma S, Mittapeeli N, Shegokar R, Mishra PR (2014) Therapeutic interventions in sepsis: current and anticipated pharmacological agents. Br J Pharmacol 171:5011–5031
32. Shim D-W, Heo K-H, Kim Y-K, Sim E-J, Kang T-B, Choi J-W, Sim D-W, Cheong S-H, Lee S-H, Bang J-K, Won H-S, Lee K-H (2015) Anti-inflammatory action of an antimicrobial model peptide that suppresses the TRIF-dependent signaling pathway via inhibition of toll-like receptor 4 endocytosis in lipopolysaccharide-stimulated macrophages. PLoS One 10:e126871
33. Wei L, Yang J, He X, Mo G, Hong J, Yan X, Lin D, Lai R (2013) Structure and function of a potent lipopolysaccharide-binding antimicrobial and anti-inflammatory peptide. J Med Chem 56:3546–3556
34. Nan YH, Lee B-J, Sin SY (2012) Prokaryotic selectivity, anti-endotoxic activity and protease stability of diastereomeric and enantiomeric analogs of human antimicrobial peptide LL-37. Bull Korean Chem Soc 33:2883–2889
35. Saravanan R, Mohanram H, Joshi M, Domadia PN, Torres J, Ruedl C, Bhattacharjya S (2012) Structure, activity and interactions of the cysteine deleted analog of tachyplesin-1 with lipopolysaccharide micelles: mechanistic insights into outer-membrane permeabilization and endotoxin neutralization. Biochim Biophys Acta 1818:1613–1624
36. Kaconis Y, Kowalski I, Howe J, Brauser A, Richter W, Razquin-Olazarán I, Inigo-Pestana M, Garidel P, Rössle M, de Tejada GM, Gutsman T, Brandenburg K (2011) Biophysical mechanisms of endotoxin neutralization by cationic amphiphilic peptides. Biophys J 100:2652–2661
37. Pini A, Falciani C, Mantengoli E, Bindi S, Brunetti J, Iozzi S, Rossolini GM, Bracci L (2010) A novel tetrabranched antimicrobial peptide that neutralizes bacterial lipopolysaccharide and prevents septic shock *in vivo*. FASEB J 24:1015–1022
38. Mas-Moruno C, Cascales L, Cruz LJ, Mora P, Pérez-Payá E, Albericio F (2008) Nanostructure formation enhances the activity of LPS-neutralizing peptides. ChemMedChem 3:1748–1755
39. Andrä J, Lohner K, Blondelle SE, Jerala R, Moriyon I, Koch MHJ, Garidel P, Brandenburg K (2005) Enhancement of endotoxin neutralization by coupling of a C12-alkyl chain to a lactoferricin-derived peptide. Biochem J 385:135–143
40. Rosenfeld Y, Lev N, Shai Y (2010) Effect of the hydrophobicity to net positive charge ratio on antibacterial and anti-endotoxin activities of structurally similar antimicrobial peptides. Biochemistry 49:853–861
41. Skovbakke SL, Larsen CJ, Heegaard PMH, Moesby L, Franzyk H (2015) Lipidated α-peptide/β-peptoid hybrids with potent anti-inflammatory activity. J Med Chem 58:801–813
42. Murugan R, Jacob B, Ahn M, Hwang E, Sohn H, Park H-N, Lee E, Seo J-H, Cheong C, Nam K-Y, Hyun J-K, Jeong K-W, Kim Y, Shin SY, Bang JK (2013) *De novo* design and synthesis of ultra-short peptidomimetic antibiotics having dual antimicrobial and anti-inflammatory activities. PLoS One 8:e80025
43. Padhee S, Smith C, Wu H, Li Y, Manoj N, Qiao Q, Khan Z, Cao C, Yin H, Cai J (2014) The development of antimicrobial α-AApeptides that suppress proinflammatory immune responses. ChemBioChem 15:688–694
44. Li Y, Smith C, Wu H, Padhee S, Manoj N, Cardiello J, Qiao Q, Cao C, Yin H, Cai J (2014) Lipidated cyclic γ AApeptides display both antimicrobial and anti-inflammatory activity. ACS Chem Biol 9:211–217
45. Cheng RP, Gellman SH, DeGrado WF (2001) β-peptides: from structure to function. Chem Rev 101:3219–3233
46. Chongsiriwatana NP, Patch JA, Czyzewski AM, Dohm MT, Ivankin A, Gidalevitz D, Zuckermann RN, Barron AE (2008) Peptoids that mimic the structure, function, and mechanism of helical antimicrobial peptides. Proc Natl Acad Sci U S A 105:2794–2799
47. Schmitt MA, Weisblum B, Gellman SH (2007) Interplay among folding, sequence, and lipophilicity in the antibacterial and hemolytic activities of α/β-peptides. J Am Chem Soc 129:417–428
48. Olsen CA, Ziegler HL, Nielsen HM, Frimodt-Møller N, Jaroszewski JW, Franzyk H (2010) Antimicrobial, hemolytic, and cytotoxic activities of selective antimicrobial β-peptoid-peptide

hybrid oligomers. ChemBioChem 11: 1356–1360
49. Yang L, Knapp KM, Yang L, Molin S, Franzyk H, Folkesson A (2013) High *in vitro* antimicrobial activity of β-peptoid-peptide hybrid oligomers against planktonic and biofilm cultures of *Staphylococcus epidermidis*. Int J Antimicrob Agents 41:20–27
50. Jahnsen R, Frimodt-Møller N, Franzyk H (2012) Antimicrobial activity of peptidomimetics against multidrug-resistant Escherichia coli: a comparative study of different backbones. J Med Chem 55:7253–7261
51. Jahnsen RD, Haney EF, Franzyk H, Hancock REW (2013) Characterization of a proteolytically stable multifunctional host defense peptidomimetic. Chem Biol 20:1286–1295
52. Mancek M, Pristovsek P, Jerala R (2002) Identification of LPS-binding peptide fragment of MD-2, a toll-receptor accessory protein. Biochem Biophys Res Commun 292:880–885
53. Jiao YL, Wu MP (2008) Apolipoprotein A-I diminishes acute lung injury and sepsis in mice induced by lipoteichoic acid. Cytokine 43:83–87
54. Ruzin A, Singh G, Severin A, Yang Y, Dushin RG, Sutherland AG, Minnick A, Greenstein M, May MK, Shlaes DM, Bradford PA (2004) Mechanism of action of the mannopeptimycins, a novel class of glycopeptide antibiotics active against vancomycin-resistant Gram-positive bacteria. Antimicrob Agents Chemother 48:728–738
55. Bucki R, Byfield FJ, Kulakowska A, McCormick ME, Drozdowski W, Namiot Z, Hartung T, Janmey PA (2008) Extracellular gelsolin binds lipoteichoic acid and modulates cellular response to proinflammatory bacterial wall components. J Immunol 181:4936–4944
56. Nell MJ, Tjabringa GS, Wafelman AR, Verrijk R, Hiemstra PS, Drijfhout JW, Grote JJ (2006) Development of novel LL-37 derived antimicrobial peptides with LPS and LTA neutralizing and antimicrobial activities for therapeutic application. Peptides 27:649–660
57. Cao L, Dai C, Li Z, Fan Z, Song Y, Wu Y, Cao Z, Li W (2012) Antibacterial activity and mechanism of a scorpion venom peptide derivative in vitro and in vivo. PLoS One 7:e40135
58. Krusong K, Pollpipat P, Supungul P, Tassanakajon A (2012) A comparative study of antimicrobial properties of crustin*Pm*1 and crustin*Pm7* from the black tiger shrimp *Penaeus monodon*. Dev Comp Immunol 36:208–215
59. Epand RF, Sarig H, Mor A, Epand RM (2009) Cell-wall interactions and the selective bacteriostatic activity of a miniature oligo-acyl-lysyl. Biophys J 97:2250–2257
60. Ruan Y, Shen T, Wang Y, Hou M, Li J, Sun T (2013) Antimicrobial peptide LL-37 attenuates LTA induced inflammatory effect in macrophages. Int Immunopharmacol 15:575–580
61. Lee S-H, Baek D-H (2012) Antibacterial and neutralizing effect of human β-defensins on *Enterococcus faecalis* and *Enterococcus faecalis* lipoteichoic acid. J Endod 38:351–356
62. Som A, Navasa N, Percher A, Scott RW, Tew GN, Anguita J (2012) Identification of synthetic host defense peptide mimics that exert dual antimicrobial and anti-inflammatory activities. Clin Vacc Immunol 19:1784–1791
63. Rietschel ET, Kirikae T, Schade FU, Mamat U, Schmidt G, Loppnow H, Ulmer AJ, Zahringer U, Seydel U, Di Padova F (1994) Bacterial endotoxin: molecular relationships of structure to activity and function. FASEB J 8:217–225
64. Williams KL (2007) Endotoxins. In: Williams KL (ed) Pyrogens, LAL-testing and depyrogenation drugs and pharmaceutical sciences, vol 167, 3rd edn. CRC Press, Boca Raton, FL
65. Morath S, von Aulock S, Hartung T (2005) Structure/function relationships of lipoteichoic acids. J Endotoxin Res 11:348–356
66. Ryu YH, Baik JE, Yang JS, Kang SS, Im J, Yun CH, Kim DW, Lee K, Chung DK, Ju HR, Han SH (2009) Differential immunostimulatory effects of Gram-positive bacteria due to their lipoteichoic acids. Int Immunopharmacol 9:127–133
67. Ziegler-Heitbrock HW, Thiel E, Futterer A, Herzog V, Wirtz A, Riethmuller G (1988) Establishment of a human cell line (Mono Mac 6) with characteristics of mature monocytes. Int J Cancer 41:456–461
68. Ralph P, Nakoinz I (1977) Antibody-dependent killing of erythrocyte and tumor targets by macrophage-related cell lines: enhancement by PPD and LPS. J Immunol 119:950–954
69. Moesby L, Hansen EW, Christensen JD (1997) Pyrogen testing of lipid-based TPN using Mono Mac 6 monocyte cell line and DELFIA. J Clin Pharm Ther 22:327–333

Chapter 28

Protocols for Screening Antimicrobial Peptides That Influence Virulence Gene Expression in *Staphylococcus aureus*

Martin Saxtorph Bojer, Mara Baldry, and Hanne Ingmer

Abstract

Compounds that inhibit virulence gene expression in bacterial pathogens have received increasing interest as possible alternatives to the traditional antibiotic treatment of infections. For the human pathogen *Staphylococcus aureus*, we have developed two simple assays based on reporter gene fusions to central virulence genes that are easily applicable for screening various sources of natural and synthetic peptides for anti-virulence effects. The plate assay is qualitative but simultaneously assesses the effect of gradient concentrations of the investigated compound, whereas the liquid assay is quantitative and can be employed to address whether a compound is acting on the central quorum sensing regulatory system, *agr*, that controls a large number of virulence genes in *S. aureus*.

Key words *S. aureus*, Anti-virulence, *agr* quorum sensing, Compound screening, Peptides, β-Galactosidase/β-lactamase reporter genes, *hla*, RNAIII, *spa*

1 Introduction

Infections by the human pathogen *S. aureus* are becoming increasingly difficult to treat due to the spread of strains that are highly resistant to antibiotics. To address such challenges, alternative treatment strategies have been proposed including anti-virulence therapy where expression or activity of virulence factors is targeted rather than pathogen viability [1]. In *S. aureus*, virulence gene expression is in part controlled by the *agr* quorum sensing system. Agr is a two-component sensory signal transduction system that at high cell density, and in response to a secreted auto-inducing peptide (AIP), induces expression of extracellular virulence factors. In contrast, expression of cell surface-associated virulence factors is repressed by induction of *agr* [2]. Among the induced gene products are a number of toxins including α-hemolysin encoded by *hla* and the *agr* effector molecule RNAIII; whereas *spa*, encoding

Paul R. Hansen (ed.), *Antimicrobial Peptides: Methods and Protocols*, Methods in Molecular Biology, vol. 1548,
DOI 10.1007/978-1-4939-6737-7_28, © Springer Science+Business Media LLC 2017

Protein A, is repressed. With knowledge of this expression pattern, both induction and repression of *agr* can easily be monitored. Based on transcriptional fusions of *hla* and *spa* promoter regions to *lacZ* encoding β-galactosidase [3], a simple plate well assay was established [4]. Here, *S. aureus* cells carrying these reporter fusions are incorporated into the agar plates also containing the β-galactosidase substrate, 5-bromo-4-chloro-3-indolyl-β-d-galactopyranoside (X-gal), and compounds or solutions to be tested are applied to wells drilled in the agar plate (Fig. 1). This assay provides qualitative data on effects on virulence gene expression of both pure compounds and complex solutions, and the diffusion of test compounds from the well allows for a gradient of concentrations to be examined concomitantly [5, 6]. Among the compounds identified by this approach is solonamide B, a cyclodepsipeptide secreted from a marine bacterium that displays striking similarity to the AIP naturally produced by *S. aureus* [7]. Interestingly, strains of *S. aureus* can be divided into at least four *agr* classes with AIPs of one class only inducing expression of *agr*-controlled genes belonging to that same class while repressing *agr* expression of other classes. Importantly, solonamide B inhibited *agr* in strains representing all four *agr* classes [8]. To assess the quantitative effects on *agr* inhibition, a liquid assay was established where the RNAIII promoter (p3) is fused to the β-lactamase reporter gene and where induction of *agr* in the reporter strain relies on externally added AIP [8]. Furthermore, to address if interference is mediated by disruption of the interactions between the AgrC sensory histidine kinase and the AIP, the liquid assay was modified to also express a constitutively active AgrC [9]. Using this variant, a compound that interferes with interactions between AIP and AgrC will be ineffective at suppressing the RNAIII promoter activity, whereas if the interference occurs downstream of

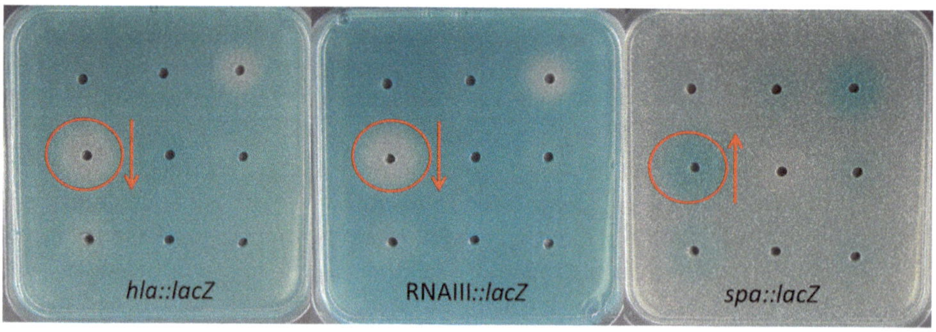

Fig. 1 Representation of the qualitative plate assay for screening compounds affecting virulence gene expression in *S. aureus*. The different virulence gene reporter strains are incorporated into the agar, and compounds of interest are added to individual wells. Following incubation the plates are visually inspected for any up- or downregulation of the genes. *Highlighted* is a compound that represses expression of *hla* and RNAIII while stimulating expression of *spa*

the *agr* regulatory cascade independently of AIP-AgrC interaction, the compound will inhibit RNAIII promoter activity. Using this assay it was demonstrated that solonamide B competitively interferes with AIP binding to AgrC [8]. The protocols described here are useful for testing any type of compound of interest and are even sensitive enough to identify small peptides with potent activity at micromolar concentrations.

2 Materials

All bacterial strains are routinely cultivated in tryptone soya broth (TSB) or on tryptone soya agar (TSA) at 37 °C and handled using aseptic techniques. *S. aureus* strain 8325-4 [10] and MOZ53 are used to generate cell-free AIP-I- and AIP III containing supernatants, respectively (*see* **Note 1**). *S. aureus* strain RN6911 (Δ*agr* derivative of 8325-4) [11] is used to generate AIP-free supernatant.

2.1 Qualitative ß-Galactosidase Reporter Plate Assay

1. *S. aureus* 8325-4-derived reporter strains PC322 (*hla::lacZ*) [3], PC203 (*spa::lacZ*) [3], and SH101F7 (RNAIII::lacZ) [12] are kept as frozen stocks in TSB with 15% glycerol at −80 °C.
2. 0.9% NaCl: Dissolve 9 g NaCl in 1 l dH_2O, autoclave, and keep at 4 °C.
3. Erythromycin stock solution: 50 mg/ml in 96% EtOH, stored at −20 °C.
4. 5-Bromo-4-chloro-3-indolyl β-d-galactopyranoside (X-gal) stored at −20 °C.
5. DMSO, molecular biology grade.
6. 70% EtOH.
7. Bunsen burner.
8. Sterile cork borer, diameter approximately 5 mm.
9. Spectrophotometer.
10. Peptides (or alternative compounds) of interest as preferred stocks.

2.2 Quantitative β-Lactamase Reporter Assay

1. *S. aureus* 8325-4-derived reporter strains RN10829 (P2-agrA; P3-blaZ)/p*agrC-I-WT* [8] and RN10829 (P2-agrA; P3-blaZ)/pEG11(*agrC-I-R238H*) [9] are kept as frozen stocks in TSB with 15% glycerol at −80 °C.
2. 0.1 M phosphate buffer: Add 0.31 g of $NaH_2PO_4 \times H_2O$ and 1.09 g of Na_2HPO_4 (anhydrous) to 100 ml dH_2O, adjust to pH 7.0 (HCl), filter sterilize (0.22 μm), and keep at 4 °C.
3. Chloramphenicol stock solution: 50 mg/ml in 96% EtOH, stored at −20 °C.

4. Nitrocefin stock solution: 500 μg/ml or 1 mM, reconstituted or prepared as recommended by supplier. Generally stable for weeks when stored protected from light at −20 °C.
5. Spectrophotometer.
6. Spectrophotometric plate reader with temperature control (37 °C), automatic shaking function, capable of reading at wavelength 486 nm.
7. Peptides (or alternative compounds) of interest as preferred stocks.

3 Methods

3.1 Qualitative β-Galactosidase Reporter Plate Assay

1. Streak strains 8325-4 and MOZ53 onto TSA and *hla*, *spa*, and RNAIII reporter strains onto TSA with erythromycin (5 μg/ml). Incubate overnight at 37 °C.
2. Inoculate each strain from single colonies into 5 ml TSB (reporter strains supplemented with 5 μg/ml erythromycin) in 15 ml test tubes, and incubate overnight (16 h) in shaking incubator at 37 °C at 225 rpm.
3. Spin down cultures of strains 8325-4 and MOZ53 (10,000 × *g*, 5 min), and pass the supernatants through 0.22 μm sterile filters. Keep sterile AIP-I- and AIP-III-containing supernatants at 4 °C (stable for at least 1 week).
4. Melt TSA (200 ml) (*see* **Note 2**) and leave to cool to 45 °C in a water bath (approximately 45 min).
5. Prepare a 10 fold dilution of the reporter strains in 0.9% NaCl (1 to 9 ml), measure OD_{600}, and adjust individual samples to $OD_{600} = 0.35$. Dilute cells another 100 fold by two tenfold dilutions (1 to 9 ml). Transfer 800 μl of the final cell suspensions to 9 cm Petri dishes marked with respective reporter strain designations.
6. Add 20 μl erythromycin stock to 200 ml preconditioned TSA (5 μg/ml final concentration). Dissolve 30 mg X-gal in 50 μl DMSO, and add to the medium while swirling (150 μg/ml final concentration). Mix thoroughly (*see* **Note 3**). Pour 20 ml of the medium into each Petri dish containing reporter strains. Swirl dishes softly to mix bacterial cells with the agar until just before solidification begins. Leave plates to solidify for 15 min, and dry plates (lid open) for approximately 45 min in an LAF bench.
7. Drill holes (up to 7, *see* **Note 4**) in each plate with a sterile cork borer and remove agar plugs. Use 70% EtOH and Bunsen burner to disinfect cork borer between plates containing different reporter strains.

8. Add 20 μl of peptide solution (to individual wells) (*see* **Note 5**). Include appropriate solvent controls and AIP-containing supernatants as preferred. Allow samples to diffuse into the agar.
9. Incubate plates at 37 °C until blue color development. Incubation time can vary between 9 and 48 h (*see* **Note 6**). Inspect plates visually for any up- and/or downregulation conferred by the samples on the different reporters (cf. Fig. 1) (*see* **Note 7**).

3.2 Quantitative β-Lactamase Reporter Assay

1. Prepare AIP-I containing and AIP-free supernatants from strains 8325-4 and RN6911, respectively, as described in Subheading 3.1.
2. Streak RN10829 reporters expressing wild-type AgrC or constitutively active AgrC(R238H) onto TSA with chloramphenicol (10 μg/ml). Incubate overnight at 37 °C.
3. Inoculate each strain from single colonies into 5 ml TSB (supplemented with 10 μg/ml chloramphenicol) in 15 ml test tubes, and incubate overnight (16 h) in shaking incubator at 37 °C at 225 rpm (*see* **Note 8**).
4. Dilute reporter strains 100× in fresh TSB (culture volume depending on number of assays to be performed), and incubate at 37 °C at 225 rpm to reach an OD_{600} of 0.4–0.5 (approximately 2 h). Note exact OD (time zero), withdraw 200 μl samples, and transfer to a 96-well microtiter plate preconditioned to −80 °C and freeze immediately at −80 °C (*see* **Note 9**).
5. Distribute 1.8 ml of the wild-type AgrC reporter into $X+2$ 15 ml test tubes, and then add 200 μl (10%) AIP-free supernatant to one culture (uninduced) and 200 μl AIP-I supernatant to the remaining (one induced and X for the X number of entities to be tested; include solvent control when relevant). Distribute the constitutively active AgrC(R238H) reporter in volumes of 2 ml into $1+X$ 15 ml test tubes. Add chemical entities to X individual cultures in desired concentrations (*see* **Note 10**), and incubate at 37 °C at 225 rpm.
6. Withdraw 200 μl samples from each culture after 30 and 60 min, and freeze in microtiter plate immediately (−80 °C). In parallel, take out 0.5 ml, mix with 0.5 ml TSB, and record OD_{600} of respective cultures (*see* **Note 11**).
7. Thaw all samples at room temperature and aliquot 50 μl volumes of each into three wells of a clear, flat-bottomed polystyrene microtiter plate (technical replicates) (*see* **Note 12**).
8. Prepare a fivefold dilution of the nitrocefin stock solution in phosphate buffer (volume depending of number of samples to be analyzed). Add 50 μl of diluted nitrocefin to each sample well using a multipipette (0.1 mM final concentration).

9. Immediately, record nitrocefin conversion in a plate reader preset to read at 486 nm every 20 s for 20 min (shaking in between readings) at 37 °C (*see* **Note 13**).
10. Analyze data by computing arbitrary β-lactamase activities of individual cultures by estimating V_{max} (slope of the conversion curves, ΔOD_{486}/time) followed by normalization of respective values to the cell density (OD_{600}) of the sample. Evaluate degree of inhibition conferred by the different compounds relative to the untreated controls (*see* **Note 14**).

4 Notes

1. Essentially any *agr*-proficient strains of known *agr* types can be used. The effect of an AIP-I supernatant may not be evident in the β-galactosidase reporter plate assay as the reporter strain endogenously produces AIP-I. The *agr* cross-inhibitory effect of an AIP-III supernatant is observed to varying degree between assays and critically depends on the incubation time.
2. 200 ml will be sufficient to prepare nine plates, i.e., three with each reporter. Adjust the protocol to the number of plates needed.
3. Local precipitation of X-gal/DMSO may occur. The substrate is added while swirling the medium flask to prevent excessive precipitation. Continued swirling for up to several minutes, while avoiding generation of air bubbles, may facilitate homogenization.
4. Adjust the number of holes to the number of samples to be analyzed. Generally, seven wells will fit into 9 cm Petri dishes. Alternatively, 12 × 12 cm squared Petri dishes can be used, which will accommodate up to 16 wells, if a large number of samples are to be analyzed. In that case, the protocol should be adjusted to use 2 ml of reporter cell suspension and 50 ml agar per plate.
5. Testing different dilutions of the same compounds will allow an evaluation of the dose dependence of any *agr*-modulatory effect. Furthermore, it will facilitate inference whether the effect occurs at concentrations near or far below the growth inhibitory concentration of the compound.
6. The incubation time may vary between assays depending, e.g., on the exact cell density of the reporter strains. Generally, the RNAIII reporter turns blue within the first 9 h of incubation, while the *hla* reporter is slightly delayed. The *spa* reporter requires 24 h (sometimes even up to 48 h) of incubation for upregulation to become apparent. For all reporters, incubation at 4 °C for another 24 h may improve color contrasts.

7. Results from the reporter plates may be supported by subsequent assays of liquid cultures by quantifying β-galactosidase activity using ONPG (ortho-Nitrophenyl-β-galactoside) on cell lysates.
8. It is recommended to include biological replicates (e.g., three) of each reporter. In that case, and if testing several different compounds, the wild-type AgrC and the constitutive AgrC(R238H) reporter assays may be performed separately to limit the number of cultures to be handled simultaneously.
9. The time zero sample is not essential for the wild-type reporter since it is uninduced at this time point and, hence, has negligible β-lactamase activity. For the constitutively active reporter, however, it is important to know the β-lactamase activity at time zero to be able to infer correctly the effect of the test compounds on the activity accumulated after addition.
10. Prior characterization of compounds of interest regarding growth inhibition is recommended to be able to evaluate if any effect on the reporter gene is related or unrelated to such activity.
11. Choose sample time points as preferred (e.g., 20, 40, and 60 min). Minimize culture volumes to spare compounds that are costly or scarce. Withdraw smaller volumes for OD measurement if necessary.
12. For samples taken at late time points, which are high in β-lactamase activity, it may be necessary to dilute the test volume (e.g., using only 25 μl sample and 25 μl phosphate buffer).
13. The reading protocol may be adjusted according to the instrument as long as the data allows the generation of a conversion curve.
14. Compounds that provide significant downregulation of both reporter strains are expected to act downstream of the agr regulatory cascade/independent of the AIP-AgrC interaction, whereas inhibition observed only for the wild-type AgrC reporter is suggestive of direct interference with the AIP-AgrC interaction. In the latter case, conclusions may be substantiated by experiments on the wild-type reporter using different AIP concentrations (e.g., 5, 13, and 20%) in which a correlation between the AIP concentration and the relative inhibition should be apparent.

References

1. Cegelski L, Marshall GR, Eldridge GR, Hultgren SJ (2008) The biology and future prospects of antivirulence therapies. Nat Rev Microbiol 6:17–27
2. Novick RP, Geisinger E (2008) Quorum sensing in staphylococci. Annu Rev Genet 42:541–564
3. Chan PF, Foster SJ (1998) The role of environmental factors in the regulation of virulence-determinant expression in *Staphylococcus aureus* 8325-4. Microbiology 144:2469–2479
4. Nielsen A, Nielsen KF, Frees D, Larsen TO, Ingmer H (2010) Method for screening

compounds that influence virulence gene expression in *Staphylococcus aureus*. Antimicrob Agents Chemother 54:509–512

5. Nielsen A, Mansson M, Wietz M, Varming AN, Phipps RK, Larsen TO, Gram L, Ingmer H (2012) Nigribactin, a novel siderophore from Vibrio nigripulchritudo, modulates *Staphylococcus aureus* virulence gene expression. Mar Drugs 10:2584–2595
6. Kjaerulff L, Nielsen A, Mansson M, Gram L, Larsen TO, Ingmer H, Gotfredsen CH (2013) Identification of four new agr quorum sensing-interfering cyclodepsipeptides from a marine Photobacterium. Mar Drugs 11:5051–5062
7. Mansson M, Nielsen A, Kjærulff L, Gotfredsen CH, Wietz M, Ingmer H, Gram L, Larsen TO (2011) Inhibition of virulence gene expression in *Staphylococcus aureus* by novel depsipeptides from a marine photobacterium. Mar Drugs 9:2537–2552
8. Nielsen A, Månsson M, Bojer MS, Gram L, Larsen TO, Novick RP, Frees D, Frøkiær H, Ingmer H (2014) Solonamide B inhibits quorum sensing and reduces *Staphylococcus aureus* mediated killing of human neutrophils. PLoS One 9:e84992
9. Geisinger E, Muir TW, Novick RP (2009) *agr* receptor mutants reveal distinct modes of inhibition by staphylococcal autoinducing peptides. Proc Natl Acad Sci U S A 106:1216–1221
10. Novick RP, Morse SI (1967) In vivo transmission of drug resistance factors between strains of *Staphylococcus aureus*. J Exp Med 125:45–59
11. Novick RP, Ross HF, Projan SJ, Kornblum J, Kreiswirth B, Moghazeh S (1993) Synthesis of staphylococcal virulence factors is controlled by a regulatory RNA molecule. EMBO J 12:3967–3975
12. Horsburgh MJ, Aish JL, White IJ, Shaw L, Lithgow JK, Foster SJ (2002) sigmaB modulates virulence determinant expression and stress resistance: characterization of a functional *rsbU* strain derived from *Staphylococcus aureus* 8325-4. J Bacteriol 184:5457–5467

Chapter 29

Methods for In Vitro Analysis of Antimicrobial Activity and Toxicity of Anti-keratitis Peptides: Bacterial Viability in Tears, MTT, and TNF-α Release Assays

Floriana Cappiello, Bruno Casciaro, Satya Sree Kolar, Hasna Baidouri, Alison M. McDermott, and Maria Luisa Mangoni

Abstract

Ease of access to the cornea makes antimicrobial peptides (AMPs) ideal candidates for topical drug application. However, before bringing them to the clinic, it is fundamental to evaluate in vitro: (1) the ability of AMPs to kill bacteria in the presence of human tears, by counting the number of surviving bacteria on agar plates; (2) the potential cytotoxicity of AMPs to mammalian cells by a colorimetric method based on the production of a colored formazan crystals by metabolically active cells; and (3) the ability of AMPs to neutralize the toxic effect of the bacterial cell wall component, lipopolysaccharide (LPS), by measuring the level of the pro-inflammatory cytokine, TNF-α, released from LPS-activated macrophages, using a sandwich enzyme-linked immunosorbent assay.

Key words Colony counts, Tear collection, Colorimetric MTT assay, Sandwich immunosorbent assay, LPS detoxification

1 Introduction

Microbial keratitis is a serious vision threatening disease characterized by deterioration of the cornea (the transparent part covering the pupil and iris) with rapid progression to inflammation, necrosis, and potentially vision loss, if not treated [1, 2]. One of the major risk factors for bacterial keratitis is contact lens wear, particularly overnight use of disposable lenses and/or extended wear modalities [3, 4]. *Pseudomonas aeruginosa* is the most common bacterial pathogen causing microbial keratitis related to contact lens wear [5]. The increasing emergence of microorganisms resistant to the available ophthalmic agents has led to a significant need for novel approaches to prevent and treat ocular infections. Antimicrobial peptides (AMPs) hold promise for the development of new antibiotics. Importantly, in order to be effective at the

Paul R. Hansen (ed.), *Antimicrobial Peptides: Methods and Protocols*, Methods in Molecular Biology, vol. 1548,
DOI 10.1007/978-1-4939-6737-7_29,

ocular surface, AMPs must be active at high salt concentration, as well as in the presence of tear film components. The tear film is the liquid layer bathing the cornea and conjunctiva and that provides lubrication, protection, and nutrients to the cornea. It contains a lipid layer in contact with the air that prevents tears from evaporating, a mucus layer that coats the corneal surface and an aqueous layer rich in salts and proteins [6]. Importantly, to be used as new therapeutics, AMPs have to be harmless to corneal epithelial cells. In addition, it is worthwhile mentioning that during bacterial division, but mostly upon antibiotic treatment, lipopolysaccharide (LPS) molecules (the most abundant components of the outer membrane in Gram-negative bacteria) are released in the surroundings [7, 8]. LPS can activate immune cells, e.g., macrophages, triggering the intracellular signaling pathway which controls the secretion of pro-inflammatory cytokines, i.e., TNF-α, whose high levels can induce serious tissue damage [9]. Ease of access to the cornea makes it an ideal candidate for topical drug application, and there have been a small number of studies investigating the efficacy of AMPs in animal models of microbial keratitis (*see* Chap. 30). However, before bringing AMPs to the clinic, it is fundamental to evaluate in vitro: (1) their ability to kill bacteria in the presence of human tears, (2) their potential cytotoxicity to mammalian cells, and (3) their ability to neutralize the toxic effect of LPS.

A brief description of the methods for in vitro studies of a potential anti-keratitis AMP, according to the aforementioned aims, is reported below.

1. With reference to the antimicrobial effect in tears, counting surviving cells in terms of colony-forming units (CFU) on plates remains the most accurate and reliable standard method [10, 11].
2. The toxic effect of AMPs on mammalian cells, in this case corneal epithelial cells, is generally investigated by the colorimetric 3(4,5-dimethylthiazol-2yl)2,5-diphenyltetrazolium bromide (MTT) assay in which the intensity of the dye is proportional to the number of viable cells. MTT is a tetrazolium salt which is reduced to a colored formazan product [12] by mitochondrial reductases which are functional only in metabolically active cells. Acidified isopropanol is then added to dissolve the insoluble purple formazan product into a colored solution. The absorbance of this colored solution can be quantified by measuring the sample at a certain wavelength (usually between 500 and 600 nm) by a spectrophotometer. MTT assays are usually performed in the dark since the MTT reagent is light sensitive [13, 14].
3. LPS detoxification can be studied by evaluating the peptide's ability to inhibit the extracellular release of TNF-α from LPS-activated immune cells, e.g., macrophages, which are the most

abundant circulating immune cells. The level of human TNF-α in samples can be determined by the sandwich enzyme-linked immunosorbent assay (ELISA), which is designed to measure the amount of the target-bound molecule (i.e., TNF-α) between a matched antibody pair. Briefly, a target-specific antibody, called the capture antibody, is pre-coated in the wells of a microplate. Samples, standards, or controls are then added into wells to bind to the immobilized capture antibody (usually a polyclonal antibody). The sandwich is formed by the addition of a second (detector) biotinylated antibody, which will recognize a different epitope of the target molecule from the capture antibody. Avidin horseradish peroxidase (HRP) is subsequently added. After incubation and washing steps to remove unbound substances, a substrate (3,3′,5,5′-tetramethylbenzidine, TMB, solution) that reacts with the enzyme-antibody-target complex is added to produce a blue color in proportion to the concentration of antigen (TNF-α) present in the sample. Finally, the stop solution changes the reaction color from blue to yellow, and absorbance of the well is read with a microplate reader at 450 and 570 nm.

2 Materials

Prepare all materials at room temperature (R.T.), and diligently follow all waste regulations when disposing waste material.

2.1 Antimicrobial Activity in the Presence of Tears

2.1.1 Collection of Human Basal Tears

1. 5 μl microcapillary glass tubes.
2. Eppendorf microcentrifuge tubes (200 μl capacity).
3. Ice.

2.1.2 Bacterial Culture

1. Luria-Bertani (LB) broth: 1% (w:v) bacto tryptone, 0.5% (w:v) yeast extract, and 0.5% (w:v) NaCl in distilled water; the final pH has to be adjusted to 7.4 with 1 N sodium hydroxide; using a graduated cylinder, dispense aliquots into glass bottles, autoclave them at 120 °C for 20 min, and store the sterilized medium at R.T.
2. LB agar Petri dishes: add 1.5% (w:v) agar to freshly prepared LB medium, autoclave at 120 °C for 20 min, and cool the medium to between 45 and 50 °C prior to pouring the plates to minimize the amount of condensation that forms; the thickness of the agar medium in the plates should be around 0.5 cm, which can be achieved by pouring 15–20 ml of medium into 100 mm dish plates; store plates at +4 °C.
3. 50 ml BD Falcon polypropylene tubes.

4. 10 μl sterile inoculating loops.
5. Phosphate buffer, pH 7.4 (PB): stock solution contains 8.2 mM Na_2HPO_4 and 1.8 mM KH_2PO_4 in water, sterilize the buffer solution by filtration (0.22 μm nitrocellulose filters) in a biological safety cabinet, and store at R.T.
6. Spectrophotometer.
7. 37 °C incubator and a centrifuge for 50 ml tubes.

2.1.3 Antimicrobial Assay

1. LB agar Petri dishes.
2. Pipette/micropipettes with disposable tips.
3. 200 μl microcentrifuge tubes.
4. Eppendorf Thermomixer Comfort; 37 °C incubator and a colony counter.

2.2 MTT Assay to Evaluate Toxicity of AMPs on Corneal Epithelial Cells

2.2.1 Cell Culture, Passaging, and Growth Medium

1. Telomerase immortalized human corneal epithelial cell line (hTCEpi) [15].
2. Keratinocyte growth medium-2 (KGM-2) bullet kit (CC-3103 and CC-4152). The kit includes 500 ml basal medium, plus supplements and growth factors.
3. 50 mg/ml normocin: use 1 ml/500 ml media, and prepare the growth media by mixing the 500 ml basal medium with the supplies, growth factors, and normocin.
4. Tissue culture-treated T25 and T75 flasks (25 cm^2 and 75 cm^2, respectively).
5. 5% CO_2 incubator for cell cultures.
6. TrypLE Express phenol red (*see* **Note 1**).
7. Phosphate-buffered saline without calcium and magnesium chloride (CMF-PBS): stock solution (10×) is prepared by mixing 1.37 M NaCl, 27 mM KCl, 100 mM Na_2HPO_4, and 18 mM KH_2PO_4 in distilled water; adjust pH to 7.4 with HCl if necessary, and autoclave at 120 °C for 20 min before storage at R.T.; working buffer is prepared by diluting one part of the stock solution with nine parts of distilled water; store at R.T.

2.2.2 Cytotoxicity Assay

1. Hank's balanced salt solution (HBSS): 136 mM NaCl; 0.34 mM Na_2HPO_4; 0.44 mM KH_2PO_4; 5.4 mM KCl; 4.1 mM $NaHCO_3$, pH 7.2, supplemented with 5.5 mM D-glucose (*see* **Note 2**).
2. MTT stock solution: dissolve 5 mg/ml in sterile HBSS, store single-use aliquots at −20 °C in dark or foil-covered bottles (the compound is light sensitive), and prepare fresh working solutions at the desired concentration (0.5 mg/ml) by dilution with HBSS (*see* **Note 3**).
3. Stop solution: acidified isopropanol by adding 0.04 N HCl.

4. Sterile 96-well polystyrene flat bottom and tissue culture-treated transparent plates.
5. 0.02 % benzalkonium chloride (BAC).
6. Pipette/micropipettes with disposable tips.
7. A microplate reader capable of measuring absorbance at 590 and 635 nm.

2.3 TNF-α Assay

2.3.1 LPS and Peptide Solution

1. Prepare LPS solution by solubilizing LPS powder (e.g., LPS from *P. aeruginosa* serotype 10), in sterile water at a final concentration of 5 mg/ml. It is recommended to prepare aliquots and store them at −20 °C. Dilute LPS to the desired concentration before usage.
2. Peptide powder is dissolved in sterile water at 2 mM (stock solution) and stored at −20 °C.

2.3.2 Macrophage Growth and Passaging

1. RAW 264.7 murine macrophage cell line.
2. Dulbecco's modified high glucose Eagle's medium (DMEM) supplemented with non-essential amino acids, sodium pyruvate (1 mM final concentration); 2 mM glutamine; 10 % heat-inactivated fetal bovine serum (FBS) and antibiotics (0.1 mg/ml of penicillin and streptomycin).
3. T25 tissue culture-treated flasks.
4. 5 % CO_2 incubator.
5. CMF-PBS (1×).
6. Cell scrapers, 1.8 cm blade.
7. Inverted optical microscope.
8. Neubauer or Burker chamber.
9. Cover glasses.
10. Pipette/micropipette with disposable tips.
11. 50 ml BD Falcon polypropylene tubes.

2.3.3 Neutralization of LPS

1. Sterile 96-well flat bottom and cell culture-treated transparent plates.
2. Multichannel pipette to measure volumes ranging from 20 to 200 μl.

2.3.4 ELISA Procedure

1. eBioscience mouse TNF-α enzyme-linked immunosorbent assay kit.
2. Microplate reader capable of measuring absorbance at 450 and 570 nm.
3. Adjustable pipettes to measure volumes ranging from 2 μl to 1 ml.

4. Phosphate-buffered saline, with $MgCl_2$ and $CaCl_2$ (PBS) sterile filtered.
5. Wash buffer (PBS + 0.05 % Tween-20, pH 7.4).
6. Stop solution: 2 N H_2SO_4 or 1 M H_3PO_4.
7. Eppendorf microcentrifuge tubes to prepare standard dilutions.
8. Plate sealer: parafilm.
9. Absorbent paper.
10. Multichannel pipette to measure values ranging from 50 to 200 μl.

3 Methods

3.1 In Vitro Antimicrobial Activity in the Presence of Human Tears

The antimicrobial activity of the peptide is studied in the presence of 50 or 70 % (v:v) human basal tears at 37 °C by a rapid, sensitive, and reliable method based on the reduction in the number of CFU compared to that of peptide-untreated samples.

3.1.1 Collection of Human Basal Tears

Obtain approval for human subject's research from the relevant regulatory body. When subject has been informed of the study and given consent, collect basal tears as described below and store them on ice during the collection process. If desired reflex tears can be collected by using an appropriate stimulus such as a cotton swab gently inserted in to the nose or onion vapors.

1. Make sure that the subjects are sitting in a comfortable position with the back of their head against a headrest.
2. The investigator (i.e., person collecting the tears) can be standing or sitting in order to collect the tears.
3. Place one capillary tube into the black bulb/holder, and collect tears by holding the collection tube perpendicular to the ocular surface.
4. Gently pull down the lower lid and collect tears from the fornix (*see* **Note 4** and Fig. 1). This method, when performed properly, rarely induces reflex tearing (*see* **Note 5**).
5. Place your finger over the hole on the back of the plunger, and expel the tears into a sterile 200 μl collection tube (previously put on ice).
6. Repeat until sufficient tears have been collected. Allow the subjects short breaks in which time they can close their lids and roll their eyes. Multiple samples from the same subject or samples from multiple subjects may be pooled as necessary.
7. Store the basal tears at −80 °C until usage.

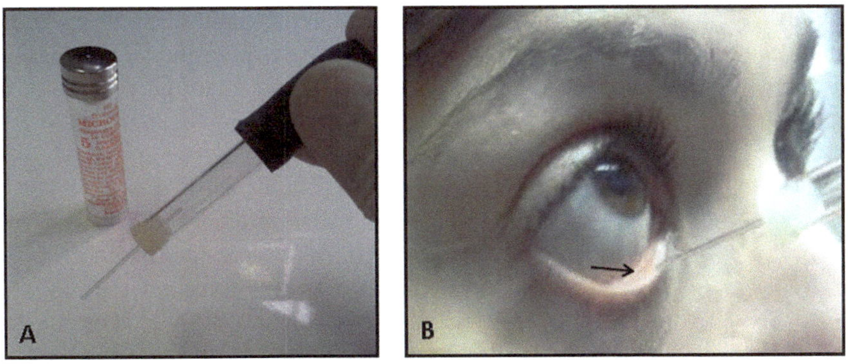

Fig. 1 Collection of basal tears. Obtain tears by inserting a 5 µl glass capillary tube (**a**) into the fornix (indicated by the arrow), in a perpendicular orientation to the ocular surface (**b**)

3.1.2 Bacterial Culture

1. Bacterial growth is started from a frozen glycerol stock. Under a biological safety cabinet class II, open the stock tube, scrape off a portion from the top of the frozen glycerol stock with a 10 µl sterile loop, and streak it onto an LB agar plate.
2. In the case of *P. aeruginosa*, the plates are incubated at 37 °C for 24 h under aerobic conditions.
3. A single bacterial colony is picked from the streak plate and then inoculated into 10 ml LB broth, previously put into a 50 ml polypropylene Falcon tube, and grown at 37 °C (use an incubator with shaking at 150 rpm), until an absorbance A_{590} 0.8 is reached (*see* **Note 6**).
4. Bacteria are harvested by centrifugation at 3000 × *g* for 10 min; the supernatant is discarded, and the pellet is washed twice with PB and resuspended in the same buffer to an optical density of approximately 1×10^7 CFU/ml.

3.1.3 Antimicrobial Assay

1. Reaction mixtures containing 50 or 70% (v:v) human basal tears are incubated with 1×10^5 CFU of *P. aeruginosa* in PB (final volume 50 µl) in microcentrifuge Eppendorf tubes (200 µl capacity), where the peptide is added at the desired concentration. Control samples, without peptide and/or tears, are also included.
2. Samples are incubated in the thermomixer, with shaking at 600 rpm, at 37 °C for different times (e.g., 30, 90, and 120 min).
3. At the corresponding time intervals, 5 µl aliquots of peptide-treated samples are withdrawn and plated on LB agar plates using 6 ml of soft agar (*see* **Notes 7** and **8**). For the other samples, a 1:10 dilution in PB is made before plating. The plates are incubated at 37 °C overnight, for counting and plotting the data (*see* **Notes 9** and **10**).
4. The results are expressed as percentage of CFU compared to the control (peptide-untreated cells).

3.2 MTT Assay to Evaluate Toxicity of AMPs on Corneal Epithelial Cells

The colorimetric measurement of viable cells using the chromophore MTT allows an automated quantification of viable cells at low cost without requiring direct reading of the soluble final product.

3.2.1 Cell Culture and Passaging

The hTCEpi cells are cultured in either T25 or T75 tissue culture flasks using KGM-2 medium supplemented with growth factors and normocin (100 μg/ml). They are incubated at 37 °C in 5% CO_2. If they do not reach 80–90% confluence in 3 days, change medium every day. When cells reach 80–90% confluence, detach cells from the flask as follows:

1. Aspirate the medium from the flask and discard it into a bottle under a biological safety cabinet class II. Rinse the cells with CMF-PBS and gently rock the vessel back and forth. Decant CMF-PBS.
2. Add an appropriate volume of pre-warmed TrypLE Express to the flask (i.e., 1 ml in a T25 flask, 3 ml in a T75 flask).
3. Gently rock the vessel, allowing the solution to coat the cells completely.
4. Incubate at 37 °C until the cells are visibly detached (observe at 5–8 min intervals).
5. Add an appropriate volume of complete growth medium, resuspend cells, and pipette repeatedly to break up any clumps that may be present.
6. Aspirate 10 μl with a micropipette tip for counting using a Neubauer or Burker chamber (*see* **Note 11**).

3.2.2 Cytotoxicity Assay

1. Seed 10,000 hTCEpi cells/well in a 96-well plate, and incubate the plate at 37 °C and 5% CO_2 for 48 h to allow the cells to attach to the plate and spread.
2. Stimulate cells with the peptide dissolved in medium at the desired concentration (50 μl final volume), in triplicate. Incubate the plate for 24 h.
3. Add 50 μl of 0.02% BAC into three wells for 15 min as a positive control.
4. Add 50 μl medium into the other wells (now total volume is 100 μl).
5. Add 10 μl MTT solution (5 mg/ml) into each well.
6. Incubate the plate at 37 °C for 3 h until purple formazan crystals are visible by light microscopy. An example of MTT assay is reported in Fig. 2.
7. Stop the reaction by adding 100 μl of acidified isopropanol to solubilize the formazan crystals. Use a multichannel pipette (*see* **Note 12**).

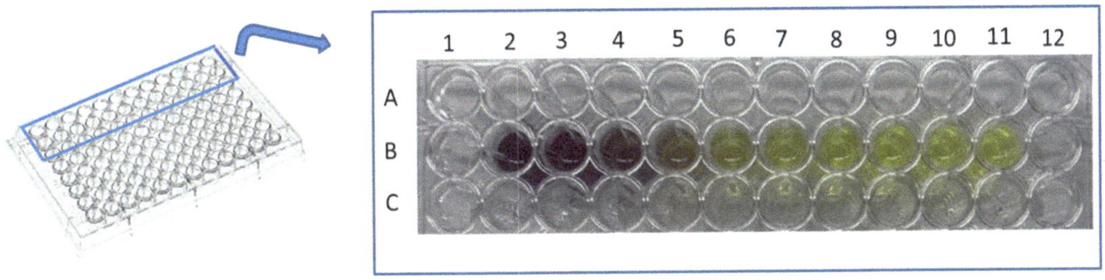

Fig. 2 Example of viability assay using the colored indicator MTT. The intensity of the purple dye (formazan) is proportional to the number of viable cells. The transparent *yellow* color in the B11 well corresponds to MTT solution without cells

8. Pipette up and down to dissolve the crystals.
9. Gently pop the bubbles using a 10 μl tip.
10. Read the plate using a multi-well reader at 590 and 635 nm as reference.
11. Subtract reference absorbance (Abs) value (635 nm) from 590 nm absorbance value and plot graph.

   Calculate the percentage of viable cells, according to the formula:

$$\frac{\text{Abs sample} - \text{Abs blank}}{\text{Abs control} - \text{Abs blank}} \times 100$$

   where the blank is given by sample without cells and not treated with the peptide.

3.3 TNF-α Assay

The level of TNF-α released in the extracellular medium upon stimulation of immune cells with LPS can be measured using murine macrophages and a mouse TNF-α enzyme linked immunosorbent assay kit according to the manufacturer's protocol. Cells that are stimulated with LPS alone and untreated cells serve as controls. All experiments should be performed in triplicate.

3.3.1 LPS and Peptide Solution

1. Dilute the peptide (from a stock solution of 2 mM in water) you wish to test in DMEM supplemented with 2 mM glutamine, 1 mM nonessential amino acids, 1 mM sodium pyruvate, and 10% heat-inactivated FBS without antibiotic (DMEMg).
2. Since each experiment is performed in triplicate on a 96-well plate, it is recommended to prepare 350 μl mixture containing DMEMg plus peptide (at the desired final concentration) and LPS (10 ng/ml).
3. In parallel, prepare a mixture without LPS.

3.3.2 Macrophage Growth and Passaging

1. Grow murine RAW macrophages in their complete culture medium in T25 flasks as described in Subheading 2.3.2.

2. Once cells reach confluence, remove the medium with a 10 ml pipette, and discard it into a bottle under the biological safety cabinet class II.
3. Wash three times with 4 ml CMF-PBS. The first two washes are very quick; for the last wash, leave the buffer for 3 min, before removing it.
4. Add 3 ml of DMEMg and detach cells with an appropriate scraper.
5. Transfer the cell suspension in to a 50 ml BD Falcon tube, and take out a small amount (10 μl) for counting in a Neubauer or Burker chamber under a microscope (*see* **Note 11**).
6. Dilute the cells to 1×10^5/ml in DMEMg.

3.3.3 Neutralization of LPS

1. One day before running the assay, seed 1×10^5 RAW 264.7 macrophages (suspended in 100 μl of DMEMg) in a 96-well plate, and incubate the plate at 37 °C and 5% CO_2.
2. The following day remove the medium using a multichannel pipette, wash each well with 100 μl of DMEMg, and replace the medium with 100 μl of the peptide mixture containing or not LPS. One hundred μl of medium (without LPS or peptide) are added to the other wells as control.
3. Incubate the plate for 4 h at 37 °C and 5% CO_2.
4. Aspirate the samples for each treatment and keep at −20 °C before ELISA test (*see* **Note 13**).
5. Perform ELISA test according to the manufacture's manual.

3.3.4 ELISA Procedure

1. Coat all wells of a 96-well plate with 100 μl of capture antibody previously diluted in coating buffer provided with the ELISA kit. Seal the plate and incubate it overnight (16–18 h) at 4 °C.
2. Aspirate and wash the plate for ~1 min with wash buffer (250 μl per well). Repeat three times. Be careful not to touch (with the plastic tip) the adjacent wells during washing and pipetting to avoid contamination.
3. To remove any residual buffer, turn the plate upside down on absorbent paper.
4. To block nonspecific binding and reduce background, add 200 μl of Assay Diluent 1× (diluted with 4 parts water) per well, seal the plate, and incubate it at R.T. for 1 h.
5. While the plate is being blocked, prepare the appropriate samples dilution (if necessary) and standards.
6. Wash the plate as in **steps 2** and **3**.
7. Using Assay Diluent 1×, dilute TNF-α standards and add 100 μl/well (at the highest concentration) in triplicate to wells A1, A2, and A3. Add 100 μl of half TNF-α concentration to the second row (B1, B2, and B3) and so on until row G (*see* **Notes 13** and

14). Do not add TNF-α to H row. Wells H1, H2, and H3 contain only media (Assay Diluent 1×) and are used as blank control. A schematic representation is shown in Fig. 3.

8. Add 100 μl/well of your samples to the appropriate wells. Seal the plate with parafilm to prevent evaporation and incubate overnight at 4 °C or at R.T. for 2 h.
9. Wash and blot the plate four times as in **steps 2** and **3**.
10. Add 100 μl/well of detection antibody diluted in Assay Diluent (as suggested by the manufacturer's procedure).
11. Seal the plate and incubate at R.T. for 1 h.
12. Wash and blot the plate four times as in **steps 2** and **3**.
13. Add 100 μl/well of avidin-HRP diluted in Assay Diluent (as suggested by the manufacturer's manual). Seal the plate and incubate at R.T. for 30 min.
14. Wash plate seven times. In this wash step, soak wells in wash buffer for 1–2 min before aspiration to minimize background.
15. Add 100 μl/well of TMB (substrate solution) to each well in the dark (use a multichannel pipette).
16. Incubate the plate at R.T. for 15 min in the dark. Positive wells should turn blue in color. It is not necessary to seal the plate during this step.
17. Add 50 μl of stop solution to each well. Positive wells should turn from blue to yellow.
18. Read absorbance at 450 and 570 nm (*see* **Notes 15** and **16**). A schematic representation is reported in Fig. 4.

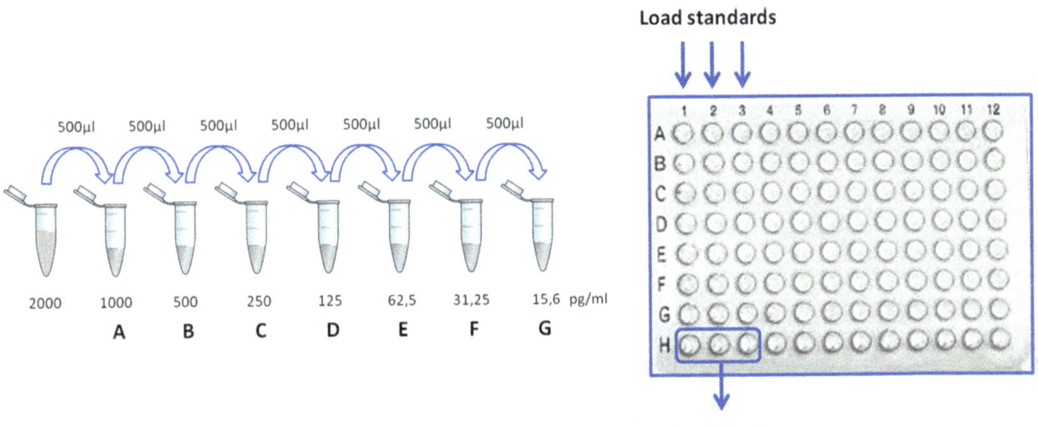

Fig. 3 Representation of TNF-α standard dilutions and their loading in a 96-well plate. Put 500 μl of Assay Diluent 1× in seven Eppendorf tubes, from A to G, and perform twofold serial dilutions by transferring 500 μl from the top standard (2000 pg/ml) to tube A, then from A to B, and so on until G. Transfer 100 μl of diluted standards in the corresponding wells (in triplicate) and load Assay Diluent 1× only in three wells H

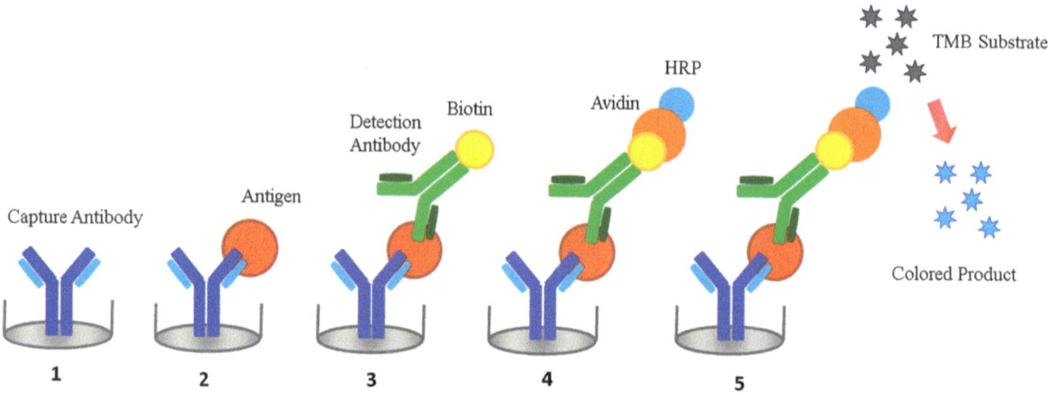

Fig. 4 Schematic representation of a sandwich enzyme-linked immunosorbent assay (ELISA). The plate is coated with a suitable capture antibody (1). Then the sample is added and the antigen present is bound to the capture antibody (2). A suitable biotin-labeled detection antibody (BDAb) which binds to the antigen is added (3). Avidin-HRP binds BDAb (4). Finally, TMB substrate is added and converted to a detectable form

4 Notes

1. TrypLE Express 10× is designed for the detachment of cells with strong adhesive properties to plastic surfaces and can be used at 10× concentration or properly diluted depending on the cell type. TrypLE Express can be substituted by trypsin-EDTA, but make sure to use media containing 10% FBS to stop the trypsin reaction.
2. If HBSS is made from scratch, sterilize it in a biological safety cabinet by filtering through a sterile filtration device fitted with 0.22 μm nitrocellulose filter. Store at +4 °C.
3. Be careful because MTT is toxic. Weigh the compound in the dark in a conical tube and solubilize it in HBSS by shaking with a vortex mixer. Sterilize it in a biological safety cabin by filtering through a sterile filtration device fitted with a 0.22 μm nitrocellulose filter and wrap the tube in aluminum foil. Fresh working solution can be stored at +4 °C for a few days.
4. The tear film can be gently collected also from the lateral canthus, while the patient can blink normally. The subject needs to look superior nasal (away from the collection site) and tilt the head toward the investigator.
5. Reflex tears are produced in response to stimulation and irritation of the cornea and conjunctiva or in response to emotions. Reflex tears are not simply a more diluted version of basal tears because the concentration of some components such as lysozyme is higher, whereas some interleukins but mainly mucins are more concentrated in basal tears than reflex.

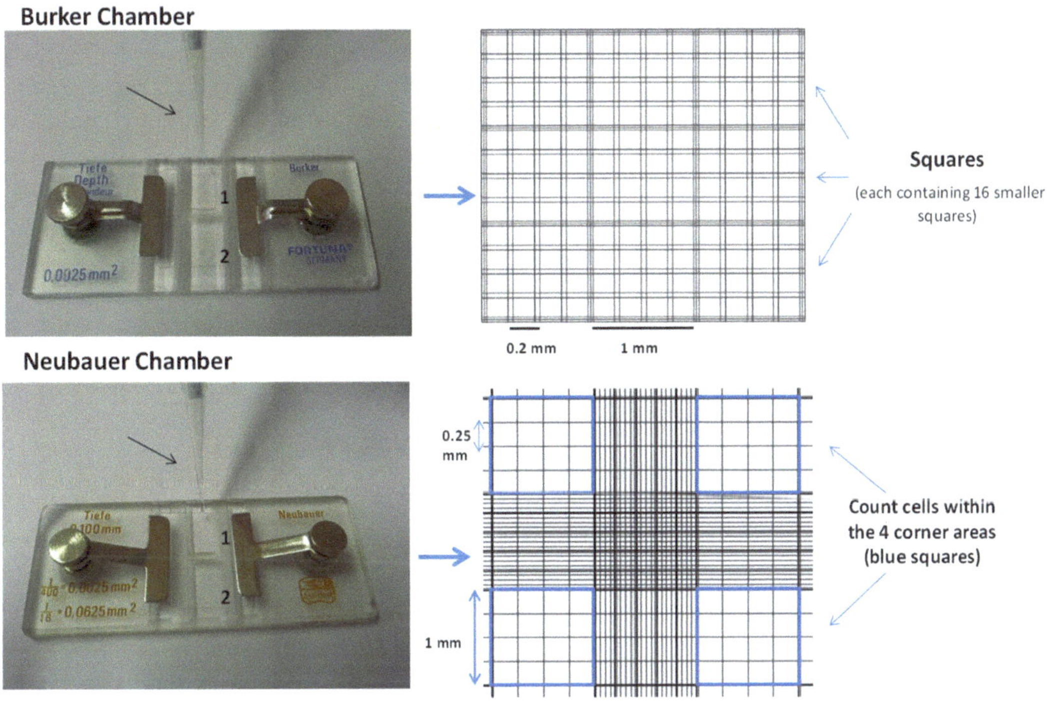

Fig. 5 Representation of a Burker and a Neubauer chamber. *Left side*: using a micropipette tip (indicated by the *arrow*), inject 10 μl of cell suspension under the cover glass (previously put over the chamber). It is recommended to count both sides (1 and 2) of the chamber. *Right side*: typical grids of Burker and Neubauer chambers, showing squares, each of which is 1 mm long and contains 16 smaller squares

6. It is important to optimize the assay with your particular strain. For example, in the case of *P. aeruginosa*, an optical density of 0.8 at 590 nm corresponds to 4×10^8 CFU/ml.
7. Soft agar differs from LB agar by having 0.75 % (w/v) agar concentration.
8. 5 μl aliquots can be added into 6 ml of soft agar (previously put into 15 ml centrifuge tubes) and then spread on top of the agar plate. Leave the plate drying at R.T. for 10–20 min; then, invert the plate and incubate as desired.
9. When enumerating CFU, a proper number of countable bacterial cells are between 50 and 300 CFU per plate. Typically, a tenfold dilution series is prepared from the original sample, using a suitable diluent, such as PBS.
10. Incubation in a closed humidified incubator will help to avoid problems with plates drying out especially when working with slow-growing colonies.
11. Place the pipette tip close to the glass cover edge, previously put above the Neubauer or Burker chamber (*see* Fig. 5); release the plunger slowly watching how the liquid enters the chamber uniformly, being absorbed by capillarity. In case of bubble

formation, rinse the chamber and repeat the loading process. If you have a very high concentration of cells, make a dilution. Start counting the cells in two big squares (each one contains 16 small squares, bounded by triple lines in Burker chamber or multiple lines in Neubauer chamber) (*see* Fig. 5). Repeat the operation for the other side of the chamber (Fig. 5). The equation used when counting cells in the big squares will be: concentration (cells/ml) = number of cells × 10,000/number of squares. In the case of a dilution, the concentration will be divided by the dilution applied.

12. Note that full solubilization may take a few hours. Incubation of the microplate at 37 °C and soft agitation for 1 h speed up the process.
13. If necessary, centrifuge cell culture supernatants to remove debris prior to analysis. After reconstitution of the lyophilized standard with Assay Diluent 1X, aliquot it into propylene tubes and store at −70 °C for up to 1 month. Avoid repeated freeze/thaw cycles.
14. Prepare TNF-α solution at a concentration of 2000 pg/ml, by dissolving the standard TNF-α provided by the kit in 10 ml of Assay Diluent 1X. Six twofold serial dilutions are recommended.
15. Read the plate within several minutes after adding the stop solution, not days/hours later. Once you have determined measurements at both 450 and 570 nm, subtract the values of 570 nm from those of 450 nm and plot the data (it reduces your background).
16. Plot the standard curve on log-log axis graph paper with TNF-α concentration on the x-axis and absorbance on the y-axis. Draw the best fit line through the standard points. Once you have determined the correct curve-fitting algorithm, you can transform the absorbance values to pg/ml. For example, if the standard curve is a linear line with the following equation $y = 0.98x + 0.032$ where y is the OD and x is the concentration of TNF-α, the unknown x values of your samples can be calculated from the corresponding OD measurements as follows: $x = (\text{OD value} - 0.0325)/0.98$. If the sample's absorbance is outside the standard curve range, the sample needs to be reanalyzed at a higher or lower dilution, as appropriate.

Acknowledgments

This work was supported grants from Sapienza Università di Roma (MLM) and NIH grant EY13175 (AMM). Part of the content of this work is object of a US patent application No. 14/506,383.

References

1. Keay L, Edwards K, Naduvilath T, Taylor HR, Snibson GR, Forde K, Stapleton F (2006) Microbial keratitis predisposing factors and morbidity. Ophthalmology 113:109–116
2. Thomas PA, Geraldine P (2007) Infectious keratitis. Curr Opin Infect Dis 20:129–141
3. Hsu HY, Nacke R, Song JC, Yoo SH, Alfonso EC, Israel HA (2010) Community opinions in the management of corneal ulcers and ophthalmic antibiotics: a survey of 4 states. Eye Contact Lens 36:195–200
4. Matthews TD, Frazer DG, Minassian DC, Radford CF, Dart JK (1992) Risks of keratitis and patterns of use with disposable contact lenses. Arch Ophthalmol 110:1559–1562
5. Evans DJ, Fleiszig SM (2013) Why does the healthy cornea resist *Pseudomonas aeruginosa* infection? Am J Ophthalmol 155:961–970.e2
6. Butovich IA (2013) Tear film lipids. Exp Eye Res 117:4–27
7. Rietschel ET, Kirikae T, Schade FU, Mamat U, Schmidt G, Loppnow H et al (1994) Bacterial endotoxin: molecular relationships of structure to activity and function. FASEB J 8:217–225
8. Trent MS, Stead CM, Tran AX, Hankins JV (2006) Diversity of endotoxin and its impact on pathogenesis. J Endotoxin Res 12:205–223
9. Rosenfeld Y, Shai Y (2006) Lipopolysaccharide (Endotoxin)-host defense antibacterial peptides interactions: role in bacterial resistance and prevention of sepsis. Biochim Biophys Acta 1758:1513–1522
10. Jett BD, Hatter KL, Huycke MM, Gilmore MS (1997) Simplified agar plate method for quantifying viable bacteria. Biotechniques 23:648–650
11. Koch AC (1994) Growth measurement. In: Gerhardt P, Murray RGE, Wood WA, Krieg NR (eds) Methods for general and molecular bacteriology. ASM Press, Washington, DC, pp 254–257
12. Kiderlen AF, Kaye PM (1990) A modified colorimetric assay of macrophage activation for intracellular cytotoxicity against Leishmania parasites. J Immunol Methods 127:11–18
13. van de Loosdrecht AA, Beelen RH, Ossenkoppele GJ, Broekhoven MG, Langenhuijsen MM (1994) A tetrazolium-based colorimetric MTT assay to quantitate human monocyte mediated cytotoxicity against leukemic cells from cell lines and patients with acute myeloid leukemia. J Immunol Methods 174:311–320
14. Gerlier D, Thomasset N (1986) Use of MTT colorimetric assay to measure cell activation. J Immunol Methods 94:57–63
15. Robertson DM, Li L, Fisher S, Pearce VP, Shay JW, Wright WE, Cavanagh HD, Jester JV (2005) Characterization of growth and differentiation in a telomerase-immortalized human corneal epithelial cell line. Invest Ophthalmol Vis Sci 46:470–478

Chapter 30

Methods for In Vivo/Ex Vivo Analysis of Antimicrobial Peptides in Bacterial Keratitis: siRNA Knockdown, Colony Counts, Myeloperoxidase, Immunostaining, and RT-PCR Assays

Satya Sree Kolar, Hasna Baidouri, Maria Luisa Mangoni, and Alison M. McDermott

Abstract

Antimicrobial peptides (AMPs) are essential components of the innate immune response. They have direct killing ability as well as immunomodulatory functions. Here, we describe techniques to identify specific AMPs involved in the protection against microbial keratitis, a vision threatening infection of the cornea of the eye which is the most serious complication of contact lens wear. Specifically we detail the use of siRNA technology to temporarily knockdown AMP expression at the murine ocular surface in vivo and then describe ex vivo assays to determine the level of bacteria, relative number of neutrophils, and levels of cytokines, chemokines, and AMPs in infected corneas.

Key words Antimicrobial peptide, Inflammation, Keratitis, Neutrophil, *Pseudomonas*

1 Introduction

Antimicrobial peptides (AMPs) are small, generally cationic peptides that can kill a range of microorganisms. These molecules are also known to modulate mammalian cell behavior and can, for example, influence the immune response and wound healing [1, 2]. Several groups of AMPs have been identified in mammals including humans, with the primary ones being defensins and cathelicidins [3, 4]. It is important to understand how AMPs function in vivo and to gain a deeper insight in to their potential as novel pharmaceutical agents to treat and prevent infection as well as modulate processes such as wound healing.

Here we describe techniques that can be utilized to study the role of endogenously expressed AMPs and efficacy of exogenously applied AMPs in a murine model of microbial keratitis. The latter refers to infection of the cornea of the eye, the transparent covering

Paul R. Hansen (ed.), *Antimicrobial Peptides: Methods and Protocols*, Methods in Molecular Biology, vol. 1548, DOI 10.1007/978-1-4939-6737-7_30,

of the pupil and iris, and it may lead to blindness or even enucleation if not treated quickly or if unresponsive to treatment. Microbial keratitis is the most serious complication of contact lens wear but may also occur after trauma to the eye [5–8]. Common agents include the Gram-negative bacterium *Pseudomonas aeruginosa*, the fungus *Fusarium solani*, and protozoa *Acanthamoeba* spp. [2, 7]. Here we use bacterial keratitis induced by *P. aeruginosa* as the model as this pathogen is the most common cause of contact lens-related microbial keratitis in Western countries [2, 5, 7]. The first assay is in vivo knockdown of AMP expression by topical application of siRNA to the ocular surface. This approach allows for the specific knockdown of one or more AMPs so that the role of endogenously produced AMPs in modulating the severity of corneal infection can be assessed directly. The development of knockout technology to eliminate genetic expression has led to many important discoveries regarding the participation of specific molecules in various biological processes. There have been a small number of AMP knockout mice created, and these have provided valuable information in regard to the involvement of AMPs including defensins mBD-1 [9–11] and mBD-3 [12, 13] and the cathelicidin CRAMP [13–18] in the recovery/protection from various infections and of β-defensins in sperm maturation [19]. However, genetic knockouts are not available for all of the AMPs one may have an interest in studying. Transient knockdown by siRNA interference is an alternative way to impair local expression of specific AMPs and hence study their role in specific infections such as keratitis. The siRNA is first administered 24 h before initiation of the infection, and then additional doses (at half the starting concentration) are administered over several days to ensure that AMP expression remains knocked down during the course of the experiment. This approach, which is described in this chapter, has been used successfully to study the involvement of defensins in experimental microbial keratitis induced by *P. aeruginosa* and *F. solani* [13, 20, 21].

The subsequent assays are viable bacterial counts and myeloperoxidase assays to determine how many pathogens and neutrophils are present in the cornea at a given time during infection. These assays provide insight regarding the underlying pathology of the response, for example, for testing the efficacy of a topically applied AMP, a reduction in the number of viable bacteria suggests that the AMP is directly killing the infecting pathogen. Quantitation of cytokines and chemokines is also informative in regard to the pathological changes, for example, high levels of infiltrating neutrophils and corneal damage may be correlated to excessive corneal levels of neutrophil-specific chemokines. Also described are methods to determine the level of AMP (mRNA and protein) at the ocular surface. This can, for example, reveal which AMPs are differentially modulated by pathogen exposure hence indirectly implicating them in protection against particular pathogens.

2 Materials

All solutions should be prepared using ultrapure water and analytical grade reagents. Reagents to be used in vivo should be kept sterile by preparation in a Class II biological safety cabinet and if necessary filtered using a 0.2 μm filter unit prior to storage or use. Use of animals requires prior approval by the relevant institutional/local regulatory body. All waste should be disposed of appropriately and experiments conducted at biosafety level 2 with relevant personal protective equipment provided to all laboratory personnel.

2.1 AMP Knockdown Using siRNA

1. AMP-specific and scrambled control siRNA.
2. 6- to 8-week-old C57BL/6 mice.
3. Ketamine.
4. Xylazine.
5. Hamilton syringe (10 μL, 26 gauge) for subconjunctival injection.
6. Surgical stereomicroscope.
7. *P. aeruginosa* (ATCC 19660) suspended at 2×10^8 colony-forming units (cfu)/mL in phosphate buffer: 8.2 mM Na_2HPO_4, 1.8 mM KH_2PO_4, pH 7.4.

2.2 Corneal Homogenate Preparation

1. Surgical equipment for corneal harvesting: 2 pairs of small scissors, 1 pair of large blunt forceps, 2 pairs of small forceps—blunt and sharp, iris scissors.
2. Sterile plastic dishes.
3. Premade sterile 10× phosphate-buffered saline (PBS)—pH 7.4, diluted 1:10 in sterile water.
4. Kinematica AG homogenizer.
5. Ultrasonic dismembrator.

2.3 Viable Bacterial Counts

1. *Pseudomonas* isolation agar: glycerol is added to isolate *P. aeruginosa* based on pigment formation. Dissolve 45 g of *Pseudomonas* isolation agar powder in 1 L of purified water, add 20 mL of sterile glycerol, then place the mixture on a stirring hot plate, boil for 1 min to dissolve the powder completely, autoclave the mixture at 121 °C for 15 min to sterilize to make the plates pour 25–30 mL of the cooled mixture into sterile 100 mm dishes, and allow to dry in a Class II Biosafety cabinet.
2. Premade sterile 10× PBS—pH 7.4, diluted 1:10 in sterile water.
3. Shaker incubator.
4. Alpha Imager Documentation System.

2.4 Myeloperoxidase Assay and ELISAs

1. Myeloperoxidase (MPO).
2. Phosphate buffer: 50 mM KH_2PO_4 at pH 6.
3. O-dianisidine dihydrochloride (O-d-d).
4. Premade sterile 10× PBS—pH 7.4, diluted 1:10 in sterile water.
5. Hexadecyltrimethylammonium bromide: prepare a 0.5% (w/v) solution by dissolving 0.5 g/100 mL in 50 mM KH_2PO_4, pH 6.
6. Table-top Eppendorf centrifuge.
7. 30% Hydrogen peroxide solution.
8. Corning 96 well plates.
9. FLUOstar Omega spectrophotometer plate reader.
10. Liquid nitrogen.
11. BCA protein assay kit.
12. ELISA kits for cytokines/chemokines of interest.

2.5 Real-Time PCR for AMP Expression

1. Aerosol-resistant (ART) barrier filter tips.
2. ToTALLY RNA total RNA isolation kit.
3. Cell lysis buffer—denaturation solution from the ToTALLY RNA kit.
4. DNase I.
5. NanoDrop 2000 spectrophotometer.
6. SuperScript III First-Strand Synthesis System.
7. SYBR green quantitative PCR Master Mix kits.
8. AMP-specific forward and reverse primers.
9. Table-top Eppendorf centrifuge.
10. Sterile 96 well polypropylene PCR plates.
11. PCR tubes.
12. DNase-/RNase-free, pyrogen safe PCR plate caps.
13. Real-time thermal cycler Mx3005p.

2.6 Immunostaining for Corneal AMP Protein Expression

1. General: Coplin jars for washing; large petri dishes, paper towels, and wooden coffee stirrers to create moist chambers; Pap pen.
2. Tissue-Tek optimal cutting temperature compound.
3. Superfrost/Plus Microscope Slides.
4. Leica CM 1950 Cryostat.
5. Sterile, cold (4 °C) premade Dulbecco's PBS.
6. Acetone.

7. Blocking solution: 5% (w/v) bovine serum albumin + 10% normal serum from relevant species + 0.1–0.3% (v/v) Triton X-100 diluted in PBS. Store at 4 °C.
8. Control serum/IgG: Immunopure Normal goat serum, rabbit serum, goat control IgG.
9. Primary AMP antibodies: goat anti-CRAMP, rabbit anti-mBD3, rabbit anti-mBD4.
10. Secondary antibodies: Donkey anti-goat-Alexa Fluor 546, goat anti-rabbit-Alexa Fluor 546.
11. Vectashield® prolong gold mounting medium.
12. Cover slips—#1, 22 × 60 mm.
13. DeltaVision Core inverted microscope.

3 Methods

3.1 In Vivo AMP Knockdown by siRNA

1. AMP-specific siRNA or scrambled control siRNA prepared as follows: specific or scrambled siRNA is diluted in sterile water to make a stock concentration of 10 μM. The stock is then diluted to 8 or 4 μM concentrations, aliquoted, and stored at −20 °C until use. Aliquots should not be subject to freeze-thaw cycles. Any excess should be disposed of and fresh aliquots used for each administration. The product is thawed on ice ready for use.
2. 24 h prior to ocular infection, 6–8-week-old C57BL/6 mice are anesthetized by intraperitoneal injection of a mix of ketamine (100 mg/kg of body weight) and xylazine (10 mg/kg) with the level of anesthesia being checked by lack of response to a toe pinch.
3. The anesthetized animal is positioned under a surgical stereomicroscope, and 5 μL of 8 μM AMP-specific siRNA or scrambled control siRNA is injected subconjunctivally into the superior-temporal quadrant of the designated eye (usually the right eye is used as the experimental eye) using a 26 g Hamilton syringe. Flush the syringe ten times with sterile water (the diluent for the siRNA) between animals to prevent cross contamination.
4. 24 h later the mouse eye is inoculated with bacteria to create the infection. The animal is again anesthetized and placed under the surgical stereomicroscope. Three parallel 1 mm scratches are made using a 27½ gauge needle on the central cornea of the experimental eye to create a central corneal wound. Five microliter (1×10^6 cfu in phosphate buffer) of *P. aeruginosa* (ATCC 19660) is then applied onto the wounded cornea using a Gilson pipette.

5. Five microliter of 4 μM AMP or scrambled control siRNA is applied topically 12 h postinfection and every 12 h thereafter for the days following infective agent challenge (*see* **Note 1**).

3.2 Ex Vivo Assays

Laboratory animals are infected with the pathogen of choice in the cornea of one eye. The precise details vary with pathogen and animal species, but as a general rule it is necessary to disrupt the corneal surface in some way (e.g., by superficial scratches on the corneal surface as described above (*see* Subheading 3.1, **step 4**)) in order to get the infection to take hold. At specific time points during the experiment, animals are euthanized as per the institutional approved method; corneas are then harvested under a dissecting microscope for analysis using the following procedures. Here we use infection with *P. aeruginosa* in mice as the model, but the procedures are readily applicable to other pathogens such as fungi and other species such as rat.

3.2.1 Preparation of Corneal Homogenates

Corneal homogenates are prepared as described below and then used to determine viable bacterial counts, MPO activity, and the levels of cytokines and chemokines (or other analytes of interest). The assays for viable counts and MPO should be performed on the same day as the homogenate is prepared. The remaining volume of homogenate can be stored frozen for determination of cytokines/chemokines at a later date.

1. Harvest corneas from three to five control or *P. aeruginosa*-19660-infected euthanized mice under a dissecting microscope.
2. Pool the corneas in 200 μL of sterile PBS/cornea in an Eppendorf tube and place on ice (*see* **Note 2**).
3. Homogenize the pooled corneas using a Kinematica AG homogenizer for 30 s at a setting of 5, pulsing it ten times by moving the tube up and down. The process is repeated two times keeping the tube of tissue on ice in between. To avoid cross contamination between samples, rinse the probe at least three times with sterile PBS between each sample. For preparation and maintenance of the probe, *see* **Note 3**.
4. Place the tube containing the sample in a small beaker filled with ice, and then sonicate at a setting of 3–4 and a rate of ten strokes for 10 s. Repeat this for a total of three times. To avoid cross contamination between samples, rinse the probe at least three times with sterile PBS between each sample. Preparation and maintenance of the probe are as for the homogenizer (*see* **Note 3**).

3.2.2 Viable Bacterial Counts

1. Using 100 μL of the homogenate prepared above, make a series of tenfold dilutions by mixing 10 μL of homogenate and

90 µL of sterile PBS. Making and plating 6–8 dilutions should ensure that at least one of the dilutions can be easily counted.

2. Plate 10 µL of each dilution in duplicate on to *Pseudomonas* isolation agar plates using bacterial spreaders. For fungal plating, *see* **Note 4**.
3. Incubate the plates for 16–18 h at 37 °C.
4. Take images of the plates with an Alpha Imager Documentation System, and then count the number of colonies by direct counting. The upper limit of detection for accurate counting is 200 cfu/plate.

3.2.3 Myeloperoxidase (MPO) Assay Protocol

Quantitation of MPO activity is used as an indicator of neutrophil (PMN) recruitment to the site of infection.

1. Add 10 µL of 0.5 % hexadecyltrimethylammonium bromide to 90 µL corneal homogenate (prepared as in Subheading 3.2.1) in an Eppendorf tube.
2. Sonicate the samples prepared in **step 1**: place the tube in a small beaker of ice, and sonicate at levels 3–4 of the ultrasonic dismembrator, two times for 5 s for a total of 10 s.
3. Freeze each sample in liquid nitrogen and then thaw. Repeat this for a total of three times (*see* **Note 5**).
4. Centrifuge the samples at 13,000 × *g* at 4 °C for 20 min in a table-top Eppendorf centrifuge.
5. Make up the MPO standards using 50 mM phosphate buffer, pH 6. Begin by preparing 0.1 U of MPO, and then make 9 twofold serial dilutions. Make sufficient volume of each standard to allow each to be run in duplicate or triplicate 10 µL volumes on the plate.
6. Add 10 µL of each standard to duplicate wells of a flat-bottomed 96 well plate and 10 µL of each sample (in triplicate).
7. Make up the ODD mixture by dissolving 0.00334 g O-dianisidine dihydrochloride (final concentration 0.0167 % w/v) in 20 mL of the 50 mM phosphate buffer, then add 1.33 µL of 30 % hydrogen peroxide (final hydrogen peroxide is 0.002 % v/v), and protect the tube from light by covering with aluminum foil.
8. Initiate the reaction by adding 90 µL of the ODD mixture (**step 8**) to each well.
9. Place the plate in the plate reader and read every 15 min for 120 min at 450 nm, and plot in comparison to the standard curve generated on the same plate.
10. Results are expressed as relative units of MPO activity per cornea. 1 MPO unit is proportional to 2×10^5 infiltrating neutrophils [22, 23].

3.2.4 Enzyme-Linked Immunosorbent Assays for Cytokines, Chemokines, and Other Analytes

1. Corneal homogenates prepared as described above (*see* Subheading 3.2.1, **steps 1–4**) are centrifuged at 4000×*g* at 4 °C for 20 min. Collect the supernatant and store frozen at −80 °C until ready to perform ELISAs (*see* **Note 6**).
2. Determine the total protein concentration in each sample by a standard protein assay kit such as the BCA kit.
3. Levels of analytes of interest such as cytokines and chemokines IL-1β, IL-6, TNF-α, MIP-2, and KC are quantitated in the corneal homogenate supernatants by specific ELISA kits. The typical sample size used is 100 μL of the supernatant and triplicate samples are plated. The assay is performed as per the kit manufacturer's instructions and data expressed as amount/mg total protein. For some analytes the supernatant may need to be diluted in order for the reading to fall within the appropriate range (above 20% but less than 80%) on the standard curve. This needs to be determined empirically for each analyte of interest.

3.2.5 Relative Quantitative Real-Time PCR for AMP Expression

1. Harvest and pool 6–8 corneas from control or infected mice in RNA lysis buffer (200 μL/cornea) on ice (*see* **Note 7**).
2. Extract total RNA from each sample using ToTALLY RNA total cellular RNA kit as per the manufacturer's instructions.

   All samples should be treated with DNase I (RNase-Free DNase Set kit) to avoid any genomic DNA contamination as follows.

3. Mix 10 μL DNase I with 70 μL Buffer RDD from the kit by gently flicking the Eppendorf tube. Pulse briefly in a microtube mini-centrifuge to collect all residual liquid from the sides of the tube. DO NOT VORTEX. Make sufficient volume for all samples to be tested.
4. Add 80 μL of the diluted DNase I to each RNA sample, and mix by flicking the tube gently. Place in a rack and leave at room temperature for 30 min.
5. Proceed with the RNA phenol-chloroform extraction as per the instructions of the ToTALLY RNA kit.
6. Quantitate the amount of RNA in 2 μL of the DNase I-treated sample using a NanoDrop spectrophotometer (*see* **Note 8**)
7. For each sample 1 μg of total RNA is reverse transcribed to make 20 μL of cDNA reaction volume using the SuperScript III First-Strand Synthesis System, which contains Moloney murine leukemia virus reverse transcriptase (RT). Each reaction tube will contain a final volume of 20 μL composed of 10 μL 2× kit buffer; 3 μL oligo dT, 1 μL RT/Block, 6 μL RNA+RNase-free water (i.e., 1 μg sample RNA+appropriate volume of RNase-free water for a final sample volume of 6 μL). Samples containing RNase-free water instead of RT serve as

negative controls. The incubation conditions for the RT step are 25 °C for 5 min, 42 °C for 45 min, and 95 °C for 5 min.

8. At the end of the RT reaction, dilute the 20 μL of reaction mix 1:1 with 20 μL sterile RNase-free water.
9. Prepare the samples for relative quantitative real-time PCR amplification performed using SYBR green PCR Master Mix kits. Gene-specific optimized primers to quantitate the mRNA expression of various murine AMPs and a relevant housekeeping gene (in this case we use RNA polymerase II, RPII) are presented in Table 1 (*see* **Note 9**). All steps and thawing of reagents are performed on ice, and the PCR plate is kept in a cold block, while the samples are being added. The preparation steps are as follows:
10. Dilute 1 μL ROX reference dye in 499 μL RNase-free water.
11. Thaw primers on ice and dilute in sterile RNase-free water, and mix to give a final concentration of 10 μM Forward (F) and 10 μM Reverse (R) in an Eppendorf tube. The premixed primers can be stored in small aliquots at −20 °C for up to 2 months.
12. Prepare the PCR reaction mix: 12.5 μL SYBR Green 2× Master Mix + 0.375 μL of the diluted reference dye + 0.5 μL of premixed gene-specific F and R primer mix + 9.625 μL RNase-free water giving a total final volume of 23 μL per reaction. These are the volumes for a single reaction—the actual volume should be scaled up based on the number of samples to be analyzed/gene of interest and accounting for performing each reaction in triplicate.
13. Pipette 23 μL of the reaction mix in to the wells of a PCR plate.
14. Add 2 μL of each cDNA to be tested in to the relevant wells to give a final PCR reaction volume of 25 μL.
15. Seal the PCR plate tightly and spin the plate briefly (30 s, 200 × *g*, room temperature) in a centrifuge equipped with a 96 well plate rotor to bring all of the contents to the bottom of the wells.
16. Place the plate in the real-time thermal cycler and run under the following conditions: 10 min denaturation at 95 °C followed by 40 cycles of amplification as follows—denaturation, 95 °C for 30 s; annealing, 56 °C for 1 min; and extension, 72 °C for 30 s.
17. Perform initial data analysis using the Stratagene Mx3005p software. Disassociation melt curve analysis should be included to ensure reaction specificity (*see* **Note 10**).
18. To determine the relative fold change in AMP expression in infected versus control mice, the data for the amplified gene products should be normalized to the housekeeper RPII and

Table 1
The table shows the sequences of primers used to determine β-defensin, cathelicidin (CRAMP), and RNA polymerase II (RPII, housekeeping gene) expression in normal and infected murine corneas [13]

Primer	Forward/reverse	Sequence (5′–3′)
mBD1	Forward Reverse	CTGGGAGTTTCACATCCTCTC CTCCATGTTGAAGGCATTTGT
mBD2	Forward Reverse	CTACCAGCCATGAGGACTCTC GTACTTGCAACAGGGGTTCTT
mBD3	Forward Reverse	GGATCCATTACCTTCTGTTTGC ATTTGAGGAAAGGAACTCCAC
mBD4	Forward Reverse	GCTTCAGTCATGAGGATCCAT CTTGCTGGTTCTTCGTCTTTT
mBD5	Forward Reverse	CCTTCTCTTTGCATTTCTCCT TTTCTCTTGCAGCAGTTGAGA
mBD6	Forward Reverse	TACCTGCTCTTTGCCTTTATCC TTCTGGCACTTATTCACATTGC
mBD14	Forward Reverse	TCTTGTTCTTGGTGCCTGCT CGACCGCTATTAGAACATCGAC
CRAMP	Forward Reverse	GCCGCTGATTCTTTTGACAT GCCAAGGCAGGCCTACTACT
RPII	Forward Reverse	CTACACCACCTACAGCCTCCAG TTCAGATGAGGTCCATGAGGAT

calibrated to uninfected samples and relative fold expression calculated using the $2^{\Delta\Delta ct}$ method [24].

3.2.6 Immunostaining for Corneal AMP Protein Expression

The antibodies and methodology described here are to stain for the presence of murine defensins or cathelicidin (CRAMP), but the method can be optimized and used to determine the expression of other molecules of interest.

1. Euthanize the mouse, then harvest the whole globe, and rinse with sterile cold Dulbecco's PBS and embed in orientation (superior, inferior, nasal, temporal) pre-marked small cryomolds containing Tissue-Tek optimal cutting temperature compound, and rapidly freeze by submerging in liquid nitrogen. Store at −80 °C until ready to use.

2. Cut 10 μm thick sections on a cryostat and mount (4–6 sections/slide) on Superfrost slides. If need be store slides at −80 °C until ready to perform the staining.
3. Preparation for staining: Prepare 1–2 L of PBS and store in fridge. Decant some acetone in to a small glass vial and chill in the fridge overnight. Prepare the blocking solution as described in Subheading 2.7—10 mL is sufficient for five slides (*see* **Note 11**).
4. If using stored frozen sections, thaw the sections for 60 min at room temperature before use.
5. Fix sections in ice-cold acetone for 3 min. Simply add sufficient acetone to cover each section. At the end of the 3 min, tip off the acetone, catching the liquid in a paper towel.
6. Rinse the sections in PBS three times, dry off carefully around sections with a Kimwipe, and then use a Pap pen to draw around each section so forming a "well" into which the antibodies can be added.
7. Place the slides in to a moist chamber (*see* **Note 12**) and add 100–200 μL (enough to cover the section completely) of blocking solution and incubate for 2 h at room temperature.
8. During the blocking step, prepare the appropriate dilutions of the desired primary antibodies in blocking solution. For polyclonal antibodies, use a nonspecific IgG from the same species as the primary antibody was raised in as the nonspecific binding control. For monoclonal antibodies, use the relevant nonspecific isotype control from the same species as the primary antibody (*see* **Note 13**). The amount of primary antibody and IgG/isotype control should be the same. For our purposes [13], we use the following dilutions of primary and control antibodies:

   Goat anti-CRAMP (0.2 μg/μL) diluted 1:50; goat IgG (0.4 μg/μL) diluted 1:100.

   Rabbit anti-mBD-3 (0.2 μg/μL) diluted 1:50; rabbit IgG (1 μg/μL) diluted 1:250.

   Rabbit anti-mBD-4 (0.2 μg/μL) diluted 1:50; rabbit IgG (1 μg/μL) diluted 1:250.
9. At the end of the blocking step, remove the blocking reagent by tipping the slide and catching the fluid in a paper towel. Then add 50–100 μL (sufficient to cover the section) of diluted primary antibody. Incubate overnight at 4 °C.
10. At the end of the incubation in primary antibody, wash the sections three times in PBS for 5 min each time.
11. Repeat the blocking step by adding 100–200 μL blocking solution, and incubate for 30 min at room temperature.

12. During the second blocking step, prepare 1:400 dilution of the relevant secondary antibodies in blocking reagent. Secondary antibodies for our purposes were goat anti-rabbit alexa-546 for mBD3 and mBD4 and donkey anti-goat Alexa-546 for CRAMP.
13. At the end of the second blocking step, remove the blocking reagent by tipping the slide and catching the fluid in a paper towel. Add 50–100 μL of diluted secondary antibody, and incubate the sections for 60 min at room temperature in the dark (cover with aluminum foil or place in a closed cupboard).
14. Sections are then washed three times in ice-cold PBS for 5 min each time. Remove the wash solution by tipping the slide and catching the fluid in a paper towel.
15. The labeled slides are then dried by using a Kimwipe to dab off any excess fluid and cover slips mounted with Vectashield® prolong gold mounting medium (*see* **Note 14**).
16. The slides are left at room temperature overnight in the dark to better bond the mounting medium and then can be stored at 4 °C (in the dark) until visualized by fluorescence microscopy.

4 Notes

1. The efficacy and specificity of siRNA silencing should be tested by performing quantitative real-time PCR and/or immunostaining for the AMP of interest. RT-PCR results showed a 60–74% decrease in the expression of AMPs compared to untreated or scrambled siRNA control 3 days postinfection [13, 21]. How long the AMP expression remains knocked down after terminating siRNA treatment has not been reported. Wu et al. [21] report a significant effect 5 days postinfection following knockdown of mBD-4 despite siRNA application being terminated 3 days postinfection. This suggests that AMPs remain knocked down for some time.
2. While pooling of multiple corneas is preferable to maximize the amount of material for analysis, if tissue is limited, repeatable data have been obtained by using as few as 1–2 corneas in 100 μL of PBS for MPO and viable pathogen count analyses ([25], unpublished data from our lab).
3. Immediately before the first sample, clean the probe of the homogenizer by rinsing in sterile water twice. Dry off with a Kimwipe and then wash twice again with sterile PBS. Between samples rinse with sterile PBS at least three times. After the last sample rinse with PBS, then sterile water, and finally 70% ethanol and allow the probe to air-dry.

4. For viable count of fungal pathogens: Plate samples of the serial dilutions on Sabouraud Dextrose agar plates in duplicate. Incubate the plates for 48–72 h at 30 °C and then image and count the number of culturable units.
5. To freeze hold the tube containing the sample with a pair of tongs and dip the end in to the liquid nitrogen for 10–20 s. To thaw remove the tube from the liquid nitrogen, place on the lab bench and allow to thaw at room temperature for about 1 min. Liquid nitrogen should be handled with care. Always use thick gloves and eye protection and work with the smallest amount possible. Always use a suitable container such as an ethylene-vinyl acetate laboratory ice bucket or Styrofoam container. Any excess should be clearly labeled and left to evaporate in a fume hood.
6. If multiple ELISAs are to be performed, then it is optimal to store multiple small volume aliquots to avoid the risks associated with freeze/thawing of larger volume samples. In order to ensure sufficient material for meaningful analysis of multiple analytes, the number of corneas harvested and pooled can be increased to 8–10/1 mL PBS. If material is limited, then measuring analytes by a cytometric bead assay (such as the Luminex platform) can enable the detection of multiple analytes in a much smaller volume of supernatant.
7. While harvesting corneas of C57BL/6 mice for analysis, special care must be taken to avoid contamination with pigmented tissue as melanin is known to interfere with polymerase activity hence RT-PCR reactions [26].
8. The purity of the RNA is assessed using the ratio of absorbance at 260 and 280 nm as read by the NanoDrop spectrophotometer. A ratio of approximately 2 is desirable. A ratio of less than 2 indicates the presence of contaminants such as protein and phenol in which case re-extraction of the RNA may be required.
9. Although the primers and methodology reported here were optimized to test mouse defensins and cathelicidin (CRAMP), this method can be optimized and used to determine the expression of other molecules of interest.
10. Sometimes it is desirable (e.g., to check transcript size) to visualize the PCR products by agarose gel electrophoresis. If this is the case, then the dissociation melt curve analysis step should be omitted and the products generated at the end of the last cycle of PCR amplification loaded directly on to a 1.5% agarose gel and stained with ethidium bromide for analysis.
11. The blocking reagent should contain 10% normal serum. The serum should be from a different species to that in which the primary antibody was raised, e.g., if the primary antibody is a rabbit polyclonal, then do NOT use normal rabbit serum in the blocking reagent.

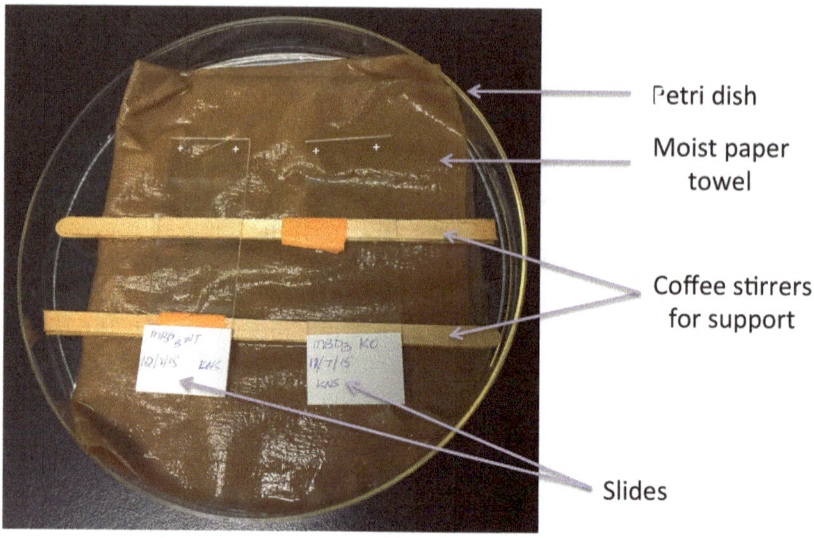

Fig. 1 The figure shows the setup of a moist chamber (lid not shown) for immunostaining tissue sections

12. To make a moist chamber (Fig. 1), take a large plastic or glass petri dish and line it with paper towel. Dampen the paper towel with water or PBS. Break off the ends from four wooden coffee stirrers if necessary so that they fit in the petri dish and bind them together in pairs with tape. Then place the wood sticks in the petri dish like train tracks. Balance the slides carefully on the wooden sticks, add the antibodies, and then cover with the lid and incubate for the specified amount of time.
13. If suitable IgG or isotype control antibody is not available, then sections incubated in the absence of any primary antibody can be used as the nonspecific binding control, but this is less than ideal.
14. If necessary the slides can be placed on a warm heating block for 5–10 min to help excess liquid evaporate.

Acknowledgment

This work was supported by NIH grant EY13175 (AMM) and grants from Sapienza Università di Roma (MLM). Part of the content of this work is the object of US patent application No. 14/506,383.

References

1. Lai Y, Gallo RL (2009) AMPed Up immunity: how antimicrobial peptides have multiple roles in immune defense. Trends Immunol 30:131–141
2. Kolar SS, McDermott AM (2011) Role of host-defence peptides in eye diseases. Cell Mol Life Sci 68:2201–2213
3. Wong JH, Ye XJ, Hg TB (2013) Cathelicidins: peptides with antimicrobial, immunomodulatory, anti-inflammatory, angiogenic, anticancer and procancer activities. Curr Protein Pept Sci 14:504–514
4. Jarczak J, Kosciuczuk EM, Lisowski P, Strazlkowska N, Jozwik A, Horbanczuk J et al (2013) Defensins: natural components of human innate immunity. Hum Immunol 74:1069–1079
5. Fleiszig SM, Evans DJ (2010) Pathogenesis of contact-lens associated microbial keratitis. Optom Vis Sci 87:225–232
6. Sharma DP, Sharma S, Wilkins MR (2011) Microbial keratitis after corneal laser surgery. Future Microbiol 6:819–831
7. Stapleton F, Carnt N (2012) Contact lens-related microbial keratitis: how have epidemiology and genetics helped us with pathogenesis and prophylaxis. Eye 26:2185–2193
8. Taneja M, Ashar JN, Mathur A, Nalamada S, Garg P (2013) Microbial keratitis following vegetative matter injury. Int Ophthalmol 33:117–123
9. Moser C, Weiner DJ, Lysenko E, Bals R, Weiser JN, Wilson JM (2002) beta-Defensin 1 contributes to pulmonary innate immunity in mice. Infect Immun 70:3068–3072
10. Morrison G, Kilanowski F, Davidson D, Dorin J (2002) Characterization of the mouse beta defensin 1, Defb1, mutant mouse model. Infect Immun 70:3053–3060
11. Tomalka J, Azodi E, Narra HP, Patel K, O'Neill S, Cardwell C et al (2015) β-Defensin 1 plays a role in acute mucosal defense against *Candida albicans*. J Immunol 194:1788–1795
12. Augustin DK, Heimer SR, Tam C, Li WY, Le Due JM, Evans DJ, Fleiszig SM (2011) Role of defensins in corneal epithelial barrier function against *Pseudomonas aeruginosa* traversal. Infect Immun 79:595–605
13. Kolar SS, Baidouri H, Hanlon S, McDermott AM (2013) Protective role of murine β-defensin 3 and 4 and cathelin-related antimicrobial peptide in Fusarium solani keratitis. Infect Immun 81:2669–2677
14. Nizet V, Ohtake T, Lauth X, Trowbridge J, Rudisill J, Dorschner RA et al (2001) Innate antimicrobial peptide protects the skin from invasive bacterial infection. Nature 414:454–457
15. Huang LC, Reins RY, Gallo RL, McDermott AM (2007) Cathelicidin-deficient (Cnlp-/-) mice show increased susceptibility to *Pseudomonas aeruginosa* keratitis. Invest Ophthalmol Vis Sci 48:4498–4508
16. Yu FS, Cornicelli MD, Kovach MA, Newstead MW, Zeng X, Kumar A et al (2010) Flagellin stimulates protective lung mucosal immunity: role of cathelicidin-related antimicrobial peptide. J Immunol 185:1142–1149
17. Kulkarni MM, Barbi J, McMaster WR, Gallo RL, Satoskar AR, McGwire BS (2011) Mammalian antimicrobial peptide influences control of cutaneous Leishmania infection. Cell Microbiol 13:913–923
18. Kovach MA, Ballinger MN, Newstead MW, Zeng X, Bhan U, Yu FS et al (2012) Cathelicidin-related antimicrobial peptide is required for effective lung mucosal immunity in Gram-negative bacterial pneumonia. J Immunol 189:304–311
19. Dorin JR (2015) Novel phenotype of mouse spermatozoa following deletion of nine β-defensin genes. Asian J Androl 17:716–719
20. Wu M, McClellan SA, Barrett RP, Hazlett LD (2009) Beta-defensin-2 promotes resistance against infection with P. aeruginosa. J Immunol 182:1609–1616
21. Wu M, McClellan SA, Barrett RP, Zhang Y, Hazlett LD (2009) Beta-defensins 2 and 3 together promote resistance to *Pseudomonas aeruginosa* keratitis. J Immunol 183:8054–8060
22. Williams RN, Paterson CA, Eakins KE et al (1982) Quantification of ocular inflammation: evaluation of polymorphonuclear leucocyte infiltration by measuring myeloperoxidase activity. Curr Eye Res 2:465–470
23. Cole N, Krockenberger M, Stapleton F, Khan S, Hume E, Husband AJ, Willcox M (2003) Experimental *Pseudomonas aeruginosa* keratitis in interleukin-10 gene knockout mice. Infect Immun 7:1328–1336
24. Livak KJ, Schmittgen TD (2001) Analysis of relative gene expression data using real-time quantitative PCR and the 2(DeltaDelta C(T)) method. Methods 25:402–408
25. Gao N, Kumar A, Guo H, Wu X, Wheater M, Yu FS (2011) Topical flagellin-mediated innate defense against *Candida albicans* keratitis. Invest Ophthalmol Vis Sci 52:3074–3082
26. Eckhart L, Bach J, Ban J, Tschachler E (2000) Melanin binds reversibly to thermostable DNA polymerase and inhibits its activity. Biochem Biophys Res Commun 271:726–730

Chapter 31

Hemolytic Activity of Antimicrobial Peptides

Alberto Oddo and Paul R. Hansen

Abstract

For antimicrobial peptides to be interesting for systemic applications, they must show low toxicity against erythrocytes. In this chapter, we describe a protocol for measuring the ability of AMPs to lyse human red blood cells, using melittin as positive control.

Key words Antimicrobial peptides, Hemolysis, Red blood cells, Melittin

1 Introduction

One requirement for AMPs to be interesting for systemic drug development is low toxicity against erythrocytes. Generally, this is measured by the ability of AMPs to lyse mammalian red blood cells. Although human red blood cells are preferred [1], the use of sheep [2], rat [3], pig, piglets [4], or rabbit red blood cells has been reported [5]. Only a few studies exist [4, 6] comparing the hemolytic activity of different peptides against red blood cells of different mammalian species. These indicate that the hemolytic activity strongly depends on the source of red blood cells. Helmerhorst et al. have shown significant differences between blood collected from different individuals and that susceptibility to hemolysis increases with storage duration of the erythrocytes [7].

Melittin, a 26-residue amphipathic peptide from honeybee, *Apis mellifera*, venom [8] is considered the "gold-standard positive control" since the peptide possesses a broad spectrum of lytic activities against a variety of cells at submicromolar concentrations and does not affect spectrophotometric reading at 414 nm [9]. The mechanism of melittin-mediated lysis is associated with the membrane poration of target cells, which causes the leakage of cell

The original version of this chapter was revised. The erratum to this chapter is available at: DOI 10.1007/978-1-4939-6737-7_32

Paul R. Hansen (ed.), *Antimicrobial Peptides: Methods and Protocols*, Methods in Molecular Biology, vol. 1548, DOI 10.1007/978-1-4939-6737-7_31, © Springer Science+Business Media LLC 2017

contents [10]. Triton-X is a viable and less expensive alternative to melittin, but working with it can be cumbersome (foaming), and it may affect the reading at 414 nm.

The AMP concentration whereby half of the red blood cells are lysed (EC_{50} or HC_{50}) constitutes one of the most used indicators of toxicity, together with the selectivity index, SI (EC_{50}/MIC) [11]. Ruiz et al. analyzed the relationship between EC_{50} and physicochemical properties of 18 peptides, having EC_{50} from <20 to >100 μM [12]. The authors were not able to correlate parameters such as length, isoelectric point, or tendency to form some secondary structures. However, they confirmed the well-known effect of charge and hydrophobicity.

Overall, more lipophilic peptides and peptides better capable of assuming a stable amphipathic secondary structure tend to be more hemolytic [13]. The amino acids which have the strongest effect on hemolytic activity either through hydrophobicity or charge are Trp, Lys, and Arg.

Tryptophan residues have been proposed to be involved in the binding of peptides to cholesterol present in biological membranes through the indole moiety [14] and therefore important in the lysis of erythrocytes. The only Trp residue of melittin has been shown to be very important for hemolytic activity [15]. Blondelle et al. reported complete set of 25 single-position synthetic Trp substitution analogues [16]. Significant increases in hemolytic activity were observed and the changes in activity of all of the analogues relative to melittin were found to be correlated to their behavior during reverse phase HPLC.

Staubitz et al. replaced all of the five Trp residues in the 13-residue peptide indolicidin, with Phe which had a significant positive effect on the hemolytic activity, while the MIC against *M. luteus* was retained [17]. Dathe and coworkers investigated a number of magainin II amide analogues with charges ranging between +3 and +7 [18]. The authors found that maintaining the hydrophobic properties while increasing the charge to +5 resulted in the best antimicrobial activity and selectivity. However, a significant increase of hemolytic activity and loss of antimicrobial selectivity were observed when charge increased beyond +5 with retention of other structural motifs.

Finally, selective d-amino acid replacements, which disrupt α-helical structure, may influence the hemolytic activity significantly while maintaining full or partial antimicrobial activity [19]. For example, the all-L peptide LKLLKKLLKKLLKLL-NH_2 and its diastereomer LKlLKkLlkKLLkLL-NH_2 showed comparable MIC values and a hemolytic activity of 100% and 0% at 100 μM, respectively [20].

Here, we describe a protocol for determining hemolysis of antimicrobial peptides against red blood cells, using melittin as positive control.

2 Materials

1. Blood: 3–4 mL of fresh 0 neg whole blood in EDTA (*see* **Note 1**).
2. PBS buffer: 137 mM NaCl, 2.7 mM KCl, 10 mM Na_2HPO_4, 1.8 mM KH_2PO_4 (*see* **Note 2**).
3. 96-Well polypropylene plates: with either V- (preferably) or U-shaped bottom.
4. 96-Well polystyrene ELISA plates: clear, flat bottom.
5. Biorad Microseal "F" film: one for each plate.
6. CappOrigami (or similar) reagent reservoirs 40 mL: at least two for each session.
7. Melittin stock solution: 1.25 mM melittin in PBS (*see* **Note 3**).
8. Eppendorf Protein LoBind tubes: two tubes are needed for each test session.
9. Incubator.
10. Spectrophotometer.
11. Plate centrifuge.

3 Methods

3.1 Preparations: Melittin, RBC Suspension, and Peptides

1. The night before the test, add 150 µL of melittin 5 µM to the positive control wells. The day of the test, discard the solution and wash the wells 2–3 times with the same volume of PBS. The day of the test, prepare a 2.50 µM solution in PBS (500 µL for each plate). *See* Fig. 1 and **Note 4**.
2. Transfer 1 mL of whole blood to a centrifuge tube.
3. Add 3 mL of PBS to the blood, mix gently, and centrifuge for 8 min at 700 × *g*.
4. Discard supernatant, add 4 mL of PBS, mix gently, and centrifuge again for 8 min at 700 × *g*.

	1	2	3	4	5	6	7	8	9	10	11	12
A	Hi Con	Hi Con	Hi Con	Hi Con	Hi Con	Hi Con	Hi Con	Hi Con	Hi Con	Hi Con	Hi Con	Hi Con
B												
C	↓	↓	↓	↓	↓	↓	↓	↓	↓	↓	↓	↓
D												
E	↓	↓	↓	↓	↓	↓	↓	↓	↓	↓	↓	↓
F												
G	Lo Con	Lo Con	Lo Con	Lo Con	Lo Con	Lo Con	Lo Con	Lo Con	Lo Con	Lo Con	Lo Con	Lo Con
H	Pos	Pos	Pos	Pos	Pos	Pos	Neg	Neg	Neg	Neg	Neg	Neg

Fig. 1 General plate layout for testing hemolysis of AMPs

5. Discard supernatant, add 4 mL of PBS, mix gently, and centrifuge for 8 min at 1000 × *g*. Discard supernatant.
6. For each plate, make 8 mL of 0.5% v/v RBC suspension: 40 μL of RBC is added to 8 mL of PBS. Please *see* **Note 5**.
7. For each peptide, prepare 500 μL of solution in PBS at double concentration. For example, dissolve 0.15 μmol of peptide in 500 μL (300 μM) of PBS to start testing down from 150 μM.

3.2 Test Procedure

Each plate can accommodate up to four peptides to be tested in triplicate.

1. Take a polypropylene plate and transfer 150 μL of peptide solution in PBS in row A as follows:

   Peptide 1: columns 1–3
   Peptide 2: columns 4–6
   Peptide 3: columns 7–9
   Peptide 4: columns 10–12
   Remember to mark each plate and to make a note of which peptide was inoculated where.
2. Pour some PBS into a CAPPOrigami.
3. Use a multichannel pipette to transfer 75 μL of PBS to all the wells from B1 to G12.
4. Transfer the same amount in the wells H7–12.
5. Transfer 75 μL of 2.50 μM melittin solution in the wells H1–6 (Fig. 2).
6. Transfer 75 μL from row A to row B, mix a couple of times.
7. Transfer 75 μL from row B to row C, mix a couple of times.
8. Transfer 75 μL from row C to row D, mix a couple of times.
9. Transfer 75 μL from row D to row E, mix a couple of times.
10. Transfer 75 μL from row E to row F, mix a couple of times.
11. Transfer 75 μL from row F to row G, mix a couple of times.
12. After mixing, discard 75 μL from row G.

	1	2	3	4	5	6	7	8	9	10	11	12
A	Pep1	Pep1	Pep1	Pep2	Pep2	Pep2	Pep3	Pep3	Pep3	Pep4	Pep4	Pep4
B	PBS	PBS	PBS	PBS	PBS	PBS	PBS	PBS	PBS	PBS	PBS	PBS
C	PBS	PBS	PBS	PBS	PBS	PBS	PBS	PBS	PBS	PBS	PBS	PBS
D	PBS	PBS	PBS	PBS	PBS	PBS	PBS	PBS	PBS	PBS	PBS	PBS
E	PBS	PBS	PBS	PBS	PBS	PBS	PBS	PBS	PBS	PBS	PBS	PBS
F	PBS	PBS	PBS	PBS	PBS	PBS	PBS	PBS	PBS	PBS	PBS	PBS
G	PBS	PBS	PBS	PBS	PBS	PBS	PBS	PBS	PBS	PBS	PBS	PBS
H	Mel	Mel	Mel	Mel	Mel	Mel	PBS	PBS	PBS	PBS	PBS	PBS

Fig. 2 General plate layout for testing four peptides with melittin and PBS as controls

By doing so the peptide is divided into a two-fold dilution series.

13. Mix the RBC suspension gently until homogeneous and then pour some into a CAPPOrigami.
14. Using a multichannel pipette, transfer 75 μL of RBC suspension into each well.
15. Mix the RBC suspension and the peptide solution from row G to A, i.e., in the direction of increasing concentration (*see* **Note 6**) (Fig. 3).

3.3 Incubating and Reading Absorbance

1. Cover the plate with a Biorad Microseal film and place in the incubator at 37 °C for 1 h (*see* **Note 7**).
2. Centrifuge the plates at 1000 × *g* for 10 min.
3. Quickly transfer 60 μL of supernatant from each well into a clear polystyrene, flat-bottom 96-well ELISA plate. Maintain the respective positions of each well (i.e., A1 to A1, A2 to A2, etc., *see* **Notes 8** and **9**).
4. Read absorbance at $\lambda = 414$ nm.

3.4 Interpretation of Results

Hemolysis values can be plotted to obtain a curve. Some peptides display a more or less linear relationship between concentration and hemolysis rate; other peptides are characterized by a sort of threshold concentration above which the hemolytic power grows steeply (Fig. 4).

First of all, negative and positive controls are averaged. Results are then normalized with respect to the averaged positive (100%) and negative (0%) controls. It is quite useful to prepare an Excel sheet to take care of these calculations automatically. The formula to apply is:

$$\%\text{Hemolysis} = \frac{\text{Abs}(\text{sample}) - \text{Abs}(\text{NegAvg})}{\text{Abs}(\text{PosAvg}) - \text{Abs}(\text{NegAvg})} \times 100$$

	1	2	3	4	5	6	7	8	9	10	11	12
A	150	150	150	150	150	150	150	150	150	150	150	150
B	75	75	75	75	75	75	75	75	75	75	75	75
C	37.5	37.5	37.5	37.5	37.5	37.5	37.5	37.5	37.5	37.5	37.5	37.5
D	18.75	18.75	18.75	18.75	18.75	18.75	18.75	18.75	18.75	18.75	18.75	18.75
E	9.4	9.4	9.4	9.4	9.4	9.4	9.4	9.4	9.4	9.4	9.4	9.4
F	4.7	4.7	4.7	4.7	4.7	4.7	4.7	4.7	4.7	4.7	4.7	4.7
G	2.35	2.35	2.35	2.35	2.35	2.35	2.35	2.35	2.35	2.35	2.35	2.35
H	1.25	1.25	1.25	1.25	1.25	1.25	0	0	0	0	0	0

Fig. 3 Plate layout showing melittin (*red*) and peptide (*blue*) concentrations

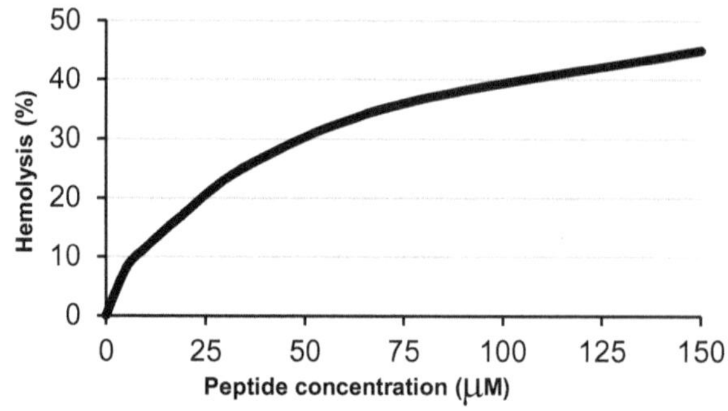

Fig. 4 Hemolytic activity of KRLFRKILKYL-NH_2 against red blood cells [1]. This shows an example of a typical graph (*see* **Note 10** for troubleshooting)

Abs(sample) = absorbance of the sample.
Abs(NegAvg) = averaged absorbance of the negative controls.
Abs(PosAvg) = averaged absorbance of the positive controls.

4 Notes

1. Blood should be stored at +4 °C and used within 2 weeks.
2. To make 500 mL of PBS, dissolve 4 g of NaCl in 400 ml of Milli-Q water. Add 0.1 g of KCl. Add 0.73 g of Na_2HPO_4 and 0.12 g of NaH_2PO_4. Add Milli-Q water up to 500 ml. Alternatively, it is possible to buy premade tablets.
3. Can be bought as 85 % pure natural product (extracted from honeybee venom).
4. Melittin is very toxic and should be handled with due care. It can bind to polypropylene surfaces, so it is important to precoat the plates and to use LoBind tubes. It is convenient to dissolve the whole amount directly inside its original glass container to prepare a stock solution. An amount of 5 mg corresponds to 1.757 μmol; adding 1.40 ml of PBS results in a 1.25 mM solution. For example, to prepare 1.5 ml of 5.0 μM solution, it is sufficient to dilute 6 μl of stock solution in 1.5 ml of PBS. Pipetting volumes in the range of a few microliters is naturally prone to random yet considerable errors. For these reasons the operational concentrations suggested in this protocol are much higher than strictly needed. We recommend to store the melittin stock solution at −20 °C and defrost only when needed.
5. RBCs are fragile and need to be handled with care. When transferring whole blood into a centrifuge tube, use a soft Pastette. When transferring the RBC pellets into PBS to create a suspension, cut off the point of the 200 μl tip in order to

widen the entrance/exit diameter. RBCs tend to precipitate from the suspension. Before pouring it into the CAPPOrigami, agitate it gently to achieve a homogeneous dispersion. They will precipitate also in the origami reservoir, so before transferring the suspension to a plate, shake it gently and use the multichannel pipette to make it homogeneous.

6. This is in order to minimize the impact of peptide transfer from well to well. Mix melittin and negative controls separately. Please note that adding the RBC suspension halves the concentration in each well.
7. Make sure that each well is individually sealed, or the content will evaporate in the incubator and concentrate the samples—invalidating the test. When preparing several test plates, plan an appropriate interval between groups of plates to be incubated based on the centrifuge's capacity.
8. This operation should be quick in order to avoid continued hemoglobin release — invalidating the test. A good technique is to let the pipette tips slide down gently along the straight left wall of the wells (for right-handed operators) until it reaches the edge of the V-bottom. Please note that, for a valid test, there must be no pellets in the melittin wells (100% hemolysis).
9. It can help to remember that the melittin half row (clearly recognizable) should be placed at the lower left corner. Be careful not to transfer or redisperse the RBC pellets at the bottom.
10. This note presents some common critical events that can invalidate the test.

RBCs are too fragile: if the blood is too old, it can be particularly sensitive to hemolysis, causing overestimation of the results. A sign can be seen during the initial washes with PBS: if after the first wash the supernatant is not clear but appears light brown, hemoglobin may already be being released. In this case the RBC suspension will contain already broken cells which have been depleted of their hemoglobin; the positive controls will thus give an unusually low absorbance.

Positive controls did not afford 100% hemolysis: after centrifugation, in the positive control wells, no RBC should be visible at the bottom. Possible reasons are that melittin has been degraded or mistakes in calculating, pipetting, and mixing have been made. Impossibility to achieve 100% makes it impossible to normalize the absorbance in the sample wells and interpret the results.

High discrepancy set to set or within a single set: it can be clearly recognized because the hemolysis values for the same peptide at the same concentration are very different between repeats or because a value at a certain concentration appears to be much higher than at higher concentrations. Unfortunately it can be very difficult to identify the reasons. Some are errors while inoculating the peptide and/or incomplete dissolution of the sample, errors while transferring it to lower rows, melittin contamination of a sample well, errors while mixing with the RBC suspension, and errors in sealing the wells before incubation causing evaporation and concentration of the solution. Please note that excessive delay between mixing and incubation start, between incubation and centrifuging and between centrifuging and transfer of the supernatant, can alter the results of the whole plate in a way that can be difficult to recognize.

High discrepancies session to session: for critical peptides of high interest, it is recommended to determine hemolysis in at least two different sessions. If individual tests appear consistent but there are discrepancies between them, the most likely reasons are errors in the preparation of the samples, delays in the critical time-sensitive operations, or major differences at the blood source. As with all manual operations, the role of the operator should also be taken into account.

References

1. Oddo A, Thomsen T, Kjelstrup S, Gorey C, Franzyk H, Frimodt-Møller N, Løbner-Olesen A, Hansen PR (2016) An all-D amphipathic undecapeptide shows promising activity against colistin-resistant strains of *Acinetobacter baumannii* and a dual mode of action. Antimicrob Agents Chemother 60:592–9
2. Ryge T, Doisy X, Ifrah D, Olsen JE, Hansen PR (2004) New indolicidin analogues with potent antibacterial activity. J Pept Res 64:171–85
3. Krishnakumari V, Sharadadevi A, Sitaram N, Nagaraj R (1999) Consequences of introducing a disulfide bond into an antibacterial and hemolytic peptide. J Pept Res 54:528–89
4. Dennison SR, Phoenix DA (2014) Susceptibility of sheep, human, and pig erythrocytes to haemolysis by the antimicrobial peptide Modelin 5. Eur Biophys J 43:423–32
5. Liu J, Jiang J, Wu Z, Xie F (2012) Antimicrobial peptides from the skin of the Asian frog, Odorrana jingdongensis: de novo sequencing and analysis of tandem mass spectrometry data. J Proteomics 75:5807–21
6. Belokoneva OS, Villegas E, Corzo G, Dai L, Nakajima T (2003) The hemolytic activity of six arachnid cationic peptides is affected by the phosphatidylcholine-to-sphingomyelin ratio in lipid bilayers. Biochim Biophys Acta 1617:22–30
7. Helmerhorst E, Reijnders I, Van't Hof W, Veerman E, Nieuw Amerongen AV (1999) Critical comparison of the hemolytic and fungicidal activities of cationic antimicrobial peptides. FEBS Lett 449:105–10
8. Yang L, Harroun TA, Weiss TM, Ding L, Huang HW (2001) Barrel-stave model or toroidal model? A case study on melittin pores. Biophys J 81:1475–85
9. Bechinger B (1997) Structure and functions of channel forming peptides: magainin, cecropins, melittin, and alamethicin. J Membr Biol 156:197–211
10. Pandey BK, Ahmad A, Asthana N, Azmi S, Srivastava RM, Srivastava S, Verma R, Vishwakarma AL, Ghosh JK (2010) Cell-selective lysis by novel analogues of melittin against human red blood cells and Escherichia coli. Biochemistry 49:7920–9

11. Munk JK, Ritz C, Fliedner FP, Frimodt-Moller N, Hansen PR (2014) Novel method to identify the optimal antimicrobial peptide in a combination matrix using anoplin as an example. Antimicrob Agents Chemother 58:1063–70
12. Ruiz J, Calderon J, Rondón-Villarreal P, Torres R (2014) Analysis of structure and hemolytic activity relationships of antimicrobial peptides (AMPs). In: Castillo LF, Cristancho M, Isaza G, Pinzón A, Rodríguez JMC (eds) Advances in computational biology, vol 232. Springer International Publishing, New York, NY, pp 253–8
13. Tossi A, Sandri L, Giangaspero A (2000) Amphipathic, α-helical antimicrobial peptides. Biopolymers 55:4–30
14. Dekruijff B (1990) Cholesterol as target for toxins. Biosci Rep 10:127–30
15. Blondelle SE, Houghten RA (1991) Hemolytic and antimicrobial activities of the twenty-four individual omission analogues of melittin. Biochemistry 30:4671–8
16. Blondelle SE, Simpkins LR, Pérez-Payá E, Houghten RA (1993) Influence of tryptophan residues on melittin's hemolytic activity. Biochim Biophys Acta 1202:331–6
17. Staubitz P, Peschel A, Nieuwenhuizen WF, Otto M, Götz F, Jung G, Jack RW (2001) Structure-function relationships in the tryptophan-rich, antimicrobial peptide indolicidin. J Pept Sci 7:552–64
18. Dathe M, Nikolenko H, Meyer J, Beyermann M, Bienert M (2001) Optimization of the antimicrobial activity of magainin peptides by modification of charge. FEBS Lett 501:146–50
19. Shai Y, Oren Z (1996) Diastereomers of cytolysins, a novel class of potent antibacterial peptides. J Biol Chem 271:7305–8
20. Papo N, Oren Z, Pag U, Sahl H-G, Shai Y (2002) The consequence of sequence alteration of an amphipathic alpha-helical antimicrobial peptide and its diastereomers. J Biol Chem 277:33913–21

ERRATUM TO

Antimicrobial Peptides: Methods and Protocols

Paul R. Hansen

Paul R. Hansen (ed.), *Antimicrobial Peptides: Methods and Protocols*, Methods in Molecular Biology, vol. 1548,
DOI 10.1007/978-1-4939-6737-7, © Springer Science+Business Media LLC 2017

DOI 10.1007/978-1-4939-6737-7_32

In chapter 11 titled "Applying Fluorescence Correlation Spectroscopy to Investigate Peptide-Induced Membrane Disruption", the following corrections are made.

1. Eq. 1: the denominator is updated as $(F)^2$ instead of (F^2).
2. Page 175, line 3 from the bottom: the correct text should be

Here, ℓ is the index of the experimental autocorrelation curves

In chapter 31 titled "Applying Fluorescence Correlation Spectroscopy to Investigate Peptide-Induced Membrane Disruption", Pg 431, the equation is updated as

$$\%\text{Hemolysis} = \frac{\text{Abs}(\text{sample}) - \text{Abs}(\text{NegAvg})}{\text{Abs}(\text{PosAvg}) - \text{Abs}(\text{NegAvg})} \times 100$$

The updated original online version for this book can be found at
http://dx.doi.org/10.1007/978-1-4939-6737-7

Paul R. Hansen (ed.), *Antimicrobial Peptides: Methods and Protocols*, Methods in Molecular Biology, vol. 1548,
DOI 10.1007/978-1-4939-6737-7_32, © Springer Science+Business Media LLC 2017

INDEX

Paul R. Hansen (ed.), *Antimicrobial Peptides: Methods and Protocols*, Methods in Molecular Biology, vol. 1548,
DOI 10.1007/978-1-4939-6737-7, © Springer Science+Business Media LLC 2017

R

S

T

V

W

Z

FSC
www.fsc.org
MIX
Papier | Fördert gute Waldnutzung
FSC® C083411

Kolibri 360 GmbH
Ferdinand-Jühlke-Straße 7
99095 Erfurt, Deutschland
produktsicherheit@kolibri360.de